T0302051

EMERGING NANOELECTRONIC DEVICES

EMERGING NANOELECTRONIC DEVICES

An Chen
GLOBALFOUNDRIES, USA

James Hutchby
Semiconductor Research Corporation, USA

Victor Zhirnov
Semiconductor Research Corporation, USA

George Bourianoff
Intel Corporation, USA

This edition first published 2015
© 2015 John Wiley and Sons Ltd

Registered office
John Wiley & Sons Ltd, The Atrium, Southern Gate, Chichester, West Sussex, PO19 8SQ, United Kingdom

For details of our global editorial offices, for customer services and for information about how to apply for permission to
reuse the copyright material in this book please see our website at www.wiley.com.

This work is supported in part by Semiconductor Research Corporation (SRC), Defense Advanced Research Projects Agency
(DARPA), and the National Science Foundation (NSF).

Library of Congress Cataloging-in-Publication Data

Emerging nanoelectronic devices / edited by Dr An Chen, Dr James Hutchby,
Dr Victor Zhirnov, Dr George Bourianoff.
      pages cm
   Includes bibliographical references and index.
   ISBN 978-1-118-44774-1 (cloth)
   1.   Nanoelectronics.   2.   Nanoelectromechanical systems.   3.   Nanostructured
materials.   I.   Chen, An (Electronics engineer), editor.   II.   Hutchby, James,
editor.   III.   Zhirnov, Victor V., editor.   IV.   Bourianoff, George, editor.
   TK7874.84.E32 2014
   621.381–dc23

                                                                2014029299

A catalogue record for this book is available from the British Library.

Set in 10/12 pt TimesLTStd-Roman by Thomson Digital, Noida, India

1   2015

To
Linda Wilson

Contents

Preface

Continued dimensional and functional[1] scaling of CMOS[2] integrated circuit technology is driving information processing[3] systems into a broadening spectrum of new applications. Many of these applications are enabled by performance gains and/or increased complexity realized by scaling. Because dimensional scaling of CMOS eventually will approach fundamental limits, several new alternative information processing devices and microarchitectures for existing or new functions are being explored to sustain the historical integrated circuit scaling cadence and reduction of cost/function in future decades. This is driving interest in new devices for information processing and memory, new technologies for heterogeneous integration of multiple functions (a.k.a. "More than Moore"), and new paradigms for systems architecture.

This book is based on the ITRS Emerging Research Device (ERD) International Technical Work Group's efforts over more than ten years to survey, research, and assess many of these new devices. As such, it provides an ITRS perspective on emerging research nanodevice technologies and serves as a bridge between CMOS and the realm of nanoelectronics beyond the end of CMOS dimensional and equivalent functional scaling. (Material challenges related to emerging research devices are addressed in a complementary ITRS chapter entitled *Emerging Research Materials*.)

An overarching goal of the ERD is to identify, assess, and catalog viable new information processing devices and systems architectures for their long-range potential and technological maturity, and to identify the scientific/technological challenges gating their acceptance by the semiconductor industry as having acceptable risk for further development. The intent is to provide an objective, informative resource for the constituent nanoelectronics communities pursuing: (1) research, (2) tool development, (3) funding support, and (4) investment, each directed to developing a new information processing technology. These communities include universities, research institutes, industrial research laboratories, tool suppliers, research funding agencies, and the semiconductor industry.

---

[1] *Functional Scaling*: Suppose that a system has been realized to execute a specific function in a given, currently available, technology. We say that system has been functionally scaled if the system is realized in an alternate technology such that it performs the identical function as the original system and offers improvements in at least one of size, power, speed, or cost, and does not degrade in any of the other metrics.

[2] Martin Hilbert and Priscila López, 2001, "The World's Technological Capacity to Store, Communicate, and Compute Information", *Science*, 332(6025), 60–65.

[3] Information processing refers to the input, transmission, storage, manipulation or processing, and output of data. The scope of this book is restricted to data or information manipulation, transmission, and storage.

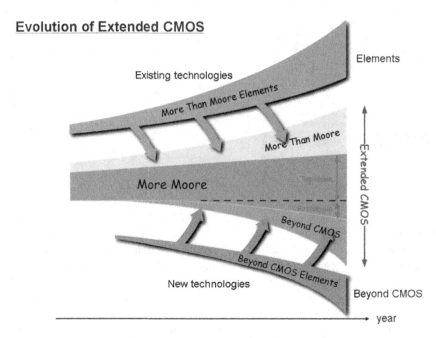

Figure P.1 Relationship between More Moore, More than Moore, and Beyond CMOS (Source: ERD, Japan)

This goal is accomplished by addressing two technology-defining domains: (1) extending the functionality of the CMOS platform via heterogeneous integration of new technologies, and (2) stimulating the invention of a new information processing paradigm. The relationship between these domains is schematically illustrated in Figure P.1. The expansion of the CMOS platform by conventional dimensional and functional scaling is often called "More Moore". The CMOS platform can be further extended by the "More than Moore" approach which is a relatively new subject. On the other hand, new information processing devices and architectures are often called "Beyond CMOS" technologies and are the main subjects addressed in this book. The heterogeneous integration of "Beyond CMOS" and "More than Moore" onto the "More Moore" platform will extend CMOS functionality to form the ultimate "Extended CMOS".

The book is partitioned into five sections: (1) Introduction, including a fundamental description of the physics of some nanodevices, (2) nanoelectronic memory devices, (3) nano-electronic logic information processing devices, (4) concepts for emerging architectures, and (5) summary, conclusions, and outlook for nanoelectronic devices. Some detail is provided for each entry regarding operation principles, advantages, technical challenges, maturity, and its current and projected performance. Also included is a device and architectural focus combining emerging research devices offering specialized, unique functions as heterogeneous core processors integrated with a CMOS platform technology. This represents the nearer-term focus of this work, with the longer-term focus remaining on the discovery of an alternate information processing technology to supplement and to eventually replace digital CMOS.

List of Contributors

Ethan C. Ahn, Department of Electrical Engineering, Stanford University, USA

Masakazu Aono, WPI Center for Materials Nanoarchitectonics, National Institute for Materials Science, Japan

Tetsuya Asai, Graduate School of Information Science and Technology, Hokkaido University, Japan

Behtash Behin-Aein, GLOBALFOUNDRIES Inc., USA

Benjamin F. Bory, Eindhoven University of Technology, The Netherlands

George Bourianoff, Components Research Group, Intel Corporation, USA

Geoffrey W. Burr, IBM, USA

An Chen, GLOBALFOUNDRIES Inc., USA

Donald M. Chiarulli, Department of Electrical and Computer Engineering, University of Pittsburgh, USA

György Csaba, University of Notre Dame, USA

Shamik Das, Nanosystems Group, The MITRE Corporation, USA

Denver H. Dash, Intel Science and Technology Center, USA

Supriyo Datta, Purdue University, USA

Vinh Quang Diep, Purdue University, USA

S. Burc Eryilmaz, Department of Electrical Engineering, Stanford University, USA

Yan Fang, Department of Electrical and Computer Engineering, University of Pittsburgh, USA

Scott Fong, Department of Electrical Engineering, Stanford University, USA

Aaron D. Franklin, Department of Electrical and Computer Engineering and Department of Chemistry, Duke University, USA

Paul Franzon, North Carolina State University, USA

Tsuyoshi Hasegawa, WPI Center for Materials Nanoarchitectonics, National Institute for Materials Science, Japan

Toshiro Hiramoto, Institute of Industrial Science, The University of Tokyo, Japan

James Hutchby, Semiconductor Research Corporation, USA

Louis Hutin, Department of Electrical Engineering and Computer Sciences, University of California, USA

Adrian M. Ionescu, Ecole Polytechnique Fédérale de Lausanne, Switzerland

Mahdi Jamali, University of Minnesota, USA

Rakesh Jeyasingh, Department of Electrical Engineering, Stanford University, USA

Alexander Khitun, Material Science and Engineering, University of California, USA

Angeline Klemm, University of Minnesota, USA

Takhee Lee, Department of Physics, Seoul National University, Korea

Steven P. Levitan, Department of Electrical and Computer Engineering, University of Pittsburgh, USA

Eike Linn, Institute of Electronic Materials II, RWTH Aachen University, Germany

Tsu-Jae King Liu, Department of Electrical Engineering and Computer Sciences, University of California, USA

Matthew J. Marinella, Sandia National Laboratories, USA

Hao Meng, Data Storage Institute, Singapore

Stephan Menzel, Peter Grünberg Institut (PGI-7), Forschungszentrum Jülich, Germany

Stefan C.J. Meskers, Eindhoven University of Technology, The Netherlands

Michael T. Niemier, University of Notre Dame, USA

Ferdinand Peper, Center for Information and Neural Networks, National Institute of Information and Communications Technology, USA

Wolfgang Porod, University of Notre Dame, USA

Mark A. Reed, Departments of Electrical Engineering and Applied Physics, Yale University, USA

Frank Schwierz, Technical University of Ilmenau, Germany

Hyunwook Song, Department of Applied Physics, Kyung Hee University, Korea

Narayan Srinivasa, Center for Neural and Emergent Systems, HRL Laboratories LLC, USA

Ken Takeuchi, Chuo University, Japan

Jian-Ping Wang, University of Minnesota, USA

Rainer Waser, Institute of Electronic Materials II, RWTH Aachen University, Germany; Peter Grünberg Institut (PGI-7), Forschungszentrum Jülich, Germany

H.-S. Philip Wong, Department of Electrical Engineering, Stanford University, USA

Victor V. Zhirnov, Semiconductor Research Corporation, USA

Acronyms

1D1R	1-Diode-1-resistor
1S1R	1-Selector-1-resistor
1T	One transistor
1T1C	1-Transistor-1-capacitor
1T1R	1-Transistor-1-resistor
2DEG	Two-dimensional electron gas
3D	Three dimensional
AD	Analog digital
AF	Anti-ferromagnetic
AIST	Silver (Ag) Indium (In) Antimony (Sb) Tellurium (Te)
ALD	Atomic layer deposition
AM	Associative memory
ASIC	Application specific integrated circuit
ASL	All-spin logic
BARITT diode	Barrier-injection transit-time diode
BBE	Brain, body, environment based interactions
BBL	Buried bit line
BDA	1,4-Benzenediamine
BDC60	Bis(fullero[c]pyrolidin-1y1)benzene
BDT	1,4-Benzenedithiol
BE	Bottom electrode
BEC	Bottom electrode contact
BEOL	Back end of line
BFO	Bismuth ferrite (BiFeO3)
BiSFET	Bilayer pseudo-spin field-effect transistor
BIST	Built-in self-test
BJT	Bipolar junction transistor
BL	Bit line
BLG	Bilayer graphene
BN	Beyond Neumann
CA	Cellular automata
CAM	Contend addressable memory

CBL	Cantilever bit line
CBRAM	Conductive-bridge random access memory
CDMA	Code division multiple access
CMIS	Current-induced magnetization switching
CMOS	Complementary metal oxide semiconductor
CNT	Carbon nanotube
CNTFET	Carbon nanotube field-effect transistor
CO	Carbon monoxide
CoFeB	Cobalt iron boron
CoPt	Cobalt platinum
CP-AFM	Conducting probe–atomic force microscopy
CPP	Current perpendicular to plane
CPU	Central processing unit
CRS	Complementary resistive switch
CTAFM	Conductive-tip atomic force microscopy
D	Density of states
DC	Direct contact
DC8 to DC12	Eight to 12 carbon atoms in alkanedithiols
dCNT	Carbon nanotube diameter
DFT	Density functional theory
DIBL	Drain induced barrier lowering
DMRG method	Density matrix renormalization group method
DNA	Deoxyribonucleic acid
DoM	Degree of match
DOS	Density of states
DRAM	Dynamic random access memory
DVD	Digital versatile disc
e	Elementary charge of an electron
EBIC	Electron beam induced current
EBJs	Electromigrated break junctions
EEPROM	Electrically erasable programmable read only memory
Eg	Energy bandgap
EM	Electromechanical
EMB	Electrochemical metallization bridge
EOT	Equivalent oxide thickness
ERD	Emerging research devices
FD	Fully depleted
FeFET	Ferroelectric field effect transistor
Fe-NAND	Ferroelectric NAND
FeRAM	Ferroelectric random access memory
FERET	Facial recognition technology
FET	Field-effect transistor
FIB	Focused ion beam
FinFET	Multi-gate MOSFET with fin-shaped active structure

FL	Free layer
FM	Ferromagnetic
FN	Fowler–Nordheim
FOPE	Fluorinated oligomer
FPGA	Field programmable gate array
FRAM	Ferroelectric random access memory
FTJ	Ferroelectric tunnel junction
FWHM	Full width at half maximum
GAA	Gate all around
GaAs	Gallium arsenide
GMR	Giant magnetoresistive
GNM	Graphene nanomesh
GNR	Graphene nanoribbon
GQ	Quantum conductance
GSHE	Giant spin Hall effect
GST	Germanium (Ge), Antimony(Sb), Tellurium(Te)
h	Planck's constant
HAO	Hf-Al-O
HDD	Hard disk drive
HEMT	High electron mobility transistor
HF	Hydrofluoric acid
HKMG	High-k metal gate
HMAX	Hierarchical memory and X
HOMO	Highest occupied molecular orbital
HP	High performance
HRS	High resistance state
HTM	Hierarchical temporal memory
IC	Integrated circuit
ICT	Information and communication technologies
IEDM	International Electron Devices Meeting
IETS	Inelastic electron tunneling spectroscopy
I-MOS	Impact-ionization MOS
InAlAs	Indium aluminum arsenide
InAs	Indium arsenide
InGaAs	Indium gallium arsenide
iNML	In-plane nanomagnet logic
InP	Indium phosphide
InSb	Indium antimonide
I_{off}	Off current
I_{on}	On current
IOP	Input–output processor
IPCM	Interfacial phase change memory
ITRS	International Technology Roadmap for Semiconductors

KJMA   Kolmogorov, Johnson and Mehl, and Avrami

L_{ch} Channel length
LEC   Lyric error correction
L_g Gate length
LLG   Landau–Lifshitz–Gilbert
LRS   Low resistance state
L_{spr} Spacer length
LtN   Less than Neumann
LUMO   Lowest unoccupied molecular orbital

m Carrier effective mass
MC   Magneto current ratio
MCBJs   Mechanically controllable break junctions
MEMS   Micro-electro-mechanical systems
MFIS   Metal ferroelectric insulator semiconductor
MFTJ   Multiferroic tunnel junctions
MG   Multigate
mHEMT   Metamorphic HEMT
MIEC   Mixed ionic electronic conduction
MIM   Metal–insulator–metal
MIT   Metal–insulator transition
MLC   Multi level cell
MO-BF   Metal oxide–bipolar filamentary
MOS   Metal oxide semiconductor
MoS_2 Molybdenum disulfide
$MoSe_2$ Molybdenum diselenide
MOSFET   Metal oxide semiconductor field-effect transistor
MO-UF   Metal oxide–unipolar filamentary
MQCA: Magnetic quantum-dot cellular automata
MRAM   Magnetic random access memory
M-SCM   Memory class storage class memory
MTJ   Magnetic tunnel junction
MtN   More than Neumann

Nc   Nanocrystal
NCC   Nano-current channel
ND   Notre Dame
NEMory   Memory cell based on a NEMS switch
NEMS   Nano-electro-mechanical systems
NIST: National Institute of Standards and Technology
NML   Nanomagnet logic
nMOSFET   n-Channel MOSFET
N-T   N-Terminal relay
NV   Nonvolatile
NVM   Nonvolatile Memory
NWFET   Nanowire field-effect transistor

ODT	Au-octanedithol junction
ONO	Oxide–nitride–oxide
OOMMF	Object oriented micromagnetic framework
OPE	Oligophenyleneethynynylene
OPI	Oligophenyleneimine
OPV	Oligophenylenevinylene
OS	Operating system
OTP	One-time programmable
OTS	Ovonic threshold switch
PABA	Persistent asynchronous background activity
PC	Phase change
PCB	Phase change bridge
PCM	Phase change memory
PCRAM	Phase change random access memory
PDMS	Polydimethylsiloxane
PE	Piezoelectric
PE:	Processing element
PET	Piezoelectronic transistor
PF	Poole Frenkel
PFM	Piezoresponse force microscopy
pHEMT	Pseudomorphic HEMT
PIDS	Process integration, devices, and structures
PLA	Programmable logic array
PLL	Phase-locked loop
PM	Permanent magnet
pMA	Pars-mercaptoaniline
PMA	Perpendicular magnetic anisotropy
PMN-PT	Lead magnesium niobate – lead titanate
pMOSFET	p-Channel MOSFET
pNML:	Perpendicular nanomagnet logic
PR	Piezoresistive
PS-MOSFET	Pseudo-spin MOSFET
PZN-PT	Lead zinc niobate–lead titanate
PZT	Lead zirconate titanate
QCA	Quantum-dot cellular automata
R	Read
R6G	Rhodamine
RA	Resistance × area
RAM	Random access memories
R_c	Contact resistance
RC	Time constant of a circuit composed of resistors and capacitors
RDF	Random dopant fluctuation
ReRAM	Redox resistive random access memory
RF	Radio frequency

RIE	Reactive ion etch
RL	Reference layer
RQ	Quantum resistance
RRAM	Resistive random access memory
RTD	Resonant tunneling diode
RWL	Read word line
SAM	Self-assembled monolayer
SB	Schottky barrier
SBM	Suspended beam memory
SBT	$SrBi_2Ta_2O_9$
SCE	Short channel effect
SCLC	Space–charge limited conduction
SCM	Storage class memory
SEM	Scanning electron microscopy
SERS	Surface-enhanced Raman spectroscopy
SET	Single electron transistor
SEU	Single event upset
SFD:	Switching field distribution
SG	Suspended gate
Si	Silicon
SLL	Super lattice like
SNDM	Scanning nonlinear dielectric microscopy
SNM	Static noise margin
SoC	System on a chip
SOI	Silicon on insulator
SPDT	Single pole double-throw
SPT	Structural phase transition
SR	Structural relaxation
SR	Stochastic resonance
SRAM	Static random access memory
SS	Subthreshold swing
S-SCM	Storage class storage class memory
SSD	Solid state disk
STD	Spin torque device
STDP	Spike timing dependent plasticity
STM	Scanning tunneling microscopy
STM-BJ	Scanning tunnel microscope controlled break junction
STMG	Spin torque majority logic gate
STNO	Spin torque nano-oscillator
STS	Scanning tunneling spectroscopy
STS	Subthreshold slope
STT	Spin transfer torque
STT-MRAM	Spin torque transfer magnetic random access memory
SWD	Spin wave devices
SWNT	Single walled carbon nanotube

TBR	Thermal boundary resistance
TE	Top electrode
TEM	Transmission electron microscopy
TER	Tunnel electro-resistance
TFET	Tunnel field-effect transistor
THz	Tera hertz
TMR	Tunneling magnetoresistance
t_{ox}	Dielectric thickness
TTR	Time domain thermo reflectance
TVS	Transition voltage spectroscopy
UHV	Ultra-high vacuum
ULSI	Ultra scale integration
UPS	Ultraviolet photoelectron spectroscopy
v	Carrier velocity
VCCPCM	Vertical chain cell type phase change memory
VCM	Valence change mechanism
VCMA	Voltage controlled magnetic anisotropy
V_{dd}	Supply voltage
V_{ds}	Drain–source bias
v_F	Fermi velocity
V_{gs}	Gate voltage
VLSI	Very large scale integration
VMR	VLSI metallic CNT removal
$V_{pull\text{-}in}$	Pull-in voltage
V_t	Threshold voltage
V_{th}	Threshold voltage
W	Write
WL	Wordline
WS2	Tungsten disulfide
WTA	Winner take all
WWL	Write word line
XANES	X-ray absorption near-edge structure
Δn	Excess carrier density
ϕ_{SB}	Schottky barrier height

Part One

Introduction

1

The Nanoelectronics Roadmap

James Hutchby
Semiconductor Research Corporation, USA

1.1 Introduction

Over the past 40 years, the global impact of the computer or digital age on most individuals, cultures, and their economies, has been nothing less than breathtaking; it has shifted the communication and human interaction paradigm and has sparked creation of several major new industries. In short, electronics technology based on semiconductor devices and integrated circuits has changed how many of us live our lives and how we do our business.

For example, semiconductor device technology of the 1960s enabled development of large, centralized, power-consuming, main frame computers using discrete transistors to replace vacuum tube binary switches. The 1970s saw the advent of e-mail, desktop and portable calculators, the earliest Internet, digital watches, and the first applications of microprocessors. As complementary metal oxide semiconductor (CMOS) gates became the dominant low-power semiconductor technology in the 1980s, its explosion following Moore's Law[1] [1] drove the development and broad application of personal and, later, laptop computers in the 1980s and 1990s. The world's capacity to process bits of information grew from 3.0×10^8 MIPS in 1986, to 4.4×10^9 MIPS in 1993, to 2.9×10^{11} MIPS in 2000, and to 6.4×10^{12} MIPS in 2007 [2].

Broadened commercial applications in the twenty-first century include Smart Phones, Global Positioning Satellite navigation systems, Digital Cameras, high definition flat panel TVs, tablet PCs, smart dining utensils (for managing food and caloric intake), and a variety of automotive sensor and information-processing electronics.

Much of the commercial success of the semiconductor industry has been driven by its ability to relentlessly follow Moore's Law in doubling the density of transistors every 2–3 years. This

---

[1] Intel co-founder Gordon Moore's bold prediction, popularly known as Moore's Law, states that the number of transistors on a chip will double approximately every 2 years. Consequently, the number of transistors on a chip of constant area will double as the transistor's linear dimension scales down by $\sqrt{2}$. This scaling enables an average annual reduction of the cost per transistor of 29%.

Emerging Nanoelectronic Devices, First Edition. An Chen, James Hutchby, Victor Zhirnov and George Bourianoff.
© 2015 John Wiley & Sons, Ltd. Published 2015 by John Wiley & Sons, Ltd.

phenomenal technology scaling not only enabled new, more complex, applications and lowered the cost per transistor by 25–30%/year [3], it provided faster circuits with lower power dissipation. In this time period, the number of transistors on a chip increased from a few hundred/cm^2 to several billion/cm^2.

The value of the semiconductor and related electronics systems industries also can be judged by their impact on the United States and global economies. For example, the annual revenues of the semiconductor-based device industry exceeded US\$ 300B in 2013, and the related electronics systems industry revenues were projected to exceed US\$ 1.55 T in 2010 [4]. Also, from 1960 to 2007, the computer industry added only 0.3% of United States value, but it contributed 2.7% of economic growth and 25% of productivity gains [5].

As these examples illustrate, the rapid and ubiquitous penetration of information-processing technologies into new domains of application and their major economic impact combine to drive aggressive reduction of the size and cost of the basic information-processing element (i.e., the transistor or some new information processing element). However, as the size of the transistor continues to shrink, its structural composition is approaching a few hundred atoms and its operation is becoming prohibitively leaky thus enabling increased power dissipation. Eventually, the semiconductor industry will need a technology to supplement and extend CMOS beyond its fundamental scaling limit.

This raises the following questions: (1) how much further can the present CMOS-based electronics technology be scaled in size and cost; (2) what (if any) new physical mechanism and technology can either supplement, or eventually, replace CMOS as a medium for continued scaling of information-processing technology?

1.2   Technology Scaling: Impact and Issues

Scaling CMOS raises some additional important issues regarding increased power dissipation of aggressively scaled metal oxide semiconductor field effect transistors (MOSFETs) leading to unwieldy power density dissipated on a silicon CMOS chip. Historically, one could reduce the threshold voltage and, thereby, the source–drain voltage and power dissipation of a scaled MOSFET transistor in a CMOS gate. However, in recent years the allowable threshold voltage of a MOSFET has reached a minimum of 0.9–1.0 V. Consequently, as the transistor is further scaled its power dissipation increases. The power dissipation on a scaled chip has two sources. One source is the scaled transistors. The other source is the system of metal interconnects.

The physical operation of electronic devices ranging from the vacuum tube, to a variety of transistors and integrated circuits, has centered on the manipulation and storage of electronic charge. The next section discusses the limits of charge confinement and relates them to a fundamental scaling limit of CMOS.

1.3   Technology Scaling: Scaling Limits of Charge-based Devices

Operation of MOSFETs is based on manipulation of electronic charge transport from source to drain controlled by another charge placed on a gate [6]. Electrons placed on the source are prohibited (for the most part) from flowing spontaneously to the drain by an energy barrier formed by the gate/source and the source/drain potentials. Lowering this energy barrier by placing a positive charge on the gate allows electrons on the source to flow to the drain when assisted by a forward bias voltage. Conversely, these electrons on the drain are blocked from

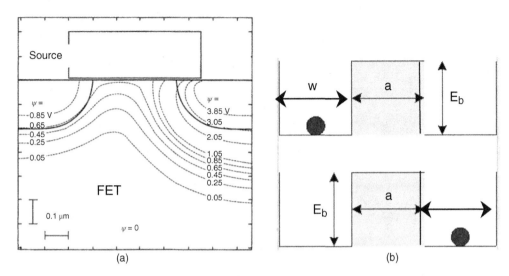

Figure 1.1 Energy model for limiting device. (a) Cross-section of a MOSFET illustrating potential distributions and their effect on the potential barrier separating the source drain potential wells. (b) Idealized model of the potential barrier separating the source drain potential wells where: $w =$ width of left-hand well (LHW) and right-hand well (RHW); $a =$ barrier width; $E_b =$ barrier energy

returning to the source by the drain/source energy barrier. The question then is: What is the minimum height (E_b) and width (a) of the gate/source potential energy barrier necessary to allow an electron to have equal probability of being either in the source or the drain?

Figure 1.1a illustrates manipulation of the energy barrier between the source and drain of a MOSFET by application of a voltage, V_{gs}, from the gate to the source, and another voltage, V_{ds}, from the drain to the source. Altering these voltages changes the height and, to a lesser extent, the width of this energy barrier and, thereby, changes the confinement of electronic charge in either of two potential wells. One is defined by the source and the other by the drain. An idealized model of this source–gate–drain potential well structure is illustrated in Figure 1.1b.

The model illustrated in Figure 1.2 is used to estimate the minimum width, a_{min}, and height, E_{bmin}, needed to provide a 50% probability that a single electron is either in the source or drain potential well. Figure 1.2 also illustrates two processes for transmitting electrons either over the barrier by thermionic emission or through the barrier by quantum mechanical tunneling. Using these criteria, a limit value of a, $a_{min} = 1.5$ nm and $E_{bmin} = kT\ln2$, where k is the Boltzmann constant, and T is absolute temperature in degrees Kelvin. An additional constraint requiring minimization of energy dissipated in changing potential wells sets $a_{min} = 5$ nm.

Since high-volume manufacturing technology for the 16 nm node is currently being ramped to full production, the 5 nm generation may go into manufacturing in 7 years or in 2020. The 8 nm generation may be in manufacturing as early as 2018. Consequently, scaling of CMOS-based information-processing technology will face fundamental limits within the next 5–7 years or by 2020. This determines a need to have any new technology for extending CMOS in place by that time. Furthermore, this analysis suggests that the search for a new information-processing technology and paradigm should consider a new "token" to replace the electronic charge or a new information-processing paradigm that does not depend on confinement of the electronic charge.

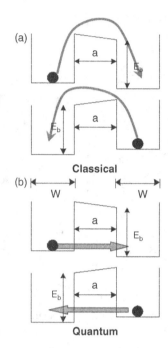

Figure 1.2 Illustration of two processes for transmitting electrons either (a) over an energy barrier by thermionic emission, or (b) through the barrier by quantum mechanical tunneling

1.4   The International Technology Roadmap for Semiconductors

Formed in 1991 by the United States Semiconductor Industry Association (SIA) assisted by Semiconductor Research Corp. (SRC) and SEMATECH, the United States National Technology Roadmap for Semiconductors (NTRS) brought together scientists and technologists from the United States microelectronics manufacturing, supplier, and academic communities and tasked them to forecast technological scaling of MOSFET integrated circuits and all related technologies 15 years into the future. The first NTRS was published in 1992.

Recognizing the value of including international microelectronics communities, the International Roadmap Committee (IRC) established the International Technology Roadmap for Semiconductors (ITRS) in 1998[2]. The ITRS was tasked to develop and maintain a 15 year assessment of the semiconductor industry's future technology requirements. These future needs drive present-day strategies for world-wide research and development among manufacturers' research facilities, universities, national laboratories, and tool manufacturers in 17 topical areas ranging from basic or emerging research materials and emerging research devices to design and systems drivers. This book is focused on Emerging Research Devices.

---

[2] The ITRS is sponsored by the European Semiconductor Industry Association (ESIA), the Japan Electronics and Information Technology Industries Association (JEITA), the Korean Semiconductor Industry Association (KSIA), the Taiwan Semiconductor Industry Association (TSIA), and the United States Semiconductor Industry Association (SIA).

1.5 ITRS Emerging Research Devices International Technology Working Group

As the size of the transistor continues to shrink, its structural composition is approaching a few hundred atoms and its operation is becoming prohibitively leaky, thereby leading to increased power dissipation. Briefly mentioned above, the semiconductor industry eventually will need a technology to supplement and extend CMOS beyond its fundamental scaling limit and one day even to replace CMOS.

As this need took on some urgency, the ITRS International Roadmap Committee (the ITRS governance committee) formed a new technical work group; this group became the Emerging Research Devices (ERD) International Technology Working Group (ITWG).

1.5.1 ERD Editorial Team

The ERD International Technology Working Group (ERD ITWG) is a large international group of approximately 80 individuals who contribute in some way to the work of the ERD Team over a 2 year cycle to produce a new ERD chapter for each updated ITRS. The ERD Editorial Team consists of approximately 12 actively engaged individuals, 8 of whom represent eight companies, 2 represent two universities, and 2 represent one consortium. This Editorial Team provides leadership for the ERD ITWG. Many ERD active contributors manage research groups that address nanoelectronics topics within their organizations and a few members of ERD are actively engaged practicing researchers contributing to this field. Much of the text in the ERD chapter is written by practicing researchers well-recognized for their leadership in the field in which they make their contributions. Their contributions are then vetted by the ERD Editorial Team and the IRC to ensure balance and accuracy of any claims or projections made for a Technology Entry.

1.5.2 Vision and Mission

The vision of the ERD is to assist the nanoelectronics research community to invent and demonstrate feasibility of new technologies to extend information processing beyond the reach of Si technology and to be a resource to the Semiconductor Industry in their evaluation of these nanodevices.

The mission of the ERD ITWG is to offer substantive input to our reader communities related to the viability of proposed nanoelectronic devices to live up to their claimed potential in heir maturity. The targeted reader communities include the global research community (including industrial, university, and government laboratories), relevant government agencies, industry technology directors and managers, and the supplier communities. ERD evaluates significant new concepts for nanodevices proposed to supplement and/or replace the silicon MOSFET for information processing. This is accomplished by critically assessing the suitability and maturity of a nanodevice concept for sustaining information processing beyond that attainable with ultimately scaled CMOS thereby leading to promising new approach(es) to memory and logic technologies to be implemented by 2020–2026.

1.5.3 Scope

The scope of ERDs activities embraces evaluating emerging research memory (including solid state storage), logic, information processing, More-than-Moore, and new nanoarchitecture technologies enabled by promising new nanodevices together with the requisite materials and process technologies chosen to fabricate the device and test structures [in collaboration with the Emerging Research Materials (ERM) work group].

Evaluation of a candidate nanodevice includes assessing its scaling potential, power dissipation, speed, operation temperature, internal gain, technological, and/or architectural compatibility with CMOS. The primary question addressed by this evaluation process is: "Assuming a device reaches its full maturity in fabrication and operation, how would it compare with ultimately scaled CMOS and to what extent can it be scaled further?" Evaluation also includes identifying the most important scientific and technological questions and issues that must be resolved to advance acceptance of the device for further attention and accelerated development. The scope includes modeling and simulation and those metrologies required by the research community to establish and demonstrate the operation and feasibility of a new device.

Evaluation is further discussed in Section 1.6.1.

1.6 Guiding Performance Criteria

1.6.1 Nanoinformation Processing

First, we need a clear understanding of just what is meant by "information." Webster's New International Dictionary defines information for broad popular use as "the communication or reception of knowledge or intelligence." For focused digital computing and digital signal-processing applications, information is defined in its most restricted technical sense "as a sequence of symbols that can be interpreted as a message." These symbols take on a specific meaning when they are encoded according to a particular algorithm. The questions addressed by the ERD work group in considering a proposed new technology include:

- In a new technology, how can a single bit of information be physically represented – what is the operative information-bearing token?" (e.g., the electron in electronic circuits or the bead in an abacus.)
- What property of the information "token" is used as the "state variable" to define and sense the state? (e.g., for an electron the commonly used state variable is charge and its presence/ absence is sensed by a voltage across the drain load capacitance.[3] For the abacus the state variable is the mass and physical position of a bead.)
- How can the state variable be manipulated to reliably change the logic state? (e.g., in a CMOS gate the voltage of the load capacitor can be increased and decreased by placement and removal of electrons on to and off of the load capacitor. For the abacus it is the movement of a bead to and from a defined position.)
- How can the state variable in one position communicate with a state variable in another position to perform a function? (e.g., in CMOS, copper metal interconnect wires conduct charge/voltage from one site another, and in an abacus, nimble fingers perform the task of moving a bead from one position to another.)
- What is the performance claimed for the new technology in terms of its power dissipation, switching speed, and so on? How does this compare with the performance of ultimately scaled silicon?
- What fabrication and manufacturing issues need to be addressed and resolved?
- What fundamental and technology issues need to be addressed and resolved to demonstrate the promise and feasibility of the proposed technology?

---

[3] The *state variable* for an electron could also be its spin, its mass, or its wave function.

1.6.2 Nanoelectronic Device Taxonomy

Information processing to accomplish a specific system function, in general, requires several different interactive layers of technology. The objective of this section is to carefully delineate a taxonomy of these layers to further distinguish the scope of this book from that of the Emerging Research Materials chapter and the Systems Design chapter in the ITRS. The scope of this book addresses the Device and the Data Representation layers shown in Figure 1.3.

One comprehensive top-down list of these layers begins with the required application or system function, leading to system architecture, micro- or nanoarchitecture, circuits, devices, and materials. As shown in Figure 1.3 a different bottom-up representation of this hierarchy begins with the lowest physical layer represented by a computational or information token (e.g., the electron) and ends with the highest layer represented by the architecture. In this more schematic representation, focused on generic information processing at the device/circuit level, a fundamental unit of information (e.g., a bit) is represented by the value of a property (e.g., presence or absence of electronic charge or the polarity of spin) of a computational information token, for example, the position of a bead in the ancient abacus calculator or the electronic charge or voltage state of a nodal capacitance in CMOS logic. A device provides the physical means of representing and manipulating a property (e.g., position of an electronic charge) of a computational information token among its two or more allowed discrete states. Eventually, device concepts may transition from simple binary switches to devices with more complex information processing functionality.

The device is a physical structure resulting from the assemblage of a variety of materials possessing certain desired properties obtained through exercising a set of fabrication processes. An important layer, therefore, encompasses the various materials and processes necessary to fabricate the required device structure, which is the domain of the ERM International Technical Work Group. The *data representation* layer is how the state variable property of the information

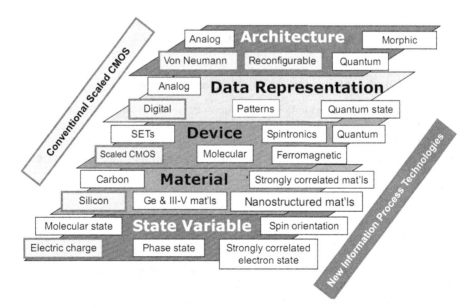

Figure 1.3 Taxonomy for nanoinformation processing devices

token is encoded by the assemblage of devices to process the bits or data. Two of the most common examples of data representation are binary digital and continuous or analog signaling. In some cases, the data are represented as arrays or images which then become the fundamental datum of computation in either primitive or compressed form. This layer is within the scope of the ERD ITWG. The architecture plane encompasses three subclasses of this taxonomy: (1) the nanoarchitecture or physical arrangement or assemblage of devices to form higher level functional primitives to represent and enable execution of a computational model, (2) the computational model that describes the algorithm by which information is processed using the primitives, for example, logic, arithmetic, memory, cellular nonlinear network (CNN), and (3) the system-level architecture that describes the conceptual structure and functional behavior of the system exercising the computational model. Subclass (1) is within the scope of the ERD Technical Work Group, and subclasses (2) and (3) are within the scope of the Design Technical Work group. The elements shown in the red-lined yellow boxes (left side of Fig. 1.3) represent the current CMOS platform technology that is based on electronic charge as a binary information token. This state token serves as the foundation for the von Neumann computational system architecture. Analog data representation also is included in the current CMOS platform technology. The other entries grouped in these five categories summarize individual approaches that, combined in some yet to be determined highly innovative fashion, may provide a new highly scalable information-processing paradigm.

1.6.3 Fundamental Guiding Principles – "Beyond CMOS" Information Processing

In considering the many disparate new approaches proposed to provide order of magnitude scaling of information processing beyond that attainable with ultimately scaled CMOS, the ERD Editorial Team proposed the following comprehensive set of guiding principles. We believe these "Guiding Principles" provide a useful structure for directing research on any information-processing technology to dramatically enhance scaling of functional density and performance while simultaneously reducing the energy dissipated per functional operation. Further this new technology would need to be realizable using a highly manufacturable fabrication process.

1.6.3.1 Computational Information Variable(s) other than Solely Electron Charge

In seeking new device opportunities, it is important to understand not only the device characteristics, but also how connected systems of these devices might be used to perform complex logic functions [7]. The basic computational element in current digital information processing systems is the binary switch. In its most fundamental form, it consists of:

1. Two states 0 and 1 (state variables), which are equally attainable and distinguishable;
2. A means to control the change of the state (WRITE);
3. A means to read the state;
4. A means to communicate with other binary switches (TALK).

The system state representation controls (WRITE), and the READ and TALK operations are all represented by a physical property of entities such as particles, quasi-particles, collections of particles, and so on. These physical entities are called *tokens*. Each token has a set of physical attributes associated with it (e.g., charge, mass, spin) and it is the physical interaction between the token attributes within the device structure that determines the operation and the resulting state of the device. In most cases, an attribute can assume several values and for this reason, we also call them *state variables*.

The estimated performance comparison of devices utilizing an alternative token (e.g., an electron, position of an atom) with an alternative "state variable" (e.g., charge and voltage, spin, current) to ultimately scaled CMOS should be made as early in a program as possible to downselect and identify key trade-offs.

1.6.3.2 Nonthermal Equilibrium Systems

These are systems that are out of equilibrium with the ambient thermal environment for some period of their operation, thereby reducing the perturbations of stored information energy in the system caused by thermal interactions with the environment. The purpose is to allow lower energy computational processing while maintaining information integrity.

1.6.3.3 Novel Energy Transfer Interactions

These interactions could provide the interconnect function between communicating information processing elements. Energy transfer mechanisms for device interconnection could be based on short-range interactions including, for example, quantum exchange and double exchange interactions, electron hopping, Förster coupling (dipole–dipole coupling), tunneling and coherent phonons.

1.6.3.4 Nanoscale Thermal Management

This could be accomplished by manipulating lattice phonons for constructive energy transport and heat removal.

1.6.3.5 Sublithographic Manufacturing Process

One example of this principle is directed self-assembly of complex structures composed of nanoscale building blocks. These self-assembly approaches should address nonregular, hierarchically organized structures, be tied to specific device ideas, and be consistent with high-volume manufacturing processes.

1.6.3.6 Alternative Architectures

In this case, architecture is the functional arrangement on a single chip of interconnected devices that includes embedded computational components. These architectures could utilize, for special purposes, novel devices other than CMOS to perform unique functions.

1.6.4 Current Technology Requirements for CMOS Extension and Beyond CMOS Memory and Logic Technologies

Some emerging nanoscale devices discussed in this book are charge-based structures proposed to extend CMOS to the end of the current roadmap. Other emerging devices offer new computational information "tokens" and will likely require new fabrication technologies.

A set of relevance or evaluation criteria, defined below, are used to parameterize the extent to which proposed "CMOS Extension" and "Beyond CMOS" technologies are applicable to memory or information-processing applications. The *relevance criteria* are: (1) scalability,

(2) speed, (3) energy efficiency, (4) gain (logic) or ON/OFF ratio (memory), (5) operational reliability, (6) operational temperature, (7) CMOS technological compatibility, and (8) CMOS architectural compatibility. Definitions of these evaluation criteria follow:

1. *Scalability*: First and foremost the major incentive for developing and investing in a new information-processing technology is to discover and exploit a new domain for scaling information-processing functional density and throughput per Joule substantially beyond that attainable by ultimately-scaled CMOS. Silicon-based CMOS has provided several decades of scaling of MOSFET densities. The goal of a new information-processing technology is to replicate this success by providing additional decades of functional and information throughput rate scaling using a new technology. In other words, it should be possible to articulate a Moore's Law for the proposed technology over additional decades.

2. *Speed*: A future information-processing technology must continue to provide (at least) incremental improvements in speed beyond that attainable by ultimately scaled CMOS technology. In addition, nanodevices that implement both logic and memory functions in the same device would revolutionize circuit and nanoarchitecture implementations.

3. *Energy Efficiency*: Energy efficiency has become the limiting factor of any beyond CMOS device using electronic charge or electric current as a computational information token used to represent a state variable. It also appears that it will be a dominant criterion in determining the ultimate applicability of alternate state variable devices. Clock speed versus density trade-offs for electron transport devices will dictate that for future technology generations; clock speed will need to be decreased for very high densities or conversely, density will need to be decreased for very high clock speeds. Nanoscale electron transport devices will best suit implementations that rely on the efficient use of multi-core processing to minimize energy dissipation.

4a. I_{on}/I_{off} *Ratio (Memory Devices)*: The I_{on}/I_{off} ratio of a memory device is the ratio of the resistance of a memory storage element in the I_{on} state to its resistance in the I_{off} state. For nonvolatile memories, the I_{on}/I_{off} ratio represents the ratio between the read current of a selected memory cell to the leakage current of an unselected cell. In cross-point memories, a very large I_{on}/I_{off} ratio is required to minimize power dissipation and maintain adequate read signal margin.

4b. *Gain (Logic Devices)*: The internal gain of nanodevices is an important limitation for presently used combinatorial logic where gate fan-outs require significant drive current and low voltages make gates more noise sensitive. New logic and low fan-out circuit approaches will be needed to use most of these nanodevices for computing applications. Signal regeneration for large circuits of nanodevices may need to be accomplished by integration with CMOS.

5. *Operational Reliability*: Operational reliability is the ability of the memory and logic devices to operate reliably within their operational error tolerance given in their performance specifications. The error rate of all nanoscale devices and circuits is a major concern. These errors arise from the difficulty of providing highly precise dimensional control needed to fabricate the devices and also from interference from the local environment. Large-scale and powerful error detection and correction schemes will need to be a central theme of any architecture and implementations that use nanoscale devices.

6. *Operational Temperature*: Nanodevices must be able to operate close to a room temperature environment for most practical applications with sufficient tolerance for higher temperature (e.g., 100 °C) operation internal to the device structure.

7. *CMOS Technological Compatibility*: The semiconductor industry has been based for the last 40 years on incremental scaling of device dimensions to achieve performance gains. The principal economic benefit of such an approach is it allows the industry to fully apply previous technology investments to future products. Any alternative technology as a goal should utilize the tremendous investment in infrastructure to the highest degree possible. Furthermore, in the near term, integratability of nanodevices with silicon CMOS is a requirement due to the need for signal restoration for many logic implementations and to be compatible with the established technology and market base. This integration will be necessary at all levels from design tools and circuits to process technology.

8. *CMOS Architectural Compatibility*: This criterion is motivated by the same set of concerns that motivate the CMOS technological compatibility, namely the ability to utilize the existing CMOS infrastructure. Architectural compatibly is defined in terms of the logic system and the data representation used by the alternative technology. CMOS utilizes Boolean logic and a binary data representation and, ideally, an alternative technology would need to do so as well.

1.7 Selection of Nanodevices as Technology Entries

Candidate Technology Entries are identified through a variety of means including a minimum of four workshops held in each 2 year cycle, extensive literature searches, personal knowledge of ERD ITWG members, ERD critiques, and so on.

These candidate Technology Entries are considered for inclusion in the ITRS ERD chapter based on their potential for accomplishing the objectives for a new technology, level of published research activity, credibility, and progress. In addition a candidate Technology Entry should have attained significant maturity in its research domain. Its continued inclusion is governed by progress in resolving those research issues gating demonstration of feasibility.

In addition to the requirements discussed above, specific criteria for including a Technology Entry in any one of the Research Devices and Architectures sections are: (1) research on the proposed Technology Entry is published by two or more groups in archival literature and peer reviewed conferences, or (2) research on the proposed Technology Entry is published extensively by one group in archival literature and in peer reviewed conferences.

1.8 Perspectives

Development and scaling of CMOS over the past 40+ years has revolutionized many aspects of our lives and has been a driver of the world-wide economy. We communicate globally at any time using a paper-thin computer–telephone, no larger than a wallet, with the computational power far exceeding that of a dated mainframe computer. Similar hand-held devices coupled to a complex satellite network system (Global Positioning System or GPS) can locate our position with an accuracy of a few feet and can navigate us through completely unknown terrain (e.g., a large metropolitan area or an arid desert). Soon automobiles will be able to talk to each other to operate an automobile safely making the human driver obsolete. These technologies have been enabled by ever-shrinking MOSFETs providing lower cost, higher speed, and lower power dissipation per transistor in silicon-based integrated circuits. Scaling of silicon MOSFETs will likely continue to the 8 nm, and possibly even to the 5 nm technology generation, at which point individual MOSFETs will have reached their fundamental scaling limit defined by the relatively small number of atoms making up a transistor.

Given this approaching limit on physical scaling of the MOSFET structure and its properties, how can the semiconductor industry continue to provide lower cost, higher density, and higher performance ICs? In the relative short-term, the design community will continue its efforts and successes to obtain performance gains and cost savings by deriving more value from current technology. Also, some important applications can be addressed by current technology for speed, but need lower power dissipation, at a lower cost per chip (e.g., a smart cell phone). Also, More-than-Moore technology, in which current IC technology is integrated with other current technologies (e.g., sensors, RF components) on a circuit board or in a package, is being used to provide many functions for new applications.

In the longer term, a new information-processing technology, capable of providing additional scaling of the primitive element's size, cost, and power dissipation together with increased data throughput, would have invaluable impact. Undoubtedly, it would set off the discovery of another round of new applications, some opening completely new possibilities such as nanomorphic microsystems used *in vivo* to diagnose and attack disease cells [8].

Perhaps Kroemer had the best perspective on application of a new technology when in his Nobel Lecture he stated [9]: "The principal applications of any sufficiently new and innovative technology always have been – and will continue to be – applications created by that technology."

So, as a community of pioneers seeking to discover a new information-processing technology, let us remain alert to new applications for our current findings.

References

1. Moore, G.E. (1965) Cramming more components onto integrated circuits. *Electronics Magazine*, **38**, 3–6.
2. Wikipedia (2014) http://en.wikipedia.org/wiki/Information_Age (accessed 16 January 2014).
3. Jefferies (2014) http://jefferies.com/CMSFiles/Jefferies.com/files/Insights/Moore%20Stress_Structural%20Industry%20Shift_09272012.pdf (accessed 16 January 2014).
4. Cellular News (2014) http://www.cellular-news.com/story/43209.php (accessed 16 January 2014).
5. Jorgenson, D.W., Ho, M., and Samuels, J. (2014) http://scholar.harvard.edu/files/jorgenson/files/02_jorgenson_ho_samuels19nov20101_2.pdf (accessed 16 January 2014).
6. Zhirnov, V.V., Cavin, R.K. III, Hutchby, J.A., and Bourianoff, G.I. (2003) Limits to binary logic switch scaling – A gedanken model. *Proceedings of the IEEE*, **91** (11), 33–36.
7. Zhirnov, V.V., Cavin, R.K. III, and Bourianoff, G.I. (2010) New State variable opportunities beyond CMOS: A system perspective, in *Emerging Technologies and Circuits* (eds A. Amara, T. Ea, and M. Belleville), Springer.
8. Zhirnov, V.V. and Cavin, R.K. III (2010) *Microsystems for Bioelectronics: the Nanomorphic Cell*, Elsevier Press.
9. Kroemer, H. in his Nobel Lecture (8 December 2000) *Quasi-Electric Fields and Band Offsets: Teaching Electrons New Tricks*, Aula Magna, Stockholm University.

2

What Constitutes a Nanoswitch? A Perspective

Supriyo Datta[1], Vinh Quang Diep[1], and Behtash Behin-Aein[2]
[1]*Purdue University, USA*
[2]*GLOBALFOUNDRIES Inc., USA*

2.1 The Search for a Better Switch

A basic element in digital logic is a switch or an inverter comprising a pair of complementary metal oxide semiconductor (CMOS) nanotransistors (Figure 2.1a) whose resistances R_1 and R_2 change in a complementary manner in response to the input voltage V_{in}. As V_{in} changes from 0 to V_{DD}, the resistance R_1 of the "NMOS" transistor gets smaller while the resistance R_2 of the "PMOS" transistor gets larger making the output voltage

$$V_{out} = V_{DD} \frac{R_1}{R_1 + R_2}$$

change from V_{DD} to 0 as shown in Figure 2.1b so that the output represents an inverted version of the input.

For some time now it has been recognized that one of the biggest obstacles to continued downscaling is the heat dissipated [1,2]. Every time a switch changes state, the charge Q stored in an input or an output capacitors gets dumped thus dissipating an energy of QV_{DD}. If there are N_{act} number of active switches switching at a frequency f per second, the power dissipated can be written as

$$P = N_{act}QV_{DD}f \tag{2.1}$$

To estimate the energy QV_{DD} dissipated per switch, we could use the numbers for the Intel® Core™ i3-530 Processor taken from their Web site at http://ark.intel.com/products/46472 [3]

Emerging Nanoelectronic Devices, First Edition. An Chen, James Hutchby, Victor Zhirnov and George Bourianoff.
© 2015 John Wiley & Sons, Ltd. Published 2015 by John Wiley & Sons, Ltd.

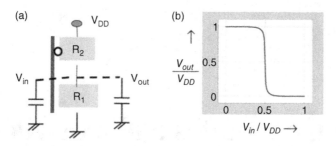

Figure 2.1 (a) CMOS inverter comprises an NMOS transistor (R_1) and a PMOS transistor (R_2). (b) Input–output characteristics of the inverter

$$P = 73 \text{ W}, f = 2.93 \text{ GHz}$$

$$N_{\text{act}} = \underbrace{559 \times 10^6}_{\substack{\text{Total number of}\\\text{transistors}}} \times \underbrace{10\%}_{\substack{\text{activity}\\\text{factor}}}$$

$$QV_{\text{DD}} \approx \frac{73 \text{ W}}{2.93 \text{ GHz} \times \left(559 \times 10^6 \times 10\%\right)} = 2785 \text{ eV} \approx 445 \text{ aJ}$$

Since the power dissipated cannot increase too much beyond 73 W we cannot increase the number of active switches N_{act} or their speed of operation f very much, unless we discover switches that dissipate less energy without compromising the speed. It was this recognition that prompted the Semiconductor Research Corporation (SRC) together with the National Science Foundation (NSF) to launch the Nanoelectronics Research Initiative (NRI) in 2005 with the objective of exploring the possibility of realizing a better switch based on any known physical mechanism.

Outline: In this chapter we would like to share our perspective on the question of what constitutes a "transistor-like switch," and how we could build one based on novel physical mechanisms and assess its performance. As an example of a radically different physical mechanism, we will focus on spintronics and nanomagnetics where there has been enormous progress in the last two decades. But we will try to phrase our discussion and conclusions in general terms so that it could be easily adapted to other phenomena as well.

The new discoveries in spins and magnets are already finding use in memory devices both to *Read (R)* information from magnets and to *Write (W)* information onto magnets. Many other new phenomena are being investigated for nanoelectronic memory as described in Part Two of this book. It seems natural to ask whether these advances in W&R units for memory devices could also translate into a new class of logic devices.

In Section 2.2 we start with a very brief and oversimplified discussion of the most common switch used to implement digital logic based on CMOS transistors stressing the key property of gain that allows us to interconnect them into complex circuits without the use of external amplifiers or clocks. To harness spins and magnets for logic applications one could either integrate them onto CMOS devices that provide the gain (see for example [4]) or try to design transistor-like spin–magnet devices having gain. It is the *latter* option that we will explore in this chapter.

We will argue that a CMOS switch can be viewed as an integrated *W–R* unit, using the word "Write" in a somewhat unconventional sense. The purpose is not to provide any new insight into CMOS, but to help understand how *W* and *R* units used for memory devices can be combined to build transistor-like switches.

In Section 2.3 we discuss the standard *W* and *R* devices used for magnetic memory devices and present one way to integrate them into a single unit where the input and output are electrically isolated, but we argue that such a unit would not provide the key transistor-like property of gain. We will then show (Section 2.4) that the recently discovered giant spin Hall effect (GSHE) could be used to construct a *W–R* unit with gain [5].

Other possibilities for transistor-like *W–R* units with gain are briefly discussed in Section 2.5 including all-spin logic (ASL) [6] along with new possibilities based on newly discovered phenomena. Indeed, with the growing research interest in STT-MRAM (spin transfer torque magnetic random access memory) for both stand-alone [7] and embedded memory applications [8,9] it is likely that many more new phenomena will be discovered that could be used to construct transistor-like *W–R* units for logic applications.

Also we should mention that there are other independent proposals like the trans-spinor [10] and m-logic [11] that could be viewed as examples of the same *W–R* paradigm for logic that we are discussing here.

In Section 2.6 we end with a brief discussion of how these alternative transistor-like switches could be evaluated in terms of possible applications. A key metric is the energy–delay product and we will argue that new materials and phenomena for *W* and *R* units are needed to provide any improvement over standard CMOS switches. On the other hand the nonvolatility and reconfigurability of switches based on magnets is a novel feature that could enable a whole new class of circuits very different from those currently possible.

2.2 Complementary Metal Oxide Semiconductor Switch: Why it Shows Gain

To understand the key characteristics of a transistor-like switch it is useful to take a brief look at a standard transistor. The simplest transistor is an NMOS or a PMOS, but we choose a CMOS switch which combines the two into a single switch that performs a logic operation, namely NOT, and has an input–output characteristic resembling those obtained from the spin switches discussed later in the chapter.

A CMOS switch is made of an NMOS and a PMOS transistor, which constitute the voltage-controlled resistors R_1 and R_2 shown in Figure 2.1a. Let us briefly describe the characteristics of an NMOS and a PMOS transistor, which can then be combined to obtain the input–output characteristics of the CMOS inverter shown in Figure 2.1b.

NMOS transistor: The resistor R_1 in Figure 2.1a is an NMOS transistor whose resistance

$$R_1 = V_{\text{out}}/I_1$$

is reduced by a positive input voltage V_{in}. For small input voltages the conductance $(1/R_1)$ increases exponentially with V_{in} (see for example [12])

$$\frac{I_1}{V_{\text{out}}} \sim e^{qV_{\text{in}}/kT}$$

Also, the resistance is not constant and ideally the current saturates for large V_{out}. We could describe this behavior approximately as (I_0: constant)

$$I_1 = I_0 e^{qV_{in}/kT} \times e^{qV_{out}/\beta kT} \times \left(1 - e^{-qV_{out}/kT}\right) \qquad (2.2)$$

With $\beta = 10^6$ the current saturates perfectly, which is what we would ideally like; but with $\beta = 10$ we have a characteristic looking more like real transistors, with the current showing an increase with V_{out} due to "drain-induced barrier lowering (DIBL)."

PMOS transistor: The other resistor (R_2) in Figure 2.1a is a PMOS transistor whose resistance is increased by a positive input voltage and we will assume that the characteristics can be described by an expression similar to Equation 2.2 but with V_{in} and V_{out} replaced by ($V_{DD} - V_{in}$) and ($V_{DD} - V_{out}$) respectively.

$$I_2 = I_0 e^{q(V_{DD}-V_{in})/kT} \times e^{q(V_{DD}-V_{out})/\beta kT} \times \left(1 - e^{-q(V_{DD}-V_{out})/kT}\right) \qquad (2.3)$$

In general the NMOS and PMOS need not be symmetric with the same constant I_0 appearing in both current expressions (see Equations 2.2 and 2.3) but we will ignore such "details", since our objective is to use the simplest model just to illustrate the main points.

Switch characteristics: The input–output characteristics of a CMOS inverter are obtained by solving Equation 2.2 for R_1 (NMOS) and Equation 2.3 for R_2 (PMOS) simultaneously. For any particular V_{in}, we adjust V_{out} numerically so as to make $I_1 = I_2$. This leads to the switch characteristics shown in Figure 2.2 for different values of the parameter β reflecting different degrees of current saturation as discussed earlier.

Gain: A key attribute of a logic unit is its *gain* defined as the change in the output voltage for a given change in the input voltage

$$Gain \equiv \frac{\Delta V_{out}}{\Delta V_{in}} \qquad (2.4)$$

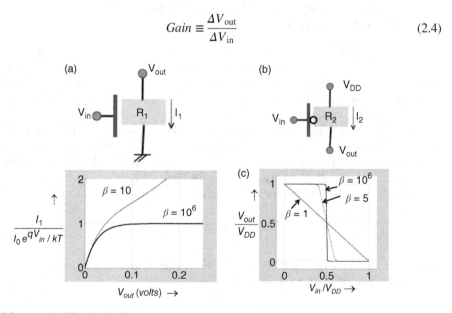

Figure 2.2 (a) NMOS, (b) PMOS and (c) Input-output characteristics of a CMOS input–output characteristics of a CMOS inverter obtained by solving Equation 2.2 for R_1 (NMOS) and Equation 2.3 for R_2 (PMOS) simultaneously for different values of β

A logic unit should have a gain >1 in order to drive another in a circuit. It is evident from Figure 2.2 that while with large values of β the inverter has a sizeable gain, the gain ≤ 1 if $\beta = 1$. Our β is a parameter introduced (see Equations 2.2 and 2.3) to account for the drain voltage dependence of the current and is usually far bigger than one for any real transistor. And so the situation with $\beta = 1$ is not of any real practical significance.

We are simply using the factor β to make the point that in order to have gain >1, one needs an input–output asymmetry whereby the current is controlled largely by V_{in}, and very little by V_{out}. This is evident if we rewrite the current in Equation 2.2 for large V_{out}

$$I_1 \approx I_0 e^{qV_{in}/kT} \times e^{qV_{out}/\beta kT}$$

showing that the factor β represents the *asymmetry* in the response of the current to the input and output voltages. With $\beta = 1$, this asymmetry is lost and so is the gain.

Switch as a Write–Read pair: Before we move on, let us point out that a CMOS switch could be viewed as a *Write–Read (W–R)* pair, if we use the word Write in a somewhat unconventional sense. This viewpoint may seem artificial and probably does not provide any insight into the operation of CMOS switches. Our reason for introducing it is as an aid to understand how Write and Read units based on spins and magnets can be combined to form switches.

We could define the state S of the complementary pair in terms of the ratio of the two resistances R_1 and R_2:

$$S = \log\left(\frac{R_2}{R_1}\right) \qquad (2.5)$$

and create two plots: the *Write (W)* characteristics showing S as a function of the input voltage V_{in} (Figure 2.3a) and the *Read (R)* characteristics showing the output voltage V_{out} as a function of the state S (Figure 2.3b).

The former describes the W operation whereby the state S of the CMOS inverter is set according to the input voltage V_{in}, while the latter describes the R operation in that the supply voltage V_{DD} results in an output voltage V_{out} depending on the state S as shown symbolically in Figure 2.3c:

$$V_{in} \rightarrow S \rightarrow V_{out} \qquad (2.6)$$

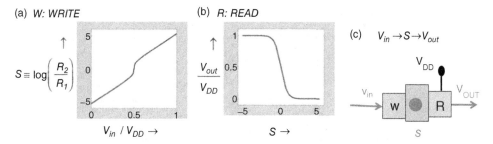

Figure 2.3 (a) "Write" operation in a CMOS inverter: State of the CMOS pair defined as $S = \log(R_2/R_1)$ as a function of the input voltage V_{in}. (b) "Read" operation: Output voltage V_{out} as a function of the state S. (c) Symbolic representation depicting the switch as a Write–Read pair

Note that we are stretching the meaning of the *Write* operation somewhat, since the state S does not persist once the input V_{in} has been removed: Unlike real memory devices, the Read operation needs to be carried out while V_{in} is present. Our purpose here is simply to connect the language of memory devices involving W and R units to that of CMOS so that we can understand and adapt the key property of gain that distinguishes logic units.

To integrate W and R into a transistor-like switch, an input–output asymmetry seems important: the gain of a CMOS switch seems intimately related to the fact that the input voltage V_{in} is far more effective in controlling the state of the switch S than the output voltage V_{out}. This input–output asymmetry and the resulting gain make a transistor very different from reversible Hamiltonian systems often discussed in the context of nanoscale systems. To harness spins and magnets for logic devices we have two broad options:

- Integrate them onto CMOS devices which provide the gain.
- Design transistor-like spin–magnet devices that have gain.

It is the latter possibility that we are discussing in this chapter.

2.3 Switch Based on Magnetic Tunnel Junctions: Would it Show Gain?

Since W and R units based on magnetic tunnel junctions (MTJs) are now well-known, it seems natural to ask whether these could be combined into a transistor-like switch. Before addressing that question let us briefly summarize how an MTJ-based W and R device works.

2.3.1 Operation of an MTJ

Figure 2.4a shows a simplified MTJ structure having one layer with a reference magnet $\hat{M}$ separated by a tunnelling barrier from a free layer magnet of nanometre scale thickness whose

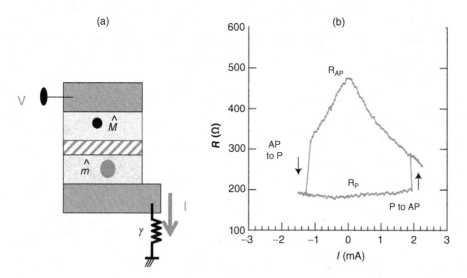

Figure 2.4 (a) A simplified schematic of a magnetic tunnel junction (MTJ). (b) A typical resistance versus current characteristic of an MTJ taken from Kubota *et al.* [13]. Reprinted by permission from Macmillan Publishers Ltd: Nature Physics, Ref. [13], copyright 2008

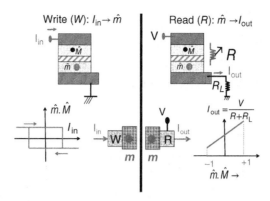

Figure 2.5   An MTJ unit can be used either as a Write (W) unit or as Read (R) unit

magnetization $\hat{m}$ represents the stored information. Figure 2.4b shows a typical resistance versus current characteristic of an MTJ taken from Kubota *et al.* [13] which illustrates the basic physical phenomena underlying both the R and W operations.

At low currents the resistance can have one of two values: The smaller one (R_P) corresponds to the ***P*** configuration with the two magnetizations *parallel*: $\hat{m}\| + \hat{M}$ while the larger one corresponds to the ***AP*** configuration with the magnetizations *anti-parallel*: $-\hat{m}\| + \hat{M}$. This phenomenon allows one to *Read* the state of the free magnet $\hat{m}$ relative to the fixed magnet $\hat{M}$ by applying a small voltage V.

On the other hand Figure 2.4b shows that at sufficiently high positive currents the free layer switches from a ***P*** to an ***AP*** configuration while at high negative currents it switches from an ***AP*** to a ***P*** configuration. This phenomenon allows one to *Write* information contained in the polarity of the current onto the magnetization of the free layer.

Figure 2.5 shows the basic characteristics of the Write and Read unit based on MTJ device and their symbolic operations. The Write unit converts the input current V_{in} into the magnetization $\hat{m}$ of the free magnet, while the Read unit converts the information stored in $\hat{m}$ into an output current V_{out} given by

$$I_{out}(\hat{m}) = \frac{V}{R(\hat{m}) + R_L} \tag{2.7}$$

where V is the supply voltage and R_L is a fixed load resistance.

2.3.2   W–R Unit with Electrical Isolation

We can now proceed to combine an MTJ *Read (R)* device with a *Write (W)* device to obtain a composite unit as shown in Figure 2.6a where the magnet $\hat{m}$ from R is coupled to the magnet $\hat{m}'$ from W through their dipolar magnetic field as indicated by a dashed line. This allows the information to propagate from input to output: An input current I_{in} switches the *Write* magnet ($\hat{m}'$) which in turn switches the *Read* magnet ($\hat{m}$) through the dipolar coupling causing a change in the output current I_{out} as described by Equation 2.7. At the same time the input is electrically isolated from the output allowing these units to be interconnected to form large circuits. This

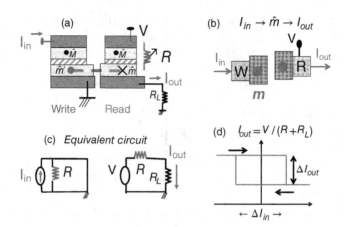

Figure 2.6   (a) A Write (W) and a Read (R) unit combined to obtain a logic unit with input–output isolation. The dashed line represents the magnetic coupling between the two free magnets. (b) Representative symbol. (c) Equivalent circuit. (d) Input–output characteristic

feature has some similarity to m-logic [11] which proposes to use exchange coupling to couple input domains to the output.

The *W–R* unit in Figure 2.6a can be modelled with an equivalent circuit of the form shown in Figure 2.6c which leads to the overall input–output characteristic shown in Figure 2.6d.

2.3.3   Does This W–R Unit Have Gain?

So far things seem straightforward, combining a *W* unit with an *R* unit using magnetic coupling to allow information transfer while maintaining electrical isolation. But can this unit exhibit gain, so that the swing in the output current will exceed that in the input current?

The swing in the output current is proportional to the voltage as in Equation 2.7 or Figure 2.6d. It seems that we could make it exceed ΔI_{in} simply by choosing a large enough supply voltage V.

The problem, however, is this. The voltages V of MTJs in the *Read* unit also give rise to a spin current I_s that acts on the *Read* magnet $\hat{m}$. Ordinarily *Read* voltages are kept small enough such that the resulting spin current I_s does not disturb the free layer whose information we are trying to read. But if we do that, the output current would be much smaller than what is needed as input to drive the next stage and we could not build circuits without using an external amplifier.

To obtain an output spin current comparable to the critical spin current needed to drive the input of the next stage we have to make the supply voltage V even larger and the resulting spin current $\vec{I}_s$ acting on the *Read* magnet $\hat{m}$ would exceed the spin current $\vec{I}'_s$ acting on the *Write* magnet $\hat{m}'$. The state of the magnet $\hat{m}$ will then be determined by the output rather than the input. This is analogous to building a CMOS inverter (Section 2.2) using transistors whose current is controlled more strongly by the drain than by the gate, described by a β lesser than one.

We need to design the magnet pair such that it "feels" the influence of the input current far more strongly than that of the output current. It may be possible to design composite magnets with different materials to achieve this, but a relatively straightforward design seems possible utilizing a relatively recent discovery, namely the giant spin Hall effect (GSHE) as we will describe next.

2.4   Giant Spin Hall Effect: A Route to Gain

The GSHE is exhibited by materials with spin–orbit coupling where the flow of charge current I is accompanied by a spin current I_s in the perpendicular direction, such that the *spin current density* equals the *charge current density* times the spin Hall angle [14]:

$$\frac{I_S}{L} = \theta_H \frac{I}{t}$$

so that

$$I_S = \theta_H \underbrace{\frac{L}{t}}_{\equiv \beta} I \tag{2.8}$$

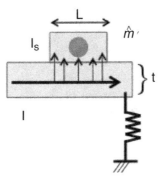

The spin Hall angle θ_H is usually quite small, but recently a number of GSHE materials have been discovered which have θ_H values as high as 0.3 [15]. More interestingly, a proper choice of geometry with $L \gg t$ can give values of β in excess of one, corresponding to a spin current I_s that exceeds the current I.

We can make use of this natural gain provided by the GSHE material, by replacing the MTJ-based Write unit in Figure 2.6a with the one shown in Figure 2.7a. To get better magnetic coupling between the *Write* and *Read* magnets it may be advisable to stack them vertically as shown in Figure 2.7b. In any case the *W–R* unit can be modelled with an equivalent circuit of the form shown in Figure 2.7c obtained by combining the *W* unit with a separate equivalent circuit for the *R* unit. Here G_m is the conductance of the MTJ device and it can be related to G and ΔG representing the sum and difference respectively of the parallel and anti-parallel conductances.

$$G_m = \frac{G}{2} + \frac{\Delta G}{2} \hat{m}.\hat{M} \tag{2.9}$$

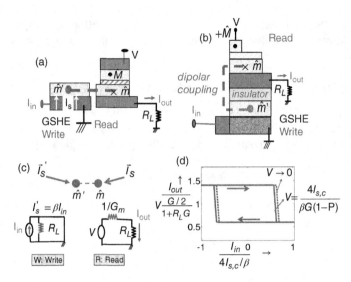

Figure 2.7 (a) A GSHE Write and a Read unit combined to build a logic unit with input–output isolation and gain. The dashed line represents the magnetic coupling between the two free magnets. (b) Better magnetic coupling between the Write and Read magnets can be obtained by stacking the units vertically rather than laterally. (c) Equivalent circuit for structure in (a). (d) Input–output characteristic

The parameters G and ΔG can be related to experimentally reported quantities like the tunnelling magnetoresistance (TMR)

$$\text{TMR} \equiv \frac{G_\text{P}}{G_\text{AP}} - 1 \qquad (2.10a)$$

or the polarization $P \equiv \Delta G/G$

$$P \equiv \frac{G_\text{P} - G_\text{AP}}{G_\text{P} + G_\text{AP}} = \frac{\text{TMR}}{\text{TMR} + 2} \qquad (2.10b)$$

The input–output characteristic (Figure 2.7d) was obtained from the equivalent circuit in Figure 2.7c using the same method as described in [5] with the spin currents coupled to the Landau–Lifshitz–Gilbert (LLG) equation for the magnet pair $\hat{m}' - \hat{m}$. For a small supply voltage $V(\rightarrow 0)$ the characteristic is symmetric about the origin but it shifts to the left as V is increased because the spin current I_s injected by the *Read* unit makes it easier to switch from +1 to −1 than to switch from −1 to +1. Indeed if the voltage V were too large we would not have a useful switch. But the GSHE allows us to use a relatively small V and still have gain.

The key point is that the gain β from the GSHE allows the use of a relatively low voltage. A *Write* unit based on an ordinary spin–torque device (like the one discussed in the last section) has a spin current that is less than the charge current I, corresponding to a β less than one and thus requires a much larger voltage V to drive the next unit.

2.4.1 Concatenability

Although the switch in Figure 2.7 exhibits gain and input–output isolation, it is not "concatenable" because the output from the *Read* unit is purely positive (assuming *V* is positive) and is not appropriate for driving the *Write* unit of the next stage which requires a bipolar input that takes on both positive and negative values. One can think of two broad approaches to addressing this concatenability issues:

- Design a *Write* unit that can be driven with purely positive voltages.
- Design a *Read* unit that produces a bipolar output.

One possible design, based on the second approach [5], is shown in Figure 2.8a. It requires a more complicated fabrication process since two MTJs are required: If they had the same resistance the output voltage would be zero since one is connected to $+V$ and one to $-V$. But the two MTJs will never have the same resistance, since their fixed magnets are antiparallel, namely $+\hat{M}$ and $-\hat{M}$. Depending on whether the free layer magnetization $\hat{m}$ is parallel to $+\hat{M}$ or $-\hat{M}$, one MTJ will be in its low resistance or P configuration while the other will be in its high resistance or AP configuration.

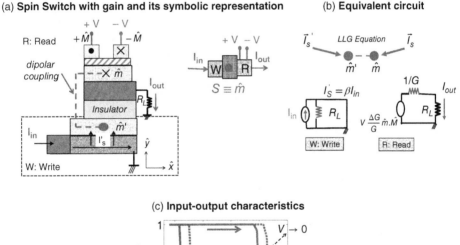

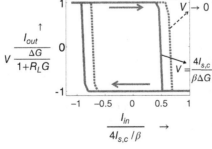

Figure 2.8 (a) An integrated *W–R* unit obtained by vertically stacking a W device based on the GSHE and a dual MTJ R device with the corresponding magnets $\hat{m}'$ and $\hat{m}$ magnetically coupled. (b) Equivalent circuit for structure in (a). (c) Input–output characteristics of the device which also implies gain and nonvolatility properties of the switch. Reprinted with permission from [5]. Copyright 2012, AIP Publishing LLC

If the MTJ connected to $+V$ is in its low resistance state then the output will be closer to $+V$ and hence positive. If the MTJ connected to $-V$ is in its low resistance state then the output will be closer to $-V$ and hence negative. This dual MTJ Read unit should thus do what we want, namely convert positive or negative magnetization into a bipolar (that is, positive or negative) output voltage; and hence a bipolar output current.

For quantitative modelling we could use the equivalent circuit shown in Figure 2.8b, where G and ΔG are the same as the ones used in Equation 2.9. This equivalent circuit shows *that the open circuit voltage is proportional to* $\hat{m}.\hat{M}$, *the component of* $\hat{m}$ *along* $\hat{M}$, giving an output current of

$$I_{\text{out}} = \frac{V\Delta G/G}{R_{\text{L}} + 1/G}\hat{m}\cdot\hat{M}$$

Figure 2.8c shows the input–output characteristic calculated using the equivalent circuit shown in Figure 2.8b and coupling the spin currents to the LLG equation for the magnet pair $\hat{m}' - \hat{m}$ [5].

The gain can be estimated by noting that from Figure 2.8c

$$\frac{\Delta I_{\text{out}}}{V(\Delta G/1 + R_{\text{L}}G)} \approx 2\frac{\Delta I_{\text{in}}}{4I_{\text{s,c}}/\beta}$$

so that

$$Gain \equiv \frac{\Delta I_{\text{out}}}{\Delta I_{\text{in}}} \approx \frac{V\Delta G}{1 + R_{\text{L}}G}\frac{\beta}{2I_{\text{sc}}} \tag{2.11}$$

For an approximate gain of ~ 2, we need a voltage of

$$V \approx \frac{4I_{\text{sc}}}{\beta\Delta G}, \quad \text{if} \quad R_{\text{L}}G \ll 1$$

which only has a minor effect on the input–output characteristic (Figure 2.8c).

2.4.2 Proof of Gain and Directionality

A good test for switches with gain and directionality is the following. It should be possible to connect an odd number of such switches to form a ring oscillator (Figure 2.9). If the voltages V on the *Read* units exceed the threshold value needed to drive the following *Write* unit, then each unit switches the next unit anti-parallel to itself. With an odd number of magnets, three in this case, in the loop, there is no way for all three units to be anti-parallel to each other and there is no satisfactory steady state. Unit 1 switches unit 2, which in turn switches unit 3, which goes on to switch unit 1 and the result is an oscillatory output as obtained from detailed simulation [5].

Such oscillations are well known using an odd number of CMOS switches, and should provide a good test for the properties of gain and directionality which ensure that a signal can propagate without losing strength or compounding errors.

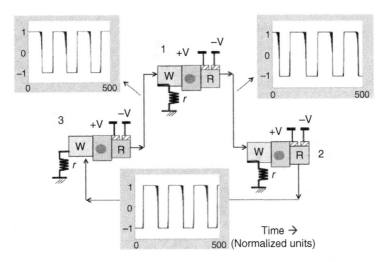

Figure 2.9 An odd number of *W–R* units with gain and directionality can be connected in a ring to form a ring oscillator. Reprinted with permission from [5]. Copyright 2012, AIP Publishing LLC

2.5 Other Possibilities for Switches with Gain

So far we have seen two examples of switches, the standard CMOS switch and a proposed one based on spins and magnets. Both can be viewed as integrated *W–R* units where the input V_{in} or I_{in} *writes* the internal state *S* which is then *read* to generate an output. As noted earlier, in the case of CMOS we are stretching the meaning of *Write* somewhat.

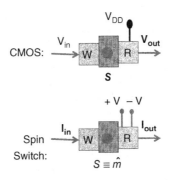

For the CMOS switch, the internal state *S* can be defined in terms of the resistances of the NMOS and PMOS transistors, while for the spin switch *S* represents the magnetizations of the magnet pair:

$$\text{CMOS Switch}: \quad S \equiv \log(R_2/R_1)$$
$$\text{Spin Switch}: \quad S \equiv \hat{m}$$

We have argued that a spin switch with gain could be implemented by combining a dual MTJ-based *Read* unit with a GSHE-based *Write* unit. However, this is by no means the only

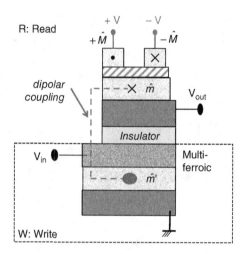

Figure 2.10   Voltage-controlled spin switch: If the Write magnet m′ is conducting and is insulated from V_{in} by an insulator (similar to the gate oxide of an MOS transistor), it may be possible to achieve electrical isolation without using a separate dipole-coupled Read magnet m so that $V_{in} \rightarrow m′ \rightarrow V_{out}$

possibility. For example, the *Write* unit could involve a voltage-controlled multiferroic [16] as shown in Figure 2.10.

Note, however, that in order for one unit to be able to drive another, the output voltage V_{out} has to be large enough to switch the multiferroic, thus requiring a minimum voltage V on the *Read* device. This voltage should not be too large, or the resulting spin current could control the magnet pair $\hat{m}′ - \hat{m}$ instead of the input voltage. As we have argued, a suitable degree of input–output asymmetry is needed and whether it can be achieved has to be assessed carefully for each individual proposal. Indeed many other phenomena like voltage-controlled magnetic anisotropy (VCMA) (see for example [17–19]) could potentially be used to design improved W units.

A key difference with the CMOS switch is that unlike the resistance of an NMOS or a PMOS transistor, the magnetization represents a nonvolatile state S. However, it may be possible to use other mechanisms for voltage-controlled resistance based on phase transition phenomena (like the Mott transition) [20] which could provide a nonvolatile internal state S like magnets.

2.5.1   All-spin Logic

Both the CMOS and the spin switch that we have described involve ordinary voltages and currents as the input and output variables and can be interconnected with ordinary wires to form circuits.

One could also envision switches based on *spin voltages* and *spin currents* as the input and output variables. An input spin current switches the *Write* magnet $\hat{m}′$ which through the dipolar coupling switches the *Read* magnet $\hat{m}$ causing the output spin current to switch (Figure 2.11). This is similar to the all-spin logic (ASL) device [6,21] with the difference that the magnets $\hat{m}′$ and $\hat{m}$ are electrically isolated in the present version. It was shown theoretically in [6,21] that switches with gain and directionality can be implemented with voltages as low as 10 mV.

Spin currents could in principle carry more information than ordinary currents and enable devices one step closer to quantum information processing. However, unlike ordinary charge currents, spin currents die out within a spin coherence length which can vary widely from tens of nanometers to tens of microns depending on the material and the temperature of operation.

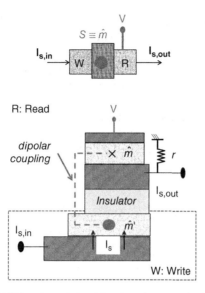

Figure 2.11 All spin logic (ASL): spin switch with spin voltages and spin currents as input and output

As it stands, this could still allow information transfer up to the first layer of interconnects (Metal layer 1, M1), but not for longer lengths (M2 and higher).

2.6 What do Alternative Switches Have to Offer?

We have tried to present a general perspective on how *Write* and *Read* units can be integrated into switches with gain. But what do these alternative switches have to offer relative to the standard CMOS switches that are widely used?

2.6.1 Energy–Delay Product

We started this chapter pointing out that a key roadblock on the path of miniaturization is the energy it takes to operate a switch. It is well-known that the switching energy can be reduced by going slow, so that the energy per se is not a fundamental property of a particular switch. It makes more sense to look at the energy–delay product. Consider for example [22] the charging of a capacitor C through a resistance R from a voltage V, for which it is well known that

$$\text{Energy}, E \sim QV$$
$$\text{Delay}, \tau \sim RC = RQ/V$$

Combining the two relations we obtain

$$\text{Energy-Delay Product}: E\tau \sim Q^2 R \tag{2.12}$$

suggesting that the energy–delay product is determined simply from two quantities:

1. How much charge Q is being switched?
2. What is the resistance R through which it is being switched?

For CMOS switches the resistance R is ~ tens of kilo-ohms, while the charge Q is more difficult to estimate. In our introduction we used system level numbers to estimate the quantity QV_{DD} as ~3000 eV, suggesting that Q ~ 3000 electrons, since V_{DD} ~1 volt. On the other hand if we look at the gate charge on an individual transistor it would be over an order of magnitude smaller. We believe the discrepancy is because the former estimate Q includes additional parasitic charges.

How does this compare with spin switches? Spin switches being all metallic structures usually have lower resistances of several tens of ohms. But the charge Q is ordinarily much larger, making Q^2R much larger. It has been shown [23] that the *minimum* charge Q needed to switch a magnet through an ordinary spin–torque mechanism (Slonczewski spin transfer torque) is given by:

$$Q \geq 2qN_s \tag{2.13}$$

where N_s is the number of spins comprising the magnet which is related to the saturation magnetization through the relation

$$N_s = \frac{M_s\Omega}{\mu_B} \tag{2.14}$$

where μ_B is the Bohr magneton, Ω is the volume of magnet.

Typical values of M_s~10^6 A/m give an N_s of about 100 spins in a volume of 1 nm^3. This means that even a magnet as small as ~$10 \times 10 \times 1$ nm has 10^4 spins, and from the inequality in Equation 2.13, the charge Q is at least 20 000 electrons well in excess of the CMOS numbers.

How fundamental is the inequality in Equation 2.13? One could understand this inequality by noting that the process of switching a magnet with a stream of incident electrons can be written as

$$\text{Incident Electrons} + N_s\hat{m} \rightarrow \text{Reflected Electrons} - N_s\hat{m}$$

During switching the magnet spin changes by $2N_s$ and it takes at least $2N_s$ electrons to conserve spin.

This argument, however, assumes that there is no other source of spin and the phenomenon of GSHE allows us to bypass this argument since the strong spin–orbit coupling provides a source of spin.

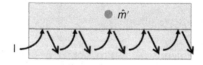

An electron on its way through the GSHE material gets deflected towards the magnet, flips its spin, has its spin randomized and then is deflected again by the spin–orbit coupling towards the magnet and so on. In other words the same electron on its way through the GSHE material is incident repeatedly on the magnet and transfers many units of spin to it.

Indeed there is experimental evidence [14] that the GSHE material allows us to switch a magnet with less number of electrons than $2N_s$. It would seem that we could reduce the charge transferred from Equation 2.13 to

$$Q \geq \frac{2qN_s}{\beta} \tag{2.15}$$

However, the GSHE gain β depends on the length of the magnet (Equation 2.8) which also makes the magnet longer and increases N_s. The possible improvement is thus useful but not unlimited. It thus seems that magnet-based switches will not provide the low power solution we are looking for unless more suitable *Write* mechanisms can be identified. Moreover, the dual MTJ *Read* unit fails with respect to another impressive characteristic of the CMOS switch: negligible standby power. These are the issues that need increased attention in the coming years.

2.6.2 Beyond Boolean Logic

Barring a major improvement is it worth pursuing alternative switches? We believe the answer is yes because of many other nonconventional applications that may be possible.

Consider for example the reconfigurable correlator shown in Figure 2.12 which should provide an output that correlates the incoming signal $\{X_n\}$ with a reconfigurable reference signal $\{Y_n\}$ stored in the m_z of the switches that could be any string of +1's and −1's of length N, N being a large number.

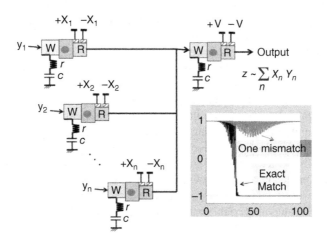

Figure 2.12 An example of a device that could be implemented by interconnecting spin switches (Figure 2.8a) which should provide an output that correlates the incoming signal $\{X_n\}$ with a reconfigurable reference signal $\{Y_n\}$ stored in the switches. Inset shows response of output magnetization as a function of time (normalized). Threshold is adjusted such that the magnet is switched only if all 20 bits of $\{Y\}$ match all 20 bits of $\{X\}$. With even one mismatch the output fails to switch. Note that no middle circuitry for signal conversion or amplification is involved. Reprinted with permission from [5]. Copyright 2012, AIP Publishing LLC

Since the output current of each Read unit is a product of $V(\sim X_n)$ and $m_z(\sim Y_n)$ it is determined by $X_n Y_n$ which are all added up to drive the output magnet. If the sequence $\{X\}$ is an exact match to $\{Y\}$, then the output voltage will be N, since every $X_n Y_n$ will equal $+1$, being either $(+1)^*(+1)$ or $(-1)^*(-1)$. If $\{X\}$ matches $\{Y\}$, in $(N - n)$, instances with n mismatches, the output will be $N - 2n$ since every mismatch lowers output by 2. If we set the threshold for the output magnet to $N - 2N_e$ then the output will respond for all $\{X\}$ that matches the reference $\{Y\}$ within a tolerance of N_e errors. The inset in Figure 2.12 shows an example with $N_e = 0$.

This is a rather unique device which would allow us to correlate an input analog signal with a stored digital code to produce an analog output. This could be useful in mobile phones for decoding CDMA signals. Moreover it has an intriguing similarity to biological systems which correlate weak analog signals from cellular processes with a digital code stored in DNA molecules.

2.7 Perspective

But let us not go too far out on a limb with speculations. The objective here is simply to present our perspective viewing switches as *Write–Read* units underlining the key role played by gain and directionality in enabling large scale circuits. Hopefully this will help guide the search for new *Write* and *Read* mechanisms that could lead to fast low energy switches allowing miniaturization to continue for many more generations. But even otherwise, additional features like reconfigurability and nonvolatility could enable new functionalities currently not available. For example, the Spin Switch [5] could provide a compact implementation for neuron and synapse [24] as well as logic gates that compute with spins and magnets [25].

2.8 Summary

This chapter presents our perspective on how W and R devices in general, spintronic or otherwise, can be integrated into switches having gain and directionality like transistors. Such switches could be interconnected to build complex circuits without external amplifiers or clocks. We start with a very brief and oversimplified discussion about CMOS transistors and argue that a CMOS switch can be viewed as an integrated W–R unit having an input–output asymmetry that give it gain and directionality. Next we discuss the standard W and R units used for magnetic memory devices and present one way to integrate them into a single unit with the input electrically isolated from the output. But this integrated W–R unit would not provide the key property of gain. We then show that the recently discovered GSHE could be used to construct a W–R unit with gain and suggest other possibilities for spin switches with gain.

We end with a brief evaluation of these alternative switches in terms of possible applications. For conventional Boolean logic, at the present magnet-based switches will not provide the low power solution over standard CMOS switches unless more suitable *Write* mechanisms can be identified. On the other hand the nonvolatility and reconfigurability of switches based on magnets is a novel feature that could enable a new class of circuits that are very different from those currently possible.

Acknowledgments

It is a pleasure to acknowledge a number of colleagues who have contributed to this work in different ways over many years: Sayeef Salahuddin, Angik Sarkar, Srikant Srinivasan, and Deepanjan Datta. S.D. and V.Q.D. are grateful for support from the Institute for Nanoelectronics Discovery and Exploration (INDEX) and the NSF-sponsored Center for Science of Information.

References

1. Haensch, W., Nowak, E.J., Dennard, R.H. *et al.* (2006) Silicon CMOS devices beyond scaling. *IBM Journal of Research and Development*, **50**, 339–361.
2. Theis, T.N. and Solomon, P.M. (2010) In quest of the 'Next Switch': Prospects for greatly reduced power dissipation in a successor to the silicon field-effect transistor. *Proceedings of the IEEE*, **98**(12), 2005–2014.
3. Augustine, C. and Srinivasan, S. (2014) Notes on energy QV$_{DD}$ dissipated per switch. See Intel® Core™ i3-530 Processor at http://ark.intel.com/products/46472 (accessed 16 January 2014).
4. Kang, S.H. and Lee, K. (2013) Emerging materials and devices in spintronic integrated circuits for energy-smart mobile computing and connectivity. *Acta Materialia*, **61**(3), 952–973.
5. Datta, S., Salahuddin, S. and Behin-Aein, B. (2012) Non-volatile spin switch for Boolean and non-Boolean logic. *Applied Physics Letters*, **101**(25), 252411.
6. Behin-Aein, B., Datta, D., Salahuddin, S. and Datta, S. (2010) Proposal for an all-spin logic device with built-in memory. *Nature Nanotechnology*, **5**(4), 266–270.
7. Everspin Technologies Inc. (2013) "Spin-Torque MRAM TECHNICAL BRIEF.".
8. Kang, S.H. (2010) Embedded STT-MRAM for Mobile Applications: Enabling Advanced Chip Architectures, Non-Volatile Memories Work, UCSD.
9. Thomas, L., Jan, G., Zhu, J. *et al.* (2013) Magnetization dynamics in perpendicular STT-MRAM cells with high spin-torque efficiency and thermal stability, 58TH Annu. Conf. Magn. Magn. Mater. 4–8 Nov. 2013 Denver, Colorado.
10. Bae, S., Zurn, S., Egelhoff, W.F. *et al.* (2001) Magnetoelectronic devices using α-Fe/sub 2/O/sub 3/bottom GMR spin-valves. *IEEE Transactions on Magnetics*, **37**(4), 1986–1988.
11. Morris, D., Bromberg, D., Zhu, Jian-Gang and Pileggi, L. (2012) mLogic: Ultra-low voltage non-volatile logic circuits using STT-MTJ devices, Des. Autom. Conf. (DAC), 2012 49th ACM/EDAC/IEEE, pp. 486–491.
12. Taur, Y. and Ning, T.H. (1998) *Fundamentals of Modern VLSI Devices*, Cambridge University Press.
13. Kubota, H., Fukushima, A., Yakushiji, K. *et al.* (2007) Quantitative measurement of voltage dependence of spin-transfer torque in MgO-based magnetic tunnel junctions. *Nature Physics*, **4**(1), 37–41.
14. Liu, L.Q., Pai, C.F., Li, Y. *et al.* (2012) Spin-torque switching with the giant spin hall effect of tantalum. *Science*, **336**(6081), 555–558.
15. Pai, C.-F., Liu, L., Li, Y. *et al.* (2012) Spin transfer torque devices utilizing the giant spin Hall effect of tungsten. *Applied Physics Letters*, **101**(12), 122404.
16. Heron, J.T., Trassin, M., Ashraf, K. *et al.* (2011) Electric-field-induced magnetization reversal in a ferromagnet-multiferroic heterostructure. *Physical Review Letters*, **107**(21), 217202.
17. Liu, L., Pai, C.-F., Ralph, D.C. and Buhrman, R.A. (Sep. 2012) Gate voltage modulation of spin-Hall-torque-driven magnetic switching, *arXiv:*1209.0962 *[cond-mat.mtrl-sci]*.
18. Amiri, P.K. and Wang, K.L. (2012) Voltage-controlled magnetic anisotropy in spintronic devices. *SPIN*, **02**(03), 1240002.
19. Zhu, J., Katine, J.A., Rowlands, G.E. *et al.* (2012) Voltage-induced ferromagnetic resonance in magnetic tunnel junctions. *Physical Review Letters*, **108**(19), 197203.
20. Zhou, Y. and Ramanathan, S. (2013) Correlated electron materials and field effect transistors for logic: A review. *Critical Reviews in Solid State and Material Sciences*, **2013**, 286.
21. Srinivasan, S., Sarkar, A., Behin-Aein, B. and Datta, S. (2011) All-spin logic device with inbuilt nonreciprocity. *IEEE Transactions on Magnetics*, **47**(10), 4026–4032.
22. Feynman, R.P. (1998) *Feynman Lectures on Computation*, Addison-Wesley.
23. Behin-Aein, B., Sarkar, A., Srinivasan, S. and Datta, S. (2011) Switching energy-delay of all spin logic devices. *Applied Physics Letters*, **98**(12), 123510.
24. Diep, V., Brian, S., Behin-Aein, B. and Datta, S. (2014) Spin Switches for compact implementation of neuron and synapse. *Appl. Phys. Lett.*, **104**, 222405.
25. Behin-Aein, B., Wang, J.-P., and Weisendanger, R. (2014) Computing with Spins and Magnets. *MRS Bulletin*, **39**, 696.

Part Two

Nanoelectronic Memories

3

Memory Technologies: Status and Perspectives

Victor V. Zhirnov[1] and Matthew J. Marinella[2]
[1]*Semiconductor Research Corporation, USA*
[2]*Sandia National Laboratories, USA*

3.1 Introduction: Baseline Memory Technologies

Memory devices are a crucial component of current and future Information and Communication Technologies (ICT). There is a common distinction between ICT *memory* and *storage*. Memory often implies devices which are random-access, fast, evanescent, and relatively expensive. Typical example are Random Access Memories (RAM) put on a microprocessor chip or installed in a computer, such as Static Random Access Memory (SRAM) and Dynamic Random Access Memory (DRAM). In contrast, storage technologies are sometimes defined as sequentially accessed, slow, permanent, and inexpensive. Classical examples are magnetic Hard Disk Drives (HDD) and flash memory based on a floating gate FET cell. It should be noted however, that the distinction between memory and storage is fluid; moreover, new emerging nonvolatile memory technologies may considerably reduce the separation between memory and storage. For example, storage-class memory is emerging as a new technology with attributes of both memory and storage devices [1].

Current baseline memory technologies include dynamic random-access memory, static random-access memory, and flash memory based on a floating gate FET structure. Because all new memory concepts attempt to mimic and improve on the capabilities of a present day memory technology, this chapter provides an outlook for key performance parameters and scaling and performance projections for the baseline memory technologies.

The memory device is viewed in this chapter as being composed of three fundamental components: (a) the "Storage node," which is usually characterized by the physics of operation of different memory devices; (b) the "Sensor" which reads the state, and is an electrical device such as transistor, and (c) the "Selector" (select device), which allows a given memory cell in an array to be addressed for read or write, and is a nonlinear element, such as transistor or diode. All three components impact performance and scaling limits for all memory devices.

Emerging Nanoelectronic Devices, First Edition. An Chen, James Hutchby, Victor Zhirnov and George Bourianoff.
© 2015 John Wiley & Sons, Ltd. Published 2015 by John Wiley & Sons, Ltd.

It should be noted that the treatment in this chapter covers the fundamentals and develops theoretical scaling limits for several memory technologies. In a number of instances, the chapter deals with idealized or simplified structures, which may be different from practical memory realizations.

3.2 Essential Physics of Charge-based Memory

The current baseline memory technologies (DRAM, SRAM, and flash) are based on storing an electron charge in a storage node. Two distinguishable states 0 and 1 are created by the presence (i.e., state 0) or absence (i.e., state 1) of electrons in a specific location (the charge storage node). In order to prevent losses of the stored charge, the storage node is defined by energy barriers of sufficient height E_b to retain charge (as shown in Figure 3.1). The properties of the barrier, that is, barrier height, E_b, and width, a, determine the retention time of a memory cell.

There are two fundamental mechanisms for the losses of the stored charge. The first is the thermal over-barrier transitions (thermionic emission), which is related to the Boltzmann probability. The electron escape frequency is given by:

$$f_{\text{therm}} = f_0 \exp\left(-\frac{E_b}{k_B T}\right) \tag{3.1a}$$

Where $f_0 \sim 10^{12}\,\text{s}^{-1}\,\text{nm}^{-2}$ is the thermal attempt frequency.

Correspondingly, the retention time for one electron in a system with cross-sectional dimension L is:

$$t_r = \frac{1}{L^2 f_{\text{therm}}} = \frac{1}{L^2 f_0} \exp\left(\frac{E_b}{k_B T}\right) \tag{3.1b}$$

For a specified t_r, the required minimum barrier height is:

$$E_{b_{\text{min}}} = k_B T \ln\left(t_r \cdot f_0 \cdot L^2\right) \tag{3.1c}$$

In the case of the "nonvolatility requirement," that is, $t_r > 10$ years, gives $E_{b_{\text{min}}} \geq 1.42\,\text{eV}$ at $T = 300\,\text{K}$. For small retention times, for example, 50–100 ms, typical for DRAM, Equation 3.1c yields $E_{b_{\text{min}}} \geq 0.8\,\text{eV}$.

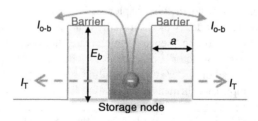

Figure 3.1 A generic electron charge-based memory element

A second source of charge loss is electron tunneling. The tunneling escape frequency for a rectangular barrier is:

$$f_T = f_0^* \cdot \exp\left(-\frac{2\sqrt{2m}}{\hbar} \cdot a \cdot \sqrt{E_b}\right) \tag{3.2a}$$

Where $f_0^* \sim 10^{13}\ \mathrm{s^{-1}\ nm^{-2}}$ is the tunneling attempt frequency.
The electron escape time due to tunneling is:

$$t_T = \frac{1}{L^2 f_0^*} \exp\left(\frac{2\sqrt{2m}}{\hbar} \cdot a \cdot \sqrt{E_b}\right) \tag{3.2b}$$

Suppose that the barrier height is large enough to suppress over-barrier escape, that is, $E_b \gg E_{b_{\min}}$, where $E_{b_{\min}}$ is given by Equation 3.1c. In this case, the store time will be determined by the tunneling time, t_T: $t_s \approx t_T$. The minimum barrier width for a specified store time, can be estimated from Equation 3.2b, for example, for $t_s = 10$ years:

$$a_{\min} = \frac{\hbar}{2\sqrt{2mE_b}} \ln\left(f_0^* \cdot t_s \cdot L^2\right) \tag{3.2c}$$

As a numerical estimate for $t_s > 10$ years, $E_{b_{\min}} \geq 1.42\ \mathrm{eV}$, $m = m_e$ and $T = 300\ \mathrm{K}$, Equation 3.2c gives $a_{\min} \sim 5\ \mathrm{nm}$.

As follows from the above, in order to obtain an electronic memory cell, sufficiently high barriers must be created to retain the charge for a long period of time. In physical systems the barrier can be created by combining materials with different properties, such as semiconductors with different doping structure, for example, n-p-n or conductor–insulator heterojunctions.

3.3 Dynamic Random Access Memory

Several schematic representations of a DRAM cell are shown in Figure 3.2: a schematic electrical diagram, a DRAM cell cross-section, and a simplified energy barrier diagram. Each DRAM cell in an array consists of a cell capacitor (Storage Node) in series with a FET (Selector).

The minimum DRAM cell capacitance should be $C_{\mathrm{cell}} \sim 25\ \mathrm{fF}$, matched to the line capacitance to enable remote sensing and soft error resilience. The capacitor needs to be charged to a minimum voltage of V_{cell} 0.4–0.5 V for a reliable reading. Thus, the number of electrons stored in DRAM is:

$$N_{\mathrm{el}} = \frac{C_{\mathrm{cell}} V_{\mathrm{cell}}}{e} \approx 10^5 \tag{3.3}$$

The energy of charging the cell capacitor is:

$$E_{\mathrm{cell}} = C_{\mathrm{cell}} V_{\mathrm{cell}}^2 = e N_{\mathrm{el}} V_{\mathrm{cell}} \approx 10^{-14}\ J \tag{3.4}$$

The capacitor insulator forms a fixed-height barrier in DRAM cell (Figure 3.2). The second (controllable height) barrier in Figure 3.2 is formed by the DRAM transistor. The

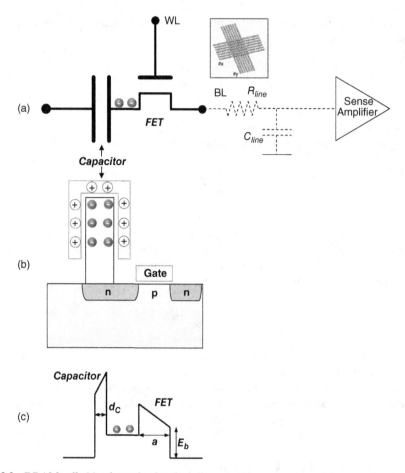

Figure 3.2 DRAM cell: (a) schematic electrical diagram, (b) cross-section, (c) energy barrier diagram

transistor also provides a means to select a given cell in the array. The maximum height of the FET barrier is limited by the bandgap of silicon (1.1 eV), which limits the retention time to a fraction of seconds. The Joint Electron Device Engineering Council (JEDEC) standards specify a minimum of 64 ms for DRAM retention time, for which the barrier height $E_{b_{min}} > 0.8\,eV$ is needed [2,3], and the corresponding FET gate control voltage should be >1 V (in practice 1.5–2.0 V).

As argued in [3], the cell capacitor is the main factor limiting DRAM scalability. The maximum capacitance is limited by two nonscalable parameters of the capacitor insulator: the thickness d_C, which is limited by tunneling leakage current between the electrodes, and the dielectric constant, which is limited by materials physics (e.g., the maximum dielectric constant which can be realized in stable materials structure is ~300 for single-crystal $SrTiO_3$). According to [3], a theoretically minimal capacitor insulator thickness calculated is $d_{C_{min}}$ ~5 nm, and the minimum external dimension of the cell capacitor are ~10 nm. (For this limiting case the capacitor must be very tall (with height $h \sim 100\,\mu m$) and the capacitor serial resistance is large, $R_C \sim 10^8$ Ohm – see Table 3.1).

Table 3.1 Resistances and capacitances in DRAM

F (nm)	90	70	60	50	40	20	10
R_C $(\Omega)^a$	210	527	928	1840	4380	1.15×10^5	1.37×10^8
R_{FET} $(\Omega)^b$	2770	3560	4150	4980	6220	12 400	22 600
R_{line} $(\Omega)^c$	144	192	228	284	374	932	2600
C_{line} (fF)^d	55	50	45	40	35	24	16
C_{cell} (fF)	25	25	25	25	25	25	25

^{a} Serial resistance of an idealized cell capacitor [3].
^{b} Channel resistance of an idealized FET in ON state [3].
^{c} Line resistance in 256×256 array (see Appendix).
^{d} Line capacitance in 256×256 array (see Appendix).

The DRAM write energy is the energy needed to "pump" N_{el} electrons through the bitline into a cell to charge the cell's capacitor, plus the energy to control the barrier of the FET. The voltage V_g is applied to the entire word line, and the corresponding portion of energy is $C_{line}V_g^2$ (for a switching cycle). Combining both terms an estimate is obtained for the energy required to write a bit of information into a memory cell in an array:

$$E_{DRAM} = C_{line}V_g^2 + (C_{line} + C_{cell})V_{cell} \sim 10^{-13} \, J \tag{3.5}$$

3.3.1 Total Energy Required to Create/Maintain the Content of a Memory Cell

An application specific parameter of a memory system is the *user's access interval* t_a, which is the average time between memory accesses [3]. This parameter enables comparison across different memory classes, including volatile and nonvolatile. In Equation 3.6, E_{tot} is the energy required to first write and then maintain the memory state, given the retention time t_r, for the cell:

$$E_{tot} = E_{DRAM} + \frac{t_a}{t_r}E_{DRAM} = E_{DRAM}\left(1 + \frac{t_a}{t_r}\right) \tag{3.6}$$

The $(t_a/t_r) \cdot E_{DRAM}$ part of Equation 3.6 represents the retention cost. As a numerical example, $E_{tot} \sim 10^{-12}$–10^{-11} J/bit for $t_a \sim 1$–10 s ($t_r = 64$ ms).

In memory systems additional energy is needed to support the peripheral circuitry. The average DRAM system-level energy per bit is $(3–6) \times 10^{-11}$ J, as shown in Figure 3.3.

3.3.2 DRAM Access Time (WRITE or READ)

The access time accumulates delays due to different resistances and capacitances present in a memory array. In a simplified form, the RC-bounded minimum access time can be written as:

$$t_{DRAM} = (R_C + R_{FETon} + R_{line}) \cdot (C_{cell} + C_{line}) \tag{3.7}$$

All characteristic resistances and capacitances in Equation 3.7 are summarized in Table 3.1 and the characteristic access times for several DRAM cell sizes are given in Table 3.2.

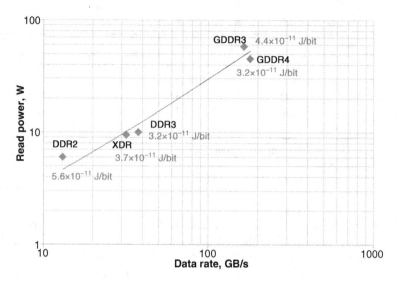

Figure 3.3 DRAM energetics (adapted from [4])

3.3.3 Energy–Space–Time Compromise for DRAM

In [3] the Space–Action metric for the evaluation of the performance of memory elements operating with different physical principles has been proposed. This metric is defined as the product of energy, space (e.g., volume) and read/write time (Action = Energy × Time). The Space–Action function of DRAM shown in Figure 3.4 has a distinct minimum suggesting that from the Energy–Space–Time metric point of view, the DRAM will achieves optimum performance in the 40–50 nm regime.

Table 3.2 Scaling and performance projections for DRAM

Parameter		Current	Minimal	Optimal[a]
Feature size F		28–45 nm	>10 nm[b]	45 nm
Access time	Practical	<10 ns	>25 ns	<10 ns
	RC limit	0.5–2 ns	25 ns[c]	0.5 ns
Retention time		64 ms	64 ms	64 ms
Write cycles		>10^{16}	>10^{16}	>10^{16}
Operating voltage		~2 V	~2 V	~2 V
Number of stored electrons		10^5	10^5	10^5
Write energy (J bit^{-1})	Cell level	10^{-14}	10^{-14}	10^{-14}
	Array level	10^{-13}	10^{-13}	10^{-13}
	System level	$(3\text{--}6) \times 10^{-11}$	>10^{-11}	$(3\text{--}6) \times 10^{-11}$

[a] Corresponds to the minimum of the Energy–Space–Time product.
[b] Limited by the dimensions of the cell capacitor; "minimal" only refers to the node size and area (in this case, the timing and energy for "minimal" is greater than "current" or "optimal" nodes.
[c] Expected RC delay at 16 nm.

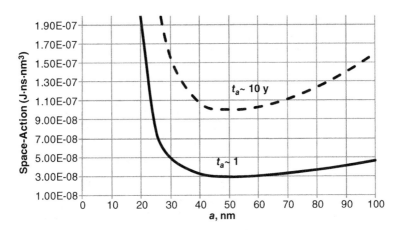

Figure 3.4 Scaling optimization for DRAM based on minimal space–action

Table 3.2 summarizes essential parameters of DRAM. The cell capacitor appears to be the main factor limiting DRAM scalability to >10 nm. Note that as the scaling limit is approached, the access time will dramatically increase due to the growing serial resistance of the capacitor.

3.4 Flash Memory

The flash memory is a nonvolatile electron charge-based memory, the basic building block of which is a floating gate cell [5], shown in Figure 3.5. The term "flash" refers to the erase operation where many cells are cleared to one state (erased) in a large block simultaneously, thereby allowing the realization of an electrical programmable and erasable cell using only

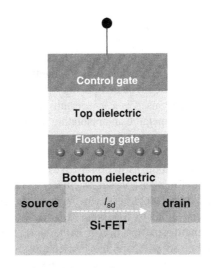

Figure 3.5 Floating gate memory

Table 3.3 Insulator material parameters and the corresponding theoretical minimum insulator thickness for floating gate nonvolatile storage

Material	Dielectric constant, K	Barrier height, E_b (with Si)	Effective electron mass, m^*	a_{min}
SiO_2	3.9	3.1 eV	$0.50m_0$	5.0 nm
Si_3N_4	7.6	2.4 eV	$0.43m_0$	6.0 nm
Al_2O_3	9	2.8 eV	$0.30m_0$	6.8 nm

one transistor, and at the same time decreases the erase time at the expense of a more complicated memory handling. In a flash memory cell, the charge storage node is either on a conductive electrode surrounded by insulators (floating gate) or in discrete traps within a defective insulator layer (charge trapping layer) or in nanocrystals embedded in the insulator layer. The fundamental physics of device operation is similar for each type of floating memory.

In order to prevent loss of the stored charge, the storage node is defined by energy barriers of sufficient height E_b to retain charge (as shown in Figure 3.1). Such barriers are formed by using layers of insulator (I), which surround a metallic storage node (M). Such an I-M-I structure forms the storage node in floating gate memory cell, the basic element of flash memory. The barrier height E_b is fixed, as it is a material-specific property. A simple estimate in Section 3.2 for the minimal barrier height and width to satisfy the "nonvolatility requirement" yielded $E_{b_{min}} \geq 1.42$ eV and a minimum width of ~5 nm. More detailed calculations in [5] that take into account different practical constrains result in the somewhat larger barrier height of 1.73 eV required to achieve 10 year retention. Such a barrier height can be achieved only with a limited number of materials, some of which are listed in Table 3.3. Note that the effective electron mass in solids is, in most cases, smaller than the free electron mass (used in the simple estimates in Section 3.2). According to Equation 3.2c the smaller mass will result in a wider barrier or thicker insulator layer (Table 3.3). The theoretical barrier width must be >5 nm for all known dielectric materials (typically >7 nm in practical devices). The corresponding practical minimum size of the floating gate cell is ~10 nm [5].

The presence of fixed barriers confining the storage node, while enabling storage, represents a considerable challenge for charge transfer across the barriers, that is, for read, write, and erase operations. For example, the sensing device, which detects presence or absence of electric charge stored in a "floating" node should be in immediate proximity to the storage node. A field effect transistor (FET) is commonly used as a sensor, thus a complete nonvolatile floating gate memory cell consists of a stack of metallic and insulating layers on the top of an FET channel, as shown in Figure 3.5.

All operations of a floating-gate memory cell correspond to different *shapes* of the storage node barriers. While for store operation, a low conductive barrier is needed, switching to a more conductive barrier is needed for a fast write. Such a switching is achieved by applying sufficiently high voltage to external electrodes, which causes the barrier shape to change from trapezoidal/rectangular (Figure 3.6a) to triangular (Figure 3.6b). Electron tunnel transport through the trapezoidal/rectangular barriers occurs in the *direct tunneling* mode, which can be very slow if both the height and width of the barrier are sufficiently large, as discussed in Section 3.2. In the case of a triangular barrier, the electron transport occurs in the

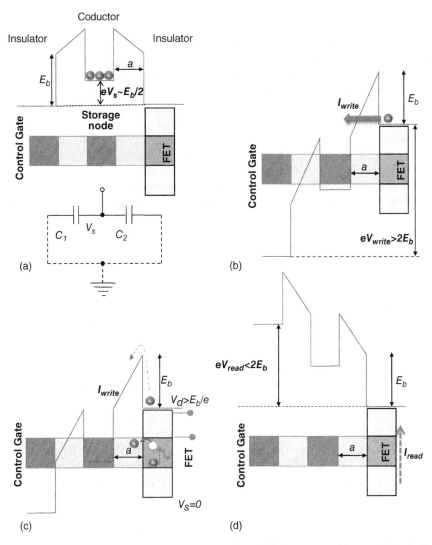

Figure 3.6 Energy barrier shapes in three basic operations of flash memory: (a) store; (b) write by F-N tunneling; (c) write by channel hot-electron injection and (d) read

Fowler–Nordheim tunneling mode, which can be sufficiently fast, resulting in a small write/ erase time. The shape of the barrier depends on the voltage difference between the two sides of the barrier, as can be seen from Figure 3.6a–d:

$$\text{Trapezoidal barrier (direct tunneling)}: \quad e\Delta V_b < E_b$$
$$\text{Triangular barrier (F-N tunneling)}: \quad e\Delta V_b \geq E_b$$

The above relations set fundamental limitations on all operations of the floating gate memory cells, as discussed in the following.

3.4.1 Store

Generally speaking, a larger number of electrons representing a state is preferable, because: (a) larger charge is easier to detect by the sensing device, (b) increased robustness to statistical fluctuations, and (c) the possibility to store more than one bit per cell. However there are limitations on the number of stored electrons due to the barrier deformations. When a number N_{el} of electrons is placed in the storage node, the node charge $q = eN_{el}$ (where e is electron charge) results in an increase of the node potential V_s (Figure 3.6a).

$$V_s = \frac{eN_{el}}{C_m} \tag{3.8}$$

C_m is the capacitance of the memory cell, formed by the capacitances C_1 and C_2 as shown in Figure 3.6a. The standard parallel-plate capacitor formula can be used to estimate the capacitance:

$$C = \frac{\varepsilon_0 K A}{a} \tag{3.9a}$$

Where $\varepsilon_0 = 8.85 \times 10^{-12}\,\mathrm{F\,m^{-1}}$, $A = L^2$ is the cross-sectional area, a is the capacitor dielectric thickness (equal to the barrier width), and K is the dielectric constant of barrier material.

Assuming a symmetrical barrier structure with $C_1 = C_2$, one obtains:

$$C_m = C_1 + C_2 = \frac{2\varepsilon_0 K}{a} L^2 \tag{3.9b}$$

Note that the potential shift V_s changes the barrier shape from rectangular to trapezoidal, when $eV_s < E_b$ (Figure 3.6a), or triangular, when $eV_s > E_b$ (Figure 3.6b). The barrier deformations considerably increase the tunneling leakage and therefore decrease the retention. In particular, the tunneling leakage dramatically increases when $eV_s > E_b$ (the triangle barrier), due to decreased tunnel distance as can be seen in Figure 3.6b. Thus, the barrier deformations limit the maximum number of stored electrons, and for long retention the requirement holds $eV_s < E_b$ and in practice $eV_s \approx 0.5\,E_b$ represents a reasonable compromise between the sense and retention requirements. For this condition, the maximum number of electrons stored in the floating gate cell from Equation 3.9 is

$$N_{el_{max}} \approx \frac{C_m E_b}{2e^2} = \frac{\varepsilon_0}{e^2} \cdot K \cdot E_b \cdot \frac{L^2}{a} \tag{3.10}$$

As an example, consider a barrier structure formed by insulating layers of SiO_2. From Table 3.3 for SiO_2, $E_b = 3.1\,\mathrm{eV}$ and $K = 3.9$. Let thickness $a = 5\,\mathrm{nm}$ and spatial dimensions $L = 20\,\mathrm{nm}$. Substituting these parameters in Equation 3.3 results in $C_1 = C_2 = 2.76 \times 10^{-18}\,\mathrm{F}$ and $N_{s_{max}} \approx 54$ electrons. The small number of stored electrons causes a severe problem of reliability degradation due to statistical variations [6].

3.4.2 Write

During the write operation (Figure 3.6b), electrons are injected into the storage node, and this requires operation in the Fowler–Nordheim (F-N) tunneling regime for faster injection. This mechanism is employed in NAND flash memory. The condition for F-N tunneling is $eV_b > E_b$, that is, the potential difference across the barrier between the storage node and the external contact must be larger than the barrier height. Since the storage node is isolated from the external contacts by two barriers (i.e., it is floating), this requires the total write voltage applied to the opposite external contacts of the memory cell to be more than the doubled barrier height: $eV_{write} > 2E_b$ as is shown in Figure 3.6b. Thus, the floating gate structure inherently requires high voltage for the write operation: for example, for SiO_2 barriers, $V_{write\ min} > 6\,V$, and the write voltage should be $>10-15\,V$ for faster ($\sim$ms to μs) operations.

Another way to perform the WRITE operation is to supply the electrons with sufficient energy to overcome the barrier (e.g., for a classical floating gate cell, the barrier between Si and SiO_2 is $E_b = 3.1\,eV$). To gain this energy, electrons are accelerated in the FET channel by applying a sufficiently high drain voltage (in general $eV_d > E_b$, typically $V_d = 4-5\,V$). This process is called channel hot electron injection (Figure 3.6c) and is typically used in NOR flash memory. Also a positive voltage has to be applied to the control gate that attracts the generated electrons. This method allows for faster write speed (in μs) at lower voltage, than the F-N tunneling mechanism. However, writing via hot electron injection is very inefficient: Typically about 10^5-10^6 channel electrons (η) are needed to inject one electron in the storage node. This requires high FET channel current in the range of $100\,\mu A$. The corresponding energy for WRITE is quite large, for example, for a number of stored electrons $N_{el} = 100$ and drain voltage $V_d = 5\,V$, the write energy can be estimated as:

$$E_{write} \sim \eta N_{el} \cdot eV_d = 10^6 \cdot 100 \cdot 1.6 \cdot 10^{-19} \cdot 5 \approx 10^{-10}\,J \qquad (3.11)$$

Note that the cell-level write energy using hot electrons Equation 3.11 is considerably larger than the cell-level F-N write energy, derived below in Equation 3.12. Also, a severe limitation on the FET channel length arises if the channel hot electron injection is used from write operation. In this case, applying a large voltage to a very short channel (e.g., 20–30 nm) will result in punchthrough or junction breakdown.

3.4.3 Read

The sensing FET is controlled by the voltage, V_{read}, applied to an external control gate electrode (Figure 3.6c). The source–drain current of the FET depends on the presence or absence of charge in the floating gate, thus the memory state can be sensed by measuring the FET current. The control gate allows modulation of the semiconductor channel of the FET by external commands, similar to the logic FET. However, unlike a conventional transistor, the degree of accessibility of the channel from the control gate is rather limited. First, the control gate is physically far from the channel, since the minimal thickness of both top and bottom dielectric layers is large due to the retention requirements, and the minimal thickness of the insulator stack is $>10\,nm$. Second, the control gates affects the channel only indirectly, as the floating gate lies between the control gate and the channel. Therefore a large read voltage must be applied to the control gate for reliable ON/OFF transitions of the sense transistor. The maximum read voltage is, however, limited by the condition for the F-N tunneling discussed above, and for the read operation is $eV_{read} < 2E_b$, for example, for SiO_2 barriers, $V_{read\ max} < 6\,V$ for a nondisturbing read. In practice, a typical read voltage is 4.5–5.0 V [5].

3.4.4 Energetics of Flash Memory

The cell-level write energy using hot electrons derived above in Equation 3.11 is $\sim 10^{-10}\,J$. For NAND flash memory, if only one isolated floating gate cell is considered, its energy of operation is remarkable low. For example, considering the write operation of Figure 3.6b, the write energy can be estimated as the sum of energy of barrier deformation (which is equivalent to the charging the stack capacitance to the voltage V_{write}) and the energy needed to "pump" N_{el} electrons through the barrier into the storage node:

$$E_{FG} = \frac{C_1 C_2}{C_1 + C_2} \cdot V_{write}^2 + e N_{el} \cdot V_{write} \qquad (3.12)$$

(Note that the stack capacitances C_1 and C_2 are connected in series in the write mode.) For the model stack parameters used for numerical calculations in Equation 3.10 and $V_{write} = 15$ V, Equation 3.11 yields $E_{FG} \sim 4 \times 10^{-16}$ J.

For flash memory arrays, due to the regular wiring, the properties of interconnecting array wires determine the operational characteristics of the memory system. A given cell in an array is selected (e.g., for read operation) by applying appropriate signals to both interconnect lines, thus charging them. The large operating voltage of flash results in rather large line charging energy, $\sim C_{line} V^2$. For $V_{write} \sim 15$ V, and the line capacitance from Table 3.1, the write energy is $\sim 10^{-11} - 10^{-12}$ J. In addition, in practical flash memory systems large energy is consumed in high-voltage peripheral circuitry, such as multiplexers and voltage pumps, which raises the system level write energy to $10^{-9} - 10^{-10}$ J bit^{-1} write [7,8]. Hence, it is important to note that the total write energy of a flash memory system is dominated by the high-voltage peripheral circuitry – which requires energy several orders of magnitude above that of the individual floating gate cells.

Table 3.4 summarizes essential parameters of flash memory. The cell stack appears to be the main factor limiting flash scalability to >10 nm. Note that the operational parameters of flash memory are not expected to improve with scaling. Moreover, as recently argued, reliability, endurance and performance could decline as further scaling continues [9].

Table 3.4   Scaling and performance projections for Flash memory

Parameter		NAND Current	NAND Minimal	NOR Current	NOR Minimal
Feature size F		16–32 nm	>10 nm	45 nm	25 nm
Access time	Write[a]	~100 μs	~100 μs	~10 μs[c]	~10 μs
	Read[b]	~10 μs[c]	~10 μs	60–120 ns[c]	~60 ns
Retention time		10 yr	<10 yr	10 yr	10 yr
Write cycles		~10⁵	<10⁴	~10⁵	~10⁵
Operating voltage	Write	15–20	15	8–10	~8
	Read	5	5	5	5
Number of stored electrons		~50	~10	~200	~100
Write energy (J bit^{-1})	Cell level	4×10^{-16}	$\sim 10^{-16}$	2×10^{-10}	$\sim 10^{-10}$
	Array level	10^{-11}–10^{-12}	$\sim 10^{-12}$	$>2 \times 10^{-10}$	$>10^{-10}$
	System level	10^{-10}–10^{-9} [d]	10^{-10}–10^{-9}	$\sim 10^{-9}$ [e]	$\sim 10^{-9}$

[a] Random write.
[b] Random read.
[c] Ref.[10].
[d] Ref. [7].
[e] Ref. [8].

3.5 Static Random Access Memory

SRAM consumes about a half or more of the area of a modern CPU chip. An SRAM cell is a bistable transistor structure typically made of two CMOS inverters connected back to back, as shown in Figure 3.7. The output potential of one inverter is applied as input of the other, forming a feedback loop that "freezes" the cell in a given state. Two access transistors in the off state are used to "isolate" the cell in the store or standby mode. The structure in Figure 3.7 represents the standard six-transistor (6 T) SRAM cell. The estimates offered in this section are

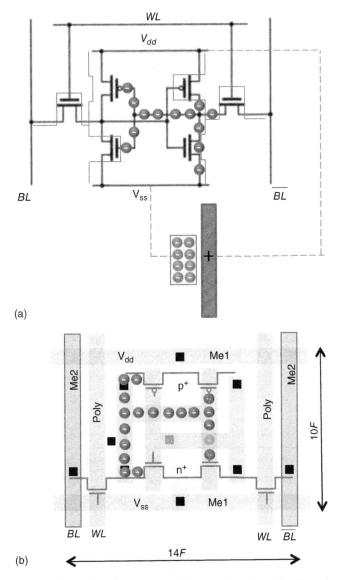

Figure 3.7 SRAM 6 T cell showing electrons stored between barriers: (a) schematic; (b) an example layout

based on simplified assumptions, and mainly serve to illustrate the basic principles of SRAM. For a more detailed analysis of SRAM operational parameters the reader can refer to a number of publications [11–14].

In order to change the state, the access transistors are turned on and external signals are simultaneously applied to inputs of both inverters from the bit line and the inverse bit line, so that the two inverters are set in the opposite states, for example, 0,1 or 1,0. If now the access transistors are turned off, the set state will be preserved, and thus the data stored.

SRAM cell looks very different from the DRAM or flash cells, for example it is built entirely from logic-type FETs and has no dedicated storage element, similar to a DRAM capacitor, or flash floating gate. However, the SRAM cell still can be represented by the generic two-barrier structure of Figure 3.1. The barriers are formed by FET channels (red lines) and by gate insulators. Since the height of the FET barrier is low, ~0.5 eV) the stored charge retention time is small – the charge leaks out the system. The lost charge in SRAM is replenished from an external source. In Figure 3.7a such a source is schematically shown as an electron reservoir formed by a galvanic cell. Thus, SRAM operates in an "instant refresh" mode and needs always be connected to a power supply. The SRAM internal storage node is formed by gate capacitances of an inverter and conductive lines of the cell.

The internal storage node capacitance is composed of gate and junction capacitances of FETs and the connecting wires within the cell (see a SRAM cell layout in Figure 3.7b)

$$C_{\text{cell}} \sim 2C_{\text{g}} + 2C_{\text{j}} + C_{\text{w}} \tag{3.13}$$

All components of the cell capacitances can be calculated using standard capacitance formulas. In these estimates the following normalized capacitance values will be used: $Cg \sim 0.8\,\text{fF}\,\mu\text{m}^{-1}$ (per FET channel width) [15], $Cj = 1\,\text{fF}\,\mu\text{m}^{-2}$ [16], $Cw = 1.9\,\text{fF}\,\text{cm}^{-1}$ (per wire length) [15].

The estimated SRAM node capacitance for several technology generations are shown in Table 3.5 along with nominal operation voltages. The corresponding number of stored electrons cam be calculated using Equation 3.3 by substituting C_{cell} and $V_{\text{cell}} = V_{\text{dd}}$ in Table 3.5. The cell-level write energy includes the energy of charging the storage node plus the energy of opening the two access transistors:

$$E_{\text{cell}} = \left(C_{\text{cell}} + 2C_{\text{g}}\right)V_{\text{dd}}^2 \tag{3.14}$$

Table 3.5 SRAM cell and array parameters for advanced technology nodes

Parameter	45 nm	32 nm	22 nm	16 nm
C_{cell}, F	1.4×10^{-15}	8.8×10^{-16}	5.3×10^{-16}	3.7×10^{-16}
V_{dd}, V	1	0.9	0.8	0.7
N_{el}	~8700	~5000	~2600	~1600
E_{cell}, J	2.2×10^{-15}	1.2×10^{-15}	5.7×10^{-16}	2.9×10^{-16}
C_{line}, F[a]	9.4×10^{-14}	7.8×10^{-14}	6.3×10^{-14}	5.2×10^{-14}
R_{line}, Ω[a]	1700	3500	4000	6300
$R_{\text{line}}C_{\text{line}}$, ns[a]	0.16	0.20	0.25	0.30

[a] 256×256 SRAM array.

Table 3.6 Scaling and performance projections for SRAM

Parameter		Current	Minimal	Optimal[a]
Feature size F		32–45 nm	Unknown	45 nm
Access time	Practical	<10 ns	<10 ns	<10 ns
	RC limit	0.2 ns	~0.3 ns[b]	0.2 ns
Write cycles		>10^{16}	>10^{16}	>10^{16}
Operating voltage		1 V	~0.7 V	1 V
Number of stored electrons		$(2–8) \times 10^3$	~10^3	~8000
Write energy (J/bit)	Cell level	~10^{-15}	~10^{-16}	~10^{-15}
	Array level	10^{-13}	~10^{-14}	10^{-13}
	System level	~10^{-12}	>10^{-13}	~10^{-12}

[a] Estimate based on $V_{\min}$ and DRV minimization, while maintaining the minimum cell area layout [20].
[b] Expected RC delay at <16 nm.

The estimated number of electrons and the cell-level write energy are also tabulated in Table 3.5.

The total dynamic energy for a write operation in array is mostly determined by the capacitances of the word and bit lines (see e.g., [17]).

$$E_{\mathrm{w}} = n \cdot C_{\mathrm{WL}} \cdot V_{\mathrm{dd}}^2 + m \cdot C_{\mathrm{BL}} \cdot V_{\mathrm{dd}}^2 \sim (n+m) C_{\mathrm{line}} \cdot V_{\mathrm{dd}}^2 \qquad (3.15)$$

where n is the number columns, m is the number of rows, and it is assumed in this simple estimate that $C_{\mathrm{WL}} \sim C_{\mathrm{BL}} \sim C_{\mathrm{line}}$. The values for C_{line} calculated using Equation 3A.4b in the Appendix, and assuming the line length factor $\beta = 10$, are shown in Table 3.5. The estimated array-level access energy is ~10^{-13} J (Table 3.6).

In practical memory systems additional energy is needed to support the peripheral circuitry. An example of recently reported SRAM system-level energy is 133 fJ per bit [18].

3.5.1 SRAM Access Time

The access time accumulates delays due to different resistances and capacitances present in a memory array. In a simplified form, the RC-bounded minimum access time can be written as:

$$t_{\mathrm{SRAM}} = (R_{\mathrm{FETon}} + R_{\mathrm{line}}) \cdot (C_{\mathrm{cell}} + C_{\mathrm{line}}) \qquad (3.16)$$

All characteristic resistances and capacitances in Equation 3.16 are summarized in Table 3.5 (see also Table 3A.2 in the Appendix) and the characteristic access times for several SRAM cell sizes are given in Table 3.6.

3.5.2 SRAM Scaling

In contrast with other electro charge-based memories, that is, DRAM and Flash, SRAM scalability has no obvious "hard" physical barriers above the FET scaling limits, that is, ~5 nm according to ITRS (Scaling limits for both DRAM and Flash are well above the 5 nm FET limit). In case of SRAM, the scaling limits are rather "soft."

A significant challenge for SRAM scaling arises for process variation impacting the fluctuation of the parameters of the transistors forming an SRAM cell, for example, different threshold voltage. The mismatch between SRAM transistors can result in the failure of the cell [19], for example, flipping the cell data during read operation or during cell data holding when the supply voltage is decreased. Important performance measures of SRAMs stability are the lowest supply voltage for reliable read/write operation, V_{min} and data retention voltage, DRV. Typically, the smallest possible transistors are used in the high-density SRAM, which makes the above challenges more difficult. For example as was argued in [20] based on V_{min} and DRV minimization, the minimum cell area cannot be maintained below 45 nm, and has to be increased as the node size decreases.

Also, single event upset (SEU) due to strikes of ionizing particles [21] is a difficult problem for SRAM scaling. SEU rate depends on the number of stored electrons which is decreasing with scaling. In addition, in advanced technology nodes the effects of multiple-node charge collection creates additional challenges [21].

Another challenge is increasing SRAM standby leakage due to increased gate, subthreshold, and junction leakage of the minimum-sized FETs so that the leakage power can be very significant in SRAM [14]. Larger-size transistors help to reduce leakage, and there is a clear trade-off between the memory cell area and the leakage power [14].

There are a number of SRAM design solutions to address above scaling challenges, including different layout topologies for the 6 T cells [22,23] and new cell design utilizing more transistors per cell, for example, 8 T and 10 T cells [24,25].

Also replacing the conventional planar CMOS with a 3-D Tri-Gate transistor technology allows to continue the SRAM scaling; recently a 22 nm SRAM was implemented in the 3-D Tri-Gate technology [26]. The viability of the Tri-Gate SRAM even at 10 nm has been confirmed by simulations [27].

3.6   Summary and Perspective

This chapter provides an outlook for the current baseline memory technologies, which include DRAM, SRAM, and Flash, all based on storing electron charge in a storage node. The physics of all charge-based memory devices have been analyzed using the barrier model, which allows derivation of the key performance parameters as well as scaling and performance projections.

Both for DRAM and flash memories, the scalability is likely to be limited to >10 nm. The main limiting factor of DRAM is the cell capacitor, while the scalability of flash cell is limited by the gate stack. In contrast with DRAM and Flash, SRAM scalability has no obvious "hard" physical barriers above the FET scaling limits ($\sim$5 nm according to ITRS). In case of SRAM, the scaling limits are rather "soft." The limiting factors include parameter variation, single event upsets, standby leakage; all of which can be addressed by design solutions, at least in principle.

This chapter has a special emphasis on the energetics of charge-based memory, including cell, array and system levels. This analysis of can be extended to new memory concepts, which would be helpful for better understanding of energetics of emerging memory including, for example, array parasitics and peripheral circuitry considerations.

Appendix: Memory Array Interconnects

In a typical memory system, the memory cells are connected to form an array, and in many instances the properties of interconnecting array wires determine the operational characteristics

Table 3A.1 Parameters for scaling analysis of array interconnects

Parameter	Numerical value	Rational
Bulk resistivity, ρ_{bulk}	2.37 μΩ·cm	Cu, $T = 400\,K$
Mean free path, λ	28 nm	Cu, $T = 400\,K$
Specularity, p	0.5	Experimental value in Cu interconnects: $p = 0.3$–0.5 [30]

of the memory system, size as speed or energy. The memory device is located at the point of intersection of *bitline* (BL) and *wordline* (WL). The length of interconnects line to accommodate N memory cells is:

$$L_{line} \sim \beta \cdot F \cdot N \tag{3A.1}$$

where F is the nominal feature size. The line length factor, β, depends on the memory types, for example, $\beta \sim 2$ for NAND flash, $\beta \sim 2$–4 for DRAM, and $\beta \sim 10$ for SRAM.

The basic parameters of array interconnects are their *resistance* R_{line} and *capacitance* C_{line} and these are directly related to L and F. The line resistance can be estimated as:

$$R_{line} = \rho \frac{L}{F^2} \tag{3A.2}$$

Now the resistivity of the metal line, ρ, is different from the bulk resistivity of a metal for small cross-sections due to dimensional effects, $\rho = f(F)$, when the wire width, w, is less than the electron mean free path, λ. The Fuchs–Sondheimer approximation [28–30] can be used to calculate resistance of nano-scale metal wires, which in the commonly used simplified form is:

$$\rho(F) = \rho_{bulk} \left[1 + \frac{3}{4} \frac{\lambda}{F} (1 - p) \right] \tag{3A.3}$$

where ρ_{bulk} is the bulk resistivity of metal wire and p is specularity, that is, the probability of an electron being scattered elastically at the side surface of the wire. The resistance parameters, used for scaling analysis of array interconnects, are given in Table 3A.1. Table 3A.2 contains example data on line resistance calculated using Equations 3A.1–3A.3 and assuming that both the width and the height of the wire's cross section is equal to the nominal feature size: $w = h = F$.

Table 3A.2 Resistances and capacitances in memory array interconnects in a 256×256 array with $L_{line} = 2_{FN}$

F, nm	90	70	60	50	40	20	10
$R_{line}\,(\Omega)^a$	151	199	238	294	383	925	2490
$r_{line}\,(\Omega/\mu m)$	3.27	5.56	7.74	11.5	18.7	90.4	486
$C_{line}\,(fF)^b$	55	50	45	40	35	24^c	160^c
$c_{line}\,(fF/\mu m)$	1.19	1.40	1.46	1.56	1.71	2.32	156

[a] Calculated using Equations 3A.1–3A.3.
[b] 3-D calculations for practical 256×256 DRAM array [31].
[c] Extrapolation of the data calculated $F = 40$–$90\,nm$ using Equation 3A.4.

The line capacitance is, in general, proportional to the wire length, $C_{line} \sim \varepsilon_0 L$. However, the capacitance calculations in a memory array are more complicated, since in practical arrays, the total capacitance is considerably larger than the 2-D capacitance of wires. The array capacitance is increased due to additional contributions arising, for example, from the junction and gate capacitances of select transistors, 3-D effects, sense amplifier capacitance, and so on. Examples of the values of DRAM bitline capacitance, derived from measurements and 3-D capacitance simulations [31] are given in Table 3.2. An extrapolation of these data by a power function yields scaling for a 256 bit line and $\beta = 2$:

$$C_{line}(F) \approx aF^k \tag{3A.4a}$$

or per unit length, using Equation 3A.1:

$$\left(\frac{C_{line}}{L_{line}}\right) = \frac{aF^k}{\beta FN} = \frac{a}{256\beta} F^{k-1} \tag{3A.4b}$$

where $a \approx 4.28$, $k \approx 0.57$; C_{line} and F are measured in fF and nm respectively. Example values of line capacitances are given in Table 3A.2.

Acknowledgments

The authors would like to acknowledge R.J. Kaplar (Sandia) for reviewing and providing useful comments on this chapter. This work was funded in part by Sandia's Laboratory Directed Research and Development (LDRD) program. Sandia National Laboratories is a multi-program laboratory managed and operated by Sandia Corporation, a wholly owned subsidiary of Lockheed Martin Corporation, for the US Department of Energy's National Nuclear Security Administration under contract DE-AC04-94AL85000.

References

1. Freitas, R.F. and Wilcke, W.W. (2008) Storage-class memory: The next storage system technology. *IBM Journal of Research and Development*, **52**, 439–447.
2. Mandelman, J.A., Dennard, R.H., Bronner, G.B. *et al.* (2002) Challenges and future directions for the scaling of dynamic random-access memory (DRAM). *IBM Journal of Research and Development*, **46**, 187–212.
3. Zhirnov, V.V., Cavin, R.K., Menzel, S. *et al.* (2010) Memory Devices: Energy-Space-Time Tradeoffs. *Proceedings of the IEEE*, **98**, 2185–2200.
4. Sekiguchi, T., Ono, K., Kotabe, A., and Yanagawa, Y. (2011) 1-Tbyte/s 1-Gbit DRAM architecture using 3-D interconnect for high-throughput computing. *IEEE Journal of Solid-State Circuits*, **46**, 828–837.
5. Zhirnov, V. and Mikolajick, T. (2012) Chapter 26: Flash memories, in *Nanoelectronics and Information Technology* (ed. Rainer Waser), John Wiley & Sons.
6. Molas, G., Deleruyelle, D., Salvo, B.De. *et al.* (2006) Degradation of floating-gate memory reliability by few electron phenomena. *IEEE Transactions on Electron Devices*, **53**, 2610–2619.
7. Grupp, L.M., Caulfield, A.M., Coburn, J. *et al.* (2009) Characterizing Flash Memory: Anomalies, Observations, and Applications, MICRO'09, Dec. 12–16 2009, New York, NY, USA, pp. 24–33.
8. Derhacobian, N., Hollmer, S.C., Gilbert, N., and Kozicki, M.N. (2010) Power and energy perspectives of nonvolatile memory technologies. *Proceedings of the IEEE*, **98**, 283–298.
9. Grupp, L.M., Davis, J.D., and Swanson, S. (2012) The bleak future of NAND flash memory, FAST'12 Proceedings of the 10th USENIX conference on File and Storage Technologies, Feb. 14–17 2012, San Jose, CA.
10. Micheloni, R. and Crippa, L. (2010) *Inside NAND Flash Memories*, Springer.
11. Evans, R.J. and Franzon, P.D. (1995) Energy consumption modeling and optimization for SRAM's. *IEEE Journal of Solid-State Circuits*, **30**, 571–579.

12. Amrutur, B.S. and Horowitz, M.A. (2000) Speed and power scaling of SRAM's. *IEEE Journal of Solid-State Circuits*, **35**, 175–185.
13. Zheng, A., Rose, K., and Gutman, R.J. (2006) Memory performance prediction for high-performance micro-porocessors at deep submicrometer technologies. *IEEE Transaction on CADICS*, **25**, 1705–1718.
14. Morifuji, E., Patil, D., Horowitz, M., and Nishi, Yoshio. (2007) Power optimization for SRAM and its scaling. *IEEE Transactions on Electron Devices*, **54**, 715–722.
15. International Technology Roadmap for Semiconductors (2014) www.itrs.net (accessed 16 January 2014).
16. Chiarella, T., Witters, L., Mercha, A. *et al.* (2010) Benchmarking SOI and bulk FinFET alternatives for PLANAR CMOS. *Solid-State Electronics*, **54**, 855–860.
17. Garg, A. and Kim, T.T.-H. (2013) SRAM array structures for energy efficiency enhancement. *IEEE Transactions on Circuits and Systems II*, **60**, 351–355.
18. Rooseleer, B., Cosemans, S., and Dehaene, W. (2012) A 65nm, 850MHz, 256 kbit, 4.3 pJ/access, ultra low leakage power memory using dynamic cell stability and a dual swing data link. *IEEE Journal of Solid-State Circuits*, **47**, 1784–1796.
19. Mukhopadhyay, S., Mahmoodi, H., and Roy, K. (2005) Modeling of failure probability and statistical design of SRAM array for yield enhancement in nanoscaled CMOS. *IEEE Transaction on CADICS*, **24**, 1859–1880.
20. Makosijej, A., Thomas, O., Amara, A., and Vladimirescu, A. (2013) CMOS SRAM scaling limits under optimum stability constraints, IEEE Intern. Symp. on Circuits and Systems (ISCAS), Beijing, China, 19–23 May 2013.
21. Black, J.D., Dodd, P.E., and Warren, K.M. (2013) Physics of multiple-node charge collection and impacts on single-event characterization and sift error rate prediction. *IEEE Transactions on Nuclear Science*, **60**, 1836–1851.
22. Mann, R.W., Hook, T.B., Nguyen, P.T., and Calhoun, B.H. (2012) Nonrandom device mismatch considerations in nanoscale SRAM. *IEEE Transactions on Very Large Scale Integration (VLSI) Systems*, **20**, 1211–1220.
23. Amat, E., Amatlle, E., Gomez, S. *et al.* (2013) Systematic and random variability analysis of two different 6T-SRAM layout topologies. *Microelectronics Journal*, **44**, 787–793.
24. Qazi, M., Stawiasz, K., Chang, L., and Chandrakasan, A.P. (2011) A 512kb 8T SRAM macro operating down to 0.57V with an AC-coupled sense amplifier and embedded data-retention-voltage sensor in 45nm SOI CMOS. *IEEE Journal of Solid-State Circuits*, **46**, 85–96.
25. Yamauchi, H. (2010) A discussion on SRAM circuit design trend in deeper nanometer-scale technologies. *IEEE Transactions on Very Large Scale Integration (VLSI) Systems*, **18**, 763–774.
26. Karl, E., Wang, Y., Ng, Y.-G. *et al.* (2013) A 4.6GHz 162 Mb SRAM design in 22nm Tri-Gate CMOS technology with integrated read and write assist circuitry. *IEEE Journal of Solid-State Circuits*, **48**, 150–158.
27. Jaksic, Z. and Canal, R. (2013) Comparison of SRAM cells for 10-nm SOI FinFETs under process and environmental variations. *IEEE Transactions on Electron Devices*, **60**, 49–55.
28. Sondheimer, E.H. (1952) The mean free path of electrons in metals. *Advances in Physics*, **1**, 1–42.
29. Blatt, Frank J. (1968) *Physics of Electron Conduction in Solids*, McGraw-Hill, New York.
30. Steinhögl, W., Schindler, G., Steinlesberger, G. *et al.* (2005) Comprehensive study of the resistivity of copper wires with lateral dimensions of 100nm and smaller. *Journal of Applied Physiology (Bethesda, Md: 1985)*, **97**, 023706.
31. Mueller, W., Aichmayr, A., Bergner, W. *et al.* (2005) Challenges for the DRAM cell scaling to 40 nm, IEEE Intern. Electron Dev. Meet., Technical Digest, pp. 347–350.

4

Spin Transfer Torque Random Access Memory

Jian-Ping Wang[1], Mahdi Jamali[1], Angeline Klemm[1], and Hao Meng[2]
[1]*University of Minnesota, USA*
[2]*Data Storage Institute, Singapore*

4.1 Chapter Overview

Over the past years, there has been a significant effort in developing the next generation of nonvolatile memory (NVM) which would be used to replace or be integrated with current technologies. Semiconductor-based memory technologies such as SRAM and Flash are predicted to soon reach their fundamental limits in term of scaling [1]; therefore, an alternative technology capable of high areal density and low-power operation which can be embedded for memory applications is highly desirable. One potential NVM, Spin Transfer Torque Magnetic Random Access Memory (STT-RAM), emerged more than one decade ago and demonstrates the potential to replace most of today's semiconductor memory technologies [2]. It provides faster and more cost-effective solutions for future applications. In this chapter, STT-RAM technology development is reviewed.

Table 4.1 summarizes characteristics of different memory technologies [1]. It compares current state of the art technologies including DRAM, SRAM, and Flash memories, as well as other prototype technologies including STT-RAM, FeRAM, and PCM. STT-RAM utilizes magnetic materials for data storage, is a nonvolatile memory, and unlike DRAM, does not require periodic refreshing of stored information. This also enables instant-on capabilities. The endurance of Flash memory is about 10^5 cycles and the writing time ranges from micro- to milli-seconds. STT-RAM is predicted to perform over 10^{15} write cycles and have writing times of less than 1 ns. Since transistors are not the main storage element of STT-RAM, its power dissipation during the "off state" of the device is much less than corresponding transistor-based memories. The fast writing speed combined with the extremely high endurance of STT-RAM meets the technical demands of CMOS-based memories. This makes STT-RAM a potential candidate for "universal memory" in future information technology.

Emerging Nanoelectronic Devices, First Edition. An Chen, James Hutchby, Victor Zhirnov and George Bourianoff.
© 2015 John Wiley & Sons, Ltd. Published 2015 by John Wiley & Sons, Ltd.

Table 4.1 Characteristics of emerging memory technologies [3]

	SRAM	DRAM	Flash (NOR)	Flash (NAND)	FeRAM	MRAM	PRAM	RRAM	STT-RAM
Nonvolatile	No	No	Yes	Yes	Yes	Yes	Yes	Yes	Yes
Cell size (F^2)	50–120	6–10	10	5	15–34	16–40	6–12	6–10	6–20
Read time (ns)	1–100	30	10	50	20–80	3–20	20–50	10–50	2–20
Write/erase time (ns)	1–100	15	1μs/ 10 ms	1ms/ 0.1 ms	50/50	3–20	50/120	10–50	2–20
Endurance	10^{16}	10^{16}	10^5	10^5	10^{12}	$>10^{15}$	10^8	10^8	$>10^{15}$
Write power	Low	Low	Very high	Very high	Low	High	Low	Low	Low
Other power consumption	Current leakage	Refresh current	None	None	None	None	None	None	None
High voltage required	No	3 V	6–8 V	16–20 V	2–3 V	3 V	1.5–3 V	1.5–3 V	<1.5 V
				Existing products				Prototype	

4.2 Spin Transfer Torque

4.2.1 Background of Spin Transfer Torque

STT-RAM utilizes the magnetization direction of a ferromagnetic layer for data storage similar to magnetic media in Hard Disc Drives. As shown in Figure 4.1a, an STT-RAM cell stack consists of a ferromagnetic free layer (FL) and a reference layer (RL) separated by a nonmagnetic space layer (SL). The self-sustained ferromagnetic free layer (FL) can maintain the data in the event of power loss and results in data nonvolatility. The changes in the magnetization configuration of the FL and RL result in a difference in the resistance of the STT-RAM cell such that it is low (high) when the two layers are parallel (antiparallel).

In 1996, Slonczewski and Berger independently predicted that a nanomagnet could be directly switched by a current flowing through a tri-layer structure [4,5]. This switching behavior is known as the spin transfer torque effect. When electrons pass through a ferromagnetic (FM) layer, FM1, as shown in Figure 4.1b, the current becomes spin polarized

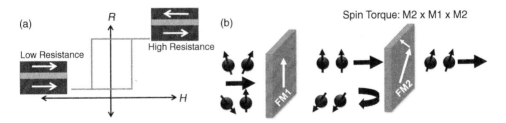

Figure 4.1 (a) Basic MRAM structure and magnetoresistive (MR) loop. (b) Spin torque transfer effect in a tri-layer magnetic structure (spacer layer is not shown)

due to the s-d interactions inside FM1. With the correct choice of the spacer layer material and thickness, the spin polarization is preserved during the traveling of the electrons from FM1 to FM2. At FM2, the polarized electrons exert a torque on the local magnetic moment due to a transfer of angular momentum. Given a sufficiently large injected spin polarized current, the spin torques can be large enough to change the magnetization of FM2.

Slonczewski derived an expression to represent this torque and put it into the Landau–Lifshits–Gilbert equation [4,6]:

$$\frac{dM_2}{dt} = \gamma\, M_2 \times \mathbf{H}_{\text{eff}} + \alpha\,\frac{M_2 \times \dfrac{dM_2}{dt}}{M_2} + \frac{Ig}{e} M_2 \times \frac{M_1 \times M_2}{M_1 M_2} \tag{4.1}$$

where g factor is defined as:

$$g = \frac{\text{sgn}(P)}{-4 + \dfrac{(1+|P|)^3(3+\cos\theta_{12})}{4|P|^{\frac{3}{2}}}} \tag{4.2}$$

M_i is magnetic moment of the spin in the ith layer ($i=1,2$), γ is gyromagnetic ratio ($\gamma = -e\hbar/m_e$), $\mathbf{H}_{\text{eff}}$ is the effective field that M_2 feels, α is the Gilbert damping constant, I is the electrical current, e is the electron charge, P is the spin polarization ratio, and θ_{12} is the angle between M_1 and M_2. The first term of Equation 4.1 represents the precession of spin about the effective field direction. The last two terms represent the damping and the current induced spin torque terms, respectively.

In 2000, Sun reported more details about the spin transfer torque mechanism [7]. He showed that there are three steps involved in the spin current induced magnetization switching: spin precession excitation, spin precession, and spin rotation. Figure 4.2a shows the model geometry for the calculation. The magnetic element has a geometric size of $a \times a \times l_m$. M represents the magnetic moment and the current flow, I, with polarized spin S is along the $-e_x$ direction. With an effective field H_{eff}, the M motion under applied spin current follows a spiral shape as shown in Figure 4.2b.

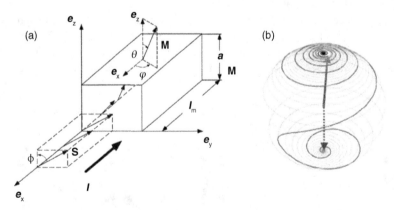

Figure 4.2 (a) Schematic showing various parameters used in the spin transfer torque critical switching current calculation. (b) Trajectory of the M under applied spin current. Switching of M is not instant, but rather it follows a spiral motion [7]

The spin torque from the spin current shifts the moment from its initial direction. Then, the moment starts to precess due to the precession torque of $M \times H_{\text{eff}}$ as shown in Figure 4.2a. The angle between M and its initial magnetization direction becomes larger by increasing the input current density. With sufficient spin current, the precession angle of M becomes large enough to switch to the opposite direction. Sun also provided the critical switching current to induce the switching [7], given by:

$$I_{\text{c}} = \frac{1}{\eta} \left(\frac{2e}{\hbar} \right) \frac{\alpha}{|\cos \varphi|} (a^2 l_{\text{m}} M)(H_{\text{k}} + 2\pi M + H) \qquad (4.3)$$

Where η is the spin polarization ratio, α is the damping constant, H_{k} is the anisotropy field, and H is the applied field. Equation 4.3 provides guidance to reduce the critical switching current. For example, the switching current reduces upon increasing the magnetic spin polarization ratios or by having a magnetic layer with a smaller damping constant.

4.2.2 Experimental Observation of Spin Transfer Torque

Spin transfer torque provides an alternative way to drive a spintronic device. Magnetization can be switched directly by spin current instead of a magnetic field. There are extensive research studies on the current induced spin transfer torque for the switching of the magnetization. Spin transfer torque effect was first observed in magnetic nanowires [8,9]. When a current is injected along a magnetic nanowire, it is polarized within a nanometer range [10]. Spin current flows along the nanowire until it reaches a domain wall where the spin current exerts torques on the magnetic moments in the domain wall and forces the local magnetic moments to follow the direction of the spin current through exchange coupling. Therefore, for a constant input current, the domain wall is pushed along the current flow direction. It was demonstrated that the domain structure remains unaltered for the current density less than 10^7 A/cm^2 where there is no breakdown in the domain wall structure [11]. Spin transfer torque effect was also observed in the point contact structure with Co/Cu multilayers and Co/Cu/Co sandwich structures [12,13] in which a sharp tip (D $\sim$ 100 nm) was used to contact a magnetic multilayer surface. The magnetic state of the multilayer is monitored by measuring the magnetoresistance of the contact. The point contact structure provides a relatively simple method to detect spin transfer torque effect. Since spin transfer occurs within, at most, a few magnetic domains beneath the tip and the domains are coupled to the continuous magnetic film, the dynamics of the spin transfer effects are affected. As a result, the current density required to observe spin transfer is an order of magnitude larger than in isolated magnetic structures, that is, nanopillar structures. There are other limitations with the structure: the point contact tip can damage the layers; the sample fabrication (i.e., contact formation) is not fully reproducible; the sample size can only be deduced from the resistance of the device.

Due to difficulties in the fabrication process, the first successful spin transfer torque switching was observed in 2000 by Katine *et al.* [14], four years after the theoretical prediction. A giant magnetoresistive (GMR) stack was prepared by sputtering the structure Cu(120 nm)/ Co(10 nm)/Cu(6 nm)/Co(2.5 nm)/Cu(15 nm)/Pt(3 nm)/Au(60 nm) onto an oxidized Si substrate. A current perpendicular to plane (CPP) nanopillar structure with 100 nm diameter was fabricated and a schematic drawing of the device is shown in Figure 4.3a. The difference in thickness for the Co layers allows the magnetization direction of the thicker layer to be held fixed so that the polarity of the current bias associated with the spin transfer excitations in the

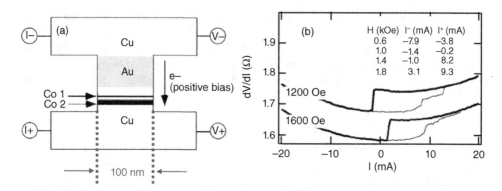

Figure 4.3 (a) Schematic drawing of a CPP GMR structure for spin transfer torque demonstration. (b) Current induced magnetization switching in CPP GMR devices [14]

thinner layer can be determined. Figure 4.3b shows the resistivity curve of the nanopillar with the applied current swept from negative to positive values. The high (low) resistance corresponds to the antiparallel (parallel) configuration between the thin and thick Co layers.

4.3 STT-RAM Operation

4.3.1 Design of STT-RAM Cells

Magnetic RAM, or MRAM, is operated by magnetic fields which are generated from write line 1 which lies along the easy axis and a bit line (write line 2) which lies along the hard axis of the magnetic elements (Figure 4.4a and b). The MRAM element selected to be written lies at the intersection of a word line and bit line. The total magnetic field generated from the two lines is strong enough to switch the magnetization of the MRAM element, while the field produced from a single line seen by other elements is not large enough to induce switching. A single word or bit line cannot produce a field sufficient to induce switching, while their sum at their intersection is larger than the switching threshold required for writing.

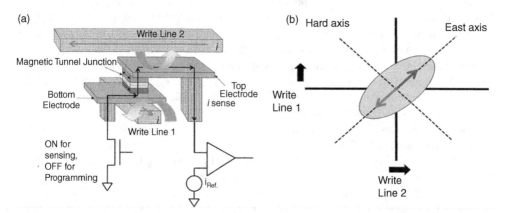

Figure 4.4 (a) Schematic drawing of MRAM. (b) Field generated by write lines 1 and 2 along the easy and hard axis

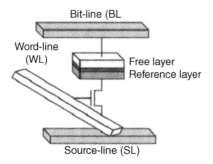

Figure 4.5 Spin transfer torque MRAM design [17]

However, this writing process makes it difficult to efficiently scale MRAM. The magnetic orientation for MRAM elements relies on the shape anisotropy, which is sensitive to geometrical parameters of the cell. Small geometry variation or defects, which are unavoidable in fabrication, could induce large switching field distributions. Other memory cells along the bit line or word line may be switched when writing a specific cell. As a result, the operation window for such MRAM designs is narrow and is known as the half-field selection problem. The switching field distribution worsens with MRAM cell size shrinking. Therefore, it is difficult to achieve high density MRAM in this writing scheme.

Spin transfer torque provides a promising scheme to solve the half-field selection problem in field driven MRAM. Figure 4.5 shows a STT-RAM design operated by spin transfer torque where no magnetic field is required. The writing current directly goes through the memory element under the word line (transistor) control. Such design reduces the unintended writing errors. In addition, the writing current has better scalability since it is proportional to the device area according to Equation 4.3. STT-RAM has attracted attention due to its advantages, including scalability, speed, and power consumption [15,16].

A typical STT-RAM cell consists of one transistor to supply the current for writing and reading and one MTJ element for the actual storage of data (Figure 4.5). The generated voltage, which is proportional to $I_{sense} \times R_{low} \times TMR$, where I_{sense}, $R_{low,}$ and TMR are the sense currents, low (parallel) resistance states, and TMR value, is then compared to a reference MTJ element to determine the state of the memory bit. In STT-RAM design, to achieve high memory density, the memory cell size can be as small as $4F^2$ based on the one MTJ and one transistor design scheme [17], where the width of transistor and MTJ is $1F$.

It imposes great challenges in STT-RAM design as the lowest J_c ($J_c = I_c/a^2$) required for writing STT-RAM (above 10^6 A/cm^2) exceeds what a transistor can supply at $1F$ width. Although drive current of a transistor is proportional to its channel width, a larger transistor is not an option due to its negative impact to the STT-RAM capacity. Thus the current density required to switch the magnetization of the FL has to be reduced. With a low J_c, the voltage across the tunneling barrier is also low, thus improving the endurance and stability of STT-RAM.

4.3.2 Key Parameters for Operation

STT-RAM utilizes MTJ structures instead of metallic giant magnetoresistive (GMR) structures. Metallic GMR devices have resistances of several ohms and small GMR ratios (<20%) [18–21]. As such, it is not a suitable structure to integrate with CMOS transistors and provide a sufficient signal to noise ratio. The MTJ structure is optimized to gain a high TMR signal as well as low product of resistance and area (RA). This is important for the MTJ to sustain a large switching current threshold and gain high data transfer rates. In 2004, the spin transfer torque effect was successfully demonstrated in a low RA MTJ structure based on an Al_2O_3 tunnel barrier [22–24]. Due to the electron incoherent tunneling mechanism in Al_2O_3, the maximum TMR value is only 70% in Al_2O_3 MTJ structures. In late 2004, a breakthrough of TMR around 200% was made by adopting a crystallized (001) MgO barrier layer with single crystal Fe (001) electrodes [25] or polycrystalline CoFe (001) electrodes [26]. Theoretical calculations show that the crystallized MgO barrier functions as a spin filter in a MTJ system. Only $\Delta 1$ states can coherently tunnel through the barrier, resulting in high spin polarization and high MR ratios. More recently, TMR values of around 600% and above have been achieved in CoFeB/MgO/CoFeB pseudo spin valve MTJs at room temperature [27–29]. Since then, MgO-based low RA MTJs with TMR above 150% and RA below $50\,\Omega\mu m^2$ [30,31] have become the most suitable structure to be used in STT-RAM.

Another consideration for STT-RAM is the size and scalability of the devices required for achieving high density STT-RAM. While sub-20 nm MTJ structures have been demonstrated [16,32], other factors such as transistor size, sense current, and thermal stability can still have an impact on the scalability of STT-RAM.

It was mentioned that STT-RAM has better scalability compared to field driven MRAM. Given a fixed J_c value, writing current amplitude keeps reducing with memory cell shrinking until it reaches the superparamagnetic limit [33,34] where the thermal energy, $k_B T$, is equal to the energy barrier and the magnetization state is no longer stable. As the memory cell size is reduced to the nanometer scale, the amplitude of the energy barrier, ΔE, between the two stable states is reduced since ΔE is defined by [34,35]

$$\Delta E = K_u V \qquad (4.4)$$

where V is the volume and the anisotropy energy is defined as

$$K_u = \frac{1}{2} H_k M_s \qquad (4.5)$$

The magnetization status change corresponds to forcing the memory elements to overcome the energy barrier E. Additionally, when $T \neq 0$, thermal fluctuation can unintentionally increase the probability of the FL switching and result in data loss. Data storage typically demands the thermally activated hopping from one state to the other to be improbable over a period of 10 years or longer. Such hopping is known as thermally activated magnetic relaxation [36–38]. The data retention period is defined by

$$\tau = \tau_0 \exp(\Delta E / k_B T) \qquad (4.6)$$

where τ_0 is the attempt frequency (typically one nanosecond). ΔE, given by Equation 4.4, has to be large enough to overcome the magnetization degradation caused by thermal fluctuation to achieve the desired data retention period. With thermal fluctuations, the magnetization of the FL is gradually changed with time as

$$M = M_0 \exp(-t/\tau) \qquad (4.7)$$

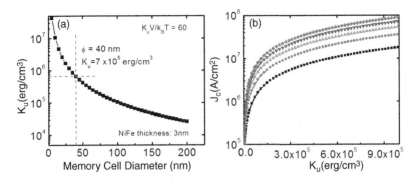

Figure 4.6 (a) Required K_u value to maintain thermal stability for high density STT-RAM. (b) Relationship between critical switching current density (J_c) and anisotropy energy (K_u). Different colors represent different values of damping constants

where t is the time and M_0 is the initial magnetization when $t = 0$. Equation 4.7 indicates that M decays with time and the rate of decay depends on τ and therefore ΔE. However, if the memory cell size is so small that the thermal energy, $k_B T$, is comparable to ΔE, τ is small and M decays very fast. To overcome this and achieve ultrahigh areal density, STT-RAM with very small volume requires a large K_u to maintain thermal stability.

As shown in Figure 4.6a, for a cell size of 30 nm, the minimum required K_u is about 7×10^5 erg/ cm^3 to maintain a thermal stability factor above 60. The required K_u has to be further increased as the memory elements are scaled down. This imposes tremendous writing difficulty for STT-RAM. According to Equation 4.3, the critical switching current density depends on the demagnetization field ($H_d = 4\pi M_s$) and anisotropy field. For a large memory cell size, J_c is dominated by H_d since H_k (K_u) is negligible compared to H_d. However, for high areal density STT-RAM, H_k (K_u) has to be greatly increased for thermal stability and results in a large J_c as shown in Figure 4.6b.

Another aspect affecting the scalability of STT-RAM is the sensing capabilities. As J_c decreases, the read current also needs to decrease in order to prevent accidental switching of the element since the switching probability depends on J_c as [39]

$$P_{sw} = 1 - \exp\{-\tau_p/\tau_0 \exp[-E/k_b T(1 - I_c/I_{c0})]\} \tag{4.8}$$

where τ_p, τ_0, E, I_c and I_{c0} are the current pulse width, attempt frequency, energy barrier, applied current, and critical switching current, respectively. In order to reduce the probability of switching, the read current needs to be reduced as the critical current for switching is reduced. Furthermore, as the MTJ devices are scaled down, variations in the device due to the fabrication process can result in reduction of the sensing current margin and affect the reliability of the device. As I_{sense} decreases, the sense margin also decreases together with device variations at small scales, and it can lead to a large bit error rate in sensing and determination of the state of the memory cell [40,41].

4.4 STT-RAM with Perpendicular Anisotropy

In the previous sections, we have discussed STT-RAM with in-plane anisotropy where magnetizations of both FL and RL are in the film plane. However, there are major drawbacks when applying in-plane anisotropy for STT-RAM. In high density STT-RAM, large shape anisotropy (high aspect ratio) is required to maintain thermal stability and eventually limits the density. In addition, high aspect ratios are not preferred for fabrication processes, since geometry variations can change magnetic characteristics and result in nonuniformity of J_c. Furthermore, the existence of the demagnetization field ($H_d = 4\pi M_s$), which does not contribute to the thermal stability, imposes difficulty on reduction of J_c for STT-RAM with in-plane anisotropy.

In 2006, Wang's group proposed and demonstrated spin transfer torque in a CPP GMR device with perpendicular anisotropy [42]. J_c in a perpendicular anisotropic system is given by

$$I_c = \frac{1}{\eta}\left(\frac{2e}{\hbar}\right)\frac{\alpha}{|\cos\varphi|}(a^2 l_m)\left(\left|2K_u - 2\pi M^2\right| + HM\right) \tag{4.9}$$

where K_u is perpendicular anisotropy, which could be interface anisotropy or crystalline anisotropy, and other parameters are similar to equation Equation 4.3. Equation 4.9 suggests that J_c depends on the effective perpendicular anisotropy $\left|2K_u - 2\pi M^2\right|$, which can be adjusted by tuning layer structure and materials. As a consequence, J_c can be further reduced. Since perpendicular anisotropy is not a function of the shape anisotropy, memory cells can be circular in shape, which is favorable in reducing J_c distribution. Thus, STT-RAM with perpendicular anisotropy has better scalability and is acknowledged as the major direction in STT-RAM development [16,43,44].

In order to generate perpendicular magnetic anisotropy, a multilayer superlattice structure of [magnetic layer/nonmagnetic layer]$_n$ is very common where the nonmagnetic layer is usually a heavy metal such as Pt or Pd [45,46]. The total magnetic anisotropy energy (k_{eff}) of the multilayer system is defined [47]

$$k_{eff}t_m = (k_{cry} - 2\pi M_s^2)t_m + 2k_s \tag{4.10}$$

where t_m is the magnetic layer thickness, k_{cry} is the crystalline anisotropy energy of the magnetic layer, $2\pi M_s^2$ is from the demagnetization field (H_d), and k_s is the perpendicular anisotropy induced by interface effects. Positive k_{eff} value corresponds to a perpendicular magnetic anisotropy. Figure 4.7 shows an example of the magnetic layer thickness dependence

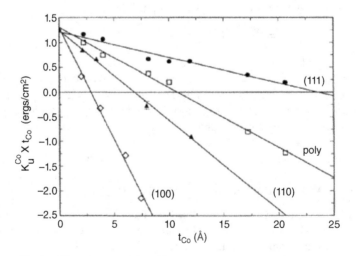

Figure 4.7   Magnetic layer thickness dependence of $K_{eff} \times t_m$ [47]

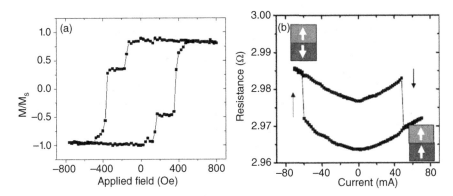

Figure 4.8 (a) *M-H* loop of a sheet film of GMR having perpendicular anisotropy. (b) Spin transfer torque in a spin valve with perpendicular anisotropy at room temperature [42]

of $K_{eff} \times t_m$, in which the magnetic layer is Co and the nonmagnetic layer is Pd [47]. The product of K_{eff} and t_m linearly depends on t_m, with a slope of $(k_{cry} - 2\pi M_s^2)$ and an intercept of $2k_s$. Magnetic anisotropy of the multilayer system is changed from perpendicular to in-plane with increasing magnetic layer thickness. Therefore, the anisotropy field (H_k) direction and amplitude of the multilayer system could be controlled by tuning the individual layer thicknesses [48,49].

Figure 4.8a shows a M-H loop of a sheet film with perpendicular anisotropy. The square loop shape indicates that the easy axis of the FL and RL are out of plane. The two-step switching indicates that the magnetizations of the FL and RL are well separated. Spin transfer torque induced magnetization switching in the nanopillar measured at room temperature is shown in Figure 4.8b. The positive current is defined along the direction from the bottom to top electrode. Two sharp resistance changes indicate current induced magnetization switching of the FL where the two different resistance states represent when the FL and RL are aligned parallel or antiparallel. Since magnetization of the FL is parallel with the direction of current flow, Figure 4.8b provides strong experimental proof that spin transfer torque is the main switching mechanism instead of the Oersted field.

Due to large J_c values (up to 10^8 A/cm^2) and low GMR ratios, spin transfer torque on all metallic CPP GMR devices with perpendicular anisotropy is still not practical for STT-RAM. In 2007, Toshiba demonstrated spin transfer switching in a MTJ device with perpendicular anisotropy. Using MgO as the barrier and TbCoFe/CoFeB as the RL and FL, J_c was reduced to 3.6×10^6 A/cm^2 with TMR around 12% [50]. In 2008, J_c was further reduced to 2.7×10^6 A/cm^2 and TMR was increased to 60% in a MTJ with MgO as the barrier. The structure used [Co/Pd]/CoFeB and FePt/CoFeB as the FL and RL, respectively [51]. However, such Pd- or Pt-based multilayer structures or alloys have a large damping constant and leads to large J_c, as shown in Equation 4.9, and low TMR due to strong spin–orbit interactions. Most recently, Ikeda *et al.* demonstrated perpendicular anisotropy in a CoFeB/MgO system, which has been widely used to generate high TMR in MTJs with in-plane anisotropy [28]. A thin but continuous CoFeB film is adopted as the FL and RL to generate perpendicular anisotropy at the interfaces with the MgO barrier layer. The anisotropy with respect to CoFeB thickness is shown in Figure 4.9 [52]. Ikeda *et al.* showed with a proper annealing process, TMR of the system is more than 120% with RA around 18 $\Omega/\mu m^2$. The average intrinsic critical current

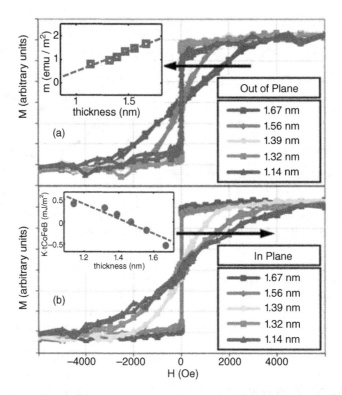

Figure 4.9 Dependence of the magnetic anisotropy on the thickness of CoFeB free layer [95]

density is 3.9×10^6 A/cm^2, which is close to the value for in-plane configuration. Successfully building perpendicular anisotropy on CoFeB/MgO systems provides a solid base for high areal density STT-RAM with high thermal stability, low critical switching current and high TMR values. However, material and structure engineering [53] is still necessary to further reduce J_c to less than 10^6 A/cm^2, the requirement to be integrated with transistors for high density STT-RAM.

4.5 Stack and Material Engineering for J_c Reduction

4.5.1 Dual Pinned Structure

Low critical switching current density and high thermal stability are two main foci in STT-RAM research and development. Theoretically, J_c can be further reduced by increasing the spin polarization ratio, reducing the damping constant, and reducing the saturation magnetization of the FL. However, these attempts offer little room for improvement due to material limitations and high TMR requirement. The research on stack layer engineering, such as dual pinned RL structures [54], provides another approach for J_c reduction. Figure 4.10 illustrates a stack structure with dual pinned reference layers. Two antiferromagnetic (AFM) pinning layers are adopted to bias the two RLs along opposite directions. The commonly used AFM materials are IrMn and PtMn in MTJ structures. The biasing direction can be set during the thermal annealing with application of a magnetic field. The FL is separated from the two RLs by two SLs which can be a metal or a tunneling barrier layer. Since the MR signal from the bottom and

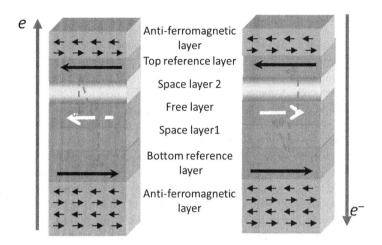

Figure 4.10 Schematics of MTJ with dual pinned reference layers [57]

top parts are opposite to each other, the two SLs cannot be identical in order to gain an output signal.

With the dual pinned RL structure, the efficiency of the spin scattering is greatly enhanced and the switching current density threshold is reduced as a result. With a high resistance value at the initial state, the magnetizations of the FL and bottom (top) RL are anti-parallel (parallel). As shown in the left diagram in Figure 4.10, polarized spin current will interact with the FL magnetization and align it with the magnetization of the bottom RL. The majority spins from the bottom RL will be reflected back by the top RL due to its opposite magnetization. Thus spin dynamic accumulation occurs between the two RLs [55,56]. When such spin accumulation exceeds a critical value, FL switching occurs. To switch the FL back, current polarity should be changed. Due to the symmetrical stack structure, spin can also be accumulated between the two FLs for both current polarities. With the spin accumulation, FL switching deviates from Equation 4.9 and a lower J_c is expected. Wang's group has successfully demonstrated that J_c amplitude can be reduced by factor of four with a dual pinned RL structure (in DC condition) [57]. Diao *et al.* further confirmed that such structure significantly reduces intrinsic J_c when excluding temperature effects [54].

Besides reduction of J_c amplitude, the dual pinned RL structure also improves the symmetry of J_c. For a MTJ with a single RL, switching of the FL is induced by either the majority spin (from high resistance to low resistance) or minority spin (from low resistance to high resistance). High spin polarization of the RL and FL is essential to ensure high TMR and reduce J_c. The two switching processes require different J_c amplitudes, resulting in asymmetrical writing current. In the dual pinned RL structure, it is the majority spins that induces switching regardless of the current polarity. Therefore, symmetry of J_c is improved with the two identical RLs.

One disadvantage of the dual pinned RL structure is the high stack resistance due to more layers introduced in the design. High voltage has to be applied to supply a large enough switching current density. Another drawback is relatively low TMR values. The total TMR signal is the sum of the TMR/GMR effects from the bottom and top parts. As magnetizations of the two RLs are opposite, magnetization of the FL is always parallel to one RL and antiparallel to the other RL so the measured resistance difference (ΔR) shrinks and results in a lower TMR signal.

4.5.2 Nanocurrent Channel Structure Design

Since a large transistor is required to supply enough current to write STT-RAM cells, the footprint of the memory cell can be quite big and can limit the density of a STT-RAM chip. For a given current amplitude, however, local current density can be increased if the current crossing area is reduced. A nanocurrent channel (NCC) integrated with a MTJ has been proposed by Wang *et al.* with J_c reduced by factor of three [58].

Figure 4.11a shows the MTJ stack with two FLs and one FL integrated with a NCC structure. A NCC can be built by lithographic patterning which is quite complicated and costly. One alternative approach is to build a NCC using direct co-sputtering. It is a feasible and cost-effective mechanism to build small NCC for STT-RAM of an effective size of 20 nm or less. Wang *et al.* has reported co-sputtering Fe and Si with oxygen doping to form granular FeSiO NCC with a diameter of about 5 nm [46,59,60] (Figure 4.11b). One highlight of co-sputtering is to form NCC or columnar (island) growth of Fe or FeSi, surrounded by SiO_2 insulating boundaries. Since Si-O has relatively higher bonding energy, Fe atoms will survive from the oxygen atmosphere and form a column structure surrounded by SiO_2 boundaries.

The FeSiO layer consists of magnetic conductive nanovolumes (Fe, Fe-Si) and insulating boundaries (SiO_2). It can function as a current filter. If a current flows through the layer, it will be confined inside the conductive columns. Due to the reduced current cross-area (total area minus

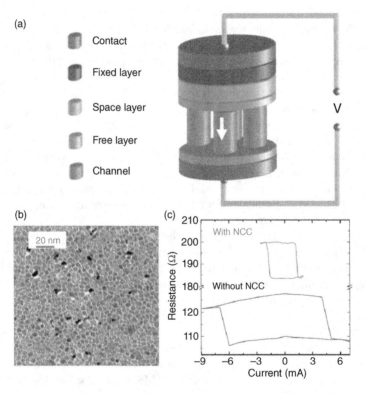

Figure 4.11 (a) MTJ stack integrated with nanocurrent channel structure. (b) Tunneling electron microscopy image of granular FeSiO nano current channel with diameter of about 5 nm [60]. (c) Switching characteristic of NCC device together with the device without NCC [58]

boundary area), the current density inside the nanocolumns will be increased. By adjusting the area ratio between the conductive areas and the boundary areas, the current density inside the conductive nanovolumes can be greatly increased. It provides a new approach to adjust the current distribution inside nanodevices that are sensitive to the current density.

Assuming the magnetizations of the FLs and RL are antiparallel at the beginning, electrons are polarized by the RL layer when they travel from the top layer to the bottom layer. Due to the current confining effect, the local current density in the current channels is increased. The magnetic moments (inside the magnetic NCC and FL 2) are reversed first and then spread the switching all over the free layer together with the help of the Fe or FeSi in FeSiO layer. After FL 2 switches, FL 1 is much easier to be switched because of the exchange coupling with FL 2 through the magnetic NCC. The reversed moments in the NCC, together with the polarized current, will switch the moment and results in switching of FL 1. In the switching process, the NCC layer is: (1) increasing the current density locally inside the nanochannels and the FL and (2) coupling with the two sub-free layers. Furthermore, since the FeSiO is "soft" due to its nanometer scale volume, its magnetic moment is much easier to be reversed (even under small polarized current). The reversed moment will help to switch FL 2. Under DC conditions, the spin transfer torque behavior of a MTJ with NCC is similar to that of a normal MTJ as shown in Figure 4.11c. As designed, the J_c of the former is significantly reduced from 1.4×10^7 to 4.2×10^6 A/cm^2 and the symmetry of J_c for the two current polarities is also improved.

The switching process discussed above was induced under scanning of quasi-static DC current, in which thermally activated spin transfer occurred. Thermal fluctuation from the outside environment helps the spin current to switch the free layer. In reality, STT-RAM is driven by nanosecond current pulses; thermal assistant energy is excluded in such short time frame. Therefore, the critical switching current density is higher. The relationship between critical switching current density, J_c, and pulse width, τ_p, is [61]

$$J_c = J_{c0} \left(1 - \frac{k_B T}{K_u V} \ln \frac{\tau_p}{\tau_0} \right) \qquad (4.11)$$

where J_{c0} is the intrinsic critical switching current density without thermal activated assistance. Equation 4.11 indicates that J_c is always smaller than J_{c0} and equals J_{c0} only at $T = 0$ K. Experimental results show that J_{c0} is normally two or three times higher than the thermally assisted critical switching current, J_c, at room temperature. Equation 4.11 also can be used to evaluate the intrinsic switching current density and the thermal stability factor [39,62–64]. Wang's group reported that J_{c0} is reduced from 2.4×10^7 A/cm^2 to 8.5×10^6 A/cm^2 with a NCC structure in MTJ.

4.5.3 Electric Field Assisted Switching

The previously mentioned methods for reducing J_c for MTJs involve changing the materials and stack structure or device design, but the magnetic properties of the material stay the same. Although there are studies that have been done using multiferroic materials [65,66], they are not commonly used in spintronic devices. It has recently been shown that the magnetic properties of ferromagnetic materials, particularly the perpendicular anisotropy, can be modified by the application of an electric field [67–69]. By tuning the magnetic properties, the energy barrier for magnetization switching can be reduced, resulting in a lower J_c for STT switching.

An electric field, which is generated by applying a voltage across a dielectric (in this case, MgO), can modify the electric properties of certain magnetic thin films by changing the electron density at the Fermi energy level. However, this effect is seen only near the interfaces between the FM and MgO and changes the interfacial anisotropy. As the ferromagnetic films become thicker, the demagnetization field and bulk crystalline anisotropy dominates over the interfacial anisotropy changes and diminishes the electric field modified anisotropy. Therefore, very thin films are required for electric field modification of the magnetic anisotropy. The perpendicular anisotropy energy density is given by [70,71]

$$E_{\text{perp}}d = \left(-\frac{1}{2}\mu_0 M_s^2 + K_u\right)d + K_{s,\text{MgO}/\text{Fe}} + \Delta K_s(V) \tag{4.12}$$

where μ_0, M_s, K_u, d, K_{s1}, K_{s2}, and $\Delta K_s(V)$ are the permeability of free space, saturation magnetization, crystalline anisotropy, thickness of the ferromagnetic film, surface anisotropy from the first interface, surface anisotropy from the second interface, and induced surface anisotropy due to an applied voltage, respectively.

The first experimental evidence showing that electric fields could be used to modify magnetic anisotropy was show in 2007 [72]. FePt and FePd thin films were epitaxially grown on MgO(001) and by the application of −0.6 V, a 4.5% and 1% change in the coercivity were observed for the FePt and FePd films, respectively.

In 2009, electric field induced magnetic anisotropy changes were demonstrated for BCC Fe(001)/MgO(001) junctions where the Fe thin film was a few monolayers thick [73] as well as for several monolayers of FeCo/MgO [74]. Using optical Kerr measurements, it was demonstrated that the anisotropy could be modified by about 40%, changing from −31.3 to −13.7 kJ/m³ with the application of 200 and −200 V [73]. By changing the applied voltage, and therefore strength of electric field at the interface between the ferromagnetic material and MgO, the magnetization of the FeCo films can be changed from in-plane to perpendicular, as shown in Figure 4.12 [74].

The application of an electric field for magnetic anisotropy modification has not only been shown in thin films, but also in MTJs. In 2011, electric field assisted switching in a CoFeB/

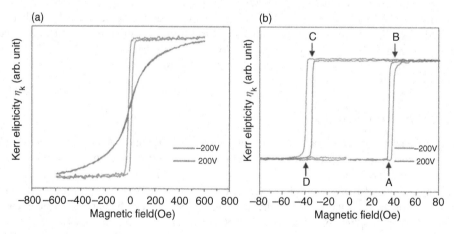

Figure 4.12 (a, b) Change in the magnetic anisotropy from in-plane to perpendicular due to the applied electric field in FeCo films [73]

MgO/CoFeB MTJ was demonstrated [70]. An appropriate bias field is applied for operation of the device and then through application of an electric field, the coercivity change results in changing the FL and RL magnetization between parallel and antiparallel. The coercivity of the FL (RL) changes from 72 Oe (115 Oe) to 20 Oe (137 Oe) with the application of a bias voltage of $V_{bias} = -870\,mV$ ($+890\,mV$). Additionally, by applying bias voltages of $-0.9\,V$ and $-1.5\,V$, the critical switching current density required for STT switching of the devices was reduced to -1.2×10^4 and $-2.4 \times 10^4\,A/cm^2$. If this electric field control of the magnetic anisotropy is integrated with STT-RAM devices, it will significantly reduce the writing current and improve the energy efficiency of the devices.

4.6   Ultra-Fast Switching of MTJs

An important issue to consider with STT-RAM is the write speed. Spin torque switching on the picosecond scale has been demonstrated for MTJs [75,76]. In order to achieve this ultra-fast switching, an orthogonal MTJ structure has been adopted, which consists of an in-plane FL and two polarizing layers, one in-plane and one orthogonal to the plane. The structure used in [77] has the following structure: bottom lead/in-plane polarizer/barrier layer/free layer/barrier/perpendicular polarizer/capping layer. The magnetoresistance value is due to the orientation of the free layer relative to the in-plane polarizing layer. In a traditional in-plane MTJ, the initial torque exerted on the free layer is small since it depends on the angle between the different magnetic layers as seen by Equation 4.1. However, the addition of an orthogonal polarizing layer results in a large initial spin torque. The magnetization trajectories for the case of a traditional in-plane MTJ and orthogonal MTJ are shown in Figure 4.13a and b. The trajectories were calculated using a macro spin approximation of the Landau–Lifshitz–Gilbert equation with the STT term. From the magnetization trajectories, we can see that the free layer magnetization of the orthogonal MTJ is quickly switched to out-of-plane and results in a faster switching speed. "To demonstrate the switching of the orthogonal MTJ, voltage pulses of varying widths and magnitudes were used and the requirements for 50% switching probability was determined. At an excitation voltage of 1.58 V, the switching time could be reduced to 120 ps without the aid of an external bias field.

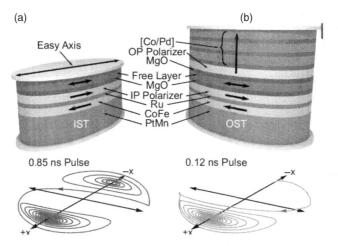

Figure 4.13   (a) Magnetization trajectories for an in-plane MTJ (b) for orthogonal MTJ [77]

In 2012, Zhao *et al.* [78] utilized a CoFeB/MgO/CoFeB MTJ structure with Fe rich CoFeB. The Fe rich CoFeB results in a strong perpendicular interface anisotropy, but the magnetization of the magnetic layers remains in-plane. They demonstrated that for a 50% switching probability, a 165 ps pulse width was required, and for a switching probability of 98%, the required pulse width was 190 ps where the excitation pulse amplitude was 1.89 V. This corresponds to switching energies of 0.16 and 0.21 pJ for 50 and 98% switching probabilities, respectively. The fast switching of the devices is attributed to a reduction in the out-of-plane demagnetization field due to the perpendicular interface anisotropy.

4.7 Spin–Orbit Torques for Memory Application

Recently, the study of current induced spin–orbit torques in ultrathin magnetic structures in the absence of a spin polarizer has attracted a strong interest among researchers due to its potential for low power magnetization switching [79–83]. Different phenomena, such as the Rashba effect [79,80] and the spin Hall effect [14,84], have been proposed to explain the current induced torques with an in-plane current. In this section, we briefly discuss the Rashba and spin Hall effects and present some of the recent progress in this area.

Although the Rashba effect was known in semiconductor materials for a long time [85], it recently has been found that the current induced Rashba field could be comparable to current induced spin transfer torque in metallic systems [79,86]. An ultra-thin film of magnetic layer ($\sim$1 nm) is sandwiched between a nonmagnetic metal and insulating layer. The nonmagnetic metal layer must have strong spin–orbit coupling such as Ta, Pt, and Pd. Due to the inversion asymmetry of multilayer thin films, the conduction electrons feel a net electric field caused by an asymmetric Coulomb potential profile. Furthermore, presence of strong spin–orbit coupling in the transition ferromagnetic metallic system interprets this electric field to an equivalent magnetic field. The magnitude of the Rashba field, H_{ra}, is given by [87,88]:

$$H_{ra} = \frac{\alpha_R P}{\mu_B M_s}(\hat{z} \times J_e),\qquad(4.13)$$

where μ_B is the Bohr magneton, M_s is the saturation magnetization, P is a parameter that represents the *s-d* electron coupling strength, α_R is the Rashba coefficient (proportional to the spin–orbit strength of the material) and J_e is the input current density. Although the Rashba field is concentrated only at the interface of the magnetic and nonmagnetic layers, its intensity could be quite large (the order of 1 Tesla for a current density of about 10^8 A/cm^2) [79,80,87].

In 2012, another proposal came out that suggested spin–orbit torque originated from the spin Hall effect [14]. In a material with a strong spin–orbit coupling, electrons with opposite spins deflect in different directions according to $\vec{J}_s \propto \vec{\sigma} \times \vec{E}$ where E is the applied electric field, σ is the electron spin direction, and J_s is the generated spin current. The direction of the accumulated spin at the boundary of the nonmagnetic layer is shown in Figure 4.14a. Since spin accumulates at each boundary of the material, the presence of a magnetic material at the interface with the nonmagnetic channel results in the absorption of the spin current and application of spin torque on the magnetic moments.

Recently, there have been some reports on the characterization of both torques in the magnetic multilayer structures [89,90]. In spite of fact that which effect is the dominant mechanism of the magnetization switching, spin orbit torque is a promising candidate for the magnetization switching. There has been a proposal of a three terminal device shown in

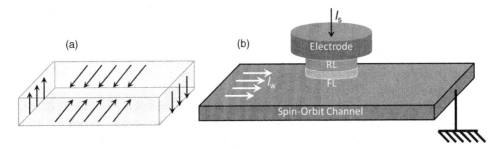

Figure 4.14 (a) Splitting of the electron spins due to the spin orbit effect. (b) Schematic of a three terminal device that operates based on the spin orbit torque

Figure 4.14b that utilizes the spin orbit torques for the magnetization writing [14]. In this device, the writing and reading have different paths. For writing of the data on the magnetic layer, a current is passed through the spin orbit channel; for the reading, a small current is passed through the MTJ stack.

The three-terminal device has several advantages over the conventional STT-RAM memory cell. Since the read and write current paths are separated, each path can be optimized for its specific purpose. By preventing the passage of large current through the tunnel barrier, there is no breakdown issue. It has been shown that the efficiency of the spin Hall induced spin torque is higher than the conventional spin transfer torque effect [81]. Furthermore, the Joule heating can be significantly reduced since the write path can have much lower resistance. In addition, the MTJ can be optimized for a large TMR. Finally, other switching mechanism such as electric field assisted switching can be easily integrated with the three-terminal device.

4.8   Current Demonstrations for STT-RAM

The first demonstration of STT-RAM was in 2005 by Sony where they demonstrated the first 4 kbit STT-RAM [91]. Hitachi and Tohuku University announced their 2 Mbit STT-RAM in 2008 [92] and later demonstrated in 2009 a 32 Mbit STT-RAM. Grandis has announced their low power STT-RAM where the writing energy is 0.16 pJ with a write current of 80 μA, write voltage of 0.4 V and write speed of 5 ns [93]. Everspin announced in November 2012 the release of their 64 Mbit STT-RAM [94].

4.9   Summary and Perspectives

STT-RAM is a competitive memory technology which is nonvolatile, fast and endurable. It offers a simpler cell structure compared to MRAM and allows for better scalability. It can be a universal memory for personal electronic devices, automobile, and space technology. After years of development, STT-RAM chips have been commercialized for embedded memory applications. The crucial issue of STT-RAM is the high critical switching current density for writing process, which requires even higher current to achieve faster write speeds.

Various cell structures and material engineering methods have been proposed and shown to successfully reduce critical switching current density without degrading the thermal stability of the STT-RAM cell. Utilization of MgO as the tunnel barrier for MTJs has brought many advantages for STT-RAM. Successful integration of MgO MTJs with spin transfer torque brings critical switching current density down to 1×10^6 A/cm^2 and TMR values of more than 150%. However, utilizing shape anisotropy will eventually limit chip capacity for in plane STT-RAM. MTJs with perpendicular anisotropy, both due to interfacial anisotropy or crystalline anisotropy, have better scalability and higher spin transfer efficiency in terms of thermal stability over critical switching current density. Spin transfer torque together with perpendicular anisotropic MTJs is a promising candidate for high areal density STT-RAM. It has also been shown that electric field assisted switching can be used to reduce the writing energy. Furthermore, recent progresses on the spin-orbit torques demonstrate an exciting approach to solve most of the STT-RAM issues and introduce a new device structure for the STT-RAM memory cell.

References

1. The International Technology Roadmap for Semiconductors (ITRS) Emerging Research Devices (2011) ITRS Online. http://www.itrs.net/Links/2011ITRS/2011Chapters/2011ERD.pdf (accessed 16 January 2014).
2. Daughton, J.M. (1997) Magnetic tunneling applied to memory (invited). *Journal of Applied Physiology*, **81**(8), 3758.
3. Future Fab (2014) www.future-fab.com/documents.asp?d_ID=4400 (accessed 16 January 2014).
4. Slonczewski, J. (1996) Current-driven excitation of magnetic multilayers. *Journal of Magnetism and Magnetic Materials*, **8853**(96), 53–58.
5. Berger, L. (1996) Emission of spin waves by a magnetic multilayer traversed by a current. *Physical Review B-Condensed Matter*, **54** (13), 9353–9358.
6. Slonczewski, J.C. (1996) Current-driven excitation of magnetic multilayers. *Journal of Magnetism and Magnetic Materials*, **8853**(96), 93–98.
7. Sun, J. (2000) Spin-current interaction with a monodomain magnetic body: A model study. *Physical Review B-Condensed Matter*, **62**(1), 570–578.
8. Partin, D.L., Karnezos, M., deMenezes, L.C., and Berger, L. (1974) Nonuniform current distribution in the neighborhood of a ferromagnetic domain wall in cobalt at 4.2K. *Journal of Applied Physiology*, **45**(4), 1852.
9. DeLuca, J.C. and Gambino, R.J. (1979) Abstract: Bias field dependence of domain drag propagated bubble domains. *Journal of Applied Physiology*, **50**(B3), 2212.
10. Togawa, Y., Kimura, T., Harada, K. *et al.* (2006) Current-excited magnetization dynamics in narrow ferromagnetic wires. *Japanese Journal of Applied Physics*, **45**(27), L683–L685.
11. Mougin, A., Cormier, M., Adam, J.P. *et al.* (2007) Domain wall mobility, stability and Walker breakdown in magnetic nanowires. *Europhysics Letters*, **78**(5), 57007.
12. Tsoi, M., Jansen, A., Bass, J. *et al.* (1998) Excitation of a magnetic multilayer by an electric current. *Physical Review Letters*, **80**(19), 4281–4284.
13. Ralls, K.S., Buhrman, R.a., and Tiberio, R.C. (1989) Fabrication of thin-film metal nanobridges. *Applied Physics Letters*, **55**(23), 2459.
14. Katine, J., Albert, F., Buhrman, R. *et al.* (2000) Current-driven magnetization reversal and spin-wave excitations in Co/Cu/Co pillars. *Physical Review Letters*, **84**(14), 3149–3152.
15. Hu, J.-M., Li, Z., Chen, L.-Q., and Nan, C.-W. (2011) High-density magnetoresistive random access memory operating at ultralow voltage at room temperature. *Nature Communications*, **2**, 553.
16. Klostermann, U.K., Angerbauerl, M., Grtining, U. *et al.* (2007) A perpendicular spin torque switching based MRAM for the 28nm technology node. *IEEE International Electron Devices Meeting, Technical Digest*, **2007**, 187–190.
17. Wang, X., Chen, Y., Li, H. *et al.* (2008) Spin torque random access memory down to 22nm technology. *IEEE Transactions on Magnetics*, **44**(11), 2479–2482.
18. Chappert, C., Fert, A. and VanDau, F F.N. (2007) The emergence of spin electronics in data storage. *Nature Materials*, **6**(11), 813–823.
19. Fert, A. and Piraux, L. (1999) Magnetic nanowires. *Journal of Magnetism and Magnetic Materials*, **200**(1), 338–358.
20. Binasch, G., Grünberg, P., Saurenbach, F., and Zinn, W. (1989) Enhanced magnetoresistance in layered magnetic structures with antiferromagnetic interlayer exchange. *Physical Review B-Condensed Matter*, **39**(7), 4828–4830.

21. Dieny, B., Humbert, P., Speriosu, V. *et al.* (1992) Giant magnetoresistance of magnetically soft sandwiches: Dependence on temperature and on layer thicknesses. *Physical Review B-Condensed Matter*, **45**(2), 806–813.

22. Huai, Y., Albert, F., Nguyen, P. *et al.* (2004) Observation of spin-transfer switching in deep submicron-sized and low-resistance magnetic tunnel junctions. *Applied Physics Letters*, **84**(16), 3118.

23. Fuchs, G.D., Emley, N.C., Krivorotov, I.N. *et al.* (2004) Spin-transfer effects in nanoscale magnetic tunnel junctions. *Applied Physics Letters*, **85**(7), 1205.

24. Meng, H., Wang, J., Diao, Z., and Wang, J.-P. (2005) Low resistance spin-dependent magnetic tunnel junction with high breakdown voltage for current-induced-magnetization-switching devices. *Journal of Applied Physiology*, **97**(10), 10C926.

25. Yuasa, S., Nagahama, T., Fukushima, A. *et al.* (2004) Giant room-temperature magnetoresistance in single-crystal Fe/MgO/Fe magnetic tunnel junctions. *Nature Materials*, **3**(12), 868–871.

26. Parkin, S.S.P., Kaiser, C., Panchula, A. *et al.* (2004) Giant tunnelling magnetoresistance at room temperature with MgO (100) tunnel barriers. *Nature Materials*, **3**(12), 862–867.

27. Ikeda, S., Hayakawa, J., Ashizawa, Y. *et al.* (2008) Tunnel magnetoresistance of 604% at 300K by suppression of Ta diffusion in CoFeB/MgO/CoFeB pseudo-spin-valves annealed at high temperature. *Applied Physics Letters*, **93**(8), 82508.

28. Ikeda, S., Miura, K., Yamamoto, H. *et al.* (2010) A perpendicular-anisotropy CoFeB-MgO magnetic tunnel junction. *Nature Materials*, **9**(9), 721–724.

29. Bai, Z., Shen, L., Wu, Q. *et al.* (2013) Boron diffusion induced symmetry reduction and scattering in CoFeB/MgO/CoFeB magnetic tunnel junctions. *Physical Review B-Condensed Matter*, **87**(1), 014114.

30. Diao, Z., Pakala, M., Panchula, A. *et al.* (2006) Spin-transfer switching in MgO-based magnetic tunnel junctions (invited). *Journal of Applied Physiology*, **99**(8), 08G510.

31. Han, X., Ali, S.S., and Liang, S. (2012) MgO(001) barrier based magnetic tunnel junctions and their device applications. *Science China Physics, Mechanics & Astronomy*, **56**(1), 29–60.

32. Jeong, J.H., Kim, Y., Lim, W.C. *et al.* (2011) Extended scalability of perpendicular STT-MRAM towards sub-20nm MTJ node, in 2011 International Electron Devices Meeting, pp. 24.1.1–24.1.4.

33. Speliotis, D.E. (1999) Magnetic recording beyond the first 100 Years. *Journal of Magnetism and Magnetic Materials*, **193**(1–3), 29–35.

34. White, R.M. (2001) Magnetic recording — Pushing back the superparamagnetic barrier. *Journal of Magnetism and Magnetic Materials*, **226–230**, 2042–2045.

35. Chappert, C., Fert, A., and F. N., VanDau. (2007) The emergence of spin electronics in data storage. *Nature Materials*, **6**(11), 813–823.

36. Neel, L. (1949) Théorie du traînage magnétique des ferromagnétiques en grains fins avec application aux terres cuites. *Annales De Geophysique*, **5**, 99–136.

37. Brown, W.J. (1963) Thermal fluctuations of a single-domain particle. *Physical Review*, **34**(1951), 1677–1686.

38. Cowburn, R.P. (2003) Superparamagnetism and the future of magnetic random access memory. *Journal of Applied Physiology*, **93**(11), 9310.

39. Yagami, K., Tulapurkar, a.a., Fukushima, A., and Suzuki, Y. (2004) Low-current spin-transfer switching and its thermal durability in a low-saturation-magnetization nanomagnet. *Applied Physics Letters*, **85**(23), 5634.

40. Chen, E., Apalkov, D., Driskill-Smith, A. *et al.* (2012) Progress and Prospects of Spin Transfer Torque Random Access Memory. *IEEE Transactions on Magnetics*, **48**(11), 3025–3030.

41. Article, R. Challenges for semiconductor spintronics, vol. i, pp. 40–44.

42. Meng, H. and Wang, J.-P. (2006) Spin transfer in nanomagnetic devices with perpendicular anisotropy. *Applied Physics Letters*, **88**(17), 172506.

43. Mangin, S., Ravelosona, D., Katine, J.a. *et al.* (2006) Current-induced magnetization reversal in nanopillars with perpendicular anisotropy. *Nature Materials*, **5**(3), 210–215.

44. Zhu, X. and Zhu, J.-G. (2006) Spin Torque and field-driven perpendicular MRAM designs scalable to multi-gb/chip capacity. *IEEE Transactions on Magnetics*, **42**(10), 2739–2741.

45. Iwasaki, S. (2012) Perpendicular magnetic recording—Its development and realization. *Journal of Magnetism and Magnetic Materials*, **324**(3), 244–247.

46. Wang, J., Shen, W., and Bai, J. (2005) Perpendicular magnetic recording, vol. 41, no. 10, 3181–3186.

47. Engel, B., England, C., Van Leewen, R.A. *et al.* (1991) Interface magnetic anisotropy in epitaxial superlattices. *Physical Review Letters*, **67**(14), 1910–1913.

48. Ding, Y., Judy, J.H., and Wang, J.-P. (2005) Magneto-resistive read sensor with perpendicular magnetic anisotropy. *IEEE Transactions on Magnetics*, **41**(2), 707–712.

49. Shi, J., Tehrani, S., and Scheinfein, M.R. (2000) Geometry dependence of magnetization vortices in patterned submicron NiFe elements. *Applied Physics Letters*, **76**(18), 2588.

50. Nakayama, M., Kai, T., Shimomura, N. *et al.* (2008) Spin transfer switching in TbCoFe/CoFeB/MgO/CoFeB/TbCoFe magnetic tunnel junctions with perpendicular magnetic anisotropy. *Journal of Applied Physiology*, **103**(7), 07A710.

51. Nagase, T., Nakayama, M., Shimomura, N. *et al.* (2008) Spin transfer torque switching in perpendicular magnetic tunnel junctions with Co based multilayer. *2008 APS March Meeting*, **53**(2), C1.00331.

52. Amiri, P. Khalili., Zeng, Z.M., Langer, J. *et al.* (2011) Switching current reduction using perpendicular anisotropy in CoFeB–MgO magnetic tunnel junctions. *Applied Physics Letters*, **98**(11), 112507.

53. Zhu, X. and Zhu, J.-G. (2007) Effect of damping constant on magnetic switching in spin torque driven perpendicular MRAM. *IEEE Transactions on Magnetics*, **43**(6), 2349–2351.

54. Diao, Z., Panchula, A., Ding, Y. *et al.* (2007) Spin transfer switching in dual MgO magnetic tunnel junctions. *Applied Physics Letters*, **90**(13), 132508.

55. Jiang, Y., Abe, S., Ochiai, T. *et al.* (2004) Effective reduction of critical current for current-induced magnetization switching by a Ru layer insertion in an exchange-biased spin valve. *Physical Review Letters*, **92**(16), 167204.

56. Parkin, S.S.P., Roche, K.P., Samant, M.G. *et al.* (1999) Exchange-biased magnetic tunnel junctions and application to nonvolatile magnetic random access memory (invited). *Journal of Applied Physiology*, **85**(8), 5828.

57. Meng, H., Wang, J., and Wang, J.-P. (2006) Low critical current for spin transfer in magnetic tunnel junctions. *Applied Physics Letters*, **88**(8), 82504.

58. Meng, H. and Wang, J.-P. (2006) Composite free layer for high density magnetic random access memory with lower spin transfer current. *Applied Physics Letters*, **89**(15), 152509.

59. Wang, J.-P., Shen, W.K., Bai, J.M. *et al.* (2005) Composite media (dynamic tilted media) for magnetic recording. *Applied Physics Letters*, **86**(14), 142504.

60. Shen, W.K., Bai, J.M., Victora, R.H. *et al.* (2005) Composite perpendicular magnetic recording media using [Co/PdSi][sub n] as a hard layer and FeSiO as a soft layer. *Journal of Applied Physiology*, **97**(10), 10N513.

61. Koch, R., Katine, J., and Sun, J. (2004) Time-resolved reversal of spin-transfer switching in a nanomagnet. *Physical Review Letters*, **92**(8), 88302.

62. Braganca, P.M., Krivorotov, I.N., Ozatay, O. *et al.* (2005) Reducing the critical current for short-pulse spin-transfer switching of nanomagnets. *Applied Physics Letters*, **87**(11), 112507.

63. Huai, Y., Pakala, M., Diao, Z., and Ding, Y. (2005) Spin-transfer switching current distribution and reduction in magnetic tunneling junction-based structures. *IEEE Transactions on Magnetics*, **41**(10), 2621–2626.

64. Yagami, K., Tulapurkar, a.a., Fukushima, A., and Suzuki, Y. (2005) Inspection of intrinsic critical currents for spin-transfer magnetization switching. *IEEE Transactions on Magnetics*, **41**(10), 2615–2617.

65. Eerenstein, W., Wiora, M., Prieto, J.L. *et al.* (2007) Giant sharp and persistent converse magnetoelectric effects in multiferroic epitaxial heterostructures. *Nature Materials*, **6**(5), 348–351.

66. Hur, N., Park, S., Sharma, P.A. *et al.* (2004) Electric polarization reversal and memory in a multiferroic material induced by magnetic fields. *Nature*, **429**(6990), 392–395.

67. Lottermoser, T., Lonkai, T., Amann, U., and Fiebig, M. (2004) Magnetic phase control by an electric field, vol. 193, no. May, pp. 541–544.

68. Ohno, H., Chiba, D., Matsukura, F. *et al.* (2000) Electric-field control of ferromagnetism. *Nature*, **408**(6815), 944–946.

69. Chu, Y.-H., Martin, L.W., Holcomb, M.B. *et al.* (2008) Electric-field control of local ferromagnetism using a magnetoelectric multiferroic. *Nature Materials*, **7**(6), 478–482.

70. Wang, W.-G., Li, M., Hageman, S., and Chien, C.L. (2012) Electric-field-assisted switching in magnetic tunnel junctions. *Nature Materials*, **11**(1), 64–68.

71. Kanai, S., Yamanouchi, M., Ikeda, S. *et al.* (2012) Electric field-induced magnetization reversal in a perpendicular-anisotropy CoFeB-MgO magnetic tunnel junction. *Applied Physics Letters*, **101**(12), 122403.

72. Weisheit, M., Fähler, S., Marty, A. *et al.* (2007) Electric field-induced modification of magnetism in thin-film ferromagnets. *Science*, **315**(5810), 349–351.

73. Maruyama, T., Shiota, Y., Nozaki, T. *et al.* (2009) Large voltage-induced magnetic anisotropy change in a few atomic layers of iron. *Nature Nanotechnology*, **4**, 158–161.

74. Shiota, Y., Maruyama, T., Nozaki, T. *et al.* (2009) Voltage-assisted magnetization switching in ultrathin Fe 80 Co 20 alloy layers. *Applied Physics Express*, **2**, 63001.

75. Sun, Z., Wu, W., Incorporated, Q., and Diego, S. (2012) A Dual-mode Architecture for Fast-Switching STT-RAM.

76. Kitagawa, E., Fujita, S., Nomura, K. *et al.* (2012) Impact of ultra low power and fast write operation of advanced perpendicular MTJ on power reduction for high-performance mobile CPU, 2012 Int. Electron Devices Meet., pp. 29.4.1–129.4.4.

77. Rowlands, G.E., Rahman, T., Katine, J.a. *et al.* (2011) Deep subnanosecond spin torque switching in magnetic tunnel junctions with combined in-plane and perpendicular polarizers. *Applied Physics Letters*, **98**(10), 102509.

78. Zhao, H., Glass, B., Amiri, P.K. *et al.* (2012) Sub-200ps spin transfer torque switching in in-plane magnetic tunnel junctions with interface perpendicular anisotropy. *Journal of Physics D-Applied Physics*, **45**(2), 25001.

79. Gambardella, P. and Miron, I.M. (2011) Current-induced spin-orbit torques. *Philosophical Transactions. Series A, Mathematical, Physical, and Engineering Sciences Royal Society (Great Britain)*, **369**(1948), 3175–3197.

80. Avci, C. Onur., Garello, K., Miron, I. Mihai. *et al.* (2012) Magnetization switching of an MgO/Co/Pt layer by in-plane current injection. *Applied Physics Letters*, **100**(21), 212404.

81. Liu, L., Pai, C.-F., Li, Y. *et al.* (2012) Spin-torque switching with the giant spin hall effect of tantalum. *Science*, **336** (6081), 555–558.

82. Ha, S.-S., Yoon, J., Lee, S. *et al.* (2009) Spin wave quantization in continuous film with stripe domains. *Journal of Applied Physiology*, **105**(7), 07D544.

83. Suzuki, T., Fukami, S., Ishiwata, N. *et al.* (2011) Current-induced effective field in perpendicularly magnetized Ta/CoFeB/MgO wire. *Applied Physics Letters*, **98**(14), 142505.

84. Liu, L., Moriyama, T., Ralph, D.C., and Buhrman, R.A. (2011) Spin-torque ferromagnetic resonance induced by the spin hall effect. *Physical Review Letters*, **106**(3), 36601.

85. Chernyshov, A., Overby, M., Liu, X. *et al.* (2009) Evidence for reversible control of magnetization in a ferromagnetic material by means of spin–orbit magnetic field. *Nature Physics*, **5**(9), 656–659.

86. Miron, I.M., Gaudin, G., Auffret, S. *et al.* (2010) Current-driven spin torque induced by the Rashba effect in a ferromagnetic metal layer. *Nature Materials*, **9**(3), 230–234.

87. Suzuki, T., Fukami, S., Ishiwata, N. *et al.* (2011) Current-induced effective field in perpendicularly magnetized Ta/CoFeB/MgO wire. *Applied Physics Letters*, **98**(14), 142505.

88. Pi, U.H., Kim, K.Won., Bae, J.Y. *et al.* (2010) Tilting of the spin orientation induced by Rashba effect in ferromagnetic metal layer. *Applied Physics Letters*, **97**(16), 162507.

89. Jamali, M., Narayanapillai, K., Qiu, X. *et al.* (2013) Spin-orbit torques in Co/Pd multilayer nanowires. *Physical Review Letters*, **111**(24), 246602.

90. Kim, J., Sinha, J., Hayashi, M. *et al.* (2013) Layer thickness dependence of the current-induced effective field vector in Ta|CoFeB|MgO. *Nature Materials*, **12**(3), 240–245.

91. Hosomi, M., Yamagishi, H., Yamamoto, T. *et al.* (2005.) Technical report. *IEEE International Electron Devices Meeting, Technical Digest*, **459**, 19–20.

92. Kawahara, T., Takemura, R., Miura, K. *et al.* (2008) 2 Mb SPRAM (SPin-Transfer Torque RAM) with Bit-by-Bit Bi-directional current write and parallelizing-direction current read. *IEEE Journal of Solid-State Circuits*, **43**(1), 109–120.

93. Grandis (2010) "Grandis Newsletter," *Gd. Newsl.*, vol. 3, no. Q2.

94. Everspin (2012) ST-MRAM Press Release.

95. Amiri, P. Khalili., Zeng, Z.M., Langer, J. *et al.* (2011) Switching current reduction using perpendicular anisotropy in CoFeB–MgO magnetic tunnel junctions. *Applied Physics Letters*, **98**(11), 112507.

5

Phase Change Memory

Rakesh Jeyasingh, Ethan C. Ahn, S. Burc Eryilmaz, Scott Fong, and H.-S.
Philip Wong
Department of Electrical Engineering, Stanford University, USA

5.1 Introduction

In the late 1960s, Stanford Ovshinsky's (1922–2012) discovery of switching and phase change phenomena in chalcogenide materials [1] seeded new possibilities in data storage understanding/application. Initially, phase change chalcogenides impacted the optical disk market, enabling DVD and Blu-ray disks. Concomitantly, innovation in the materials and solid-state memory device research has led to Phase Change Memory (PCM) as one of the potential candidates for future nonvolatile memory technology [2]. The traditional memory hierarchy has a major bottleneck for improving the overall system performance due to increasing performance gap between the main memory and hard disk storage. PCM has the potential to combine DRAM-like features such as bit alteration, fast read and write, and good endurance and Flash-like features such as nonvolatility using a simple device structure. Thus introduction of PCM in the memory hierarchy would enable a seamless and versatile data exchange between the processor and storage [3]. PCM is also expected to be a highly scalable technology extending beyond the scaling limit of existing memory devices [4].

In this chapter, we focus on one of the mature emerging memory technologies – PCM – by summarizing the important material and device learning in recent years [5–8], with a focus on how fundamental physics interact with device properties and the device scaling potential of PCM. We start with a description of the basic device operation in Section 5.2. The properties of the phase change material, reviewed in Section 5.3, are of fundamental importance to device optimization for the targeted application (such as finding the best speed, retention, and endurance tradeoff) as well as the scalability of PCM. Any new semiconductor technology, including PCM, must be scalable for many generations. The potential for the PCM to scale to nanoscale dimensions is explored in Section 5.4. The vision of a high-density memory attained via multi-bit operation of the memory cell and three-dimensional stacking of the memory array is reviewed in Section 5.5. In Section 5.6, we review two of the most promising applications of PCM. Finally, we offer a view of the future and conclude in Section 5.7.

Emerging Nanoelectronic Devices, First Edition. An Chen, James Hutchby, Victor Zhirnov and George Bourianoff.
© 2015 John Wiley & Sons, Ltd. Published 2015 by John Wiley & Sons, Ltd.

5.2 Device Operation

One of the most common structures of a PCM cell, called the T-cell or a mushroom cell is shown in Figure 5.1a. The name "mushroom" stems from the fact that the programming region is in the shape of a "half dome." The phase change material forms a "T" shape with a highly resistive, narrow structure called a "heater" element.

PCM uses the large resistivity contrast between the crystalline (low resistivity) and amorphous (high resistivity) state of a phase change material in order to store the information. This phase (resistivity) change is achieved by applying sufficient thermal energy to change the phase of the material. In order to reset the PCM cell into its amorphous state, a short electrical pulse (typically<50 ns) is applied to the bottom electrode contact (BEC). The amplitude of the pulse is such that the heat generated in the phase change material–heater interface causes a region of the phase change material to melt. The thermal pulse is quenched rapidly to cause the molten region to cool to its amorphous state. For the case of set programming, the PCM cell is applied an electrical pulse that is sufficient to increase the temperature of the programming region above the crystalliza-tion temperature and a time period sufficiently long to crystallize the phase change material. To read the state of the cell a small electric pulse is applied to measure the cell resistance such that it does not disturb the state of the cell. The different pulse shapes are summarized in Figure 5.1b.

The current–voltage characteristics of the set and reset states are shown in Figure 5.2. It can be seen that there is a large resistance contrast between the set and the reset state for voltages below the threshold switching voltage (V_{th}). The cell in the reset state has a high resistance below V_{th} and shows electronic threshold switching behavior at V_{th}, that is, a negative differential resistance. For voltages below V_{th} (sub-threshold region) the PCM returns to its original amorphous state after removing the electrical stimulus. However for voltages above V_{th}, when the electrical stimulus is retained for a sufficient amount of time, the PC material undergoes memory switching into a low-resistivity, crystalline state. The programming of the PCM cell critically depends on this electronic threshold switching process [6], the physics of which is yet to be fully understood. The conduction of high current through the amorphous region is enabled only by this switching process in order for the crystallization process to occur.

Reset programming consumes the largest power since the cell needs to reach the melting temperature of about 600 °C. The reset current required for the cell to reach this melting

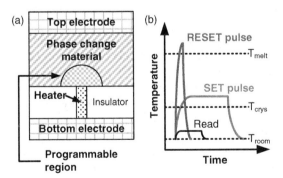

Figure 5.1 (a) Cross-section schematic of a conventional phase change memory cell. The electrical current passes through the phase change material between the top electrode and heater. Current crowding at the "heater" to phase change material contact results in a programmed region illustrated by the mushroom boundary. This is typically referred to as the mushroom cell. (b) PCM cells are programmed and read by applying electrical pulses to change the temperature inside the PCM cell accordingly. © IEEE. Reprinted with permission, from [7]

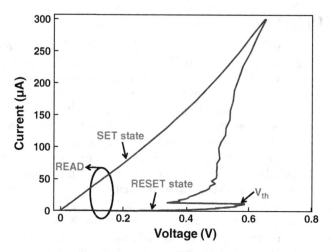

Figure 5.2 I–V characteristics of the SET and RESET states. The RESET state shows switching behavior at the threshold switching voltage (V_{th}). The RESET state stays in the high resistance state below V_{th} (sub-threshold region) and switches to the low resistance state at V_{th}

temperature depends on a number of factors such as the electrical resistivity and thermal conductivity of the materials used as well as the specific device structure (which determines the thermal environment of the device). The set programming however limits the operating speed of the PCM as it takes longer time to crystallize the amorphous region.

5.3 Material Properties

The active material of a PCM cell is the phase change material that determines the majority of the device characteristics. These materials have three major requirements: bistability of the amorphous and crystalline states, ability to rapidly switch between the two states, and high electrical resistance or optical reflectivity contrast between the two states. While most known phase change materials can exist in both a glassy and crystalline state, few have both the biphasic stability and suitable transition dynamics between both phases required for memory applications. More specifically, crystallization time of materials varies significantly, but only values in the nanosecond regime are viable candidates for commercial applications. These requirements significantly limit the range of viable materials for PCM to those mostly in the chalcogenide family (Group 16 elements, primarily S, Se, and Te). Despite the discovery of the switching phenomenon in the 1960s [1] the technological success of optical storage based on phase change materials was only possible after the discovery of a new class of materials that fulfilled the requirements of this technology. The chalcogenide alloys along the GeTe-Sb_2Te_3 pseudo-binary line (Figure 5.3) are among the most commonly used materials because of their large optical and electrical contrast and faster crystallization times. The discovery of this class of chalcogenide alloys led to a very successful re-writable storage technology with its third generation 100 GB capacity Blu-ray disks [9]. In addition, the excellent electrical and scaling properties of these materials sparked a new interest in early 2000 for these materials to be used in nonvolatile memory applications [2]. The following sections describe in detail the different aspects of the phase change material properties that make PCM a potential candidate for the future nonvolatile memory technology.

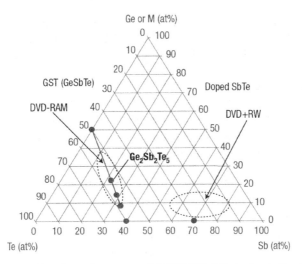

Figure 5.3 The most popular phase change material GST 225 falls in the tie-line between GeTe and Sb_2Te_3. © Nature. Reprinted with permission, from [8]

5.3.1 Electrical and Phase Transformation Properties

The electrical resistivity contrast between the crystalline and amorphous (as-deposited) state in phase change materials can be up to five orders of magnitude [10]. Figure 5.4 shows the resistivity change with temperature for different class of phase change materials. It can be seen that there is a sudden drop in the resistivity at a particular temperature and this temperature is commonly referred to as crystallization temperature (T_{crys}). Materials need to be chosen such

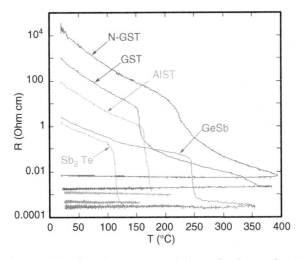

Figure 5.4 Resistivity as a function of temperature during a heating cycle at $1\,K\,s^{-1}$ for initially amorphous, as-deposited 50 nm films of various phase change materials. Initially, the thin films have a high resistance that drops sharply when the crystallization temperature is reached, and it stays low upon cooling. GST – $Ge_2Sb_2Te_5$, N-GST – 7 at% N doped GST, GeSb – Ge:Sb ratio 15:85, AIST – Sb_2Te doped with 7 at% Ag and 11 at% In. © Journal of Applied Physics. Reprinted with permission, from [10]

that the T_{crys} is high enough to ensure proper retention of the amorphous state for 10 years at the operating temperature of the PCM cells, that is, 85 °C for embedded memory applications and 150 °C for automotive applications. In contrast to the above requirement, the materials must undergo phase transformation in the nano-second regime to meet the memory application requirements. These are very conflicting requirements and a number of different alloy compositions and doping techniques [11] have been studied to identify the desired material composition that meets the specifications of both the retention and speed for nonvolatile memory applications. Recently, a N-doped GST alloy [12] achieved retention at 120 °C for 10 years that meets the requirements for industrial and some automotive applications.

5.3.1.1 Conduction Mechanism and Threshold Switching

The electrical conductivity in the amorphous phase can be described by thermally activated hopping transport [13] as shown in Figure 5.5. A Poole–Frenkel (PF) transport of carriers through traps leads to a current which is linear with voltage for very small voltages and exponential for high voltages. The trap energies are randomly distributed in the band gap of the amorphous material and the PF conduction happens through preferred filamentary paths that have the least overall activation energy [14]. At a certain material-dependent threshold field on the order of 10–100 V μm^{-1} [15] the resistivity of the amorphous phase change material suddenly decreases by orders of magnitude, negative differential resistance is observed, and so-called threshold switching occurs (Figure 5.2). The mechanism behind threshold switching is still not fully understood and several models have been proposed as a possible mechanism. The Joule Heating model [16] attributes the threshold switching to thermal runaway within the chalcogenide layer. This model is based on a simple observation that the current through the phase change material increases exponentially due to the temperature-dependent conductivity of the phase change material as temperature increases. However, since the speed of switching is faster than the thermal time constant, electronic mechanisms are favored over purely thermal mechanisms [17]. In the Impact Ionization model, for higher electric fields the current rises exponentially due to secondary carrier generation in the amorphous region. The electrons get trapped in the donor-like traps. But as the field increases, all the trap states are filled and the free carrier density increases. The impact ionization overcomes the carrier recombination process, the traps are filled and the voltage snap back occurs. As a result, higher current can flow through the device at a lower voltage and the impact ionization process still happens but with a lower multiplication rate [18–20]. In another electronic model, threshold switching is attributed to energy gain of electrons in a high electric field leading to a voltage–current instability [6]. In the field-induced nucleation model [21],

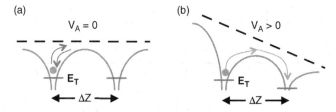

Figure 5.5 (a) Schematic showing the carrier trapped in the localized states of the amorphous phase change region at the trap energy E_T. The inter-trap spacing is Δz (b) On applying a bias voltage V_A across the amorphous region the barrier height to the carrier hopping is reduced enabling them to hop from one trap to the other easily, resulting in sub-threshold conduction [139]

threshold switching is explained based on the electric field assisted formation of crystalline nuclei across the programming region. These crystalline nuclei disappear if the electric field is removed prematurely and would go back to its original amorphous form. There is no experimental demonstration as to which of the above models correctly explains the physics behind the threshold switching. The dominant switching mechanism may depend on the material system and more than one mechanism might be required to explain the experimental observations at different operating conditions or device dimensions. The electrical conduction in the crystalline phase of the phase change material can be described straightforwardly with the drift diffusion behavior of a doped semiconductor resulting in Ohmic behavior at low voltages [18]. NonOhmic behavior of PCM cells for higher voltages can be attributed to Joule heating by the current causing the temperature-dependent resistance to change.

5.3.1.2 Resistance Drift

The phase transformation to the amorphous state is achieved either by ion implantation [22] or by melt-quenching with an electrical/optical pulse. The amorphous state resistance of a PCM cell, measured at a given read voltage is determined by the nature and volume of the amorphous region formed during the reset programming. The amorphous resistance does not remain constant but rather drifts logarithmically with time to a higher value immediately after the reset pulse is removed following the power law $R = R_0 \times (t/t_0)^\nu$ where R and R_0 are the present and initial resistances, t and t_0 the present and initial times, and ν the drift exponent [23]. This phenomenon is called resistance drift. Figure 5.6 shows the change in the amorphous resistance of a PCM cell over time for different reading temperatures immediately after reset programming [24]. The drift is also a strong function of the temperature [25] and the volume of the amorphous region [26]. Higher temperatures and larger amorphous volumes have larger drift coefficient. The threshold voltage also shows a similar behavior with time [27]. While resistance drift to higher values increases the on/off ratio for two-level memory, it is a major problem for the multi-level PCM implementation [28] because the different resistance levels overlap as time progresses and is no

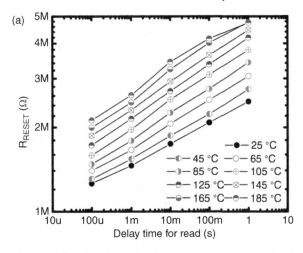

Figure 5.6 Reset resistance as a function of time after reset programming for various annealing temperatures (T_A). Cells are programmed and read at room temperature, 25 °C [27]. © IEEE. Reprinted with permission, from [140]

more distinguishable from each other. There is currently no consensus on what causes drift, but there are two major theories that explain as to why drift occurs in PCM. According to the Structural Relaxation (SR) model, drift is explained as a reduction in the number of structural defects (distorted bonds, wrong bonds, vacancies, etc.) that causes an increase in the band gap and of the activation energy for conduction [23,24]. In other words, the density of traps decreases with time due to reduction of the structural defects, causing the trap spacing to increase, which in turn raises the barrier for PF conduction through the trap states. Alternatively, resistance drift is explained as the relaxation of the compressive stress developed in the amorphous phase at solidification [29]. In fact, the density of the amorphous phase is 6.5% lower than the crystalline phase, thus possibly causing a compressive stress after reset programming in PCM cells where the amorphous region is surrounded by a crystalline region. Experimental results on nanowire PCM devices indicate that the drift coefficient decreases with the size of the phase change nano-wire [29]. The results also show that the drift exponent is higher for nanowires encapsulated with nitride or oxide, presumably due to the enhancement of the local stress by mechanical confinement. However, a recent study on PCM thin films and devices subjected to external mechanical stress [30] suggests that the resistance drift is primarily due to structural relaxation and not by any mechanical stress relaxation. There is still a lack of proper systematic study explaining the direct impact of mechanical stress on PCM devices and whether there is any correlation between the two theories.

5.3.1.3 Mechanism of Crystallization

When the electric pulse is applied for a sufficient amount of time after the threshold-switching event, the amorphous region is transformed into its crystalline phase provided the temperature reaches above the crystallization temperature, but below the melting temperature; this process is called crystallization. The nucleation and growth processes govern the crystallization kinetics of PC materials. Nucleation involves the formation of small crystalline nuclei in the amorphous matrix, and growth involves the subsequent expansion of the phase front separating the amorphous and crystalline regions. The driving force for crystallization is the gain in free energy below the melting temperature. While similar in bonding structure, common phase change materials are traditionally divided into two categories based on their crystallization kinetics: growth-dominated materials and nucleation-dominated materials [31]. (Figure 5.7) Growth dominated materials are generally related to the family of AgIn doped Sb_2Te (AIST) materials, and are characterized by slow nucleation, but fast growth of crystalline domains. In contrast, the prototypical nucleation dominated material is $Ge_2Sb_2Te_5$ (GST) where nucleation events happen

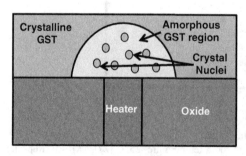
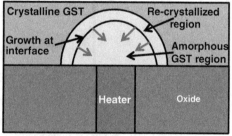

Figure 5.7 Nucleation and growth dominated phase change materials

quickly during crystallization. Nucleation-dominated materials generally fall on the tie-line between GeTe and Sb_2Te_3 whereas growth-dominated materials contain much more antimony. (Figure 5.3) In case of the growth dominated materials, the time it takes for crystallization (t_{crys}) depends on the size of the amorphous region, while it is independent in case of nucleation dominated materials [31]. It has been shown that, in the case of highly scaled PCM devices, the growth dominated phase change materials enable faster crystallization times because of the smaller amorphous volume [33]. The dynamics of crystallization (phase change) in GST and related systems have been fairly well interpreted in terms of the classical nucleation and growth model developed by Kolmogorov, Johnson and Mehl, and Avrami (KJMA model) [32]. The KJMA model gives the macroscopic evolution of the transformed phase under isothermal annealing condition and has been extended to analyzing transformations under a constant heating rate. Recently [34], a modification involving the introduction of fractal geometry and the Meyer–Neldel rule [132] for thermally activated processes was introduced into the KJMA theory to account for the deviations in the measured physical parameters like Avrami exponent (n) and frequency factor (ν). Further a number of other crystallization models [35,36] have been proposed that considers the effect of percolation effects and local temperature effects in practical PCM devices. For the same target resistance, the type of the programming scheme used can also result in different crystallization times due to the variations in the nature of the amorphous volume formed during the programming [37]. Furthermore, fast crystallization times can be correlated to the structural properties of the phase change materials. Fast switching materials often show a simple cubic or rocksalt structure with random atomic distributions that require little atomic movement to change from the amorphous to the crystalline state [38]. In addition it was found that resonance bonding [39] plays an important role for fast switching of phase change materials [40]. A low degree of ionicity and low tendency towards hybridization is typical of fast switching phase change materials [41]. Recently, it was shown that a small incubation electric field could be used to produce rapid crystallization within a few hundred ps, up to an order of magnitude faster than the conventional crystallization times of $>10\,\text{ns}$ [42]. This electrical controllability of the crystallization speed is attributed to the temperature rise due to the small incubation field which is sufficient to cause a prestructural ordering of the local cluster of atoms, but still small to start the crystallization process. Recent studies [43] also show that decreasing the grain size of the deposited phase change material can significantly reduce the crystallization time of the cell because of the presence of a large number of broken or loosely bonded atomic structures at the grain boundaries. The process of crystallization has also been studied in different phase change materials [44,45] using density functional simulations to understand different bonding reconfigurations that happens at the atomic level. Hence, understanding the crystallization process and related parameters can guide us to explore novel materials with improved retention and switching characteristics.

5.3.1.4 Mechanism of Amorphization

The process of amorphization is traditionally considered to involve melting the phase change material and rapidly quenching it to the room temperature. This results in the formation of a large number of dislocations or vacancies in the programmed region and thereby resulting in high resistance. However, some of the recent studies [46–48] show that it is possible to bring about crystalline to amorphous phase transformation without actually going through the molten state. Ultra-fast photo excitation methods were employed to provide sufficient energy to trigger the rupture of the sacrificial (resonant) bonds that are inherently present in the crystalline phase of the GST to collapse to its amorphous phase without actually melting the material [47]. This photo-induced lattice instability requires about five times less power than that of the conventional melting process. Studies using nanowire PCMs [48] show that amorphization can also occur as electrical fields drive the movement of dislocations in the crystalline phase to move in the direction of the E-field to be collected as a narrow amorphous region resulting in a sudden increase in the device resistance. Further, the recent demonstration of low-power amorphous switching in superlattice-based interfacial PCM [40] and small grain PCM devices [43] can be explained based on the fact that these devices have a large number of interfaces/grain boundaries in their programmed volume where the resonant bonds/dislocations play a major role in the phase transformation. The above studies are only the first step in understanding the physics behind the phase transformation of these unique nanoscale class of materials and further research in this area can help us engineer highly energy efficient and ultra-fast phase change memory devices.

5.3.2 Thermal and Mechanical Properties

The programming of PCM is achieved by Joule heating of the phase change layer to temperatures above the melting temperature. The current required to achieve this temperature depends strongly on the thermal properties of the various materials that form the PCM device namely the phase change layer, heater electrode, top electrode and the surrounding dielectrics. The thermal resistance arising from carrier energy scattering in the bulk of the material and the thermal boundary resistance (TBR) arising from scattering in the interface region are the two main mechanisms that contribute to the overall thermal network of the PCM cell [49]. Hence, understanding thermal conduction in thin film phase change materials, thin film electrode materials, and at their interfaces is essential for reducing the programming energy.

The most common thin film thermal conductivity measurement techniques for phase change materials are the 3ω method [50], nanosecond transient thermoreflectance (TTR) [51], and picosecond time domain thermoreflectance [52,53]. Measurements on the common phase change material GST show thermal conductivities at room temperature in ranges of 0.14–0.29, 0.28–0.55, and 0.83–1.76 $\mathrm{W\,m^{-1}\,K^{-1}}$ in the amorphous, rocksalt, and hexagonal phases, respectively [54]. The rocksalt phase thermal conductivity exhibits a slow increase with temperature consistent with other highly defective crystalline materials [52,53]. Recently, high density nanostructured materials have exhibited thermal conductivities as low as 0.05 $\mathrm{W\,m^{-1}\,K^{-1}}$ [55]. Nanostructured phase change regions [40,43] have the potential to offer dramatically reduced programming currents through exceptionally low thermal conductivities. Another key challenge is extending thermal conductivity measurements to the melting temperature, which is notoriously difficult due to the volatility of many phase change materials at high temperatures. These measurements will shed light on the relative electron and phonon contributions in conduction at device operating conditions, informing better material selection. There have been many studies [51,53] of the thermal conductivity normal to GST films (k_{normal}), but the in-plane thermal conductivity ($k_{lateral}$) or the conductivity anisotropy can also be an important factor that determines the programming current. Thermal conductivity anisotropy in thin films can be due to electron or phonon scattering on film interfaces, partially oriented grains or inhomogeneous material quality [56]. Lateral thermal conduction in GST films can also negatively affect the thermal cross talk in high density cells and the programming current of PCM devices, the latter being more pronounced for novel lateral designs. The anisotropy ratio ($k_{lateral}/k_{normal}$) depends on the annealing time and temperature of the deposited phase change films [57] and can reduce the programming current by up to 40% [56].

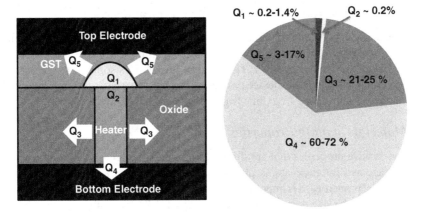

Figure 5.8 Percentage of energy dissipated in different regions of PCM [60]. The useful energy spent in bringing the phase transformation (Q1, Q2) is much smaller compared to the rest of the energy that is dissipated through the surroundings

Much of the heat generated during GST cell switching occurs in the GST bulk and at the bottom electrode interface [58,59]. Coupled electrical–thermal simulations revealed that the programming current of a PCM device decreases by as much as 30% for GST-electrode TBR values of $50 \, \mathrm{m^2 \, K \, GW^{-1}}$ [59]. In many PCM designs, significant heat loss occurs through the electrode [60–62] due to the large thermal conductivity of the heater material such as TiN $\sim 8 \, \mathrm{W \, mK^{-1}}$ compared to the GST $\sim 0.4 \, \mathrm{WmK^{-1}}$ (Figure 5.8). This problem can be alleviated by introducing alternative electrode materials such as C and W-WN$_x$ [62] or by inserting low thermal conductivity TaN electrodes within a TiN electrode [63]. Engineering the electrode to have acceptable electrical conductivity and very low thermal conductivity is a key step toward reducing programming current. Composite electrodes may leverage TBR to increase the device effective thermal resistance.

The large current density and temperature excursion exceeding 600 °C in the phase change layer gives rise to a more pronounced effect of the thermoelectric properties such as the Seebeck effect on the programming properties of a PCM cell. Recent measurements provided evidence of thermoelectric transport in PCM cells through observation of a modification in the amorphous region [64] and the programming condition [65] with the bias polarity. The Seebeck coefficient of $Ge_2Sb_2Te_5$ films shows strong dependence on temperature and temperature history as governed by the phase purity. It scales with film thickness due to varying degrees of crystallization and phase purity in crystalline GST films [66]. Recent electrothermal simulations indicated nearly a 45% increase and 16% decrease in the peak temperature and programming current, respectively, if thermo-electric effects are considered [67].

From this discussion it is clear that the search for the best phase change material is a multi-parameter optimization process with some seemingly contradictory requirements such as high stability of the amorphous phase at operating temperature, but very fast crystallization of the amorphous phase at switching temperature. Many material parameters will also change with size of the phase change material when devices are scaled to nanoscale dimensions. Much research is still required to understand the fundamental relationship between material composition and structure, and phase change properties for a physics/chemistry-based design of new phase change materials.

5.4 Device and Material Scaling to the Nanometer Size

One of the major advantages of PCM over the conventional FLASH technologies is its scalability to sub-10 nm dimensions. However it is unclear how properties would change both at the material and at the device level. This section summarizes some of the recent works on how the material physics and the device properties change as we enter technology nodes on the order of the grain size or the trap dimensions of phase change materials.

5.4.1 Materials Scaling Properties

Nanomaterials have properties that are different from bulk materials of the same composition because surface and interface atoms play an increasingly important role in determining the different material properties. It is important to know how phase change material properties change with size in order to be able to evaluate the scalability of PCM technology. It is predicted that as the volume of phase change material gets smaller, the phase stability characteristics will change. Since PCM relies on the bi-stability of the amorphous and crystalline phases, the technology will only be reliable as long as both states are distinct. It is conceivable that there is a limit to the minimum number of atoms required for the two phases to be distinct, and that there is a minimum size at which both phases are stable under the conditions required for commercial applications.

Scaling studies of phase change materials have been done on thin films, nanowires, nanoparticles, and PCM devices (see [68–70] for overviews). It was found that many properties of the phase change materials do depend on size, in particular below the 10 nm range. These changing properties include crystallization temperatures and times, related activation energies for crystallization, melting temperatures, resistances, and optical and thermal properties. Crystallization temperatures can vary up to 200 °C and can be increased or decreased for very thin phase change films depending on the interface material [71,73] which in turn also significantly affects the crystallization time of the material [73]. Melting temperatures are reduced for thinner films [69], hence potentially reducing the power to melt-quench the material. Electrical resistances on the other hand are increased as film thickness is reduced [72].

While these dependencies increase the complexity of materials optimization they also enable us to tune interfacial properties in such a way that desirable switching properties are obtained. Nanostructured materials such as the multi-layered superlattice-like (SLL) structures have switching properties that are significantly different from the bulk materials [74]. In PCM, higher heterogeneous crystallization rates can be achieved in the nanostructured phase change materials due to their high surface area to volume ratios [75]. Nanoscale active device regions or volume of phase change materials can also have lower phase-transition temperatures than the bulk materials due to greater phonon softening effects [76,77]. Materials with SLL structures have lower thermal conductivities than the bulk materials with the same composition [78]. They also have favorable thermal confinement properties due to the phonon scattering effects at the interfaces [79]. Phase change materials with SLL structures can be formed by alternating layers of Sb_2Te_3 and GeTe, which have fast switching speed and good data retention, respectively [78,80].

Recently, a superlattice-like structure of GeTe/Sb_2Te_3 of thicknesses between 5 Å and 40 Å was used to fabricate a PCM cell called Interfacial PCM (IPCM) [40], (Figure 5.9) that shows excellent endurance properties and requires an order of magnitude lower switching energy than the conventional single layer GST cells. These excellent switching and endurance properties of IPCM is due to confinement of the atomic movements to a single dimension, that is, the interface between GeTe and Sb_2Te_3 controls the local atomic switching of Ge atoms resulting in a phase transition with substantially reduced entropic losses. The ultrathin, uniform stratum

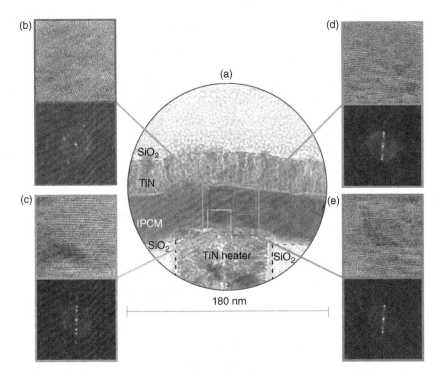

Figure 5.9 TEM image of a $(GeTe)_2(Sb_2Te_3)_4$ Interfacial Phase Change Memory (IPCM) in the RESET state. The diffraction data (b–e) at different locations above the electrode shows that (b) the layered structure of the IPCM is preserved even after RESET, implying that the layers did not undergo any melting during the programming operation. © Nature. Reprinted with permission, from [40]

of Sb_2Te_3 and GeTe lowers the entropy of the covalent amorphous state by restricting the number of atomic configurations that can exist. For the SET operation, IPCM-based devices use only 12% of the energy required by similar GST-based devices. This increase in efficiency means less energy is dissipated during the phase transition, which in turn leads to more than an order of magnitude improvement in the SET–RESET cyclability. Reducing the entropic losses has further ramification of producing highly repeatable device characteristics. Indeed, the resistance of the IPCM during the SET–RESET cycle is identical over a million cycles.

Engineering the grain-size of the as-deposited phase change material can also result in improvements in switching speed and reduced switching energy. GST with small grain sizes have high interface area to volume ratios, which can promote hetero-crystallization (interfacial growth) at the grain interfaces or boundaries [81]. These can alter the phase change mechanism of GST from a nucleation-dominated mechanism to a growth-dominated mechanism, enabling faster crystallization. The increased number of grain boundaries also decreases the thermal conductivity of GST [56]. This increases the thermal confinement in cells, thus reducing the power needed for RESET. The large surface to volume ratio in these materials also facilitates stress relaxation after RESET and hence results in a lower resistance drift compared to the bulk GST. Overall, the scaling properties of nanostructured phase change materials are promising for the development of the fastest and the most efficient phase change memory devices that will be required to meet the demands of the future technology.

Phase change nanoparticles offer another intriguing way of understanding the scaling properties of many chalcogenide materials. A number of techniques have been used to synthesize nanoparticles, including pulsed laser ablation [82,83], electron-beam lithography [10], selfassembly-based lithography techniques using sputter deposition [84,85] or spin-on phase change materials [86], and solution-based chemistry [87]. Large nanoparticles show properties similar to bulk, but the small nanoparticles below 10 nm show size-dependent properties such as higher crystallization temperature and reduced melting temperature. Both are beneficial for PCM applications and illustrate the favorable scaling properties of phase change materials. GeTe nanoparticles can be synthesized in the amorphous phase and can be crystallized by heating them over their crystallization temperature (remarkably increased compared to bulk) for nanoparticle sizes as small as 1.8 nm [87]. As such, GeTe nanoparticle is a convenient material for exploring the ultimate size limit of PCM. Figure 5.10 shows transmission electron microscope images of these GeTe nanoparticles of various sizes. Down to these small sizes, phase change materials still show phase transformation. These nanoparticles are as small as about two to three times the

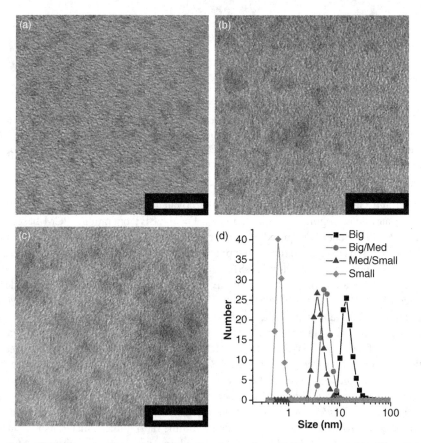

Figure 5.10   TEM images of size-selected samples of amorphous GeTe nanoparticles. All scale bars are 10 nm. (a) Small nanoparticles of 1.8 ± 0.44 nm. (b) Medium nanoparticles of 2.6 ± 0.39 nm. (c) Large nanoparticles of 3.4 ± 0.74 nm. (d) Dynamic Light Scattering results of another instance of size-selected nanoparticles. See http://pubs.rsc.org/en/content/requestpermission?msid=b917024c. © Journal of Materials Chemistry. Reprinted with permission, from [87]

Table 5.1 Scaling rules of PCM cell for both isotropic and aggressive scaling. $k \sim 1/F$ with F being the feature size

Parameters	Scaling factor	
	Isotropic	Aggressive
Heater contact area	$1/k^2$	$1/k^2$
Vertical dimensions	$1/k$	1
Electrical/thermal resistances	k	k^2
Power dissipation	$1/k$	$1/k^2$
Current	$1/k$	$1/k^2$
Voltage	1	1
Current density	k	1

lattice constant, so this will be close to the ultimate scaling limit of phase change technology as far as the phase change materials themselves are concerned.

The future challenges for the material scientists from the technological standpoint will include exploring phase change materials that do not exhibit void formation or elemental segregation, tailoring the increasingly important interfaces that support high cyclability, good data retention, and fast switching, and continuing the study of scaling properties of phase change materials as dimensions shrink to the few-nanometer length scale.

5.4.2 Device Scaling Properties

While it has been established that phase change materials (e.g., GeTe) can exist in two stable phases, it is still necessary to explore how the memory device characteristics may behave when PCM are scaled to nanometer size. The active device area of the phase change memory needs to be scaled every year in order to meet the demands of the increasing device density. Furthermore, device scaling has the advantage of reduced programming power. Table 5.1 summarizes the scaling rules for both isotropic and aggressive scaling. There are two major device properties that are significantly affected by the scaling of the active device, namely the threshold switching voltage and the reset current. Several device structures and methodologies have been used to understand how scaling affects these device parameters and will be discussed below.

5.4.2.1 Threshold Voltage Scaling

The present models for threshold switching are based on the inter-trap distance [13] or the minimum distance that the carriers have to travel to cause impact ionization [18]. It is not clear how threshold switching properties will change when the film thickness becomes comparable to either of these distances. A number of device structures ranging from phase change bridge cells to electrodes of carbon nanotubes have been used to study how the threshold voltage scales as the size of the active programming region decreases. In [88], an additional top electrode was introduced in the GST layer at a specific height from the bottom electrode of the PCM cell to confine the programming volume to a specified thickness. It has been shown that the threshold switching voltage linearly increases with GST layer thickness with a nonzero offset. In the case of nanowire PCM devices [75,89] the threshold switching voltage scaling changed from constant field to constant voltage scaling with the length of the amorphous region below 10 nm. The phase change bridge (PCB) device is another cell design uniquely suited to study the electrical scaling characteristics of phase change materials [90]. It comprises a narrow

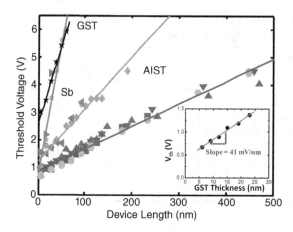

Figure 5.11 Scaling of threshold voltage with the length of the active phase change region in a phase change bridge device (Device Length). Results are shown for four different phase change materials. The slopes of the linear fits are used to determine the threshold fields for each material. © Journal of Applied Physics. Reprinted with permission, from [15]. Interestingly all these materials have nonzero threshold voltages. Inset shows similar threshold voltage scaling on vertical PCM devices using GST. The linear threshold voltage scaling applies to thicknesses as small as 6 nm. © IEEE. Reprinted with permission, from [88]

line of thin phase change material bridging two underlying electrodes. The PCB structure is ideally suited for a variety of characterization experiments [15,91]. For instance, in Figure 5.11 the critical field necessary for threshold switching for different materials has been obtained by plotting the measured threshold voltage of the PCB device as a function of the device length. The threshold fields for $Ge_{15}Sb_{85}$, Ag- and In-doped Sb_2Te, $Ge_2Sb_2Te_5$ and 4 nm thick Sb devices are 8.1, 19, 56, and 94 V μm^{-1}, respectively. Recently, carbon nanotubes (CNTs) with diameters of ~ 1 to 6 nm were used as electrodes to reversibly induce phase change in nanoscale GST bits [92]. The study also demonstrates reversible switching with programming currents from 0.5 to 8 µA, two orders of magnitude lower than state of the art PCM devices. The active device area is the small nanoscale gap formed in the middle of the CNTs through electrical breakdown [133]. By controlling the breakdown voltage, gaps ranging from 20 to 300 nm was formed. Threshold voltages scale proportionally to the size of the nanogap, at an average field of ~ 100 V μm^{-1} for GST material.

5.4.2.2 Reset Current Scaling

The large programming current is still a key issue that limits the adoption of PCM in many applications. Furthermore, a large programming current in PCM imposes a stringent requirement on the current delivered by the memory cell selector integrated in series with the PCM. In order to provide the current required to switch the states of PCM, the area of the memory cell selector may not be scaled down as fast as the memory cell itself, thus the size of the cell selection device becomes the limiting factor for device density and annihilates the small size advantage of PCM technology. Therefore, reducing the programming current is necessary for achieving both high density and low power consumption of PCM.

To decrease the reset current, one way is to increase the heater thermal resistance by reducing the contact area [4]. The feature size of the conventional mushroom structure

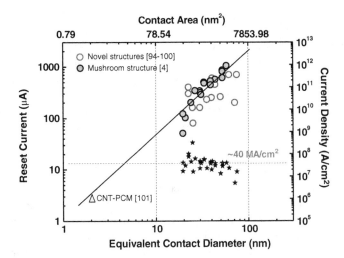

Figure 5.12 Reset current as a function of equivalent contact diameter, showing a linear scaling trend with the effective contact area as the device feature size goes down. A constant ~40 MA cm^{-2} current density is required to program the PCM cell. © IEEE. Reprinted with permission, from [7]

(Figure 5.1) of PCM is limited by lithography and process capability. This was recognized early on and many innovative device structures have been explored to reduce the effective bottom electrode contact (BEC)/GST interface to the sub-lithographic regime. Figure 5.12 shows the reset current reduction as a function of the equivalent diameter of a circular contact and the effective contact area for different cell structures, such as conventional mushroom type [93], edge contact type [94], μTrench [95], cross-spacer PCM [96], "wall" structure [97], "pore" structure [98], ring type contact [99], and dash-type confined cell [100]. We can clearly see that the reset current scales with the effective contact area of the PCM and that a constant current density ~40 MA cm^{-2} is required to program an average PCM cell. Those with carefully engineered cell structures and materials can be programmed using ~10 MA cm^{-2}. Recent demonstrations of PCM devices using CNT as the electrodes [92,101] show extreme scalability of the reset current down to 1 μA range. These PCM devices have an active area of only a few nm^2 because CNT diameters are only about 1–2 nm. They demonstrate typical electrical switching characteristics that are observed in larger devices. This shows that, the scalability of the phase change material properties can also be observed in practical device configurations [101] and hence providing for a viable technology that can extend the semiconductor memory device scaling for several generations.

5.5 Multi-Bit Operation and 3D Integration

The advancements of the phase change memory cell itself does not guarantee the success of this technology for practical applications. In order for PCM to be commercialized in a large scale we have to increase the available storage capacity per unit area of memory. Multi-level operation and multi-layer 3D stacking are two possible ways of achieving this goal. In this section we will see some of the major advances and the challenges that we face in this field and also discuss the issues that we need to consider as we scale to large array sizes.

5.5.1 Multiple Bits per Element

Multilevel cell (MLC) storage is essential for reducing the cost per bit of PCM technology and for increasing its potential for market acceptance. The large resistivity contrast between the crystalline and the amorphous state of the phase change materials and the ability to access the intermediate states in a reliable way have made possible the realization of multi-level implementation in PCM. Some of the earlier methods of MLC implementation are based on engineering the properties of the chalcogenide layer by doping [102,103] or by stacking multiple layers with differing electro-thermal properties as the storage medium [104–106] or by using parallel combination of multiple resistances [107,108]. These demonstrations thus depended on precisely engineering the cell structure to achieve the intermediate resistance states. However, even conventional cell structures such as the common "mushroom" phase change element could be programmed to store multi-bit data by altering the size of the amorphous region in a continuous manner to achieve different resistance levels. The size changes as a function of the amplitude of the write pulse, as quantified by the cell's programming curve shown in Figure 5.13a. However, process and material variations give rise to resistance levels with broad distributions when single programming pulses are applied. A common solution is to employ iterative programming schemes, in which a sequence of write and verify steps is used in a feedback loop to minimize the error between the programmed and a specified target resistance level. A basic iterative scheme is shown in Figure 5.13b. It was shown that by varying the amplitude or slope of the trailing edge of the programming pulses to control the evolution of temperature in the cell, up to 16 intermediate levels could be programmed in a cell, thus demonstrating a four-bit cell [109]. These programming techniques are based on the fact that the cell resistance R_{cell} can be increased by applying programming pulses of larger amplitudes that result in melting of larger volumes of the amorphous region (partial RESET), or can be decreased by applying pulses of lower amplitude (partial SET), or sequences of annealing pulses of appropriate magnitude to crystallize and shrink the size of the amorphous volume [110]. A novel iterative programming scheme that uses both partial-SET and partial-RESET pulses is depicted in [111]. Operation starts from the partial-SET regime

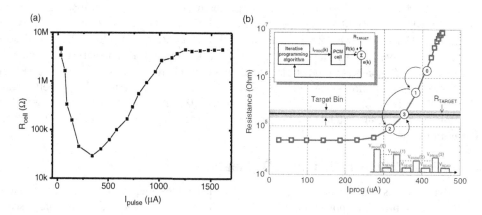

Figure 5.13 (a) Typical R_{cell}–I_{pulse} curve obtained by varying the current pulse amplitude for a mushroom phase change memory cell. © IEEE. Reprinted with permission, from [103]. (b) Schematic illustrating the basic iterative programming concept using a sequence of adaptive write and verify steps. © IEEE. Reprinted with permission, from [112]

and either terminates there if the target resistance is reached, or switches to the partial-RESET regime if the programmed resistance drops below the target level. This hybrid method combines the low energy dissipation of the partial-SET regime with the flexibility of achieving higher or lower resistance level using the partial-RESET regime.

Regardless of how much efficient the MLC scheme can be in producing tight and dense resistance distributions, the resistance drift adversely affects the reliability of MLC storage in PCM, because the distance between adjacent levels is small and stochastic fluctuations of the resistance are more likely to cause level overlap over time than in binary storage. Currently there is no known material solution to mitigate the effect of drift, as the physics of drift in itself is not well understood. However, a novel drift-resilient MLC PCM state metric has been proposed for MLC programming that is largely robust to the impact of drift [112]. Traditionally, the low field resistance is used as a metric to define the different MLC levels. This resistance value is a strong function of the activation energy (E_a) and suffers a large change as E_a changes with time during the drift phenomenon [28]. Instead, what is required for drift-tolerant MLC is a cell state metric that has minimal dependence on the activation energy, but has a linear relationship to the amorphous thickness. To achieve this, the read voltage is progressively increased until a certain predefined current level I_R, is reached. This current level is chosen to be a safe value much below the current needed for threshold switching to occur. The time needed to reach I_R can be considered as a measure, M of the programmed state. It is shown that M is proportional to the effective amorphous thickness [112], yet it is a weak function of the activation energy (nearly linear as opposed to exponential), suggestive of a significant tolerance to drift. Experimental results suggest that the new metric has an order of magnitude smaller drift coefficient than the standard resistance metric. In another approach, the effect of drift on the selected cells was predicted by using reference cells that have similar thermal history and undergo same drift behavior [113]. Another way of mitigating the impact of drift on the resistance levels is by using an additional annealing pulse to accelerate the drift immediately after the reset programming [114,138]. This additional pulse quickly anneals out many defects (that are responsible for the drift) and provide a drift-free period that enlarges the read window. This technique not only reduces the read latency but also enables a more reliable read window for multi-level applications.

5.5.2 3D Stackable Memory

A vision for high density memory is the cross-point architecture with a memory cell integrated with a cell selector within a $4F^2$ footprint that can be stacked in the third dimension Figure 5.14. To realize this, the selector should have a $4F^2$ footprint that can be scaled with the bitline/wordline pitch [134,135], a small off current, and an on-state current that is sufficient for programming the memory (the reset current for the case of PCM). Furthermore, the fabrication process should be compatible (e.g., material and process temperature) with CMOS BEOL and the memory cell. The trend for the reset current in Figure 5.12 shows an average of $40\,MA/cm^2$ with the lower bound of about $10\,MA/cm^2$. This large current is fairly difficult to achieve even for single crystal silicon diodes. A number of recent efforts have been made to integrate a memory cell selector with a phase change memory cell. Sasago et al. [93], used a low-thermal budget process to fabricate a $4F^2$ poly-Si diode with a drive current capability in excess of $8\,MA\,cm^{-2}$ and on-off ratio more than 10^4. Kau et al. [115] used a stackable cross-point phase change memory utilizing the Ovonic threshold switching (OTS) property of chalcogenide materials to make the memory cell selector. Since both the memory device and the selector exhibit threshold switching behavior, the programming voltage conditions have to be carefully

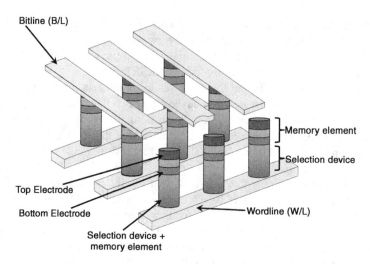

Figure 5.14 Cross-point memory with a memory cell integrated with a cell selector within a 4F² footprint. This structure can be stacked in the third dimension. © IEEE. Reprinted with permission, from [7]

chosen to avoid disturbing the state of the cells in the un-selected bit and word lines. Gopalakrishana *et al.* [116], proposed an access device based on Cu ion motion in novel Cu containing Mixed Ionic Electronic Conduction (MIEC) materials. These MIEC-based devices can be fabricated at <400 °C, are scalable, and can conduct very high current densities (up to 50 MA/cm²) – making them suitable for stacking of multilayer PCM arrays in the BEOL. Sasago *et al.* [117] proposed another stackable access device structure in which the phase change layer is directly deposited on top of the channel poly-Si layer (Figure 15.5a). This contactless, simple memory-cell configuration enables a poly-Si MOS-driven 4F² stackable memory array with low fabrication cost and smaller programming current. In an effort to demonstrate a true 3D PCM memory, Kinoshita *et al.* [120], showed a scalable 3D vertical chain cell type phase change memory (VCCPCM) with 4F² poly-silicon diode, as shown in Figure 5.15b. The VCCPCM features formation of memory holes in multi-layered stacked

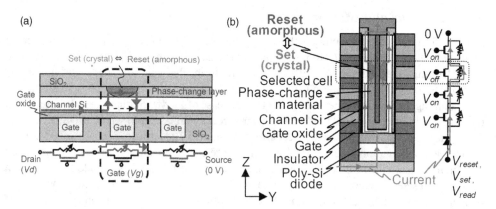

Figure 5.15 (a) Device structure and operation of PCM driven by Poly-Si MOS. © The Japan Society of Applied Physics. Reprinted with permission, from [117]. (b) Device structure and operation of VCCPCM. © IEEE. Reprinted with permission, from [120]

gates by using a single mask and a memory array without a selection transistor. Each memory cell consists of a poly-silicon transistor and a phase change layer connected in parallel. The memory cells are connected serially in the vertical direction. The excellent scalability of the new phase change material used in VCCPCM while retaining a reasonable crystallization temperature makes it possible to reduce the cell size beyond the scaling limit of flash memory.

5.6 Applications

Phase Change Memory with its ability to achieve multiple resistance levels, fast access time, high endurance, scalability, CMOS compatibility and 3D stacking has the potential to be become the next generation memory technology for a variety of applications. In this section we will explore the use of phase change memory in two different applications namely, as storage class memories for conventional computing systems and as an electronic synaptic element for building brain-like systems for future computing applications.

5.6.1 PCM as a Storage Class Memory

Historically, device scaling has resulted in the benefit of both cost reduction and performance improvement for the computing systems and hence the unrelenting focus on the processor performance and logic device scaling was justified. However, due to power constraints in the processor and the diminishing returns of the performance advantage of logic device scaling, it is important to seek other areas of improvement through which we can continue to enhance the overall system performance. Of these, the most important one is the memory storage hierarchy. The major bottleneck today is the large gap in the access time between the magnetic HDD and the DRAM main memory [121,122]. There are no major technologies on track to improve the access times of the HDD in the near future. Because critical computing applications are becoming more data-centric than computer-centric, a high-performance, high-density, and low-cost nonvolatile memory technology whose access time falls between that of an HDD and the dynamic memory located near the processor would significantly improve the overall system performance while keeping the cost low. This class of memories is referred to as the Storage Class Memories (SCM) [121]. Figure 5.16 shows the different memory hierarchies that form a

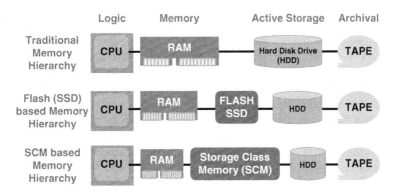

Figure 5.16 Memory hierarchy in computing systems. The performance gap between the main memory and the hard disk drive can be significantly reduced by using storage class memories in the memory hierarchy [140]

part of the computing system. The SCMs can fit in the memory hierarchy between the main memory and the HDD.

Of all the other emerging technologies, PCM is in the most advanced stages of meeting all the requirements for the SCM that companies are considering to include it in the memory hierarchy to improve the performance of their high-end systems [123]. The access time, endurance and retention properties of the PCM [124] are well within the requirements of the SCM. As compared to Flash, PCM can do in-place updates (it is bit-alterable) and there is no need to first erase a whole block before one can write new data. Instead of 10^5 (or even lower) write endurance cycles, PCM has more than 10^9 write endurance cycles. All these add up to reducing a lot of the overhead in the memory controller to keep track of where the written bits are, doing garbage collection, and wear leveling that is required for Flash. A prototype high-performance solid state drive using PCM [136] is shown to be 72–120% faster than Flash for small write and for all reads and incurs 20–51% less CPU overhead per IOP (Input Output Processor) for small requests. In another hybrid main memory system using phase change memory in conjunction with a small DRAM buffer [137] it was found that for a 16-core 8 GB DRAM baseline system, this hybrid PCM design reduces page faults by 5× and speeds up the system by 3×.

In addition, the major issues of having a $4F^2$ access device and reducing the programming current are also being addressed. Recently, there were demonstrations of several large scale Gb array demonstrations of PCM using $4F^2$ poly-Si diode [118,119]. If this trend in the technology improvement of PCM continues, then in the next few years we may be able to see commercial computing and mobile systems that contain PCM chips as a part of their memory hierarchy.

5.6.2 PCM as an Electronic Synapse

The ability to gradually program the cell into its SET or RESET state or in other words the ability to change the conductance of the device gradually makes PCM an excellent candidate to mimic the synaptic elements in the brain. Human brain predominantly consists of two main units, namely the neurons and the synapses that connect these neurons (Figure 5.17a). These synaptic elements have a certain behavior called Spike Timing Dependent Plasticity (STDP) wherein the conductance of the synaptic element changes based on the difference in the arrival time of the pulses or spikes from the post- and presynaptic neurons; that is, the conductance of the synapse increases when the prespike occurs before the post-spike and the conductance decreases when vice versa, as shown in Figure 5.17b. This particular behavior of the synapse is found to be one of the major factors that help in the process of learning and perception in brain [125]. PCM devices can be programmed in such a way that the pre- and postspike arriving at the two opposite terminals of the device can modulate the conductance of the device [126–128]. Thus PCM can act as an electronic synapse that connects neuron circuit elements (built with CMOS for instance) in a system architecture that closely resembles the brain. One implementation is to use a crossbar-like architecture that connects the different synaptic elements to the neuron circuits, with the top and bottom metal lines providing the pre- and postspike. Kuzum *et al.* [126] proposed a specific spiking scheme to translate the spike arriving timing relationship into PCM programming pulses. This enables the PCM to mimic the synaptic STDP behavior. The spiking scheme and the TEM cross-section of the nanoelectronic synapse used are shown in Figure 5.17c,d. Another pulsing scheme was also proposed in [127], where single pulses are used as opposed to multiple pulses as in Figure 5.17c to reduce complexity of neuron circuits and potentially energy consumption. The eventual goal is to

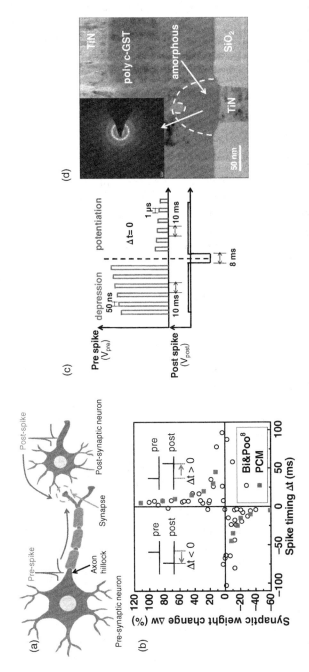

Figure 5.17 (a) A biological neuron and synapse. The synapse has the post- and prespikes applied on either side. © Journal of Neuroscience. Reprinted with permission, from [125]. (b) The asymmetric STDP curve for the synapse (both biological [125] and electronic [126]) showing the change in the conductance value with respect to the timing difference between the pre- and postspike. © American Chemical Society. Reprinted with permission, from [126]. (c) Pre- and postspikes that were used to program the electronics synapse [126]. The overlap of the two spikes at different time instants results in a different cell conductance due to the difference in the overall voltage difference across the cell. (d) TEM cross-section of the electronic synapse that is programmed in its lowest conductance (amorphous) state. Reprinted with permission from [126]. Copyright 2012 American Chemical Society

build an array of synapses and neurons to emulate different learning functions. Another application example of PCM in cognitive computing was shown by Suri *et al.* [128]. A 2-PCM configuration was used in which the depression of the synapse is emulated by gradual SET of one cell, and potentiation is implemented by gradual SET in the other cell. Hence the energy consumption on the system level can be reduced since the RESET pulses are replaced by SET pulses, which consume less energy. Besides, this can be used when RESET process is abrupt due to device design or material composition required, given that gradual SET can be performed. However, this configuration occupies larger area than a one-cell synapse and requires a separate control to RESET both cells regularly while keeping the synaptic weight. Besides single synaptic device level experimental demonstrations, array level brain-like associative learning has also been experimentally implemented with phase change memory devices [129]. It was shown that for cognitive computing applications, resistance variations can be tolerated at the expense of energy consumption [130].

PCM technology has the promise to build human-scale intelligent systems which has not been possible so far due to lack of energy and area efficient hardware. Energy efficient artificial neural networks that can perform complex tasks with large amounts of inputs and variables can be made possible due to low energy consumption and nano-scale feature sizes of PCM technology. In a possible implementation, PCM devices can be deposited on top of CMOS neuron circuits. Several challenges such as process integration, RC delay, thermal disturbances and architectural problems should be addressed for such a configuration. Device models suitable for neuromorphic architectures are still needed for application-specific performance evaluations of these systems.

5.7 Future Outlook

The last decade has seen tremendous advances of the technology and fundamental research in phase change materials. Multi giga-bit phase change memory chips on par with conventional Flash technology were demonstrated in 20 nm technology [119]. Phase Change Memory is maturing from being a mere research topic and is now moving to large-scale manufacturing following the recent announcement [131] by Micron and Nokia where PCM multi-chip packages will be used in their smart phones to enable faster boot time, longer battery life, and better reliability. A number of industrial joint ventures such as Intel-Micron and IBM-Hynix have been formed to accelerate the PCM development for both mobile and high-end server market. Table 5.2 summarizes key PCM parameters that have been demonstrated and projected.

In terms of the fundamental research, the resonant bonding phenomenon observed in the crystalline phase of phase change materials is an important factor that governs their unique properties in terms of reflectivity/resistivity change. Therefore, it is now possible to derive structure–property maps based on an understanding for which compositions resonance bonding can be expected in the crystalline state. Such maps are not only useful because they identify candidates for storage applications. More importantly, the understanding of the bonding characteristics contained in such data will provide insight into systematic property changes, which can help tailor materials for specific storage applications. The phase transition from the crystalline to the amorphous phase has traditionally been achieved by melt-quenching the phase change material. However, the recent demonstration of Interfacial PCM and other ultra-fast optical switching studies indicate that the phase transformation can be brought about by bond rearrangements at substantially reduced

Table 5.2 Summary parameter table of Phase Change Memory

Parameter	Demonstrated/projected	Value for the device in the chapter
Bottom electrode contact area	Demonstrated	$165\,nm^2$ [119]
	Projected[1]	$2.5\,nm^2$ [101]
Feature size F	Demonstrated	20 nm [119]
	Projected	8 nm [144]
Cell size[2]	Demonstrated	$4F^2$ [119]
	Projected	$4F^2$
Density[3]	Demonstrated	$59.5\,Gb\,cm^{-2}$ [119]
	Projected	$6.25\,Tb\,cm^{-2}$ [119] (4 bit/cell)
Read time	Demonstrated	12 ns [145]
	Projected	<10 ns [144]
Write (reset) time	Demonstrated	100 ns [145]
	Projected	400 ps [146]
Write (Set) time	Demonstrated	150 ns [143]
	Projected	500 ps [42]
Read operation voltages	Demonstrated	1.2 V [145]
	Projected[4]	<0.5 V [147]
Write operation voltages	Demonstrated	3 V [145]
	Projected[4]	<0.5 V [147]
Reset current	Demonstrated	90 μA [119]
	Projected	<2 μA [92,101]
Programming current density	Demonstrated	$10\,MA\,cm^{-2}$ [7]
	Projected[4]	$3.3\,MA\,cm^{-2}$ [147]
Switching energy	Demonstrated	$6E\text{-}12\,J\,bit^{-1}$ [145]
	Projected[5]	$<1E\text{-}15\,J\,bit^{-1}$ [92]
Endurance	Demonstrated	1E-12 [119]
	Projected	1E-15 [148]
Retention	Demonstrated	>10 yr
	Projected	>10 yr
Binary throughput	Demonstrated	$40\,MB\,s^{-1}$ [143]
	Projected[6]	$100\,MB\,s^{-1}$ [148]

The "Demonstrated" numbers are values from devices in multi-GB PCM chips.
The "Projected" numbers are values or estimates based on single device demonstrations.

[1] The smallest active programming area demonstrated using a PCM device with CNT electrodes.
[2] Including a diode-based cell selector device.
[3] The numbers are based on a single device dimension. The effect of peripheral circuits and other interconnects are not included.
[4] Based on $GeTe/Sb_2Te_3$ superlattice memory.
[5] Estimated based on a minimum pulse width of 1 ns.
[6] Based on the requirements of Storage Class Memories.

entropic losses. This can lead to the development of fast and efficient switching devices based on nano-structured materials that make use of this entropy controlled switching. The future PCM devices would see more of nano-structured phase change materials obtained via superlattice formation, dopant addition, or chemically synthesized nanoparticles. These provide an efficient way of tailoring some of the fundamental material properties such as crystallization time, crystallization temperature, melting temperature, and threshold field.

Understanding the electrical properties of sub-threshold conduction and threshold switching continues to be a topic of investigation as there is no consensus on the physical mechanism that governs these processes. It is important to understand these phenomena, especially as we approach device dimensions and material thicknesses that are close to the inter-trap distances. In terms of resistance drift, the constant debate continues as to whether the change in resistance with time is due to the structural relaxation of the decaying trap states or a mechanical stress release of the compressed programming volume. Understanding this drift process is a key to engineer materials and device structures that have minimal drift coefficient and to enable multi-level programming in phase change memories. However it is encouraging to see that, this problem of drift has been mitigated to some extent by using clever read out metrics that are inherently drift-tolerant, albeit comes with the cost of additional energy and area requirements.

In terms of device scaling, the recent demonstrations of carbon nanotube-based PCM devices, have given more confidence in that, the PCM technology will continue to scale even to a single-digit nanometer scale and that the programming power can be significantly reduced at such highly scaled devices as expected from continued scaling of the electrode area. But the question remains however as to whether it would be possible to build large arrays of phase change memory at such scaled dimensions, because the influence of the interconnects would then play a major role in determining the speed, power and reliability of the overall system [137]. The increasing wire delays as the technology scales down, electro-migration effects at higher current densities, reduction in the read/write margin for increasing wire resistances will demand a more rigorous device and interconnect co-optimization for cross-bar arrays in the sub-10 nm node [132]. Possible solutions also include using of wires with better conductivity and scalability (e.g., graphene and CNTs), memory arrays with smaller partition sizes, strapping the bitlines/wordlines with wider wires, and memory elements with larger resistance values and ratios.

The demonstration of $4F^2$ access devices for large arrays could lead to 3D stackable arrays in the future that form the basis of ultra high density phase change memory arrays. Achieving multi-bit, multi-layer phase change memory would be a key step in the large scale commercialization of PCM in the storage market. In addition to the nonvolatile memory applications, phase change memory has the potential to be used for neuromorphic applications as artificial synapses to emulate the learning behavior. Such brain-like systems are being contemplated for augmenting conventional computing because of the limitations of the traditional von Neumann architecture in terms of its energy efficiency in application areas such as performing cognitive operations. It is also possible to use phase change memories to perform accumulation operations in either optical or electrical domain, exploiting the gradual changes that take place in programming volume as one goes from the amorphous phase to the crystalline phase [133]. With a simple accumulator we can perform all the basic arithmetic processes of addition, subtraction, multiplication, and division while simultaneously storing the result at the same physical location.

It has almost been over four decades since the discovery of the switching effect in chalcogenides by Stanford Ovshinsky, but these materials still remain a mystery in terms of their fundamental understanding and new theories are developed every year and new applications are being proposed. One can only say that, with all that these materials have to offer to us, "There is plenty of room in the disorder."

5.8 Summary

The field of phase change memory research has gained momentum in the last decade because of its interesting device and material properties that make them an excellent candidate for future nonvolatile memory applications. This chapter gives an overview of the basic device structure, the physical mechanisms underlying the various processes of phase change, the different device metrics that are important in evaluating the success of PCM technology and some of the possible future applications. There is however a vast body of literature in the field of phase-change materials research and its application in optical disk storage that are not covered in this chapter. We highlight some of the recent innovations in materials and device engineering that has enabled low power programming, ultimate device scaling and large-scale device integration.

Acknowledgments

This work is supported in part by the National Science Foundation (NSF, ECCS 0950305), Intel through the Global Research Collaboration (GRC) of the Semiconductor Research Corporation, the MSD Center of the Focus Center Research Program (FCRP), a Semiconductor Research Corporation subsidiary, and the member companies of the Non Volatile Memory Technology Research Initiative (NMTRI) at Stanford University. Collaborations with Prof. Kenneth Goodson (Stanford), Prof. Mehdi Asheghi (Stanford), Dr. Simone Raoux (IBM), Dr. Chung Lam (IBM), Dr. G.A.M. Hurkx (NXP), Dr. B.J. Bae (Samsung), and Dr. Delia Milliron (LBNL Molecular Foundry) are greatly appreciated. We thank the contributions of graduated students and post-doctoral researchers to this review: Dr. SangBum Kim, Dr. Marissa Caldwell, Dr. Duygu Kuzum, and Dr. Jiale Liang.

References

1. Ovshinsky, S.R. (1968) Reversible electrical switching phenomena in disordered structures. *Physical Review Letters*, **21**, 1450.
2. Lai, S. (2003) Current status of the phase change memory and its future. Electron Devices Meeting, 2003. IEDM '03 Technical Digest. IEEE International, pp. 10.1.1–10.1.4.
3. Burr, G.W., Kurdi, B.N., Scott, J.C. *et al.* (2008) Overview of candidate device technologies for storage-class memory. *IBM Journal of Research and Development*, **52**, 449–464.
4. Pirovano, A., Lacaita, A.L., Benvenuti, A. *et al.* (2003) Scaling analysis of phase-change memory technology. Electron Devices Meeting, 2003. IEDM '03 Technical Digest. IEEE International, pp. 29.6.1–29.6.4.
5. Burr, G.W., Breitwisch, M.J., Franceschini, M. *et al.* (2010) Phase change memory technology. *Journal of Vacuum Science and Technology B*, **28**, 223.
6. Ielmini, D. (2008) Threshold switching mechanism by high-field energy gain in the hopping transport of chalcogenide glasses. *Physical Review B*, **78**, 035308.
7. Wong, H.-S.P., Raoux, S., Kim, S. *et al.* (2010) Phase change memory. *Proceedings of the IEEE*, **98**(12), 2201–2227.
8. Lankhorst, M.H.R., Ketelaars, B.W.S.M.M., and Wolters, R.A.M. (2005) Low-cost and nanoscale non-volatile memory concept for future silicon chips. *Nature Materials*, **4**, 347–352.
9. Yamada, N., Kojima, R., Nishihara, T. *et al.* (2009) 100 GB rewritable triple-layer optical disk having Ge-Sb-Te films. Europ. Phase Change and Ovonic Science Symp, Aachen, Germany, pp. 23–28.
10. Raoux, S., Rettner, C.T., Jordan-Sweet, J.L. *et al.* (2007) Direct observation of amorphous to crystalline phase transitions in nanoparticle arrays of phase change materials. *Journal of Applied Physics*, **102**, 094305-8.
11. Raoux, S., Cabrera, D., Devasia, A. *et al.* (2011) Influence of Dopants on the Crystallization Temperature, Crystal Structure, Resistance, and Threshold Field for $Ge_2Sb_2Te_5$ and GeTe Phase Change Materials. European Phase Change and Ovonics Symposium.

12. Cheng, H.Y., Wu, J.Y., Cheek, R. *et al.* (2012) A Thermally Robust Phase Change Memory by Engineering the Ge/N Concentration in (Ge,N)$_x$Sb$_y$Te$_z$ Phase Change Material. Electron Devices Meeting, 2012. IEDM '12 Technical Digest. IEEE International, pp. 31.1.1–31.1.4.
13. Ielmini, D. and Zhang, Y. (2007) Analytical model for subthreshold conduction and threshold switching in chalcogenide-based memory devices. *Journal of Applied Physics*, **102**, 054517-13.
14. Fugazza, D., Ielmini, D., Lavizzari, D.S., and Lacaita, A.L. (7–9 Dec. 2009) Distributed-Poole-Frenkel modeling of anomalous resistance scaling and fluctuations in phase-change memory (PCM) devices. Electron Devices Meeting (IEDM), 2009 IEEE International, pp. 1–4.
15. Krebs, D., Raoux, S., Rettner, C.T. *et al.* (2009) Threshold field of phase change memory materials measured using phase change bridge devices. *Applied Physics Letters*, **95**, 082101.
16. Owen, A.E., Robertson, J.M., and Main, C. (1979) The threshold characteristics of chalcogenide-glass memory switches. *Journal of Non-Crystalline Solids*, **32**, 29–52.
17. Adler, D., Henisch, H.K., and Mott, S.D. (1978) The mechanism of threshold switching in amorphous alloys. *Reviews of Modern Physics*, **50**, 209–220.
18. Pirovano, A., Lacaita, A.L., Benvenuti, A. *et al.* (2004) Electronic switching in phase-change memories. *Electron Devices, IEEE Transactions on*, **51**, 452–459.
19. Adler, D., Shur, M.S., Silver, M., and Ovshinsky, S.R. (1980) Threshold switching in chalcogenide glass thin films. *Journal of Applied Physics*, **51**, 3289–3309.
20. Adler, D., Shur, M.S., Silver, M., and Ovshinsky, S.R. (1984) Reply to comment on threshold switching in chalcogenide glass thin films. *Journal of Applied Physics*, **56**, 579–580.
21. Karpov, V.G., Kryukov, Y.A., Savransky, S.D., and Karpov, I.V. (2007) Nucleation switching in phase change memory. *Applied Physics Letters*, **90**, 123504–123504-3.
22. Raoux, S., Cohen, G.M., Shelby, R.M. *et al.* (April 2010) Amorphization of crystalline phase change material by ion implantation. Mater. Res. Soc. Spring Meeting, San Francisco.
23. Pirovano, A., Lacaita, A.L., Pellizzer, F. *et al.* (2004) Low-field amorphous state resistance and threshold voltage drift in chalcogenide materials. *Electron Devices, IEEE Transactions on*, **51**, 714–719.
24. Boniardi, M., Redaelli, A., Pirovano, A. *et al.* (2009) A physics-based model of electrical conduction decrease with time in amorphous Ge2Sb2Te5. *Journal of Applied Physics*, **105**, 084506.
25. Ielmini, D., Lavizzari, S., Sharma, D., and Lacaita, A.L. (2008) Temperature acceleration of structural relaxation in amorphous Ge2Sb2Te5. *Applied Physics Letters*, **92**, 193511.
26. Karpov, I.V., Mitra, M., Kau, D. *et al.* (2007) Fundamental drift of parameters in chalcogenide phase change memory. *Journal of Applied Physiology*, **102**, 124503–124509.
27. Kim, S., Lee, B., Asheghi, M. *et al.* (2011) Resistance and threshold switching voltage drift behavior in phase-change memory and their temperature dependence at microsecond time scales studied using a micro-thermal stage. *Transactions on Electron Devices*, **58**, 584–592.
28. Ielmini, D., Lavizzari, S., Sharma, D., and Lacaita, A.L. (10–12 Dec. 2007) Physical interpretation, modeling and impact on phase change memory (PCM) reliability of resistance drift due to chalcogenide structural relaxation. Electron Devices Meeting, 2007. IEDM 2007. IEEE International, pp. 939–942.
29. Mitra, M., Jung, Y., Gianola, D.S., and Agarwal, R. (2010) Extremely low drift of resistance and threshold voltage in amorphous phase change nanowire devices. *Applied Physics Letters*, **96**, 222111–222113.
30. Rizzi, M., Spessot, A., Fantini, P., and Ielmini, D. (2011) Role of mechanical stress in the resistance drift of Ge2Sb2Te5 films and phase change memories. *Applied Physics Letters*, **99**, 223513–223515.
31. Zhou, G.F., Borg, H.J., Rijpers, J.C.N., and Lankhorst, M. (2000) Crystallization behavior of phase change materials: comparison between nucleation- and growth-dominated crystallization. Optical Data Storage, 2000. Conference Digest, pp. 74–76.
32. Avrami, M. (1939) Kinetics of phase change. I general theory. *Journal of Chemical Physics*, **7**(12), 1103–1112.
33. Bruns, G., Merkelbach, P., Schlockermann, C. *et al.* (2009) Nanosecond switching in GeTe phase change memory cells. *Journal of Applied Physiology*, **95**, 043108.
34. Shimakawa, K. (2012) Dynamics of crystallization with fractal geometry: Extended KJMA approach in glasses. *Physica Status Solidi (b), Special Issue: Phase-change memory: Science and applications*, **249**(10), 2024–2027.
35. Peng, C., Cheng, L., and Mansuripur, M. (1997) Experimental and theoretical investigations of laser-induced crystallization and amorphization in phase-change optical recording media. *Journal of Applied Physiology*, **82**(9), 4183.
36. Senkader, S. and Wright, C.D. (2004) Models for phase-change of Ge$_2$Sb$_2$Te$_5$ in optical and electrical memory devices. *Journal of Applied Physiology*, **95**(2), 504–511.

37. Ahn, C., Lee, B., Jeyasingh, R.G.D. *et al.* (2011) Crystallization properties and their drift dependence in phase-change memory studied with a micro-thermal stage. *Journal of Applied Physiology*, **110**, 114520.
38. Yamada, N. (2009) Phase change materials: science and applications, in: *Development of Materials for Third Generation Optical Storage Media*, Springer, Berlin.
39. Shportko, K., Kremers, S., Woda, M. *et al.* (2008) Resonant bonding in crystalline phase-change materials. *Nature Materials*, **7**, 653–658.
40. Simpson, R.E., Fons, P., Kolobov, A.V. *et al.* (2011) Interfacial phase change memory. *Nature Nanotechnology*, **6**, 501–505.
41. Lencer, D., Salinga, M., Grabowski, B. *et al.* (2008) A map for phase-change materials. *Nature Materials*, **7**, 972–977.
42. Loke, D., Lee, T.H., Wang, W.J. *et al.* (2012) Breaking the speed limits of phase-change memory. *Science*, **336**(6088), 1566–1569.
43. Wang, W.J., Loke, D., Law, L.T. *et al.* (2012) Engineering Grains of Ge2Sb2Te5 for Realizing Fast-Speed, Low-Power, and Low-Drift Phase-Change Memories with Further Multilevel Capabilities. Electron Devices Meeting, 2012. IEDM '12 Technical Digest. IEEE International, pp. 31.3.1–31.3.4.
44. Akola, J., Jones, R.O., Kohara, S. *et al.* (2009) Experimentally constrained density-functional calculations of the amorphous structure of the prototypical phase-change material Ge2Sb2Te5. *Physical Review B-Condensed Matter*, **80**, 020201.
45. Matsunaga, T., Akola, J., Kohara, S. *et al.* (2011) From local structure to nanosecond recrystallization dynamics in AgInSbTe phase change materials. *Nature Materials*, **10**, 129.
46. Kolobov, A.V., Krbal, M., Fons, P. *et al.* (2011) Distortion-triggered loss of long-range order in solids with bonding energy hierarchy. *Nature Chemistry*, **3**, 311.
47. Fons, P., Osawa, H., Kolobov, A.V. *et al.* (2010) Photoassisted amorphization of the phase-change memory alloy $Ge_2Sb_2Te_5$. *Physical Review B-Condensed Matter*, **82**, 041203–041206.
48. Nam, S.-W., Chung, H.-S., Lo, Y.C. *et al.* (2012) Electrical wind force–driven and dislocation-templated amorphization in phase-change nanowires. *Science*, **336**(6088), 1561–1566.
49. Reifenberg, J., Pop, E., Gibby, A. *et al.* (2006) Multiphysics Modeling and Impact of Thermal Boundary Resistance in Phase Change Memory Devices. Thermal and Thermomechanical Phenomena in Electronics Systems, 2006. ITHERM '06. The Tenth Intersociety Conference on, May 30 2006–June 2 2006, pp. 106–113.
50. Risk, W.P., Rettner, C.T., and Raoux, S. (2008) *In situ* 3 omega techniques for measuring thermal conductivity of phase-change materials. *Review of Scientific Instruments*, **79**, 14–19.
51. Reifenberg, J.P., Panzer, M.A., Kim, S. *et al.* (2007) Thickness and stoichiometry dependence of the thermal conductivity of GeSbTe films. *Applied Physics Letters*, **91**, 111904-1–111904-3.
52. Reifenberg, J.P., Chang, K.W., Panzer, M.A. *et al.* (2010) Thermal boundary resistance measurements for phase-change memory devices. *IEEE Electron Device Letters*, **31**, 56–58.
53. Lyeo, H.K., Cahill, D.G., Lee, B.S. *et al.* (2006) Thermal conductivity of phase-change material Ge2Sb2Te5. *Applied Physics Letters*, **89**, 151904–151904-3.
54. Reifenberg, J.P. (Mar. 2010) Thermal phenomena in phase change memory. Ph.D dissertation: Stanford University.
55. Goodson, K.E. (2007) Ordering up the minimum thermal conductivity of solids. *Science*, **315**, 342–343.
56. Li, Z., Lee, J., Reifenberg, J.P. *et al.* (2011) Grain boundaries, phase impurities, and anisotropic thermal conduction in phase change memory. *IEEE Electron Device Letters*, **32**, 961–963.
57. Lee, J., Li, Z., Reifenberg, J.P. *et al.* (2011) Thermal conductivity anisotropy and grain structure in $Ge_2Sb_2Te_5$ films. *Journal of Applied Physiology*, **109**, 084902.
58. Small, E., Sadeghipour, S.M., Pileggi, L., and Asheghi, M. (2008) Thermal analyses of confined cell design for phase change random access memory (PCRAM). Thermal and Thermomechanical Phenomena in Electronic Systems, 2008. ITHERM 2008. 11th Intersociety Conference on, pp. 1046–1054.
59. Reifenberg, J.P., Kencke, D.L., and Goodson, K.E. (2008) The impact of thermal boundary resistance in phase-change memory devices. *Electron Device Letters, IEE*, **29**, 1112–1114.
60. Sadeghipour, S.M., Pileggi, L., and Asheghi, M. (2006) Phase change random access memory, thermal analysis. Thermal and Thermomechanical Phenomena in Electronics Systems, 2006. ITHERM '06. The Tenth Intersociety Conference on, May 30 2006–June 2 2006, pp. 660–665.
61. Russo, U., Ielmini, D., Redaelli, A., and Lacaita, A.L. (2008) Modeling of programming and read performance in phase-change memories—part I: cell optimization and scaling. *Electron Devices IEEE Transactions on*, **55**, 506–514.

62. Bozorg-Grayeli, E., Reifenberg, J.P., Panzer, M.A. *et al.* (2011) Temperature-dependent thermal properties of phase-change memory electrode materials. *Electron Device Letters, IEEE,* **32**, 1281–1283.

63. Wu, J.Y., Breitwisch, M., Kim, S. *et al.* (2011) A low power phase change memory using thermally confined TaN/TiN bottom electrode. Electron Devices Meeting (IEDM), 2011 IEEE International, pp. 3.2.1–3.2.4.

64. Castro, D.T., Goux, L., Hurkx, G.A.M. *et al.* (2007) Evidence of the Thermo-Electric Thomson Effect and Influence on the Program Conditions and Cell Optimization in Phase-Change Memory Cells. Electron Devices Meeting, 2007. IEDM 2007. IEEE International, 10–12 Dec. 2007, pp. 315–318.

65. Suh, D.-S., Kim, C., Kim, K.H.P. *et al.* (2010) Thermoelectric heating of Ge2Sb2Te5 in phase change memory devices. *Applied Physics Letters,* **96**(12), 123115.

66. Lee, J., Kodama, T., Won, Y. *et al.* (2012) Phase purity and the thermoelectric properties of Ge2Sb2Te5 Films down to 25nm Thickness. *Applied Physics Letters,* **112**, 014902–014902-6.

67. Lee, J., Asheghi, M., and Goodson, K.E. (2012) Impact of thermoelectric phenomena on phase- change memory performance metrics and scaling. *Nanotechnology,* **23**(20), 205201–205201-7.

68. Raoux, S., Burr, G.W., Breitwisch, M.J. *et al.* (2008) Phase-change random access memory: A scalable technology. *IBM Journal of Research and Development,* **52**, 465–479.

69. Raoux, S., Shelby, R.M., Jordan-Sweet, J. *et al.* (2008) Phase change materials and their application to random access memory technology. *Microelectronic Engineering,* **85**, 2330–2333.

70. Raoux, S. (2009) Scaling properties of phase change materials, in *Phase Change Materials: Science and Applications* (eds S. Raoux and M. Wuttig), Springer, Berlin, pp. 99–124.

71. Raoux, S., Cheng, H.-Y., Jordan-Sweet, J.L. *et al.* (2009) Influence of interfaces and doping on the crystallization temperature of GeSb. *Applied Physics Letters,* **94**, 183114-3.

72. Wei, X., Shi, L., Chong, T.C. *et al.* (2007) Thickness-dependent nano-crystallization in Ge2Sb2Te5 and its effect on devices. *Japanese Journal of Applied Physics,* **46**, 2211–2214.

73. Cheng, H.-Y., Raoux, S., Munoz, B., and Jordan-Sweet, J. (2009) Influence of interfaces on the crystallization characteristics of Ge2Sb2Te5. Non-volatile Memory Technol. Symp., Portland, OR.

74. Loke, D., Shi, L., Wang, W. *et al.* (2011) Ultrafast switching in nanoscale phase-change random access memory with superlattice-like structures. *Nanotechnology,* **22**(25), 254019.

75. Lee, S.-H., Jung, Y., and Agarwal, R. (2007) Highly scalable non-volatile and ultra-low-power phase-change nanowire memory. *Nature Nanotechnology,* **2**, 626–630.

76. Wautelet, M. (1991) Estimation of the variation of the melting temperature with the size of small particles, on the basis of a surface-phonon instability model. *Journal of Physics D-Applied Physics,* **24**, 343–346.

77. Khonik, V.A., Kitagawa, K., and Morii, H. (2000) On the determination of the crystallization activation energy of metallic glasses. *Journal of Applied Physiology,* **87**, 8440–8443.

78. Chong, T.C., Shi, L.P., Zhao, R. *et al.* (2006) Phase change random access memory cell with superlattice-like structure. *Applied Physics Letters,* **88**, 122114.

79. Chen, Y., Li, D., Lukes, J.R. *et al.* (2005) Minimum superlattice thermal conductivity from molecular dynamics. *Physical Review B-Condensed Matter,* **72**, 174302.

80. Tominaga, J., Fons, P., Kolobov, A. *et al.* (2008) Role of Ge switch in phase transition: approach using atomically controlled GeTe/Sb2Te3 superlattice. *Japanese Journal of Applied Physics,* **47**, 5763–5766.

81. Wang, W.J., Loke, D., Shi, L.P. *et al.* (2012) Enabling universal memory by overcoming the contradictory speed and stability nature of phase-change materials. *National Science Report,* **2**, 360.

82. Park, G.S., Kwon, J.H., Kim, M. *et al.* (2007) Crystalline and amorphous structures of Ge-Sb-Te nanoparticles. *Journal of Applied Physics,* **102**, 70–72.

83. Choi, H.S., Seol, K.S., Takeuchi, K. *et al.* (2005) Sythesis and size-controlled Ge2Sb2Te5 nanoparticles. *Japanese Journal of Applied Physics,* **44**, 7720–7722.

84. Zhang, Y., Wong, H.S.P., Raoux, S. *et al.* (2007) Phase change nanodot arrays fabricated using a self-assembly diblock copolymer approach. *Applied Physics Letters,* **91**, 013104 3.

85. Zhang, Y., Raoux, S., Krebs, D. *et al.* (2008) Phase change nanodots patterning using a self-assembled polymer lithography and crystallization analysis. *Journal of Applied Physics,* **104**, 20–22.

86. Milliron, D.J., Raoux, S., Shelby, R., and Jordan-Sweet, J. (2007) Solution-phase deposition and nanopatterning of GeSbSe phase-change materials. *Nature Materials,* **6**, 352–356.

87. Caldwell, M.A., Raoux, S., Wang, R.Y. *et al.* (2010) Synthesis and size-dependent crystallization of colloidal germanium telluride nanoparticles. *Journal of Materials Chemistry,* **20**, 1285–1291.

88. Kim, S., Bae, B.-J., Zhang, Y. *et al.* (2011) One-dimensional thickness scaling study of phase change material (Ge2Sb2Te5) using a Pseudo 3-terminal device. *IEEE Transactions on Electron Devices*, **58**(5), 1483–1489.

89. Yu, D., Brittman, S., Lee, J.S. *et al.* (2008) Minimum voltage for threshold switching in nanoscale phase-change memory. *Nano Letters*, **8**, 3429–3433.

90. Chen, Y.C., Rettner, C.T., Raoux, S. *et al.* (2006) Ultra-Thin Phase-Change Bridge Memory Device Using GeSb. Electron Devices Meeting, 2006. IEDM '06. International, pp. 1–4.

91. Krebs, D., Raoux, S., Rettner, C.T. *et al.* (2009) Characterization of phase change memory materials using phase change bridge devices. *Journal of Applied Physics*, **106**, 90–92.

92. Xiong, F., Liao, A.D., Estrada, D., and Pop, E. (2011) Low-power switching of phase-change materials with carbon nanotube electrodes. *Science*, **332**(6029), 568–570.

93. Sasago, Y., Kinoshita, M., Morikawa, T. *et al.* (2006) Cross-point phase change memory with 4F^2 cell size driven by low-contact-resistivity poly-Si diode. VLSI Technology, 2009 Symposium on, 2006, pp. 24–25.

94. Ha, Y.H., Yi, J.H., Horii, H. *et al.* (2003) An edge contact type cell for Phase Change RAM featuring very low power consumption. VLSI Technology, 2003. Digest of Technical Papers. 2003 Symposium on, 2003, pp. 175–176.

95. Pellizzer, F., Pirovano, A., Ottogalli, F. *et al.* (2004) Novel m-trench phase-change memory cell for embedded and stand-alone non-volatile memory applications. VLSI Technology, 2004. Digest of Technical Papers. 2004 Symposium on, 2004, pp. 18–19.

96. Chen, W.S., Lee, C., Chao, D.S. *et al.* (2007) A Novel Cross-Spacer Phase Change Memory with Ultra-Small Lithography Independent Contact Area. Electron Devices Meeting, 2007. IEDM 2007. IEEE International, pp. 319–322.

97. Pirovano, A., Pellizzer, F., Tortorelli, I. *et al.* (2007) Self-aligned μTrench phase-change memory cell architecture for 90nm technology and beyond. Solid State Device Research Conference, 2007. ESSDERC 2007. 37th European, pp. 222–225.

98. Breitwisch, M., Nirschl, T., Chen, C.F. *et al.* (2007) Novel Lithography-Independent Pore Phase Change Memory. VLSI Technology, 2007 IEEE Symposium on, 2007, pp. 100–101.

99. Song, Y.J., Ryoo, K.C., Hwang, Y.N. *et al.* (2006) Highly Reliable 256Mb PRAM with Advanced Ring Contact Technology and Novel Encapsulating Technology. VLSI Technology, 2006. Digest of Technical Papers. 2006 Symposium on, 2006, pp. 118–119.

100. Im, D.H., Lee, J.I., Cho, S.L. *et al.* (2008) A unified 7.5nm dash-type confined cell for high performance PRAM device. Electron Devices Meeting, 2008. IEDM 2008. IEEE International, pp. 1–4.

101. Liang, J., Jeyasingh, R.G.D., Chen, H.-Y., and Wong, H.-S.P. (2012) An ultra-low reset current cross-point phase change memory with carbon nanotube electrodes. *IEEE Transactions on Electron Devices*, **59**(4), 1155–1163.

102. Liu, B., Zhang, T., Xia, J.L. *et al.* (2004) Nitrogen-implanted Ge2Sb2Te5 film used as multilevel storage media for phase change random access memory. *Semiconductor Science and Technology*, **19**, L61–L64.

103. Nirschl, T., Phipp, J.B., Happ, T.D. *et al.* (2007) Write Strategies for 2 and 4-bit Multi-Level Phase-Change Memory. Electron Devices Meeting, 2007. IEDM 2007. IEEE International, pp. 461–464.

104. Lai, Y.F., Feng, J., Qiao, B.W. *et al.* (2006) Stacked chalcogenide layers used as multi-state storage medium for phase change memory. *Applied Physics a-Materials Science & Processing*, **84**, 21–25.

105. Rao, F., Song, Z., Wu, Z.M.L. *et al.* (2007) Multilevel data storage characteristics of phase change memory cell with doublelayer chalcogenide fims (Ge2Sb2Te5 and Sb2Te3). *Japanese Journal of Applied Physics*, **46**, 80–82.

106. Zhang, Y., Feng, J., Zhang, Y. *et al.* (2007) Multi-bit storage in reset process of Phase-change Random Access Memory (PRAM). *Physica Status Solidi-Rapid Research Letters*, **1**, R28–R30.

107. Yin, Y., Ota, K., Higano, N. *et al.* (2008) Multilevel storage in lateral top-heater phase-change memory. *IEEE Electron Device Letters*, **29**, 876–878.

108. Oh, G.H., Park, Y.L., Lee, J.I. *et al.* (2009) Parallel multi-confined (PMC) cell technology for high density MLC PRAM. VLSI Technology, 2009 Symposium on, 2009, pp. 220–221.

109. Bedeschi, F., Fackenthal, R., Resta, C. *et al.* (2009) A bipolar-selected phase change memory featuring multi-level cell storage. *IEEE Journal of Solid-State Circuits*, **44**, 217–227.

110. Nakayama, K., Takata, M., Kasai, T. *et al.* (2007) Pulse number control of electrical resistance for multi-level storage based on phase change. *Journal of Physics D-Applied Physics*, **40**, 5061–5065.

111. Papandreou, N., Pozidis, H., Pantazi, A. *et al.* (2011) Programming Algorithms for Multilevel Phase-Change Memory. Proc. ISCAS, pp. 329–332.

112. Papandreou, N., Pozidis, H., Mittelholzer, T. *et al.* (2011) Drift-tolerant Multilevel Phase-Change Memory. Proc. IMW, pp. 147–150.

113. Hwang, Y.N., Um, C.Y., Lee, J.H. *et al.* (2010) MLC PRAM with SLC write-speed and robust read scheme. Proc. Symposium VLSI Circuits, pp. 201–202.

114. Lin, Y.Y., Chen, Y.C., Lee, F.M. *et al.* (2012) A Simple New Write Scheme for Low Latency Operation of Phase Change Memory. Symposium on VLSI Tech, Digest of Tech Papers, pp. 51–52.

115. Kau, D., Tang, S., Karpov, I.V. *et al.* (2009) A stackable cross point phase change memory. Electron Devices Meeting, 2009. IEDM '09 Technical Digest. IEEE International, pp. 617–620.

116. Gopalakrishnan, K., Shenoy, R.S., Rettner, C.T. *et al.* (2010) Highly-scalable novel access device based on Mixed Ionic Electronic conduction (MIEC) materials for high density phase change memory (PCM) arrays. VLSI Technology (VLSIT), 2010 Symposium on, 15–17 June 2010, pp. 205–206.

117. Sasago, Y., Kinoshita, M., Minemura, H. *et al.* (2011) Phase-change memory driven by poly-Si MOS transistor with low cost and high-programming gigabyte-per-second throughput. VLSI Technology (VLSIT), 2011 Symposium on, 14–16 June 2011, pp. 96–97.

118. Lee, S.H., Park, H.C., Kim, M.S. *et al.* (5–7 Dec. 2011) Highly productive PCRAM technology platform and full chip operation: Based on 4F^2 (84nm pitch) cell scheme for 1Gb and beyond. Electron Devices Meeting (IEDM), 2011 IEEE International, pp. 3.3.1–3.3.4.

119. Kang, M.J., Park, T.J., Kwon, Y.W. *et al.* (5–7 Dec. 2011) PRAM cell technology and characterization in 20nm node size. Electron Devices Meeting (IEDM), 2011 IEEE International, pp. 3.1.1–3.1.4.

120. Kinoshita, M., Sasago, Y., Minemura, H. *et al.* (2012) Scalable 3-D vertical chain-cell-type phase-change memory with 4F2 poly-Si diodes. VLSI Technology (VLSIT), 2012 Symposium on, 12–14 June 2012, pp. 35–36.

121. Burr, G.W., Kurdi, B.N., Scott, J.C. *et al.* (2008) Overview of candidate device technologies for storage-class memory. *IBM Journal of Research and Development*, **52**(4), 720–722.

122. Freitas, R.F. and Wilcke, W.W. (2008) Storage-class memory: The next storage system technology. *IBM Journal of Research and Development*, **52**(4), 521–528.

123. Infoworld (2013) http://www.infoworld.com/d/hardware/ibm-turbocharge-more-servers-accelerators-364 (accessed 16 July 2013).

124. Ahn, Su Jin, Song, Yoonjong, Jeong, Hoon *et al.* (5–7 Dec. 2011) Reliability perspectives for high density PRAM manufacturing. Electron Devices Meeting (IEDM), 2011 IEEE International, pp. 12.6.1–12.6.4.

125. Bi, G.-Q. and Poo, M.-M. (1998) Synaptic modifications in cultured hippocampal neurons: dependence on spike timing and synaptic strength. *Nature*, **18**, 10464–10472.

126. Kuzum, D., Jeyasingh, R., Lee, B., and Wong, H.-S.P. (2012) Nanoelectronic programmable synapses based on phase change. *Nano Letters*, **12**(5), 2179–2186.

127. Kuzum, D., Jeyasingh, R., Yu, S., and Wong, H.-S.P. (2012) Low-energy robust neuromorphic computation using synaptic devices. *IEEE Transactions on Electron Devices*, **59**(12), 3489–3494.

128. Suri, M., Bichler, O., Querlioz, D. *et al.* (5–7 Dec. 2011) Phase change memory as synapse for ultra-dense neuromorphic systems: Application to complex visual pattern extraction. Electron Devices Meeting (IEDM), 2011 IEEE International, pp. 4.4.1–4.4.4.

129. Eryilmaz, S.B., Kuzum, D., and Jeyasingh, R.G.D. *et al.* (9–11 Dec. 2013) Experimental demonstration of array-level learning with phase change synaptic devices. Electron Devices Meeting (IEDM). 2013 IEEE International, pp. 25.5.1–25.5.4.

130. Eryilmaz, S.B., Kuzum, D., and Jeyasingh, R.G.D. *et al.* (2014) Brain-like associative learning using a nanoscale nonvolatile phase change synaptic device array. *Frontiers in Neuroscience*, **8**(205).

131. EE Times (2013) http://www.eetimes.com/design/memory-design/4403221/Micron-ships-phase-change-memory-for-Nokia-phones (accessed 16 July 2013).

132. Jiale, Liang, Yeh, S., Wong, S.S., and Wong, H.-S.P. (20–23 May 2012) Scaling Challenges for the Cross-Point Resistive Memory Array to Sub-10nm Node – An Interconnect Perspective. Memory Workshop (IMW), 2012 4th IEEE International, pp. 1–4.

133. Wright, C. David, Liu, Yanwei, Kohary, Krisztian I. *et al.* (2011) Arithmetic and biologically-inspired computing using phase-change materials. *Advanced Materials*, **23**(30), 3408–3413.

134. Meyer, W. and Neldel, H. (1937) Fractal geometry and the Meyer–Neldel rule. *Zeitung Technische Physik*, **12**, 588.

135. Collins, P.G., Arnold, M.S., and Avouris, P. (2001) Engineering carbon nanotubes and nanotube circuits using electrical breakdown. *Science*, **292**(5517), 706–709.

136. Liang, J. and Wong, H.-S.P. (2010) Cross-point memory array without cell selectors – device characteristics and data storage pattern dependencies. *IEEE Transactions on Electron Devices*, **57**(10), 2531–2538.

137. Liang, J., Yeh, S., Wong, S.S., and Wong, H.-S.P. (2013) Effect of wordline/bitline scaling on the performance, energy consumption, and reliability of cross-point memory array. *ACM Journal on Emerging Technologies in Computing Systems (JETC)*, **9**(1), 9:1–9:14, Article 9.

138. Akel, A., Caulfield, A.M., Mollov, T.I. *et al.* (2011) Onyx: a prototype phase change memory storage array. Proc. of the 3rd USENIX conference on Hot topics in storage and file systems, pp. 2.

139. Qureshi, M.K., Srinivasan, V., and Rivers, J.A. (2009) Scalable high performance main memory system using phase-change memory technology. ISCA '09 Proc. of the 36th annual international symposium on Computer architecture, pp. 24–33.

140. Kim, S., Lee, B., Asheghi, M. *et al.* (2–6 May 2010) Thermal disturbance and its impact on reliability of phase-change memory studied by the micro-thermal stage. Reliability Physics Symposium (IRPS), 2010 IEEE International, pp. 99–103.

141. Jeyasingh, R.G.D., Kuzum, D., and Wong, H.-S.P. (2011) Investigation of trap spacing for the amorphous state of phase-change memory devices. *Electron Devices, IEEE Transactions on*, **58**(12), 4370–4376.

142. Freitas, R. (June (2009)) Storage Class Memory: Technology, Systems and Applications. 35th SIGMOD International Conference on Management of Data.

143. Choi, Y., Song, I., Park, M.-H. *et al.* (19–23 Feb. 2012) A 20nm 1.8V 8Gb PRAM with 40MB/s program bandwidth. Solid-State Circuits Conference Digest of Technical Papers (ISSCC), 2012 IEEE International, pp. 46, 48.

144. ITRS (2011) http://www.itrs.net/Links/2011ITRS/Home2011.htm, Emerging Research Devices, Table 2.

145. Sandre, G.De., Bettini, L., Pirola, A. *et al.* (2011) A 4 Mb LV MOS-selected embedded phase change memory in 90nm standard CMOS technology. *Solid-State Circuits, IEEE Journal of*, **46**(1), 52, 63.

146. Wang, W.J., Shi, L.P., Zhao, R. *et al.* (2008) Fast phase transitions induced by picosecond electrical pulses on phase change memory cells. *Applied Physics Letters*, **93**, 043121.

147. Takaura, N., Ohyanagi, T., Kitamura, M. *et al.* (2013) "Charge Injection Super-Lattice Phase Change Memory for Low Power and High Density Storage Device Applications." paper T9.1, Symp. VLSI. Tech.

148. Ahn, D.H., Cho, S.L., Park, J.H. *et al.* (2012) "Key Metrics and Reliability Prospects for High Performance PRAM," E\PCOS.

6

Ferroelectric FET Memory

Ken Takeuchi[1] and An Chen[2]
[1]*Chuo University, Japan*
[2]*GLOBALFOUNDRIES Inc., USA*

6.1 Introduction

Ferroelectric materials have been utilized in commercial memories, for example, ferroelectric random access memory (FRAM), where a transistor-accessed ferroelectric capacitor store information in polarization directions [1–3]. Ferroelectric dielectrics can also be integrated in the gate stack of a field effect transistor (FET) whose channel conductance can be switched by modulation of the ferroelectric polarization, also known as a "Ferroelectric FET" (FeFET) [3–9]. Recently, the concept of ferroelectric tunnel junction (FTJ) was proposed as a memory element based on a very thin layer of ferroelectric tunnel barrier between two metal electrodes [10,11]. The tunneling current can be switched by ferroelectric reversal of the thin tunnel barrier. Figure 6.1 illustrates the three types of ferroelectric memories. Both FRAM and FTJ as memory devices combine an access transistor and a storage node, that is, 1-transistor-1-capacitor (1T1C) for FRAM and 1-transistor-1-resistor (1T1R) for FTJ. However, FeFET is an 1T memory, similar to Flash memory devices. FTJ will be discussed in detail in Chapter 9; this chapter focuses on FeFET.

FeFET has attracted a lot of attention for both memory and logic applications. The nonvolatile memory applications of FeFET are based on the remnant polarization in absence of an external field and the reversibility of polarization direction under an applied field. By coupling the ferroelectric polarization directly to the channel of a FET, FeFET enables a capacitor-less ferroelectric memory with simplified device design and nondestructive readout. However, analysis has also shown that FeFET retention can be degraded by the presence of a depolarization field and gate leakage [7]. The performance of FeFET is strongly affected by the interface between ferroelectric gate and semiconductor channel. Insertion of a buffer layer (e.g., high-κ), that is, a metal/ferroelectric/insulator/semiconductor, may improve the interface properties.

For stand-alone memory applications, a ferroelectric NAND (Fe-NAND) flash memory has been developed to achieve low power consumption, high reliability, and high scalability

Emerging Nanoelectronic Devices, First Edition. An Chen, James Hutchby, Victor Zhirnov and George Bourianoff.
© 2015 John Wiley & Sons, Ltd. Published 2015 by John Wiley & Sons, Ltd.

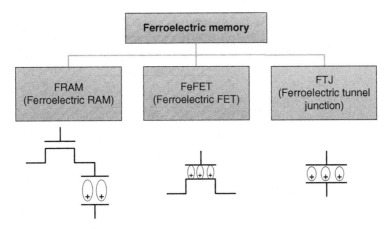

Figure 6.1 Three types of ferroelectric memories

[17–21]. The Fe-NAND flash memory is composed of serially connected MFIS (Metal Ferroelectric Insulator Semiconductor) transistors.

For logic applications, it has been suggested that a FET with a ferroelectric gate stack may achieve negative capacitance and a sub-threshold slope below 60 mV/dec [12–15]. A FeFET-based SRAM with self-adjusted V_{TH} was also developed to improve the static noise margin in a scaled SRAM with low V_{DD} [16].

This chapter is organized in three main sections. Fe-NAND will be discussed in detail as an example of FeFET applications in standalone memory in Section 6.2. FeFET-based SRAM will be covered in Section 6.3. Section 6.4 addresses some system-level considerations with Fe-NAND memory in solid state drive (SSD) systems.

6.2 Ferroelectric FET for Flash Memory Application

Recently, SSD based on NAND flash memories was adopted in mobile and PC applications. SSD provides exceptional bandwidth, random I/O performance superior to a hard disk drive (HDD), power saving, and improved system reliability. SSD is also promising in enterprise applications including data centers [22,23]. In the last five years, as the data traffic on the internet rapidly increased, the power consumption at data centers in the United States doubled to a level corresponding to the output of five nuclear power plants. The Fe-NAND flash memory is suitable for enterprise SSD owing to its low power operation and high endurance.

The Fe-NAND flash memory is made up of serially connected MFIS, as shown in Figure 6.2a. The MFIS device in Figure 6.2b utilizes a ferroelectric $SrBi_2Ta_2O_9$ (SBT) and Hf-Al-O (HAO) buffer layer in the gate stack to overcome the retention loss issues in FeFET [6]. The SBT/HAO interface is chemically stable and gate leakage is suppressed to below 1 nA/cm^2. The HAO/Si interface has high quality as the channel of the transistor. Data retention is measured up to 33 days.

6.2.1 Fe-NAND Flash Memory Operation

Ferroelectric FETs are in principle scalable below 10 nm to crystal unit-cell size because data are stored in the polarization directions of a ferroelectric gate insulator. The program and erase

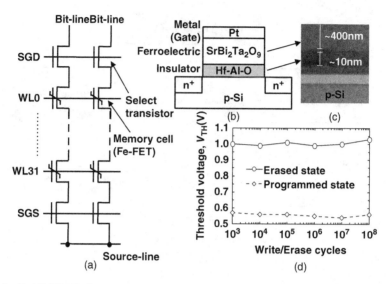

Figure 6.2 Fe-NAND flash memory [17]. (a) Fe-NAND made up of serially connected MFIS transistors. (b) MFIS transistor structure. (c) TEM cross-section of the MFIS transistor. (d) Measured programming and erasing endurance characteristics. IEEE ©. Reprinted with permission, from [19]

operation conditions are illustrated in Figure 6.3c. With program/erase pulses of 6 V amplitude and 10 µs width, a 0.5 V V_{TH} window is realized (Figure 6.3). This is significantly lower than the operation voltage of a conventional floating-gate NAND (FG-NAND) flash ($\sim$ 20 V), which contributes to the lower operation power of Fe-NAND. Fe-NAND also achieved much

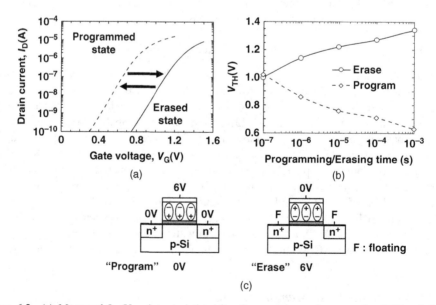

Figure 6.3 (a) Measured I_D–V_G characteristics after the program and erase pulse (6 V and 10 µs). (b) Measured program and erase characteristics with a 6 V pulse. (c) Bias conditions of program and erase operations of FeFET. IEEE ©. Reprinted with permission, from [19]

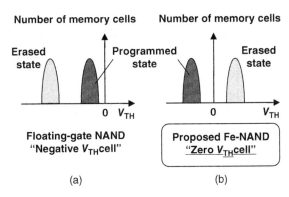

Figure 6.4 (a) Negative V_{TH} memory cell scheme for the floating-gate NAND flash memory [24]. (b) Zero V_{TH} memory cell scheme for the Fe-NAND flash memory. IEEE ©. Reprinted with permission, from [19]

longer cycling endurance, for example, 10^8 cycles in Figure 6.2d in comparison to $\sim 10^4$ cycles in FG-NAND flash [17].

While the lower operation voltages of Fe-NAND helps to reduce operation power, the difference between the program voltage and the read voltage also becomes smaller and consequently read/program disturbance may increase. To reduce disturbance, a conventional NAND flash memory could adopt a negative V_{TH} cell scheme [24], where the middle between the V_{TH}s of erased and programmed cells is negative, as shown in Figure 6.4a. However, if the negative V_{TH} cell scheme is applied to the Fe-NAND flash memory, the data retention drastically degrades due to the depolarization field in the ferroelectric layer. Therefore, a zero V_{TH} memory cell scheme has been proposed for Fe-NAND flash memory to achieve both long retention and strong resistance against read/program disturbance, as shown in Figure 6.4b. By adopting this scheme, the measured V_{TH} shift due to read disturb, program disturb, and retention loss is reduced by 32, 24 and 10%, respectively [19]. In comparison, the negative V_{TH} cell scheme results in V_{TH} shift as much as 192% due to retention loss. The positive V_{TH} cell scheme, where the middle between the V_{TH}s of erased and programmed cells is positive, suffers severe read and program disturbance.

6.2.2 Nonvolatile Page Buffer

A Fe-NAND flash memory with a nonvolatile (NV) page buffer is also proposed [18]. A critical problem of SSD is slow random write. The write unit in a NAND flash memory is a page with typical size of 4–8 KB, which is usually composed of memory cells sharing a word-line. A page is written only once to avoid a program disturb. The large page size is acceptable for digital camera, MP3 player, and camcorder application because their data size is typically over 1 MB and multiple pages are sequentially programmed. However, for a PC and data center application, the minimum write unit of the operating system (OS) is a sector of 512 Bytes. 50% of data written by OS are less than eight sectors, that is, 4 KB. A random write of data smaller than a page size frequently happens. In the case when an OS writes one sector in SSD, the remaining 80% of the page becomes garbage. As garbage accumulates, a garbage collection is performed to increase workable memory capacity. The garbage collection takes as much as 100 ms [25], which is 100 times longer than a page programming time of $\sim 800\,\mu s$ and thus causes serious performance degradation. As memory cell size scales down, more cells are

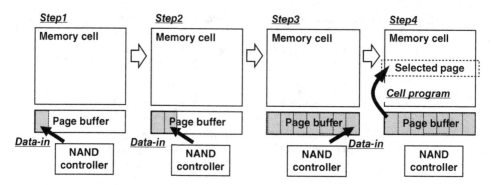

Figure 6.5 Batch write algorithm for SSD. IEEE ©. Reprinted with permission, from [18]

connected to each word-line and the page size increases, which widens the discrepancy between the page size and sector size and causes more SSD performance degradation.

To solve the random write issue, a batch write algorithm is proposed, as shown in Figure 6.5. A page buffer in the Fe-NAND flash memory temporarily stores program data. To avoid a random write, the memory programming starts only after the data to be programmed accumulate to the page size in a page buffer. In Step 1, when the OS issues one sector (512 Bytes) "write" command to the SSD, data are stored in page buffers without programming to memory cells. The NAND controller reports to the OS that the write is completed although no memory cell programming actually occurs. In Step 2, when the OS issues the second sector "write" command, the second data are also temporarily stored in the page buffer. The process continues until the data in the page buffer reach the page size (Step 3), and then memory cell programming starts in Step 4. As the logical address issued by the OS and the actually written NAND physical address are different, the NAND controller updates the logical–physical address mapping table. The batch write algorithm eliminates the data fragmentation of a page during the random write. Considering the sequential write of the SSD is over ten times faster than the random write [26], the batch write algorithm can significantly increase the SSD speed.

One problem with the batch write algorithm is data loss during a power outage. If a power outage happens in Step 4 of Figure 6.5, all data in the page buffers are lost since page buffers are volatile latch circuits. The computer system would fail because the OS recognizes that the program in steps 1–3 has been completed successfully. High reliability against a power outage is essential in an enterprise SSD. To solve this problem, a nonvolatile (NV) page buffer is proposed. The NV-page buffer consists of a volatile latch and a NV-latch, as shown in Figure 6.6. The NV-latch consists of one NMOS and one ferroelectric NMOS. The NV-latch is realized with no additional process because it has the same gate stack structure as memory cells. The area penalty for the NV-latch is less than 1% of the die size as only two transistors are added. In addition, unlike memory cells where the well is biased at 6 V, the well of the ferroelectric transistor in the NV-latch is fixed at 0 V and thus is located in the same well as the other NMOS. In normal operations, the memory cell is programmed based on data in the latch, similar to the conventional NAND flash memory. When the power outage occurs, data in the latch is transferred to the NV-latch to avoid a data loss. When power is restored, the data in the NV-latch is copied to the latch and the memory cell programming is performed based on the data in the latch. An important benefit of the Fe-NAND is that the NV-page buffer can be implemented in the peripheral circuits without any additional process cost because the Fe-FET structure is already implemented for the memory cells.

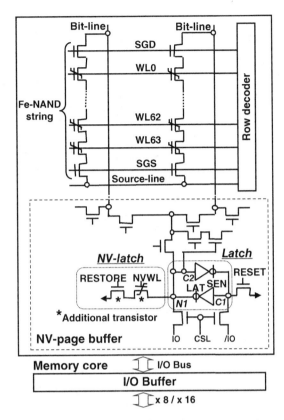

Figure 6.6 Ferroelectric NAND flash memory with nonvolatile page buffer. IEEE ©. Reprinted with permission, from [18]

6.3 Ferroelectric FET for SRAM Application

The active power consumption of digital circuits is proportional to $f \times C \times V_{DD}^2$; therefore, decreasing supply voltage (V_{DD}) is essential for reducing the power consumption of CPU and system on chip (SoC). To realize a low voltage/power CPU and SoC, a very low supply voltage (e.g., 0.5 V) is required, which however presents a challenge for SRAM noise margin. As shown in Figure 6.7, the static noise margin (SNM) of SRAM is represented by the diagonal length of the largest square in the butterfly curves of two inverters in a SRAM cell. As V_{DD} decreases, SNM is also reduced. At a high V_{DD}, SNM is so large that the hold/read of the SRAM is stable. As the feature size of SRAM decreases to sub-30 nm, the V_{TH} variation due to random dopant fluctuation (RDF) significantly increases. At V_{DD} of 0.5 V, the SNM of conventional SRAM decreases to almost zero and the conventional SRAM can no longer operate as shown in Figure 6.7 [27]. Because decreasing V_{DD} of SRAM is difficult in the nanoscale CMOS, the V_{DD} of CPU and SoC cannot be further lowered.

To overcome this problem, a 6T-SRAM with Fe-FETs has been proposed [16]. It has the same structure as a conventional SRAM but uses six FeFETs instead. The Fe-FET has the same structure as the MFIS in Section 6.2 with a ferroelectric SBT layer integrated in a gate stack of standard CMOS transistors with the metal gate (Pt) and a high-κ HfAlO buffer layer between the SBT and the Si substrate.

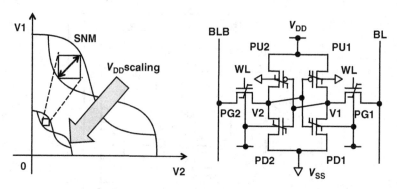

Figure 6.7 Left: Static noise margin (SNM) decreases as V_{DD} decreases. Right: Schematic of a ferroelectric 6T-SRAM. The bodies of NMOS and PMOS are biased to V_{DD} and V_{SS}. The diode current to the body is suppressed at 0.5 V V_{DD}. β ratio = 2. Since there is no extra body contact, the cell area of the ferroelectric 6T-SRAM cell is the same as the conventional 6T-SRAM. IEEE ©. Reprinted with permission, from [16]

The V_{TH} of Fe-FETs changes by controlling the electric field between the gate and the body/channel, as shown in Figure 6.8. For example, V_{GB} of 0.5 V applied on a NMOS forces positive polarization near the channel and decreases V_{TH}, and a reversed V_{GB} would increase V_{TH}. The opposite trend V_{TH} self-adjustment also occurs in PMOS.

By biasing the bodies of the NMOS and the PMOS to V_{DD} and V_{SS}, respectively, the V_{TH} of Fe-FETs automatically shifts to increase SNM. Figure 6.9 illustrates SRAM storing data "0" (a) and data "1" (b), with the directions of V_{TH} adjustment of the pull-up and pull-down transistors marked. The V_{TH} adjustments shift the I-V characteristics of the left and right inverters. For example, when the SRAM holds data "0," the trip point of the right inverter decreases and that of the left inverter increases. As a result, the storage nodes V1 and V2 become more likely to hold 0 V and 0.5 V, respectively. Therefore, holding data "0" become more stable, that is, SNM is improved. A similar analysis can be done on data "1."

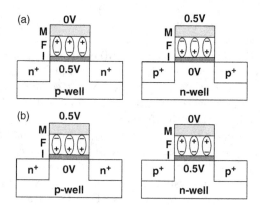

Figure 6.8 The V_{TH} self-adjustment of pull-up and pull-down FeFETs. The electric field between the gate and channel/body shifts V_{TH} (a) higher and (b) lower with proper bias on n-channel and p-channel FeFETs. IEEE ©. Reprinted with permission, from [16]

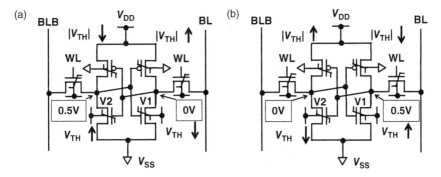

Figure 6.9   The V_{TH} of Fe-FETs of (a) "0" datum and (b) "1" datum. The V_{TH} automatically shifts to increase SNM. For (a) "0" datum and (b) "1" datum, the $|V_{TH}|$ of pull-up and pull-down FETs changes in an opposite direction to increase SNM. IEEE ©. Reprinted with permission, from [16]

Monte Carlo simulation shows 60% increase of SNM, as shown in Figure 6.10 [16]. The measured SNM on a fabricated FeFET SRAM chip is plotted in (a) for data (V1, V2) = (Low, High) and (b) for data (V1, V2) = (High, Low). A large SNM of 1.46 V is demonstrated, which is even larger than the 1.27 V SNM of an ideal inverter with infinitely steep slope [16].

The on-current of the read NMOS determines the read speed. During the read operation, the V_{TH} of the read transistor is automatically set to the low value. As a result, the read cell current increases to enhance the read speed, as shown in Figure 6.11.

The sub-threshold current of the off-state FETs shown in Figure 6.12 determines the leakage of the SRAM cell. During the stand-by operation, the V_{TH} of these transistors is automatically set to high values to reduce leakage current. Leakage reduction of 42% is demonstrated by measurement [16].

The enlarged SNM in the ferroelectric SRAM realizes the V_{DD} scaling of 0.11 V and decreases the active power by 32% [16]. Since the transistor count is minimized to six, similar to a conventional SRAM, the ferroelectric SRAM cell also achieves the smallest cell size.

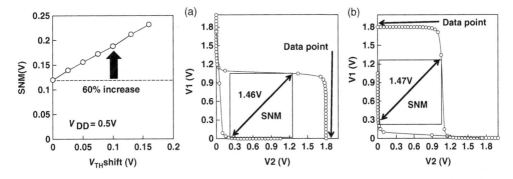

Figure 6.10   Simulated read SNM without the V_{TH} variation. SNM increases by 60% from 0.11 V of the conventional SRAM to 0.19 V with 0.1 V V_{TH} shift. Measured SNM for (a) data (V1, V2) = (Low, High) and (b) data (V1, V2) = (High, Low). IEEE ©. Reprinted with permission, from [16]

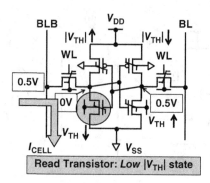

Figure 6.11 High-speed read. The V_{TH} of the read NMOS is automatically set low to increase current for high speed reading. IEEE ©. Reprinted with permission, from [16]

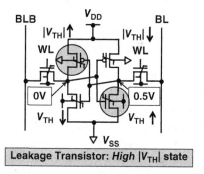

Figure 6.12 Stand-by leakage current reduction. The $|V_{TH}|$ of both leakage NMOS and PMOS is automatically set to high, suppressing the leakage current. IEEE ©. Reprinted with permission, from [16]

6.4 System Consideration: SSD System with Fe-NAND Flash Memory

The write speed in SSD can be enhanced by increasing the number of NAND chips (N_{NAND}) written in parallel; however, N_{NAND} is limited by the chip power consumption in the given SSD power budget because current is also increased in proportion to N_{NAND} [23]. Therefore, minimizing NAND power consumption is essential for realizing high-speed SSD. As the feature size of NAND flash memory decreases, the total bit-line capacitance increases and contributes significantly to the power consumption. The best strategy to decrease power consumption is by reducing V_{CC}. However, in conventional FG-NAND, decreasing V_{CC} below 2 V results in an increasing total power consumption because more charge pump stages are needed (to boost V_{CC} to 20 V for programming). Increasing power consumed by charge pumps surpasses the power saving by lowering V_{CC}.

Fe-NAND can be programmed with a much lower voltage, which helps to suppress the power consumption of the charge pump. As shown in Figure 6.13, N_{NAND} of the Fe-NAND is maximized around $V_{CC} = 1.0$ V. Higher V_{CC} increases the bit-line charging current and the total power consumption, which results in lower N_{NAND} under power constraint. Lower V_{CC} increases the required charge pump stages and the total power consumption, which also reduces N_{NAND}. The optimal N_{NAND} of the Fe-NAND is 6.9 times higher than that of the

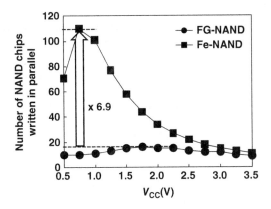

Figure 6.13 Number of NAND chips written in parallel (N_{NAND}) as a function of V_{CC}. The 6.9 times higher write performance is obtained in the Fe-NAND at $V_{CC} = 1.0\,\text{V}$. IEEE ©. Reprinted with permission, from [28]

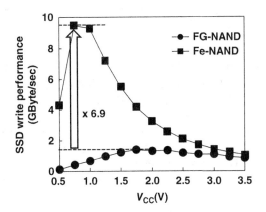

Figure 6.14 SSD write performance of the FG-NAND and the Fe-NAND. The Fe-NAND can achieve 9.5 GB/s at $V_{CC} = 1.0\,\text{V}$ which is 6.9 times higher than the FG-NAND at $V_{CC} = 1.8\,\text{V}$. IEEE ©. Reprinted with permission, from [28]

FG-NAND. Notice that the N_{NAND} of FG-NAND optimizes at higher V_{CC} because of the higher programming voltage of FG-NAND.

The improved N_{NAND} in FeNAND translates directly into a better performance of the FeNAND-based SSD. As shown in Figure 6.14, Fe-NAND Flash SSD achieves a throughput of 9.5 GB/s at $V_{CC} = 1.0\,\text{V}$, significantly higher than that of FG-NAND SSD (below 2 GB/s).

6.5 Perspectives and Summary

FeFET has demonstrated promising potential for both standalone flash memory applications and embedded logic/SRAM applications. The Fe-NAND flash memory discussed in this chapter achieves a much lower operation voltage ($\sim 6\,\text{V}$) than that of conventional FG-NAND ($\sim 20\,\text{V}$), which not only reduces power consumption but also improves reliability (from 10^4 to 10^8 cycles). For SRAM/logic applications, the ferroelectric SRAM has a unique configuration

Table 6.1 Summary of key FeFET memory parameters

Parameter		Value
Device structure		1-Transistor (1 T)
Feature size F	Demonstrated	28 nm [31]
	Projected	The same as CMOS transistor
Cell size	Demonstrated	$4F^2$ [32]
	Projected	$4F^2$
W/E speed	Demonstrated	10 ns [33]
	Projected	Ferroelectric switching time < 100 ps
W/E operation voltage	Demonstrated	± 5 V (20 ns pulse) [31]
	Projected	—
Write energy per bit	Demonstrated	1 fJ [34]
	Projected	0.1 fJ
Cycling endurance	Demonstrated	10^{12} [6]
	Projected	$> 10^{12}$
Retention	Demonstrated	33 d [35]
	Projected	> 10 yr

to adjust V_{TH} by biasing the bodies of NMOS and PMOS to V_{DD} and V_{SS}. During the read and the hold, the V_{TH} of Fe-FETs automatically changes to increase SNM by 60%. In the standby mode, the V_{TH} of the FeFET SRAM cell increases to decrease the leakage current by 42%. During reading, the V_{TH} of the read transistor decreases to increase the cell read current for a fast reading speed. During writing, the V_{TH} of the SRAM cell dynamically changes and assists the cell data to flip. The enlarged SNM enables V_{DD} reduction by 0.11 V, which decreases the active power by 32%. The ferroelectric SRAM also minimizes footprint.

SSD based on Fe-NAND benefits significantly from the low operation voltage of Fe-NAND. A lower supply voltage can be used without incurring higher power consumption from charge pumps. The 86% power reduction in Fe-NAND increases the number of NAND chips written in parallel in SSD by 6.9 times and enhances the SSD performance up to 9.5 GB/s.

The performance of FeFET devices and systems hinges on high-quality materials and integration technologies. The SBT/HAO gate stack utilized in the MFIS and ferroelectric SRAM structures in this chapter plays a critical role to enable functional FeFETs with good data retention. Integration of ferroelectric materials in a standard CMOS process has always been a challenge. A recent discovery of ferroelectric property in Si-doped HfO_2 allows fabrication of FeFETs at a highly scaled node in the state of the art CMOS process [30]. A 28 nm FeFET with TiN/Si:$HfO_2/SiO_2/Si$ gate stack has been demonstrated [31]. A 0.9 V memory window was achieved with ± 5 V 20 ns program/erase pulses. This fabrication-friendly material may provide a promising solution for ferroelectric devices including FeFET.

Table 6.1 summarizes some key parameters of FeFET memories reported in the literature.

References

1. Scott, J.F. and De Araujo, C.A.P. (1989) Ferroelectric memories. *Science*, **246**, 1400–1405.
2. Chung, Y. (2002) Experimental 128-kbit ferroelectric memory with 10^{12} endurance and 10-year data retention. *IEE Proceedings-Circuits Devices and Systems*, **149**(2), 136–142.
3. Kang, Y.M. *et al.* (2006) World smallest 0.34 μm^2 COB cell 1T1C 64Mb FRAM with new sensing architecture and highly reliable MOCVD PZT integration technology, Symp. VLSI Tech., pp. 124–125 Jun.

4. Heyman, P.M. and Heilmeier, G.H. (1966) A ferroelectric field effect device. *Proceedings of IEEE*, **54**(6), 842–848.

5. Wu, S.Y. (1974) A new ferroelectric memory device, metal–ferroelectric–semiconductor transistor. *IEEE Transactions on Electron Devices*, **ED-21**(8), 499–504.

6. Sakai, S. and Ilangovan, R. (2004) Metal–ferroelectric–insulator–semiconductor memory FET with long retention and high endurance. *IEEE Electron Device Letters*, **25**(6), 369–371.

7. Ma, T.P. and Han, J.P. (2002) Why is nonvolatile ferroelectric memory field-effect transistor still elusive? *IEEE Electron Device Letters*, **23**(7), 286–388.

8. Hoffman, J., Pan, X., Reiner, J.W., Walker, F.J., Han, J.P., Ahn, C.H., and Ma, T.P. (2010) Ferroelectric field effect transistors for memory applications. *Advanced Materials*, **22**, 2957–2961.

9. Kaneko, Y., Tanaka, H., Ueda, M., Kato, Y., and Fujii, E. (2011) A dual-channel ferroelectric-gate field-effect transistor enabling NAND-type memory characteristics. *IEEE Transactions on Electron Devices*, **58**(5), 1311–1318.

10. Tsymbal, E.Y. and Kohlstedt, H. (2006) Tunneling across a ferroelectric. *Science*, **313**(5784), 181–183.

11. Chanthbouala, A. *et al.* (2012) Solid-state memories based on ferroelectric tunnel junctions. *Nature Nanotechnology*, **7**, 101–104.

12. Salahuddin, S. and Datta, S. (2008) Use of negative capacitance to provide voltage amplification for low power nanoscale devices. *Nano Letters*, **8**(2), 405–410.

13. Salvatore, G.A., Bouvet, D., and Ionescu, A.M. (Dec. 2008) Demonstration of subthrehold swing smaller than 60mV/decade in Fe-FET with P(VDF-TrFE)/SiO2 gate stack, IEDM Tech. Dig.

14. Rusu, A., Salvatore, G.A., Jimenez, D., and Ionescu, A.M. (Dec. 2010) Metal-ferroelectric-metal-oxide-semiconductor field effect transistor with sub-60mV/decade subthreshold swing and internal voltage amplification, IEDM Tech. Dig., pp. 395–398.

15. Bratkovsky, M. and Levanyuk, A.P. (2006) Depolarizing field and "real" hysteresis loops in nanometer-scale ferroelectric films. *Applied Physics Letters*, **89**(25), 253108-1-3

16. Tanakamaru, S., Hatanaka, T., Yajima, R., Takahashi, M., Sakai, S., and Takeuchi, K. (Dec. 2009) A 0.5 V operation, 32% lower active power, 42% lower leakage current, ferroelectric 6T-SRAM with V_{TH} self-adjusting function for 60% larger static noise margin, IEDM Tech. Dig., pp. 283–286.

17. Sakai, S., Takahashi, M., Takeuchi, K., Li, Q.H., Horiuchi, T., Wang, S., Yun, K.Y., Takamiya, M., and Sakurai, T. (May. 2008) Highly Scalable Fe(Ferroelectric)-NAND Cell with MFIS(Metal-Ferroelectric-Insulator-Semiconductor) Structure for Sub-10nm Tera-Bit Capacity NAND Flash Memories, Non-Volatile Semiconductor Memory Workshop (NVSMW), pp. 103–104.

18. Hatanaka, T., Yajima, R., Horiuchi, T., Wang, S., Zhang, X., Takahashi, M., Sakai, S., and Takeuchi, K. (2010) Ferroelectric(Fe)-NAND flash memory with batch write algorithm and smart data store to the nonvolatile page buffer for data center application high-speed and highly reliable enterprise solid-state drives. *IEEE J. Solid-State Circ.*, **45**(10), 2156–2164.

19. Hatanaka, T., Takahashi, M., Sakai, S., and Takeuchi, K. (Sep. 2009) A zero V_{TH} memory cell ferroelectric-NAND flash memory with 32% read disturb, 24% program disturb, 10% data retention improvement for enterprise SSD, European Solid-State Device Research Conference (ESSDERC), pp. 225–228.

20. Yajima, R., Hatanaka, T., Takahashi, M., Sakai, S., and Takeuchi, K. (Oct. 2009) A Negative Word-line Voltage Step-Down Erase Pulse Scheme with $\Delta VTH = 1/6 \Delta VERASE$ for Enterprise SSD Application Ferroelectric(Fe)-NAND Flash Memories, Inter. Conf. Solid State Dev. and Mater. (SSDM).

21. Noda, S., Hatanaka, T., Takahashi, M., Sakai, S., and Takeuchi, K. (Oct. 2009) A 1.2V Operation 2.43 Times Higher Power Efficiency Adaptive Charge Pump Circuit with Optimized V_{TH} at Each Pump Stage for Ferroelectric (Fe)-NAND Flash Memories Inter. Conf. Solid State Dev. and Mater. (SSDM).

22. Takeuchi, K. (2008) NAND Memories for SSD IEEE Inter. Solid-State Circ. Conf. (ISSCC), Tutorial T7 Feb.

23. Takeuchi, K. (Jun. 2008) Novel co-design of NAND flash memory and NAND flash controller circuits for sub-30nm low-power high-speed solid-state drives (SSD), Symposium on VLSI Circuits, pp. 124–125.

24. Takeuchi, K., Satoh, S., Tanaka, T., Imamiya, K., and Sakui, K. (1999) A negative V_{th} cell architecture for highly scalable, excellently noise-immune, and highly reliable NAND flash memories. *Journal of Solid State Circuits*, **34**(5), 675–684.

25. Takeuchi, K. *et al.* (Feb. 2006) A 56nm CMOS 99mm^2 8Gb multi-level NAND flash memory with 10MB/s program throughput, IEEE Inter. Solid-State Circ. Conf. (ISSCC) pp. 144–145.

26. Agrawal, N., Prabhakaran, V., Wobber, T.d., Davis, J.D., Manasse, M., and Panigrahy, R. (2008) Design Tradeoffs for SSD Performance, USENIX Technical Conference.

27. Itoh, K. (Feb. 2009) Adaptive Circuits for the 0.5-V Nanoscale CMOS Era, ISSCC, pp. 14–20.

28. Miyaji, K., Noda, S., Hatanaka, T., Takahashi, M., Sakai, S., and Takeuchi, K. (2010) A 1.0 V Power Supply, 9.5 GByte/sec Write Speed, Single-Cell Self-Boost Program Scheme for Ferroelectric NAND Flash SSD, IEEE International Memory Workshop, pp. 42–45.

29. Miyaji, K., Noda, S., Hatanaka, T., Takahashi, M., Sakai, S., and Takeuchi, K. (2011) A 1.0 V power supply, 9.5 GByte/sec write speed, single-cell self-boost program scheme for ferroelectric NAND flash SSD. *Solid-State Electronics*, **58**(1), 34–41.

30. Böscke, T.S., Müller, J., Bräuhaus, D., Schröder, U., and Böttger, U. (Dec. 2011) Ferroelectricity in hafnium oxide: CMOS compatible ferroelectric field effect transistors, IEDM Tech. Dig., pp. 547–550.

31. Müller, J. *et al.* (Jun. 2012) Ferroelectricity in HfO$_2$ enables nonvolatile data storage in 28 nm HKMG, Symp. on VLSI Tech., pp. 25–26.

32. Zhang, X., Takahashi, M., Takeuchi, K., and Sakai, S. (2011) First 64 kb ferroelectric-NAND flash memory array with 7.5 V program, 10^8 endurance and long data retention, Inter. Conf. Solid State Dev. and Mater. (SSDM), pp. 975–976.

33. Kaneko, Y., Nishitani, Y., Ueda, M., Tokumitsu, E., and Fujii, E. (2011) A 60nm channel length ferroelectric-gate field-effect transistor capable of fast switching and multilevel programming. *Applied Physics Letters*, **99**(18), 182902 1-3.

34. Fu, W., Xu, Z., Bai, X., Gu, C., and Wang, E. (2009) Intrinsic memory function of carbon nanotube-based ferroelectric field-effect transistor. *Nano Letters*, **9**(3), 921–925.

35. Takahashi, M. and Sakai, S. (2005) Self-aligned-gate metal/ferroelectric/insulator/semiconductor field-effect transistors with long memory retention. *Japanese Journal of Applied Physics*, **44**(25), L800–L802.

7

Nano-Electro-Mechanical (NEM) Memory Devices

Adrian M. Ionescu
Ecole Polytechnique Fédérale de Lausanne, Switzerland

7.1 Introduction and Rationale for a Memory Based on NEM Switch

In parallel with the tremendous progress of ultra-large scale integration (ULSI) of silicon CMOS, the engineering of Micro-Electro-Mechanical Systems (MEMS) for analog, Radio Frequency (RF), and sensing applications arrived to maturity because of the demand and growth of the portable electronic devices (smart phones, tablets and notebooks) [1]. Highly reliable accelerometers and gyroscopes exploiting MEMS structures are today in our portable devices and commercial versions of these devices, with low power consumption have revolutionized smart phone applications and created markets with exponential growth [2]. Radio-Frequency MEMS (RF MEMS) [3,4] offer improved circuit performance (reconfigurability and compactness) at high frequency (MHz to 100 GHz) combined with low-power static consumption and affordable technology costs. They also form another domain where similarly major progress has been experienced in strong connection with new portable applications. The convergence of NEMS and CMOS technology (fabrication platforms using compatible materials and process steps) was an essential factor for this success and it will be even more critical and interesting for future logic and memory applications based on the scaled version of the MEMS called nano-electro-mechanical systems (NEMS) [5,6].

It is accepted that the mechanical information processing has two different forms: (i) *multistate logic*, with the logic states dictated by a spatial configuration of movable objects (the ones mainly discussed in this chapter), and (ii) *vibrational modes* of mechanical elements (with complex equivalent functionality offered by mechanical resonators) [7]. Moreover, devices with small nanometer size generate supplementary interest for massive collective information processing and for building devices addressing the main limitation of dense digital computation with solid-state MOSFET switches. Therefore, the main rationale for a NEM switch is based on the *added value in terms of functionality and performance for computing and storage* compared to the solid-state nanometer MOSFET switch.

Emerging Nanoelectronic Devices, First Edition. An Chen, James Hutchby, Victor Zhirnov and George Bourianoff.
© 2015 John Wiley & Sons, Ltd. Published 2015 by John Wiley & Sons, Ltd.

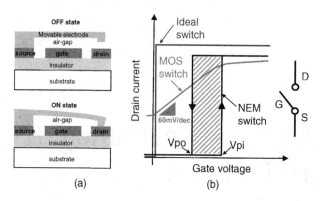

Figure 7.1 Current–voltage characteristics of a three-terminal NEM switch (a possible embodiment of the device is sketched on the left-hand side of the figure) compared to the MOS switch

The question then is what are the specific features of a NEM relay motivating such a choice? One can easily understand the unique characteristics of this device by analyzing the current–voltage characteristics of an electrostatically actuated three terminal NEM relay compared to the MOSFET switch, as depicted in Figure 7.1:

- Quasi-zero off current I_{off} (zero static power) due to the in-series air gap in the off state spatial configuration,
- Quasi-infinite transition between off and on states [no thermal limit of 60 mV/decade for the subthreshold swing, $S = dV_G/dlog(I_D)$] at room temperature) due to the abrupt switching under nonequilibrium between electrostatic and mechanical spring forces,
- High I_{on}, as the device channel is a metal (or highly doped semiconductor), assuming that a low contact resistance can be engineered,
- Hysteresis in the I-V characteristics due to different values of pull-in, V_{pi} (off–on transition) and pull-out, V_{po} (off–on transition) voltages. This last unique property suggests the possibility to use this device as a memory cell, with appropriate control of the hysteresis window and actuation voltages.

NEM switch applications are diverse and put different constraints on the switch design and operation conditions, they can be categorized as it follows: (i) low-cycle NEM switch for power management and reconfigurable circuits, (ii) intermediate-cycle NEM switch for static and nonvolatile memories, and (iii) high-cycle NEM switch for high-performance logic circuit applications. The most demanding in combined scaling and performance is the logic switching, while the memory NEM switch specifications are of interest when addressing some of the open challenges of advanced nonvolatile memories.

This chapter reviews and discusses recent progress in nano-electro-mechanical (NEM) memory devices and their main figures of merit, promises, and challenges. We concentrate on NEM structures that exploit the specific features of movable structures (like their spatial stable reconfiguration and hysteresis) and, therefore, show a clear potential as nonvolatile (NV) memory cells.

A remarkable NEM memory feature is their extremely low power consumption. Compared to their competitors like the NOR FLASH, PCM, and ionic memory, the NEM memories are able to operate with program/erase energies sub-10^{-16} J/operation [9]. For some NEM memory architectures the operation voltages can be scaled into sub-2 V and with read times in the order of ns (corresponding to NEM cells based on suspended nano-structures with resonance frequencies higher than 1 GHz). The lowest achieved voltage operation in a memory cell has been recently reported to be smaller than 1 V [10].

Note that the embodiment of NEM memories may or may not include a storage layer (combined with the mechanical switching of a movable part and injection mechanisms). Nonvolatility in storage-free layer NEM devices can be achieved by the use of other special mechanisms such as adhesion forces and bistability.

A NEM memory exploiting nonvolatile storage defined by a spatial configuration can offer extremely long term storage capability (much beyond the 10 year benchmark), which opens new specific application domains. Additionally, NEM memories can support logic information computing and storage that can operate under extreme conditions: high radiation doses (larger than MRads) and high temperature (larger than 200 °C), which makes them very interesting for applications in space, airborne, automotive and control instrumentation fields. This is one of the unique features of NEM memories because it can address ranges of temperature and harsh environment conditions where other conventional charge-based memories fail to retain data. Furthermore, in a recent work [8], the maximum operation temperature was reviewed for various NV memory technologies and concluded that M/NEM is the unique technology offering storage performance at temperatures higher than 200 °C, Figure 7.2.

It is worth noting that many reported NEM memory cells show a high co-integration capability with silicon CMOS (both in front-end and back-end of the line) because the materials used for their movable layers are conventional conductive materials used in the semiconductor industry (highly doped semiconductors or metals) and the fabrication processes are based on semiconductor compatible sacrificial layers and etching. Some of the reported NEM structures combine movable (bi-stable) electrodes with specific storage layers [oxide–nitride–oxide (ONO), nanodots, ferroelectric, etc.] inspired by the existing NV memories to ensure the cell nonvolatility and by exploiting some advantages (like low off currents) of the mechanical structures.

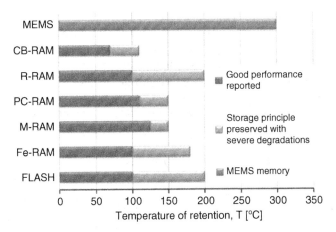

Figure 7.2 Maximum operation temperature of various NV memory technologies pointing out that M/NEM memories are the only solution for high temperature (> 200 °C) applications. © IEEE. Reprinted with permission from [5]

Other novel ideas combining NEM structures and embedded active devices in transistor architectures with movable gates or body have generated a lot of interest; their potential for highly scalable hybrid memory embodiments is reported in another section of the chapter. Particularly such structures open new design space in terms of functionality and trade-off between solid-state device operation and electro-mechanical operation. For instance, switchable-body transistors can avoid the limitations on in-series mechanical contacts (subject to contact degradation and failure when operated in hot switching conditions) and are compatible with CMOS front-end technology.

A special section of this chapter is dedicated to discussing some more advanced yet speculative memory concepts, based on movable nanostructures [like semiconducting nanowires (NW) and carbon nanotubes (CNTs)] that can exploit nano-scale specific balance of elasto-static and van der Waals energies to define the stable states and have intrinsically smaller size. The recent emergence of novel 2D materials like graphene is creating further design opportunities for nano-scale memories.

The open challenges of the NEM memories are listed and discussed in the last section of the chapter. Beyond the memory cell scaling and the required electrical properties related with nonvolatile storage, the reliability and packaging of electro-mechanical devices are adding further constraints for their successful take-up in industrial applications. These devices have failure mechanisms (such as stiction or contact resistance degradation) of their contacts that depend on both the quality and nature of the surfaces in contact, the current densities and the environmental conditions. Therefore specific contact engineering and control of environment are needed. As for the achieved endurance (cycling), current reports mention experimental investigations from tens up to 10^5 cycles under diverse condition (hot and cold switching). Significant progress has been made in the field of process control for the 3D integration of micro-electro-mechanical devices and their wafer level packaging enabling the integration of future NEM memory cells as a back-end of line (BEOL) module on silicon CMOS, with high packaging density and low fabrication costs.

7.2 NEM Relay and Capacitor Memories

Historically, the first integrated nonvolatile memory cell based on a micro-electro-mechanical structure was proposed in 1990 [11]; it consisted of a thin micro-machined bridge elastically deformed in such a way that it has two stable mechanical states to which the logical levels "0" and "1" could be assigned. The state of the bridge is changed using electrostatic forces and it may be read out by sensing the corresponding capacitance (see Figure 7.3a). The author observed that such cell is completely immune to electromagnetic fields and mechanical shocks, and the stored data could be retained for an unlimited time. However, this demonstrator device was large (hundreds of μm^2) and high required operation voltages of the order of 30 V.

A more sophisticated MEMS memory cell based on mechanical *bistability* and a *read-out integrated transistor* read was proposed [12–14]) based on a μm down to sub-μm-long NEMS structure, combined with nanocrystalline (nc) Si quantum dots. A basic concept of nonvolatile memory based on the mechanical bistability of a micro-machined bridge has been reported and demonstrated. This NEMS memory features a suspended SiO_2 beam formed in the cavity as a floating gate, which incorporates nc-Si dots for charge storage (Figure 7.1b,c). Once an electron is stored in such a small Si nanodot, another electron transfer probability into the dot is strongly reduced owing to the Coulomb blockade effect even at room temperature. The authors

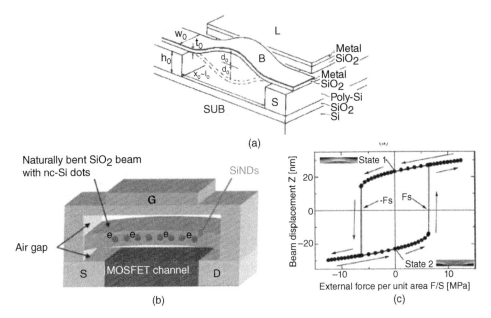

Figure 7.3 (a) Halg's [11] and (b) Oda's [12] bistable MEMS memory concepts; (c) mechanical hysteresis exploited in memory device (b). © IEEE. Reprinted with permission from [11] and [12]

reported that the amount of charges in the beam is determined by the number of nc-Si dots and, with ~8 nm nc-Si dots, densities of about 10^{11}–10^{12} cm^{-2} in a monolayer can be achieved.

In order to inject electrons into the nc-Si dots, the authors applied a high voltage to the gate electrode for initialization. Then, the electrons are injected into the nc-Si dots through the sidewalls, or the gate electrode contacts the floating gate and the electrons are injected through the SiO$_2$ of the floating gate. The beam is buckled either upward or downward, and both ends are clamped at the cavity sidewalls. When the gate voltage is applied, the charged beam moves in the cavity via electrostatic interactions between electric field in the cavity and the charge stored in the beam.

The write/erase addressing scheme is realized by a combination of gate voltage and substrate voltage (Figure 7.4a). In write/erase operation the gate voltage is positive/negative, respectively. Substrate voltage is negative because of the n type transistor. In the write operation, both the memory cell's gate and substrate voltages are applied to switching the state. In erase operation, only a gate voltage is required to switch the state. A positional displacement of the beam changes the surface potential of the metal oxide–semiconductor field effect transistor (MOSFET) placed underneath and is therefore sensed as a shift of its threshold voltage. The conditions for the read-out addressing scheme are selected by a combination of gate voltage and drain voltage that can be scaled into the sub-5 V (Figure 7.4b). The memory state is dictated by the drain current value. Write and erase operations of the NEMS memory are not associated with charge tunneling via the gate oxide and, therefore, do not cause any gate oxide deterioration as in conventional flash memory.

Another MEMS memory concept is the so called NEMory cell proposed in 2007 [15,16]. The NEMory cell consists of four elements (Figure 7.5): a read word line (RWL) which serves as the upper electrode, a suspended bit line (BL) which is a movable mechanical beam, a

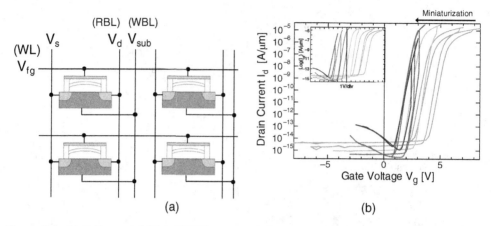

Figure 7.4 (a) Cell array of Oda's NEMS memory. WL (gate) and RBL (drain) are used for readout selection. WL (gate) and WBL (substrate) are used for write/erase selection. (b) $I_d - V_g$ characteristics for various L at 50–1500 nm. Inset shows $I_d - V_g$ characteristics shifted to negative bias direction. Shift amount increases 1 V per curve. © IEEE. Reprinted with permission

dielectric oxide/nitride/oxide (ONO) stack which serves to store a fixed amount of charge, and a write word line (WWL) which serves as the lower electrode. Information is stored in the form of the BL position: If the BL is pulled down then the cell is in the "0" state; if the BL is released it is in the "1" state. It should be noted that the role of the charge-storage layer in a NEMory cell is distinctly different from its use in conventional Flash memory. The charge stored within the

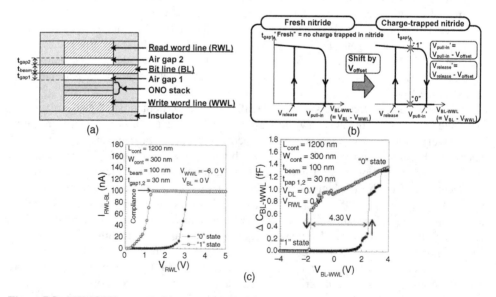

Figure 7.5 NEMORY concept: (a) cross-section of the device architecture featuring airgap, ONO stack and read, bit and write word lines, (b) qualitative characteristics of the fresh and charge-trap nitride, and (c) experimental confirmation of both "fresh" and "charge-trapped" states

ONO layer is *not used to store information*; rather, it is used to shift the hysteretic gap-closing behavior of the BL in order to achieve nonvolatile storage, that is, two stable states under zero applied voltage. The amount of charge stored in the ONO layer is constant and can be introduced during or after fabrication (i.e., with a single charging operation). Based on simulations, the authors of NEMory cell reported 100 nm cantilever beams operating at 3 V within 0.4–4.0 ns response time and demonstrated that the most of the memory energy is consumed as electrical energy being much less than the one required in program/erase in conventional Flash memory cells.

Some of the most scaled NEM memory demonstrations were reported [17] in 2007. Two types of titanium nitride (TiN) based NEM hysteretic switches, among the smallest ever made by typical top-down CMOS fabrication technology were demonstrated. NEM cantilever and clamp switches, with 30 nm TiN beam and 20 nm air gap were successfully fabricated and electrically characterized (see Figure 7.6a). The fabricated switches showed excellent on/off current characteristics with ultra-low off current (due to the in-series air gap), experimental sub-threshold slope <3 mV/decade, and on/off current ratio $>10^5$. The reported endurance is in the order of several hundred switching cycles under dc and ac bias conditions, in ambient air. The memory cell consists of a cantilever BL (CBL) connected to a buried BL (BBL) through direct contact (DC), a WL for read (RWL), and a WL for write (WWL), including a charge trapped oxide/nitride/oxide (ONO) layer (Figure 7.6b). The charge is trapped in the ONO layer using an ion implantation process during the fabrication or Fowler–Nordheim (FN) tunneling by

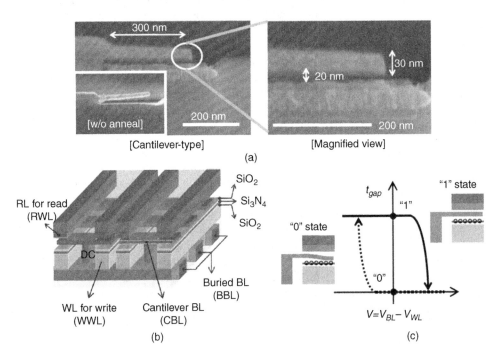

Figure 7.6 (a) Fabricated TiN suspended-beam memory cells. (b) Schematic diagram of the proposed cell array structure. (c) Hysteresis curves (applied voltage versus air gap thickness) of the SBM cell with the charge trapped in ONO layer, showing the operation mechanism of the SBM cell. © IEEE. Reprinted with permission

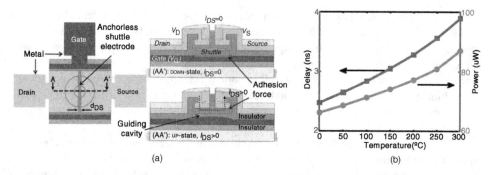

Figure 7.7 (a) Principle, design and operation of the Shuttle NEM NV memory cell. (b) Memory delay and power consumption versus temperature, showing high robustness of operation up to 300 °C

applying the high voltage between the WWL and the CBL after finishing the fabrication. A shift of the typical hysteresis curve by the charge trapped in ONO layer is obtained.

Recently, a very interesting NEM NV memory cell based on the switching of a free electrode between two stable states has been proposed [8,18,19]; the anchorless design (Figure 7.7a) of this particular memory cell is very compact and scalable. The device principle exploits the switching of a free electrode between two stable states; this shuttle electrode has no mechanical anchors and smoothly switches two stable positions being guided inside an insulator pod. Adhesion forces between the shuttle and fixed electrodes serve to hold the shuttle in stable positions.

The authors optimized the design and proposed various control schemes that are discussed elsewhere [19] together with their performance analysis. For instance, Figure 7.7 shows the read power and delay using a 1.8 V supply. This memory cell consumes tens of μW and is very robust with respect to temperature; when the temperature changes from 0 to 300 °C, the read delay increases from 2.5 to 3.8 ns (52%). The proposed memory can operate at a maximum read frequency of around 250 MHz.

7.3 NEM-FET Memory

Another conceptual direction of NEM memories consists in the so-called suspended gate (SG) MOSFET operated as one-transistor memory cell [20,21] and having full compatibility with CMOS front-end. The SG-MOSFET memory principle exploits the electro-mechanical hysteresis in the drain-voltage characteristics of a movable-gate transistor to define a volatile memory cell (Figure 7.7c) or the storage of charge in a storage layer included in the gate stack for nonvolatile operation (Figure 7.7d). The specific current transitions form low to high states and vice versa are subthermal, with voltage swings that are much lower than 60 mV/decade at room temperature. Fabricated 1 T SG-FET memory cells have been reported at micrometer scale (Figure 7.8) and simulations have been used to show the scalability of this device into sub-100 nm size, with operation voltages less than 2 V. A remarkable feature of the 1 T SG-MOSFET memory has been experimental demonstration of 10^5 cycling without major degradation of the characteristics. Further investigations [22] highlighted the important role of the oxide charging in SG-FET and the need of selecting an appropriate storage material in the gate stack, like nitride or ferroelectric layers instead of oxides to improve the long-term storage capability.

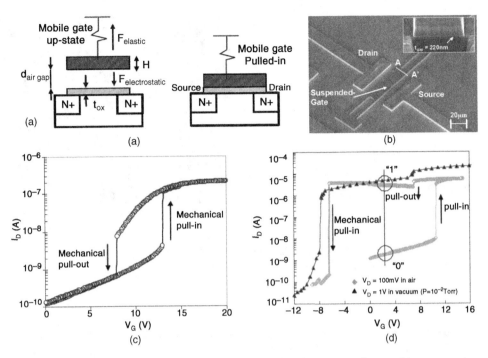

Figure 7.8 1 T SG-FET memory: (a) cross sections depicting the concept of a movable gate-transistor, (b) fabricated SG-FET with metal gate electrode, (c) device characteristics and hysteresis without storage layer in gate stack, (d) device characteristics and hysteresis when storage takes place in the gate dielectric

A higher performance version of the SG-MOSFET consists of designing this device architecture on Silicon On Insulator (SOI) with a switchable body and two lateral gates separated from the body by nanometer size air gap. The very first demonstration of switchable body double-gate MOSFET at micrometer scale [23] exploited the hysteretic characteristic with abrupt transitions to propose a new circuit principle for data transmission.

Later, the monolithic integration of NEMS and CMOS for a mechanically flip-flopped fin memory transistor was demonstrated [24,25] at nanometer scale. A FinFET with two independent gates using air gap insulators defines the fin flip-flop actuated channel that can be used for a 1 T NEM memory cell. Figure 7.9 shows the mechanical states of the proposed memory device. The fin is located at the center, being straight in the initial state. After a write operation, the fin can come into contact with gate G1 for bit "1" (pull-up) or gate G2 for bit "0" (pull-down). The SEM images from Figure 7.9b clearly show the binary mechanical memory states. The drain current modulation is practically used to define the stable memory states. The fabricated switchable fin NEM devices featured operation for few hundred cycles before failure.

Reference [26] reports other work exploiting the lateral switching of suspended cantilever for nonvolatility in a nRAM cell via work–function engineering. This eliminates the need for cell selection devices in a crossbar array which uses a displacement current-based read scheme. The configuration of the nRAM is such that *the elastic potential energy due to the beam bending is reversibly used for switching*, which enables the combination of ultralow operation voltages with high switching speed. However, this interesting principle was not fully

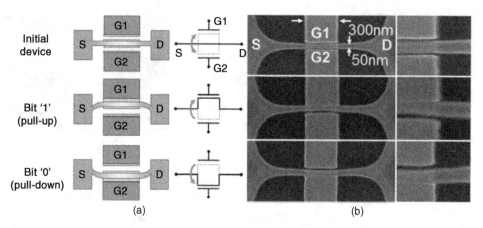

Figure 7.9 Illustration of Fin mechanical states for 1 T memory application: (a) device schematic and symbol, (b) SEM images. The fin remains straightened at the initial state. Bit "1" and bit "0" are distinguished when the fin is in contact with G1 and G2, respectively. © IEEE. Reprinted with permission after [24]

experimentally demonstrated; it was suggested that the manufacturing challenge for nRAM is the precise adjustment of pullout voltage ($V_{po} < E_G/2q$) with a minimal standard deviation (ΔV_{po}) across the wafer and between different process runs.

7.4 Carbon-based NEM Memories

Carbon nanotubes (CNTs) and graphene have attracted major interest for building NEM devices, due to their outstanding electrical and mechanical properties in 1D and 2D structures, respectively. For instance high stiffness (Young modulus, E, near 1 TPa), low-density, defect-free structure, nm cross-section and outstanding electrical thermal conductivity can be simultaneously achieved in a carbon nanotube, making it electro-mechanically unmatchable by a traditional semiconductor nanostructure based on silicon. For NEM memories these properties are of interest because they can be exploited by scaling to obtain high density and ultra-low power operation.

Nantero Inc. proposed some of the first electromechanical memory arrays using nanotube ribbons [27,28]. Their nanotube-based random access memory (NRAM) was composed of movable nanotube bridges, freely suspended between source and drain electrodes, in the vicinity of a gate. The van der Waals (vdW) interactions act between the CNT bridge and the oxide. Electrical charge can be induced in the suspended CNT bridge by a voltage applied to the gate electrode. The resulting capacitive force between the CNT bridge and gate bends the CNT bridge and brings the CNT bridge into the vdW contact above the gate electrode. Figure 7.10 shows the simple schematics of the NRAM operations and energy diagram. In the total energy plot obtained from the superposition of the elasto-static and vdW energies, the point of inflection is found in the total energy. When the vdW energy between the CNT and the oxide increases, the well depth of the potential energy also increases and an increased nonvolatility of the NRAM is achieved. However, despite its high initial attractiveness, progress was very slow to transform it in an industrial memory technology with clearly demonstrated experimental figures of merit.

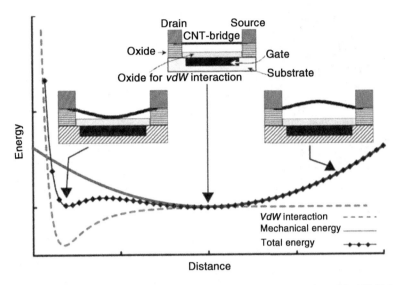

Figure 7.10 NRAM schematics originated by Nantero. The simple schematics of the NRAM operations and the energy diagram: the solid, dashed, and symbolic lines indicate the elasto-static energy, van der Waals energy, and total energy, respectively

Another CNT NEM memory concept was proposed by the University of Cambridge [29]. They reported a nano-electro-mechanical switched capacitor structure based on vertically aligned multi-walled carbon nanotubes in which the mechanical movement of a nanotube-relative to a carbon nanotube-based capacitor defines "ON" and "OFF" states (see Figure 7.11). Their carbon nanotubes were grown with controlled dimensions at predefined locations on a silicon substrate in a process that could be made compatible with existing silicon technology. The vertical orientation of the device allows for a significant decrease in cell area over conventional devices. The authors were able to write data to the structure and it should be possible to read data with standard dynamic random access memory sensing circuitry.

Based on modeling of the capacitance in the CNT NEM DRAM cell (CNT diameter, 60 nm; length, 1.6 μm; SiNx, 40 nm), the authors derived a value of 0.59 fF, which would give an available potential of 2.4 mV for bit line sensing in a conventional DRAM design. However, with minimum acceptable values of capacitance and bias difference for gigabit-level DRAMs of 10–15 fF and 60–80 mV, respectively, a challenge was made to reach these values in the CNT NEM-DRAM cell by replacing the SiNx dielectric layer with ultra-thin (10 nm) layers of high-k dielectrics.

7.5 Opportunities and Challenges for NEM Memories

In the last decade, NEM memories experienced major progress in terms of experimental proof of concepts, embodiments in CMOS compatible technologies, validation of new ideas and experimental demonstration of promising figures of merit. They stand particularly high as memory solutions for harsh environments (high temperature and high doses of radiation). However, the majority of the reported architectures and studies remain at the device level or basic memory blocks.

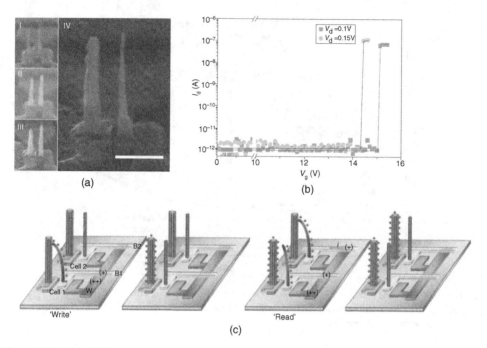

(a)

(b)

(c)

Figure 7.11 (a) CNT NEM switch based on vertically aligned multiwalled carbon nanotubes and corresponding I-V characteristics. (b) WRITE and READ operations to a NEM memory cell. (c) Data is written through the voltage levels of the bit lines and the word lines. When the write voltage is removed, the nanotube returns to a vertical position. Cell 1 now stores a "1" and cell 2 stores a "0"

Overall, NEM memories (especially the silicon-based embodiments) show clear promise for:

- *Power savings* beyond the capabilities of all known nonvolatile memory competitors (PCM, FLASH, ionic memory, RRAM). The zero-I_{off} current of hysteretic NEM relays or the shift induced in the NEM hysteretic I-V characteristics by various charge storage mechanisms coupled to a NEM structure can be smartly exploited in bi-stable memory architectures to achieve energies per bit program/erase close to 10^{-17} J, which is lower than for any other NV memory candidates.
- *High robustness in harsh environments* (high temperature and high radiation). By their nature a nonvolatile M/NEM memory cell is one of the most robust memory architectures for preserving the stored information in very harsh environments; their figures of merits in this respect are unmatched.
- *Co-integration with silicon CMOS* (in both front-end and back-end of the line). The various surface micromachining techniques developed for N/MEMS technology are applicable and have enabled today the merging of advanced CMOS with sub-100 nm movable structures. Their process temperature is low and compatible with CMOS. The 3D integration of NEM memory with CMOS for future embedded memory applications is possible.
- *Voltage scaling below 2 V and switching time below 10 ns*. The voltage scaling of electrostatic actuation imposes aggressive scaling of the actuation air gaps down to tens of nm. However, the recent nanostructure cells based on movable fins show that reasonable

voltage scaling can be experimentally achieved in structures with gaps higher of the order of 10 nm. In bi-stable structures switching voltage scaling down to 2 V was suggested with realistic technology and switching times sub-10 ns.

Among the open challenges of NEM one can cite the following (many of these are generic issues of MEMS and NEMS):

- *Limited cycling and endurance.* Very few of the reported NEM proof of concepts have been tested in terms of industrial requirements for endurance and cycling. In some of the reported devices (like Fin-FACT) some design trade-off could exist between the retention time and the endurance. However, some of the reported NEM memory cells have been cycled up to tens of thousands of times, showing promising characteristics. In general, problems are reported after multiple cycling with electrical charging (shift of electrical characteristics) or with contact failures, not concerning mechanical fatigue or mechanically specific failures.
- *Reliability.* The understanding of the main mechanisms for failures and specific accelerated reliability tests should be further defined and investigated for NEM devices in general. However, significant technical progress in the field of NEM switches, showing high reliability after specific surface engineering could be also applied to the NEM memory cells. This issue can be solved in the near future.
- *Need of adapted packaging (zero-level and wafer-level) with controlled ambience (hermetic).* Many of the reported NEM memory devices have been tested in laboratory conditions (vacuum of ambient air) and would require aprotective thin film (zero-level) and wafer-level packaging for real applications. The integration of NEM memory as a back-end of line module would simplify greatly the packaging limitations in the future; this issue can be solved in the near future.

In general, it important to note that NEM memories can have better speed and much lower energy than conventional flash memory, being co-integrable with CMOS. This observation could lead to the identification of some interesting future opportunities and applications of NEM memories in mobile communications or internet of things nodes, as a generic domain, where energy-efficient technologies offering substantial power savings could prevail.

References

1. Bryzek, J. (April 2 2013) Roadmap for the Trillion Sensor Universe, iNEMI Spring Member Meeting and Webinar Berkeley, CA.
2. Paris Tech Review (2013) http://www.paristechreview.com/2013/12/09/the-mems-revolution/ (accessed 16 September 2013).
3. Nguyen, C.T.-C. (1999) Frequency-selective MEMS for miniaturized low-power communication devices. *IEEE Transactions on Microwave Theory and Techniques*, **47**(8), 1486–1503.
4. Nguyen, C.T.-C. (2007) MEMS technology for timing and frequency control. *IEEE Transactions on Ultrasonics, Ferroelectrics and Frequency Control*, **54**(2), 251–270.
5. Roukes, M.L. (2004) Mechanical Computation, Redux?, Technical Digest of IEEE International Electron Devices Meeting, 2004, IEDM, pp. 539–542.
6. Roukes, M.L. (2007) Nanoelectromechanical systems: Potential, progress, & projections, IEEE 20th International Conference on Micro Electro Mechanical Systems, MEMS 2007, pp. 93–94.
7. Ionescu, A.M. (2013) Chapter "Nanoelectromechanical Systems (NEMS)", in *MEMS-Based Circuits and Systems for Wireless Communication* (eds A. Kaiser and C. Enz), Springer, New York.
8. Pott, V., Geng Li, Chua, Vaddi, R. *et al.* (2012) The shuttle nanoelectromechanical nonvolatile memory. *IEEE Transactions on Electron Devices*, **59**(4), 1137–1143.

9. Liu, T.J. King. (October 9th 2009) Micro-Electro-Mechanical Memory Technology for Energy-Efficient Electronics, EE298-12 Seminar, UC Berkeley.

10. Lee, J.O., Song, Y.-H., Kim, M.-W. *et al.* (2013) A sub-1-volt nanoelectromechanical switching device. *Nature Nanotechnology*, **8**, 36–40.

11. Halg, B. (1990) On a micro-electro-mechanical nonvolatile memory cell. *IEEE Transactions on Electron Devices*, **37**(10), 2230–2236.

12. Tsuchiya, Y., Takai, K., Momo, N. *et al.* (2004) Proc. IEEE Silicon Nanoelectronics Workshop, p. 101.

13. Tsuchiya, Y., Takai, K., Momo, N. *et al.* (2006) Nanoelectromechanical nonvolatile memory device incorporating nanocrystalline Si dots. *Journal of Applied Physics*, **100**, 094306.

14. Nagami, T., Mizuta, H., Momo, N. *et al.* (2007) Three-dimensional numerical analysis of switching properties of high-speed and nonvolatile nanoelectromechanical memory. *IEEE Transactions on Electron Devices*, **54**(5), 1132–1139.

15. Choi, W.Y., Osabe, T., and Liu, T.-J.King. (2008) Nano-electro-mechanical nonvolatile memory (NEMory) cell design and scaling. *IEEE Transactions on Electron Devices*, **55**(12), 3482–3488.

16. Jang, W.W., Lee, J.O., and Yoon, J.B. (2007) A DRAM-like mechanical non-volatile memory, Proceedings of Conf Solid-State Sens & Actuat, 153–156.

17. Jang, W.W., Yoon, J.B., Kim, M.S. *et al.* (2008) NEMS switch with 30 nm-thick beam and 20 nm-thick air-gap for high density non-volatile memory applications. *Solid-State Electronics*, **52**, 1578–1583.

18. Vaddi, R., Pott, V., Geng Li, C. *et al.* (2012) Design and scalability of a memory array utilizing anchor-free nanoelectromechanical nonvolatile memory device. *IEEE Electron Device Letters*, **33**(9), 1315–1317.

19. Anh Tuan, Do., Jayaraman, K.G., Pott, V. *et al.* (2013) An improved read/write scheme for anchorless NEMS-CMOS non-volatile memory, IEEE International Symposium on Digital Circuits and Systems, ISCAS. 2013, pp. 1456–1459.

20. Abele, N., Fritsch, R., Boucart, K. *et al.* (2005) Suspended-gate MOSFET: bringing new MEMS functionality into solid-state MOS transistor, Technical Digest of IEEE International Electron Devices Meeting, IEDM, pp. 479–481.

21. Abele, N., Villaret, A., Gangadharaiah, A. *et al.* (2006) 1T MEMS Memory Based on Suspended Gate MOSFET, Technical Digest of IEEE International Electron Devices Meeting, IEDM, pp. 1–4.

22. Molinero, D., Abele, N., Castaner, L., and Ionescu, A.M. (2008) Oxide charging and memory effects in suspended-gate FET, IEEE 21st International Conference on Micro Electro Mechanical Systems, MEMS, pp. 685–688.

23. Grogg, D., Meinen, C., Tsamados, D. *et al.* (2008) Double gate movable body Micro-Electro-Mechanical FET as hysteretic switch: Application to data transmission systems, 38th European Solid-State Device Research Conference, ESSDERC, pp. 302–305.

24. Han, J.-W., Ahn, J.-H., Kim, M.-W. *et al.* (2009) Monolithic integration of NEMS-CMOS with a Fin Flip-flop Actuated Channel Transistor (FinFACT), 2009 IEEE International Electron Devices Meeting (IEDM), pp. 1–4.

25. Han, J.-W., Ahn, J.-H., and Choi, Y.-K. (2010) FinFACT - fin flip-flop actuated channel transistor. *IEEE Electron Device Letters*, **31**(7), 764–766.

26. Akarvardar, K. and Wong, H.-S.P. (2009) Ultralow voltage crossbar nonvolatile memory based on energy-reversible NEM switches. *IEEE Electron Device Letters*, **30**(6), 626–628.

27. Segal, B.M., Block, D.K., and Thomas, R. (2004) Electromechanical memory array using nanotube ribbons and method for making same, US Patent submission number 2004–850100 (www.nantero.com).

28. Kanga, J.W., Leeb, J.H., Leeb, H.J., and Hwanga, H.J. (2005) A study on carbon nanotube bridge as a electromechanical memory device. *Physica E*, **27**, 332–340.

29. Jang, J.E., Cha, S.N., Choi, Y.J. *et al.* (2008) Nanoscale memory cell based on a nanoelectromechanical switched capacitor. *Nature Nanotechnology*, **3**, 26–30.

8

Redox-based Resistive Memory

Stephan Menzel[2], Eike Linn[1], and Rainer Waser[1,2]

[1]*Institute of Electronic Materials II, RWTH Aachen University, Germany*
[2]*Peter Grünberg Institut (PGI-7), Forschungszentrum Jülich, Germany*

8.1 Introduction

Redox-based resistive memories (ReRAM) are a highly promising class of emerging memories [1]. ReRAM cells offer nonvolatile data storage in terms of at least two different resistances, a high resistive state (HRS) and a low resistive state (LRS). By applying appropriate voltage pulses one can switch between these resistance states. Intermediate states are feasible, too, enabling multilevel operation modes. ReRAM cells consist of a simple to fabricate Metal–Isolator–Metal (MIM) stack (see Figure 8.1), in which the insulator is an ion conductor at high fields and/or high temperature. These stacks enable 3D stackable crossbar array architectures. However, simple 1R array implementations suffer from the sneak path problem. Thus ReRAMs cells either require a two-terminal select device (1S1R or 1 CRS) in a passive crossbar architecture configuration, or a select transistor (1T1R) in an active memory matrix (Figure 8.1). The individual ReRAMs cells offer a high scaling potential below feature sizes of $F < 10\,\mathrm{nm}$, retention times >10 years, and excellent endurance properties ($>10^{12}$ cycles). In this chapter we will give an overview of the present knowledge of the physical switching mechanism, the modeling of ReRAMs and the state of the art of device performance.

Although all ReRAM cells are based on redox processes the actual device behavior strongly depends on the applied materials [2]. Hence, it is reasonable to define three subclasses (compare Figure 8.2) [3]:

- Electro-chemical metallization (ECM) cells
- Valency change mechanism (VCM) cells
- Thermo-chemical mechanism (TCM) cells.

ECM cells are dominated by the electrodes, offering an active Cu or Ag electrode on one side and an inert (e.g., Pt, TiN) electrode on the other. In contrast, VCM and TCM cells are

Emerging Nanoelectronic Devices, First Edition. An Chen, James Hutchby, Victor Zhirnov and George Bourianoff.
© 2015 John Wiley & Sons, Ltd. Published 2015 by John Wiley & Sons, Ltd.

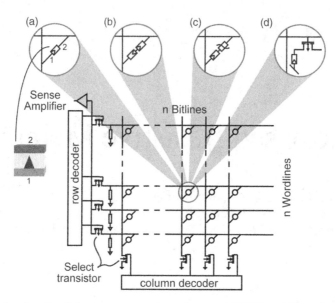

Figure 8.1 Redox-based resistive memory matrix and external CMOS periphery (decoders and sense amplifiers). In the memory array either (a) a single ReRAM cell (1R), (b) a complementary resistive switchs (1CRS), (c) a selector plus ReRAM cell (1S1R), or (d) a transistor plus ReRAM cell (1T1R) is present at each junction of word and bit lines

dominated by the sandwiched material, that is, the chalcogenide (e.g., TaO_2, $SrTiO_3$). From an application perspective, the polarity of *I-V* characteristic – either bipolar (ECM, VCM) or unipolar (TCM) – is also suited for classifying ReRAM devices.

For practical memories one will either use ECM or VCM cells since power consumption is much lower compared to TCM cells and endurance properties are superior. Hence, we will restrict our considerations to ECM and VCM cells in this chapter. In Section 8.2 we describe the physical fundamentals of redox memories and highlight the common generic properties

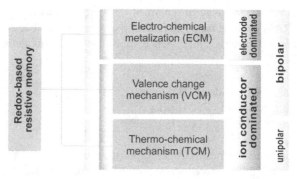

Figure 8.2 Classification scheme for redox-based resistive memory. The memory cells are either dominated by the sandwiched material (VCM and TCM) or by the electrode material (ECM). An alternative classification is based on the *I-V* characteristic which can be unipolar (TCM) or bipolar (VCM, ECM) Reproduced with permission from [3], © 2012 Wiley-VCH

of all redox devices in terms of conduction mechanism, generic switching properties, and dynamical device modeling. In Sections 8.3 and 8.4 we then emphasize the specific features of ECM and VCM, respectively. Finally, exemplary devices offering the excellent performance in the categories feature size, cell area, switching time retention time, write cycles, and multilevel feasibility are presented in Section 8.5.

8.2 Physical Fundamentals of Redox Memories

ECM and VCM cells are based on the same basic physical processes, which are discussed in this section. The most relevant processes in describing/modeling the resistive switching behavior are the conduction mechanisms in HRS and LRS as well as the processes determining the switching kinetics. In addition, ReRAMs show some generic switching characteristics that are related to the nonlinear switching kinetics. These physical fundamentals and switching characteristics build up the basis for modeling ReRAMs. A critical overview of the various modeling activities is given at the end of this section.

8.2.1 Electronic Conduction Mechanisms

The I-layer in ReRAMs is a mixed ionic electronic conductor (MIEC). Typically, the electronic conductivity dominates and accounts for the resistance states of the ReRAM. This is always true in the LRS, whereas the ionic conductivity can be higher in the HRS in some ECM systems, for example, in Cu/SiO$_2$/Pt cells [4]. The ionic conductivity is very low, but the motion of ions triggers the resistive switching effect. In this subsection, we focus on the electronic current contribution. The possible electronic conduction processes in MIM structures are illustrated in Figure 8.3. It can be distinguished between interface and bulk dominated conduction processes. The former are: (i) electron tunneling into the conduction band, (ii) thermionic emission, (iii) direct tunneling from electrode to electrode, and (iv) electron injection into trap states.

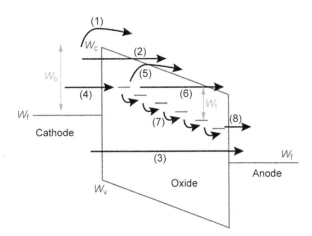

Figure 8.3 Possible electronic conduction processes in MIM structures. (1) Thermionic emission. (2) Electron tunneling into the conduction band. (3) Direct tunneling from electrode to electrode. (4) Electron injection into trap states. (5) Thermionic emission from trap to conduction band. (6) Direct tunneling from trap to conduction band. (7) Electron hopping or electron tunneling from trap to trap. (8) Tunneling from trap to anode

As bulk related processes (v) electron hopping, (vi) electron tunneling from trap to trap, (vii) band conduction, or (viii) Poole–Frenkel emission might occur. In a real MIM device various conduction processes are always present simultaneously with different balance, which depends also on temperature, applied electric field, and voltage polarity. During the resistive switching process this balance changes. The dominating conduction mechanism might change, for example, from thermionic emission to ohmic conduction. Alternatively, the dominating conduction mechanism might stay the same and only the current density level changes. For example, the thermionic emission barrier and in turn the current density level is altered by the variation of the doping concentration close to the interface due to the Schottky effect. In VCM cells oxygen vacancies (or cation interstitials) can be shallow donors or create deep trap level within the band gap. Changing the amount or distribution of oxygen vacancies thus modifies the cell resistance. The bulk related resistance or electronic transport mechanism is related to oxygen vacancy concentration, which is linked to the conduction center concentration within the insulator. Reference [5] discriminates between five different transport mechanisms. If the oxygen vacancy concentration and in turn the conduction center concentration is below a percolation threshold, the conduction centers are separated from each other and the oxide is in an insulating state. By increasing the oxygen vacancy concentration a percolation threshold is reached increasing the conductivity. By a further increase of conduction centers the conduction mechanism changes to strongly localized nearest neighbor hopping, weakly localized variable range electron hopping, and finally to a metallic conduction mechanism.

In general the electronic transport in MIM cells can be either locally confined along so called conducting filaments or spread over the whole sample area. Certainly, a leakage current parallel to a conducting filament is always present and their balance again depends on electric field strength, voltage polarity, and temperature. In addition, the conducting filaments can be so small that quantized conduction effects appear.

8.2.2 Ionic Conduction Mechanism

Ions move by hopping from one site to another vacant site within the insulating layer. The ions thus have always to overcome a migration barrier ΔG_{hop}. The ionic transport is described mathematically by the Mott–Gurney law of ion hopping

$$j_{hop} = 2\, zecaf \exp\left(-\frac{\Delta G_{hop}}{k_B T}\right) \sinh\left(\frac{aze}{2k_B T}E\right), \qquad (8.1)$$

where a is the mean hopping distance, f the attempt frequency, z the charge number of the hopping ion and c the ion concentration. Due to the intrinsic landscape of ionic crystals the hopping distance is in the range of inter atomic distances. The electric field E is the external electric field and not the local Lorentz field [7]. If the electric field E is lower than the characteristic field $E_0 = 2k_B T/aze$, the ionic current density depends linearly on the electric field. In contrast, the dependence becomes exponential when $E > E_0$. The ionic current density also increases exponentially if the temperature increases. Thus, the ionic transport can be either temperature or field enhanced. Note that the ion velocity cannot exceed the speed of sound.

8.2.3 Switching Kinetics

All types of ReRAMs exhibit strongly nonlinear switching kinetics for SET and RESET operation, that is, if a voltage pulse with a certain amplitude and length can be used to switch a cell, a change in the voltage amplitude will highly over proportionally change the pulse length required for switching. Here, we discuss the processes which may be the origin of this nonlinearity of the kinetics. In general, several of these processes are present in a ReRAM device, but the kinetics are determined by the slowest one. Depending on the electrical (applied voltage) and ambient conditions, the rate-limiting process can change also in a single ReRAM device.

Since the movement of ions triggers the resistive switching effect, the *ionic transport* may be a potential rate limiting step. As discussed in the previous subsection the ionic transport can be either temperature or electric field enhanced (cf. Equation 8.1).

Another rate-limiting step can be an *electron transfer reaction* which occurs at the metal/insulator interface. This process depends on the overpotential η_{et} and can be mathematically described by the Butler–Volmer equation:

$$j_{et} = j_{0,et} \left[\exp\left(\frac{(1-\alpha)ze}{k_B T} \eta_{et} \right) - \exp\left(-\frac{\alpha ze}{k_B T} \eta_{et} \right) \right]. \tag{8.2}$$

Here, z denotes the number of electrons in the electron transfer and α the charge transfer coefficient. The exchange current density $j_{0,et}$ is strongly temperature dependent according to

$$j_{0,et} \propto \exp\left(-\frac{\Delta G_{et}}{k_B T} \right), \tag{8.3}$$

where ΔG_{et} is the free activation energy required for the electron transfer reaction. The right term in Equation 8.2 describes the reduction, whereas the left term corresponds to the oxidation reaction. For low overpotentials $|\eta_{et}| \ll k_B T/ze$ the current becomes linearly dependent on η_{et} and exponentially dependent when $|\eta_{et}| \gg k_B T/ze$. Thus, the electron transfer reaction can also be field or temperature enhanced.

A *nucleation process* starts the formation of a new phase, for example, the electrocrystallization process of a metal on a foreign substrate. In order to permit further growth of the new phase, the nucleus has to achieve a critical size consisting of an integer number of atoms N_c. For the nucleation process to take place an activation energy ΔG_{nuc} has to be overcome. The nucleation time under an applied bias (nucleation overpotential η_{nuc}) is calculated according to

$$t_{nuc} = t_{0,nuc} \exp\left(\frac{\Delta G_{nuc}}{k_B T} \right) \exp\left[-\frac{(N_c + \alpha)ze}{k_B T} \eta_{nuc} \right]. \tag{8.4}$$

Accordingly, the nucleation process can be field and temperature enhanced. Note that the nucleation can only be a rate-limiting process if a critical nucleus needs to be formed.

Phase transformations occurring in the I-layer can also be rate-limiting. The velocity of a phase transformation is related to an activation energy and depends strongly on temperature. Thus, the kinetics can be strongly nonlinear when local Joule heating occurs. In ReRAMs phase transformations can be induced by ion depletion and local redox reactions. For instance, a phase transition from TiO_2 to a Magnèli phase as Ti_4O_7 can occur if oxygen vacancies move within a TiO_2 layer [8]. Note that a nucleation process is necessary if a phase transition occurs.

In summary, the switching kinetics can be electric field or temperature enhanced. The electric field is directly connected to the applied voltage. Due to local Joule heating the

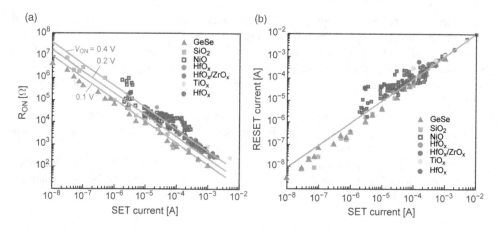

Figure 8.4 (a) ON resistance R_{ON} as a function of SET current compliance current for various material systems. (b) RESET current (maximum current before OFF switching) as a function of SET current compliance current for various material systems. Reproduced with permission from [9], © 2012 Wiley-VCH

temperature increase is also connected to the applied voltage. Thus, in both cases a faster switching process is expected for increasing voltage.

8.2.4 Generic Switching Properties

All types of ReRAMs exhibit generic SET and RESET characteristics. During SET operation the ON resistance can be programmed over many orders of magnitude by limiting the maximum current as illustrated in Figure 8.4a. The relation between programming current I_{cc} and ON resistance R_{ON} obeys the empirical law $R_{ON} = V_{ON}/I_{cc}$, whereas V_{ON} is a system inherent constant voltage. The physical origin of this universal SET characteristic lies in the nonlinear switching kinetics. As soon as the current compliance is reached any further decrease in resistance is accompanied by a voltage drop. Since the switching kinetics depends exponentially on the applied voltage, the driving force for further resistance change is suppressed, although it is still continuous. Accordingly, the ON voltage is adjusted.

It is experimentally observed that the RESET current scales linearly with the programming SET current (cf. Figure 8.4b). This empirical relation can be explained by the nonlinear switching kinetics and the almost linear I-V characteristic in the ON state. The RESET driving force is strongly voltage-dependent and almost independent of the programmed ON state. This results in a nearly constant RESET voltage V_{RES} for a distinct ReRAM under the same external switching conditions. In combination with the linear ON state we can write

$$I_{RES} = \frac{V_{RES}}{R_{ON}} = \frac{V_{RES}}{V_{ON}} I_{cc} = A I_{cc}. \tag{8.5}$$

Here, A is a system inherent constant. It should be noted that these empirical laws are independent of the actual switching mechanism as long as the system exhibits nonlinear switching kinetics. It has been shown by simulation that a thermally assisted lateral growth of a low resistive filament as well as the electrochemical change of a tunneling gap leads to these empirical generic switching characteristics.

8.2.5 Modeling Redox Memories for Circuit Simulations

The basic requirement for a behavioral model is to reproduce the device properties for arbitrary excitation signals and connection to further electronic components, for example, series resistors or further devices. Having this requirement in mind we review existing approaches and try to categorize them.

8.2.5.1 Quasi-Static Models

Quasi-static models have been the basic approach to model ReRAMs cells [10] and are extracted from quasi-static *I-V* characteristics. The advantage of this approach is the direct accessibility from measurement data which can directly fitted to obtain a usable model. The limitations of this approach directly result from absence of sweep rate/pulse rate dependency and the unfeasibility of any gradual multilevel properties.

8.2.5.2 Initial Dynamical Models

The development of dynamical models has obeyed a great advance by the initial memristive device model of Strukov *et al.* [11]. In this approach, the device is considered a dynamical system offering a readout and state equation of the form [12]:

$$
\begin{aligned}
I &= G(x, V, t) \cdot V \\
\dot{x} &= f(x, V, t)
\end{aligned}
\tag{8.6}
$$

or

$$
\begin{aligned}
V &= R(x, I, t) \cdot I \\
\dot{x} &= f(x, I, t)
\end{aligned}
\tag{8.7}
$$

The relevance of a certain dynamical model depends strongly on the applied differential equation, that is, the detailed understanding of the device physics. Starting from the basic linear model [11], several kinds of window functions were applied to enter more physical behavior to the model, for example [13–15]. But, this approach is limited since actual device properties cannot be modeled accurately.

8.2.5.3 Empirical Dynamical Models

In a next step, the concepts of the dynamical model were applied to specific devices and their dynamical behavior. In this approach, generic nonlinear equations (e.g., sinh equations) as well as certain physics-based equations (e.g., tunneling equations) are used to build-up dynamical models whose parameters are fitted to the experimental data, for example [16,17].

8.2.5.4 Generic Physics-based Dynamical Models

In this approach physical considerations are the starting point for modeling rather than experimental observations, leading to models for generic device classes, for example, ECM or VCM cells. In deference to the empirical approach, the physics-based approach tries to implement the internal physics as accurately as possible, avoiding piecewise definitions among others. For example, the fact that a ReRAM cell is mixed-ionic-electronic models – thus offer at least two current paths – is considered in this approach, see for example [18].

Sections 8.3 and 8.4 discuss selected empirical as well as generic physics-based modeling approaches for ECM and VCM cells, respectively.

8.3   Electrochemical Metallization Memory Cells

Electrochemical metallization memory (ECM) cells, also known as programmable metallization memory cell (PMC), conductive bridge random access memory (CBRAM) or atomic switch, rely on the electrochemical growth and dissolution of a conducting filament within the insulating layer [19,20]. They consist of an active Cu or Ag electrode, a cation conducting insulating layer, and an inert counter electrode such as Pt, W, or Ir [21]. The ECM effect has been demonstrated in typical insulators such as Ta_2O_5 [22], TiO_2 [23], HfO_2 [24], or SiO_2 [25] as well as solid electrolytes (e.g., AgI [26], GeS [27]). ECM cells exhibit multibit data storage capability [28], scalability almost down to the atomic level [29], and very low programming power [2]. Moreover, potential back end of line (BEOL) compatible integration has been demonstrated [30] making ECM cells of high interest for future nonvolatile memory.

8.3.1   Physical Switching Mechanism

8.3.1.1   Electroforming and SET Process

The electroforming and the SET process in ECM cells involve the same electrochemical steps as illustrated in Figure 8.5.

When a positive voltage is applied to the Cu or Ag active electrode, that electrode is oxidized electrochemically by a charge transfer process. The corresponding cation is dissolved into the insulating layer and migrates under the applied electric field towards the inert cathode (step B). At the cathode a redox reaction takes place and the cations are reduced. After formation of a stable nucleus a metallic Cu or Ag filament starts to grow towards the active electrode (step C). As the filament approaches the anode an electron tunneling current sets in and finally the preset current compliance is reached (step D). Any further resistance change forces the voltage to decrease and further filamentary growth is suppressed [18]. The formation of the metallic filament is accompanied with mechanical stress. Hence, a prerequisite for proper ECM operation is supposed to be the existence of extended defects such as nanopores or grain boundaries within the insulating layer. As illustrated in Figure 8.6a, the initial switching voltage during the electroforming cycle is higher than in the successive switching cycles [25].

Moreover, the switching voltage during the electroforming cycle is thickness-dependent, whereas it is almost thickness independent for all following cycles. Probably, the formation of the filament during the initial cycles involves mechanical stress, which makes the filamentary growth the limiting process. After RESET a template for fast ionic transport has been formed leading to the almost constant switching voltage.

The strength of the electric contact is determined by the preset current compliance level during the SET process according to $R_{ON} \propto I_{cc}^{-1}$. In this way the ON resistance can be adjusted over 10 orders of magnitude (see Figure 8.4).

The origin of the multilevel states has been attributed to varying tunneling gaps between the active electrode and the growing filament for ON resistances higher than $R_{ON} > 12.9\,k\Omega$ [18]. This resistance value corresponds to a single atomic contact conductance, that is, $1/G_0 = 12.9\,k\Omega$ [31].

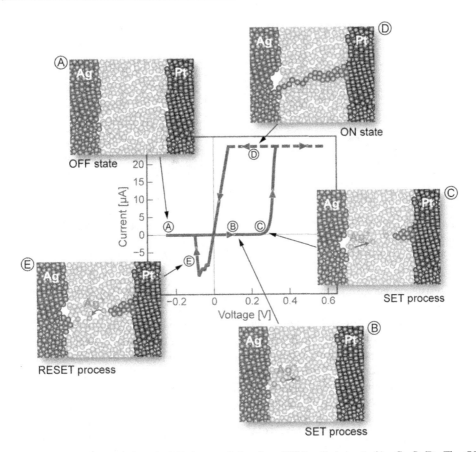

Figure 8.5 A typical quasi-static *I-V* characteristic of an ECM cell, here Ag/Ag-Ge-Se/Pt. The ON resistance is determined by the SET current compliance. The insets A to E show the different process steps: Step A – OFF state, steps B, C – SET process, step D – ON state, step E – RESET process. Reproduced with permission from [9], © 2012 Wiley-VCH

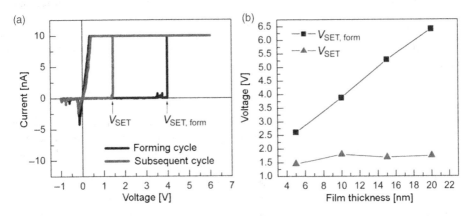

Figure 8.6 (a) Depending on the actual material system, in the initial forming cycle much larger voltages than the subsequent cycles are required, as shown here for Cu/SiO_2/Pt. (b) The initial SET forming voltage shows a clear thickness dependency while the subsequent SET voltages show no SiO_2 film thickness dependency. Reprinted with permission from [25]. Copyright 2009, American Institute of Physics

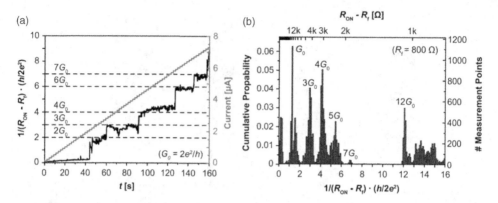

Figure 8.7 (a) In ECM (here: AgI-based) cells, single atomic contacts can occur, leading to quantized conductance offering conductance which is a multiple of G_0. (b) Cumulative statistics of the ECM cell conductivity. Reproduced with permission from [26], © 2012 IOP

In this resistance regime quantized conduction steps have been observed [22,26,32] as illustrated in Figure 8.7.

This gives an indirect proof of the existence of a remaining tunneling gap for $R_{ON} > 1/G_0$. Below this boundary the on resistance decreases further first by quantized steps and then more continuous by filament widening. This filament widening can be very small since the filament resistivity is diameter dependent according to Fuchs–Sondheimer theory [33]. Note that a further lateral growth is accompanied by mechanical stress.

ECM cells show strongly nonlinear switching kinetics. Based on the switching mechanism each physical process can be the rate limiting step. Depending on the materials involved and thermodynamic conditions, this can be: (i) the nucleation process prior to filamentary growth, (ii) the electron-transfer reaction occurring at the metal/insulator interfaces, and (iii) the ion transport within the electrolyte/insulator thin film (cf. Equations 8.1–8.3) [26,34,35]. Figure 8.8 shows the SET switching kinetics for a AgI system, which exhibits three distinct regimes [35]. By comparison to a simulation model these regime could be identified as (i) nucleation limited regime, (ii) electron transfer limited regime, and (iii) limitation by electron transfer and ion transport.

This demonstrates that the rate-limiting process also depends on the voltage regime. Due to the low currents involved during SET switching a temperature enhancement of the switching kinetics is rather unlikely.

8.3.1.2 RESET Process

The RESET mechanism strongly depends on the type of ON state. As previously discussed a tunneling gap can remain for low current compliances, whereas a true galvanic contact is achieved for high current compliances. In the former case the electrochemical cell is still present and the RESET is simply achieved by applying a negative potential to the active electrode [18]. The cell operation is inherently bipolar. If a galvanic contact is established during SET operation the conducting filament bridges the two metal electrodes. Due to the low ON resistance Joule heating will occur at the narrowest part of the filament [36,37]. Here, surface tension at the metal surface (temperature assisted ad-atom diffusion leading to Ostwald ripening) may contribute to the rupture. Due to temperature increase the electron transfer

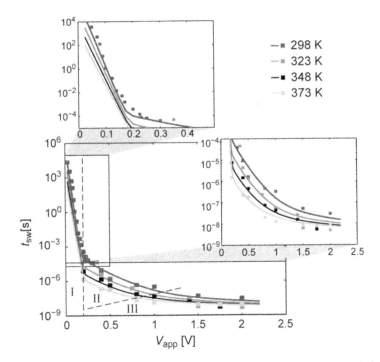

Figure 8.8 SET time as a function of pulse height for AgI-based cells. The symbols denote the experimental data and the solid lines simulation results. Reproduced by permission of the PCCP Owner Societies [35]

reactions occurring at the boundary are thermally enhanced (cf. Equation 8.2). This leads to a self-dissolution of the filament, that is, oxidation and reduction occur at the filament rather than at the filament and the active electrode [38]. After filament rupture the electrochemical RESET further proceeds as in the remaining tunneling gap case. Due to the self-dissolution of the bridging filament and the thermal nature of the RESET, the RESET process can be polarity-independent as reported in experiment [38].

The RESET switching kinetics have been rarely investigated. According to the study of Kozicki *et al.* the RESET time is a function of the ON state resistance and is also strongly voltage dependent [36]. Here, the ON resistance is very low suggesting a galvanic contact. In this regime first a lateral dissolution has to occur, which might explain the dependence of the RESET time on the ON state. For higher resistance state corresponding to a remaining tunneling gap in ON state, the RESET time is presumably independent on the ON resistance. In contrast to the SET switching kinetics the limiting factors are only the electron transfer reaction and the ionic transport, whereas a nucleation process does not occur. If a galvanic contact is present, the RESET switching is also thermally enhanced.

8.3.2 Modeling

Several compact behavioral models for ECM cells have been published. Basically, two different approaches have been pursued: (i) vertical [18,39,40] and (ii) lateral filamentary growth [28,36,41]. For the first approach the filament length or the tunneling gap between growing filament tip and counter electrodes are used as state variable, whereas in the latter the state variable is the filament radius.

A first self-consistent model for vertical filamentary growth has been published by Menzel *et al.* [42], which has been later extended to multilevel switching [18] and to model the SET switching kinetics including all relevant processes [35]. The dynamic state equation describing the tunneling gap x is given by Faradays law

$$\frac{dx}{dt} = -kj_{ion},$$ (8.8)

where k is a constant. The equivalent circuit diagram is shown in Figure 8.9a. Note that the change in the state variable only depends on the ionic current and not the electronic current as in the standard memristor-like models (cf. Section 8.2.5). The model accounts for the nonlinear switching kinetics, *I-V* characteristics, and the multilevel switching by modulation of the tunneling gap (see Figure 8.9b,c).

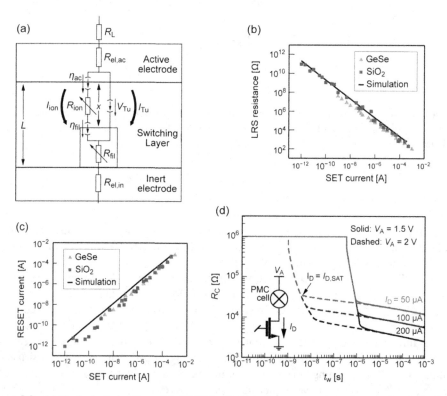

Figure 8.9 (a) Equivalent circuit model of ECM cell. Note the two current paths, one for the ionic and one for the electronic current. Reprinted with permission from [18]. Copyright 2012, American Institute of Physics. (b,c) The vertical ECM model accounts for nonlinear switching kinetics and reproduces experimental findings of SET current compliance dependencies for RESET current and actual LRS resistance. Reprinted with permission from [18]. Copyright 2012, American Institute of Physics. (d) In a 1T1R structure, the drain current I_D limits the maximum device current, thus plays the role of a current compliance. Depending on I_D a certain resistance level can be written. © 2009 IEEE. Reprinted, with permission, from [28]

The change of the filament diameter was first used to model the RESET in ECM cell by Russo *et al.*, and was supplemented by the SET modeling later [28,36] (see Figure 8.9d). This model has been also adapted by Yu *et al.* [41]. In this model Joule heating is also included due to the high currents involved. The state equation for radial dissolution and growth reads

$$\frac{dr}{dt} = \pm k j_{ion},\qquad(8.9)$$

where k is a constant. Also in this model the state equation depends only on the ionic current. Note that the algebraic sign in Equation 8.9 has to be changed for SET and RESET and the dynamic state equations are not self-consistent as in the vertical model. The lateral growth model has been used to describe multilevel switching and the switching kinetics. According to the considerations in the previous section the lateral growth model is restricted to the small resistance regime in which the filament bridges both electrodes, whereas the vertical growth model is valid until a galvanic contact is established. Both models could be thus easily combined. The occurrence of quantized conduction steps, however, is not covered by either model.

8.4 Valence Change Memory Cells

In valence change memory (VCM) cells the resistive switching effect relies on the motion of oxygen vacancies (respectively oxygen ions) and a subsequent redox reaction on the nanoscale. This leads to a valence change of the cation sublattice, which in turn changes the local conductivity. Due to the involved ion migration the device operation is inherently bipolar. The VCM effect has been observed in binary oxides as well as perovskites, where the insulating layer is a mixed ionic electronic conductor. Typically, the MIEC is sandwiched between an Ohmic electrode (OE) and an active electrode (AE) consisting of a high work function/less oxygen affinity metal [43]. Typically, the resistive switching takes place in front of the latter electrode. Many variants of VCM switching are known such as filamentary- or interface-type switching or in *p*- or *n*-conducting MIEC oxides. Here, we will focus on the filamentary type switching in *n*-conducting MIEC oxide, which is the common type of VCM switching. VCM cells represent a highly promising NVM technology because of its recent performance demonstrations: for example, very high endurance ($>10^{12}$ [44]), ultrafast switching (<1 ns [45]), high scalability (down to <10 nm [46]), and low power operation (<0.1 pJ [47]). In addition, multiple resistance states in a single ReRAM cell can be achieved by proper programming [48,49].

8.4.1 Physical Switching Mechanism

8.4.1.1 Electroforming Process

In general, an initial electroforming process is required prior in order to switch the MIM cell repetitively. The electroforming voltage is significantly higher than the SET/RESET voltage in the following switching cycles. Depending on the MIM stack an electroforming into the LRS as well as the HRS is possible, whereas the former is the more common process. In a MIM cell with identical metal electrodes the electroforming process induces an asymmetry and defines active and ohmic electrodes. During the electroforming process the oxide material is reduced and the MIM stack becomes conducting. Typically, this process is nonuniform and leads to the formation of *n*-conducting filaments or even induces a phase transformation, for example, a Magnéli phase Ti_4O_7 in TiO_2-based VCM cells [50]. Due to the localized currents and the relatively high voltages very high local temperatures are achieved, which accelerates the forming process. The microscopic mechanism is described using the process of forming the HRS as an example. Let us consider a cell with one low work function metal electrode, for example, Ti, and one high work function metal electrode such as Pt. By application of a positive potential to the active electrode oxygen is released at the anode leaving oxygen vacancies behind. The positively charged oxygen vacancies migrate under the applied electric field towards the cathode. They pile up at the cathode increasing locally the conductivity and the electric field in this region is reduced. Subsequently, a virtual cathode grows towards the anode until the current compliance is reached [3,51].

Alternatively, the forming process in HfO_x-based VCM cells has been described in terms of moving oxygen interstitials [52]. In contrast the oxygen vacancies are assumed to be immobile and to serve as deep trap states facilitating trap assisted tunneling current. Due to the applied electric field and local Joule heating this model suggests that the Hf–O bond breaks and an oxygen vacancy is created. The interstitial oxygen vacancy diffuses away from the vacancy and avoids recombination. These additional oxygen vacancies increase the trap assisted tunneling current and a self-accelerating process sets in leading to the formation of a highly reduced filament. Many aspects within this model remain unresolved at the current state of knowledge.

The electroforming voltage depends on the MIEC thickness. As a consequence, forming-free MIM cells can be fabricated using very thin oxide films (2–3 nm) [53]. By building in oxygen gradients during deposition of the oxide film (e.g., by deposition of bilayer structures such as TiO_x/TaO_x [54]), the electroforming voltages might be reduced. For WO_x-based VCM cell it is reported that the as-deposited film is highly deficient of oxygen and thus highly conducting [55]. Hence, the resistance needs to be increased during electroforming to allow resistive switching.

8.4.1.2 SET and RESET

Figure 8.10 illustrates a typically filamentary SET and RESET mechanism in VCM cells.

In this model, the HRS conducting filamentary region consists of a highly oxygen-deficient region and an oxygen-rich region near the active electrode. The corresponding band diagram exhibits an electrostatic barrier which defines the electric current. This band diagram facilitates an asymmetric *I-V* characteristic. By application of a negative potential to the active electrode the oxygen vacancies move towards the active electrode. A local redox reaction (i.e., the reduction of Zr ions in the disc; cf. Figure 8.10) occurs and the local conductivity is increased. This means that the barrier height and width are reduced enabling high tunneling currents and the cell switches to the LRS. Eventually, the electron barrier conduction is increased so much that the cell exhibits a linear, that is, a metal-like *I-V* characteristic. By reversing the polarity the oxygen vacancies are pushed backed and the HRS is re-established. In the picture of moving interstitial oxygen ions the switching polarity is the same. To enable RESET switching the oxygen ions have to recombine with the oxygen vacancies and the SET process is equivalent to the described electroforming process [6,52,56]. The above described switching polarity is called counter-eight-wise. In contrast, the switching polarity is called eightwise when the switching occurs at the same electrode but with opposite polarities for SET and RESET.

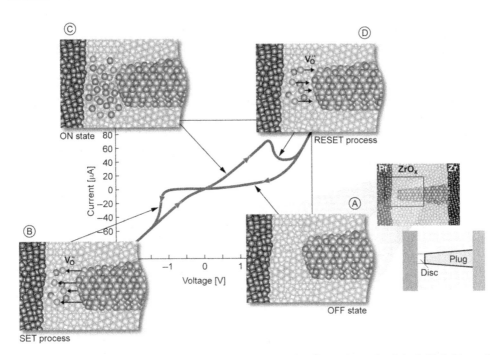

Figure 8.10 Typical *I-V* characteristic of a VCM cell. Pt is the active electrode of the Pt/ZrO$_x$/Zr stack while Zr is a purely ohmic contact. Light gray spheres indicate oxygen vacancies while dark spheres stand for lower-valence Zr ions. Note: only oxygen vacancies are mobile! The steps A to D show the switching process: step A – OFF state, step B – SET process, step C – ON state, step D – RESET process. Reproduced with permission from [9], © 2012 Wiley-VCH

Resistive switching with both polarities has been experimentally observed. It can also occur in the same device by proper electrical stimuli [43,53]. This behavior can be attributed, for example, to the change of the active switching interface [43], which means it is still counter-eight-wise switching. In contrast the counter-eightwise switching in SrTiO$_3$-based VCM cells has been attributed to a filamentary switching mechanism, while the eightwise switching occurs on a larger area at the same electrode [57]. Ultimately, complementary switching within a cell can occur, where the two different interfaces correspond to the two anti-serially connected bipolar switches of complementary resistive switch [58].

VCM cells show highly nonlinear SET and RESET switching kinetics. Here, the ion migration is supposed to be the limiting step due to the very low mobility of oxygen vacancies at room temperature and low electric fields. According to the Mott–Gurney law of ion hopping (cf. Equation 8.1) the ion transport can be either field or temperature enhanced. Due to the high current densities, involved in VCM switching, Joule heating is expected. The exact balance between these two effects depends on the specific MIM cell and the considered voltage regime. For very fast switching high currents and voltages are involved, supposing temperature as the dominating factor. During quasi-static switching very low programming power has been observed indicating that the electric field may be the dominating factor in this regime.

8.4.2 *Modeling*

In this section, typical dynamical models for VCM devices are highlighted. We restrict our considerations to models which are written as a dynamical system or at least can simply be rewritten in this manner (compare Equations 8.6 and 8.7).

First, we want to mention the TaO_x-VCM model presented by Hur *et al.* [59]. This model is based on the work of Strukov *et al.* [11] and keeps the linear dependency of R on the state variable x which corresponds to the length of a low resistive domain (i.e., a vertical filament) w here

$$V = \left[R_{ON} \frac{w}{L} + R_{OFF} \left(1 - \frac{w}{L} \right) + R_0 \right] \cdot I \qquad (8.10)$$

To model physical boundary conditions a window function is applied for calculation of the state variable

$$\dot{w} = C \cdot I \cdot \underbrace{w \cdot (L - w)}_{\substack{\text{window} \\ \text{function}}} \qquad (8.11)$$

Furthermore, the model accounts for the active electrode by introducing a Schottky barrier and includes field induced barrier lowering. The equivalent circuit model is depicted in Figure 8.11a and the simulation results in Figure 8.11b. The model is able to reproduce quasi-static *I-V* characteristics very well, but due to lack of nonlinearity in the state variable, dynamic properties of fast pulses for example will not be feasible by this model. Furthermore, as in the basic memristor models [14,15] the physical plausibility of the window function is only weak.

Next, we consider the TiO_x model from Pickett [16,60], another further development of the initial "memristor" model [11]. The basic modeling idea is depicted in Figure 8.12a showing a conductive channel which is modeled as a series resistor R_s and a gap which is described by a tunneling equation, and the state variable x which is considered the tunneling barrier width w. The advantages of this model are the nonlinear dependency on the state variable (tunneling equation) and the avoidance of an explicit window function. Hence, the model is physically

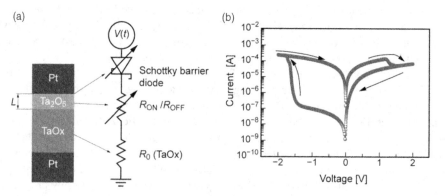

Figure 8.11 (a) Equivalent circuit model of a TaO_x VCM-type cell. (b) Simulated *I-V* curve which fits well to experimental data. Reprinted figure with permission from [59]. Copyright (2010) by the American Physical Society

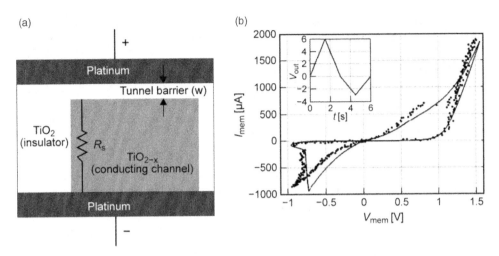

Figure 8.12 (a) Equivalent circuit model of a TiO$_x$ VCM-type cell. (b) Measured *I-V* and fitted simulation curve. © 2011 IEEE. Reprinted, with permission, from [60]

motivated, but the total approach is still mainly empirical, trying to fit simulations to an experimental quasi-static *I-V* curve (Figure 8.12b). Furthermore, the existence of two current paths – one electronic and one ionic – is not reflected by this model at all.

For VCM cells, there are also models assuming a lateral growth of a filament [61], thus the state variable *x* corresponds to the diameter of the filament. Furthermore, temperature could be implemented as an additional state variable, too.

In [62], the importance of the temperature as an additional state variable is highlighted. Here, a typical VCM cell (Ti/SrTiO$_3$/STO:Nb) is considered, which consists of an active Ti electrode, an ohmic STO:Nb bottom electrode, a highly conducting plug (created during electroforming) and a disc region where the switching takes place (see Figure 8.13a). Since the kinetics of the switching are determined by the drift velocity of the oxygen vacancies, both temperature

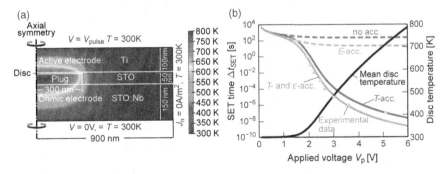

Figure 8.13 (a) Cross-section of a SrTiO$_3$ VCM cell using finite element (FEM) simulations. For a 5 V voltage pulse, the active disc region heats up to 800 K. (b) The SET time as a function of the applied pulse height. To fit the experimental data, strong temperature acceleration is needed. Pure field acceleration is not sufficient to explain the observed experimental nonlinear SET kinetics. Reproduced with permission from [62], © 2011 Wiley-VCH

and *E*-field acceleration could have significant impact. However, from the simulations in [62], the major contribution is due to temperature acceleration (see Figure 8.13b), making temperature one of the most important internal state variables, and thus should be considered in accurate VCM modeling.

In general, the challenge for VCM device modeling is the lack of detailed physical understanding of these devices, which makes physics-based modeling so difficult. Thus, in contrast to ECM modeling no generic physics-based dynamical models is available for VCM cells so far.

8.5 Performance

There are several metrics which can be applied to evaluate the performance of emerging memory devices [1]. Here we will present the most relevant criterions and corresponding best results from literature.

8.5.1 Minimum Feature Size

In terms of scalability the minimum feature size is the most significant figure of merit. In theory, feature sizes down to $F = 4$ nm are projected [63], and features sizes of $F = 30$ nm [44] or even $F = 10$ nm [46] have been shown for experimental setups for VCM cells (compare Figure 8.14). For ECM cells feature sizes down to 20 nm have been reported on [64,65]. Ultimatively, ECM operation has been demonstrated with only a few atoms involved [29].

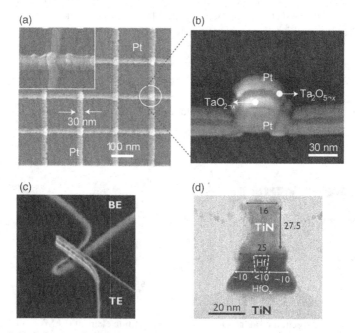

Figure 8.14   (a,b) A feature size of $F = 30$ nm were reported on for TaO$_x$ VCM devices in a crossbar array by Lee *et al.* [44]. Reprinted by permission from Macmillan Publishers Ltd: Nature Materials [44], copyright (2011). (c,d) Single crosspoint structures down to $F = 10$ nm were shown for HfO$_x$-based VCM devices. © 2011 IEEE. Reprinted, with permission, from [46]

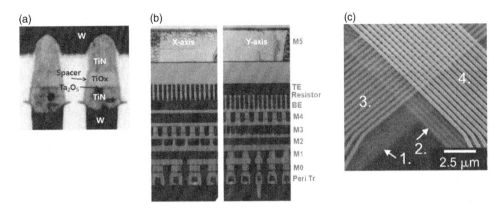

Figure 8.15 (a,b) A fully integrated TiO_x/Ta_2O_5-based crossbar array was presented in [66]. © 2012 IEEE. Reprinted, with permission, from [66]. (c) The feasibility of 3D stacking for crossbar arrays (here four layers) is depicted on the SEM picture (half pitch 200 nm). © 2009 IEEE. Reprinted, with permission, from [67]

8.5.2 Minimum Cell Area

The minimum cell array is highly relevant for area efficient memory layout. Due to the two-terminal nature of ReRAM devices the minimum cell area within crossbar arrays, which is $4F^2$, is feasible (Figure 8.15a). By applying 3D stacking techniques, for 1R, 1D1R or 1CRS devices even $4F^2/n$ can be achieved (Figure 8.15b), where n is the number of stacked layers, in a crossbar array configuration.

In principle, the scaling of ReRAM cells can be driven into the few nanometer range. For VCM cells, the redox-based resistive switching has been demonstrated at the exits of dislocations of $SrTiO_3$ single crystals with a lateral extension of less than 2 nm [68]. For ECM cells, the atomic switch concept with the Landauer conductance steps indicates contacts of few metal atoms [69]. Figure 8.15a shows a VCM device (TiO_x/Ta_2O_5) while Figure 8.15b shows a ECM array (Ag-MSQ).

8.5.3 Minimum Switching Time

The minimum switching time is directly related to the nonlinear device kinetics described above; compare also [62]. In [70], fast pulse analysis of TiO_x VCM nano-crossbars revealed (Figure 8.16a) feasible operation speed of <5 ns. In [45] an experimental coplanar waveguide test setup was used to demonstrate even faster switching properties of VCM-type TaO_x ReRAM cells (picoseconds), see Figure 8.16b. For ECM devices no coplanar waveguide test setup measurements were performed yet, and minimum observed switching is below 10 ns [71].

8.5.4 Retention Time

Data retention is another important issue for memories. Using elevated temperature experiments, the time to fail can be estimated (compare Figure 8.17). VCM-type TaO_x-based ReRAM cells were investigated and the time to fail for the tail bits of a 256 kbit array (1T1R) was estimated to be larger than 10 years at 85 °C [72] (Figure 8.17a). For Ag-chalcogenide ECM cells retention times larger 10 years were also estimated (Figure 8.17b).

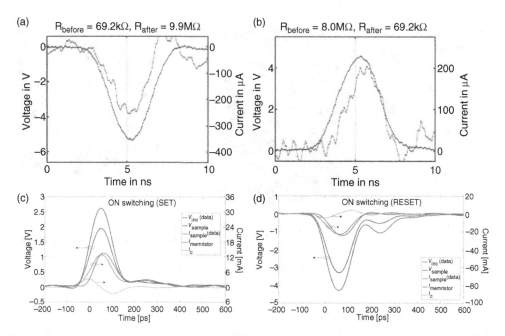

Figure 8.16 (a, b) Fast switching in TiO$_2$-based VCM cells. © 2011 IEEE. Reprinted, with permission, from [70]. (c, d) Sub-nanosecond ON and OFF switching could be shown for TaO$_x$ VCM cells. Reproduced with permission from [45], © 2011 IOP

8.5.5 Write Cycles

Besides retention, the maximum number of write cycles is important for the feasibility of long-life memories. Up to 10^{12} write cycles subsequent RESET-read-SET-Read-cycles could be performed for TaO$_x$-based VCM cells (see Figure 8.18a). For Ag-GeSe ECM devices 10^{10} write cycles are feasible (see Figure 8.18b). Especially, the endurance data for TaO$_x$-based VCM cells is very promising and can be considered a milestone on the way to future ReRAM memories.

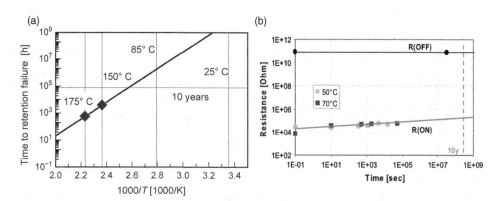

Figure 8.17 (a) Arrhenius plot visualizing the retention behavior at elevated temperatures and the extrapolated trend for lower temperatures. © 2012 IEEE. Reprinted, with permission, from [72]. (b) Increase in ON-state resistance with time. The retention is extrapolated. © 2005 IEEE. Reprinted, with permission, from [64]

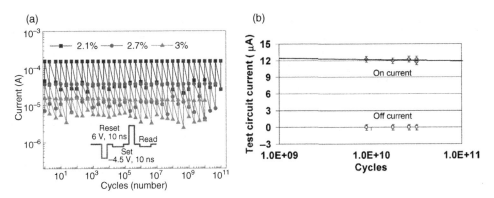

Figure 8.18 (a) TaO$_x$ VCM devices offer promising endurance properties. Up to 10^{12} write cycles could be shown in [44]. Best results were obtained with an oxygen content of 3%. Reprinted by permission from Macmillan Publishers Ltd: Nature Materials [44], copyright (2011). (b) Ag-GeSe ECM devices offer endurance of 10^{10} write cycles. © 2005 IEEE. Reprinted, with permission, from [73]

8.5.6 Multilevel

To compete with FLASH memories multiple bits should be stored in each ReRAM cell. For both VCM as well as ECM cells, multi-bit properties could be shown (see Figure 8.19).

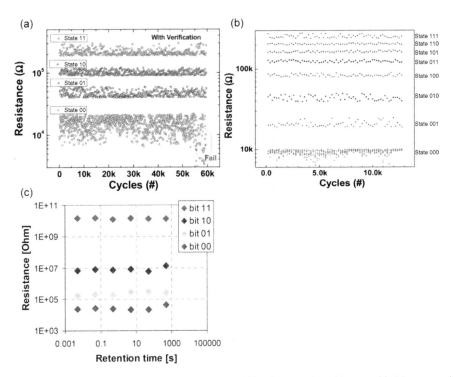

Figure 8.19 (a,b) WO$_x$-based VCM cells offering multilevel properties, either two bit (a) or even three bit (b). A read–verify technique was applied. © 2011 IEEE. Reprinted, with permission, from [49]. (c) Ag-based ECM cells showing two-bit multilevel behavior. © 2007 IEEE. Reprinted, with permission, from [complete publication information]

Table 8.1 Multi-bit properties of ReRAM cells

Performance measure	Best value		Material system		References	
	VCM	ECM	VCM	ECM	VCM	ECM
Minimum feature size F	10 nm	<10 nm	HfO$_x$	Ag-chalcogenide	[46]	[29,64]
Minimum cell area	$4F^2$	$4F^2$	Any	Any	See, for example, [66]	See, for example, [75]
Minimum switching time	< 200 ps	<10 ns	TaO$_x$	Ag-MSQ	[45]	[71]
Retention time	>10 yr (extrapolated)	>10 yr (extrapolated)	TaO$_x$	Ag-chalcogenide	See, for example, [72]	[64]
Write cycles	>10^{12}	>10^{10}	TaO$_x$	Ag-GeSe	[44]	[73]
Multilevels	8	4	WO$_x$	Ag-chalcogenide	[49]	[74]

In [49] two bits (Figure 8.19a) and up to three bits per cell (Figure 8.19b) were feasible (VCM), while in [74] two bits per cell (ECM) could be stored. The results are summarized in Table 8.1.

8.6 Summary

Redox-based resistive nonvolatile memories are among the most promising emerging memory devices. The devices can be classified by means of the mechanisms either as ECM, VCM or TCM. Both ECM and VCM devices offer the good properties for future memory applications, while ECM cells are much better understood than VCM devices. This issue is reflected in the availability of physics-based device models which are much more elaborated for ECM cells.

For modeling, the fact that ReRAMs are MIEC devices must be taken into account, thus models should offer at least two current paths. Moreover window functions should be used with caution, since the physical accuracy is often low. To model a restricted device domain, for example, the quasi-static behavior, this approach may be valid, but one should be aware of the limitation. A good first check for device model consistency is if the device models can reproduce the frequency dependency of t_{SET}/t_{RESET} as well as the multilevel behavior, that is, offer the correct inherent nonlinear switching kinetics.

In terms of actual device performance one can say that VCM cells in general offer more stable mode of operation, while ECM cells are more power-efficient. Both types of cells show excellent scaling potential, and in combination with a suitable selector, large-scale $4F^2$ structures are feasible.

References

1. ITRS (2011). "The International Technology Roadmap for Semiconductors - ITRS 2011 Edition".
2. Waser, R. and Aono, M. (2007) Nanoionics-based aresistive switching memories. *Nature Materials*, **6**, 833–840.
3. Waser, R., Dittmann, R., Staikov, G., and Szot, K. (2009) Redox-based resistive switching memories - nanoionic mechanisms, prospects, and challenges. *Advanced Materials*, **21**, 2632–2663.
4. Tappertzhofen, S., Menzel, S., Valov, I., and Waser, R. (2011) Redox processes in silicon dioxide thin films using copper microelectrodes. *Applied Physics Letters*, **99**, 203103.
5. Goldfarb, I., Miao, F., Yang, J.J. *et al.* (2012) Electronic structure and transport measurements of amorphous transition-metal oxides: observation of Fermi glass behavior. *Applied Physics A-Materials Science & Processing*, **107**, 1–11.

6. Wong, H.-S.P., Lee, H.-Y., Yu, S. *et al.* (2012) Metal–Oxide RRAM. *Proceedings of the IEEE*, **100**, 1951–1970.
7. Meuffels, P. and Schröeder, H. (2011) Comment on *Exponential ionic drift: fast switching and low volatility of thin-film memristors* by D.B. Strukov and R.S. Williams in *Applied Physics A*, 94, 515–519 (2009). *Applied Physics A: Materials Science and Processing*, **105**, 65–67.
8. Szot, K., Rogala, M., Speier, W. *et al.* (2011) TiO_2 – a prototypical memristive material. *Nanotechnology*, **22**, 254001/1–254001/21.
9. Waser, R., Menzel, S., and Bruchhaus, R. (2012) *Nanoelectronics and Information Technology*, 3rd edn, Wiley-VCH.
10. Mustafa, J. and Waser, R. (2006) A novel reference scheme for reading passive resistive crossbar memories. *IEEE Transactions on Nanotechnology*, **5**, 687–691.
11. Strukov, D.B., Snider, G.S., Stewart, D.R., and Williams, R.S. (2008) The missing memristor found. *Nature*, **453**, 80–83.
12. Chua, L.O. and Kang, S.M. (1976) Memristive devices and systems. *Proceedings of the IEEE*, **64**, 209–223.
13. Joglekar, Y.N. and Wolf, S.J. (2009) The elusive memristor: properties of basic electrical circuits. *European Journal of Physics*, **30**, 661–675.
14. Biolek, Z., Biolek, D., and Biolkova, V. (2009) SPICE model of memristor with nonlinear dopant drift. *Radioengineering*, **18**, 210–214.
15. Benderli, S. and Wey, T.A. (2009) On SPICE macromodelling of TiO_2 memristors. *Electronics Letters*, **45**, 377–379.
16. Pickett, M.D., Strukov, D.B., Borghetti, J.L. *et al.* (2009) Switching dynamics in titanium dioxide memristive devices. *Journal of Applied Physics*, **106**, 074508.
17. Sheridan, P., Kim, K.H., Gaba, S. *et al.* (2011) Device and SPICE modeling of RRAM devices. *Nanoscale*, **3**, 3833–3840.
18. Menzel, S., Böttger, U., and Waser, R. (2012) Simulation of multilevel switching in electrochemical metallization memory cells. *Journal of Applied Physics*, **111**, 014501.
19. Valov, I., Waser, R., Jameson, J.R., and Kozicki, M.N. (2011) Electrochemical metallization memories-fundamentals, applications, prospects. *Nanotechnology*, **22**, 254003.
20. Hasegawa, T., Terabe, K., Tsuruoka, T., and Aono, M. (2012) Atomic switch: Atom/ion movement controlled devices for beyond von-neumann computers. *Advanced Materials*, **24**, 252–267.
21. Valov, I. and Kozicki, M.N. (2013) Cation-based resistance change memory. *Journal of Physics D: Applied Physics*, **46**, 074005.
22. Tsuruoka, T., Hasegawa, T., Terabe, K., and Aono, M. (2012) Conductance quantization and synaptic behavior in a Ta_2O_5-based atomic switch. *Nanotechnology*, **23**, 435705.
23. Yang, L., Kügeler, C., Szot, K. *et al.* (2009) The influence of copper top electrodes on the resistive switching effect in TiO_2 thin films studied by conductive atomic force microscopy. *Applied Physics Letters*, **95**, 13109.
24. Wang, Y., Liu, Q., Long, S. *et al.* (2010) Investigation of resistive switching in Cu-doped HfO_2 thin film for multilevel non-volatile memory applications. *Nanotechnology*, **21**, 245202.
25. Schindler, C., Staikov, G., and Waser, R. (2009) Electrode kinetics of $Cu-SiO_2$-based resistive switching cells: Overcoming the voltage-time dilemma of electrochemical metallization memories. *Applied Physics Letters*, **94**, 072109.
26. Tappertzhofen, S., Valov, I., and Waser, R. (2012) Quantum conductance and switching kinetics of AgI based microcrossbar cells. *Nanotechnology*, **23**, 145703.
27. Kozicki, M.N., Balakrishnan, M., Gopalan, C. *et al.* (2005) Programmable metallization cell memory based on Ag-Ge-S and Cu-Ge-S solid electrolytes. *Proceedings of the Non-Volatile Memory Technology Symposium (NVMTS)*, **20**, 83–89.
28. Russo, U., Kamalanathan, D., Ielmini, D. *et al.* (2009) Study of multilevel programming in programmable metallization cell (PMC) memory. *IEEE Trans Electron Devices*, **56**, 1040–1047.
29. Terabe, K., Hasegawa, T., Nakayama, T., and Aono, M. (2005) Quantized conductance atomic switch. *Nature*, **433**, 47–50.
30. Bernard, Y., Renard, V.T., Gonon, P., and Jousseaume, V. (2011) Back-end-of-line compatible conductive bridging RAM based on Cu and SiO_2. *Microelectronic Engineering*, **88**, 814–816.
31. Scheer, E., Agrait, N., Cuevas, J. *et al.* (1998) The signature of chemical valence in the electrical conduction through a single-atom contact. *Nature*, **394**, 154–157.
32. Jameson, J.R., Gilbert, N., Koushan, F. *et al.* (2012) Quantized conductance in $Ag/GeS_2/W$ conductive-bridge memory cells. *IEEE Electron Device Letters*, **33**, 257–259.
33. Sondheimer, E.H. (1952) The mean free path of electrons in metals. *Advances in Physics*, **1**, 1–42.

34. Valov, I., Sapezanskaia, I., Nayak, A. *et al.* (2012) Atomically controlled electrochemical nucleation at superionic solid electrolyte surfaces. *Nature Materials*, **11**, 530–535.

35. Menzel, S., Tappertzhofen, S., Waser, R., and Valov, I. (2013) Switching kinetics of electrochemical metallization memory cells. *PCCP*, **15**, 6945–6952.

36. Kamalanathan, D., Russo, U., Ielmini, D., and Kozicki, M.N. (2009) Voltage-driven on-off transition and tradeoff with program and erase current in programmable metallization cell (PMC) memory. *IEEE Electron Device Letters*, **30**, 553–555.

37. Tsuruoka, T., Terabe, K., Hasegawa, T., and Aono, M. (2010) Forming and switching mechanisms of a cation-migration-based oxide resistive memory. *Nanotechnology*, **21**, 425205.

38. Menzel, S., Adler, N., van den Hurk, J. *et al.* (2013) Simulation of polarity independent RESET in electrochemical metallization memory cells, *Proceedings of the 5th IEEE International Memory Workshop (IMW)*.

39. Jameson, J.R., Gilbert, N., Koushan, F. *et al.* (2011) One-dimensional model of the programming kinetics of conductive-bridge memory cells. *Applied Physics Letters*, **99**, 063506.

40. Lin, S., Zhao, L., Zhang, J. *et al.* (2012) Electrochemical Simulation of Filament Growth and Dissolution in Conductive-Bridging RAM (CBRAM) with Cylindrical Coordinates, *IEEE International Electron Devices Meeting - IEDM '12*.

41. Yu, S. and Wong, H.-S. (2011) Compact modeling of conducting-bridge random-access memory (CBRAM). *IEEE Transactions on Electron Devices*, **58**, 1352–1360.

42. Menzel, S., Klopstra, B., Kügeler, C. *et al.* (2009) A simulation model of resistive switching in electrochemical metallization memory cells. *Materials Research Society symposium proceedings*, **1160**, 101–106.

43. Yang, J.J., Borghetti, J., Murphy, D. *et al.* (2009) A family of electronically reconfigurable nanodevices. *Advanced Materials*, **21**, 3754–3758.

44. Lee, M.J., Lee, C.B., Lee, D. *et al.* (2011.) A fast, high-endurance and scalable non-volatile memory device made from asymmetric Ta_2O_{5-x}/TaO_{2-x} bilayer structures. *Nature Materials*, **10**, 625–630.

45. Torrezan, A.C., Strachan, J.P., Medeiros-Ribeiro, G., and Williams, R.S. (2011) Sub-nanosecond switching of a tantalum oxide memristor. *Nanotechnology*, **22**, 485203.

46. Govoreanu, B., Kar, G.S., Chen, Y.-Y. *et al.* (2011) $10 \times 10 \, nm^2$ Hf/HfOx Crossbar Resistive RAM with Excellent Performance, Reliability and Low-Energy Operation, *IEDM Technical Digest*, pp. S31.6.

47. Cheng, C.H., Chin, A., and Yeh, F.S. (2011) Ultralow switching energy Ni/GeOx/HfON/TaN RRAM. *IEEE Electron Device Letters*, **32**, 366–368.

48. Ielmini, D. (2011) Modeling the universal set/reset characteristics of bipolar RRAM by field- and temperature-driven filament growth. *IEEE Transactions on Electron Devices*, **58**, 4309–4317.

49. Chien, W.-C., Lee, M.-H., Lee, F.-M. *et al.* (2011) A Multi-Level 40 nm WO_X Resistive Memory with Excellent Reliability, *IEEE International Electron Devices Meeting - IEDM '11*.

50. Kwon, D.-H., Kim, K.M., Jang, J.H. *et al.* (2010) Atomic structure of conducting nanofilaments in TiO_2 resistive switching memory. *Nature Nanotechnology*, **5**, 148–153.

51. Yang, J.J., Miao, F., Pickett, M.D. *et al.* (2009) The mechanism of electroforming of metal oxide memristive switches. *Nanotechnology*, **20**, 215201.

52. Bersuker, G., Gilmer, D.C., Veksler, D. *et al.* (2011) Metal oxide resistive memory switching mechanism based on conductive filament properties. *Journal of Applied Physics*, **110**, 124518/1–124518/7.

53. Bruchhaus, R., Hermes, C.R., and Waser, R. (2011) Memristive switches with two switching polarities in a forming free device structure. *MRS Online Proceedings Library*, **1337**, 73–78.

54. Yang, J.J., Zhang, M., Pickett, M.D. *et al.* (2012.) Engineering nonlinearity into memristors for passive crossbar applications. *Applied Physics Letters*, **100**, 113501.

55. Chen, Y., Chien, W., Lee, M. *et al.* (2012) Evaluation of the WO*x* film properties for resistive random access memory application. *Japanese Journal of Applied Physics*, **51**, 191–202.

56. Gao, B., Kang, J., Liu, L. *et al.* (2011) A physical model for bipolar oxide-based resistive switching memory based on ion-transport-recombination effect. *Applied Physics Letters*, **98**, 232108.

57. Müenstermann, R., Menke, T., Dittmann, R., and Waser, R. (2010) Coexistence of filamentary and homogeneous resistive switching in Fe-doped $SrTiO_3$ thin-film memristive devices. *Advanced Materials*, **22**, 4819.

58. Nardi, F., Balatti, S., Larentis, S. *et al.* (2013) Complementary switching in oxide-based bipolar resistive-switching random memory. *IEEE Transactions on Electron Devices*, **60**, 70–77.

59. Hur, J.H., Lee, M.-J., Lee, C.B. *et al.* (2010) Modeling for bipolar resistive memory switching in transition-metal oxides. *Physical Review B*, **82**, 155321.

60. Abdalla, H. and Pickett, M.D. (2011) SPICE Modeling of Memristors, *2011 IEEE International Symposium on Circuits and Systems*, pp. 1832–1835.

61. Ielmini, D. (2012) Evidence for voltage-driven set/reset processes. *IEEE Transactions on Electron Devices*, **59**, 2049–2055.
62. Menzel, S., Waters, M., Marchewka, A. *et al.* (2011) Origin of the ultra-nonlinear switching kinetics in oxide-based resistive switches. *Advanced Functional Materials*, **21**, 4487–4492.
63. Zhirnov, V.V., Meade, R., Cavin, R.K., and Sandhu, G. (2011) Scaling limits of resistive memories. *Nanotechnology*, **22**, 254027.
64. Kund, M., Beitel, G., Pinnow, C.U. *et al.* (2005) Conductive bridging RAM (CBRAM): an emerging non-volatile memory technology scalable to sub 20nm. *IEDM Technical Digest*, pp. 754–757.
65. Aratani, K., Ohba, K., Mizuguchi, T. *et al.* (2007) A novel resistance memory with high scalability and nanosecond switching. *IEDM Technical Digest*, 783.
66. Lee, H.D., Kim, S.G., Cho, K. *et al.* (2012) Integration of $4F_2$ selector-less crossbar array 2Mb ReRAM based on transition metal oxides for high density memory applications. *Japanese Journal of Applied Physics*, **51**, 151–152.
67. Meier, M., Rosezin, R., Gilles, S. *et al.* (2009) A multilayer RRAM nanoarchitecture with resistively switching Ag-doped spin-on glass, *Proceedings of 10th International Conference on Ultimate Integration on Silicon*, pp. 143–146.
68. Szot, K., Speier, W., Bihlmayer, G., and Waser, R. (2006) Switching the electrical resistance of individual dislocations in single-crystalline $SrTiO_3$. *Nature Materials*, **5**, 312–320.
69. Sakamoto, T., Kaeriyama, S., Sunamura, H. *et al.* (2004) A nonvolatile programmable solid electrolyte nanometer switch. *IEDM Technical Digest*, **2004**, 290–529.
70. Hermes, C., Lentz, F., Waser, R. *et al.* (2011) Fast pulse analysis of TiO_2 based RRAM nano-crossbar devices. *Nature Materials*, **10**, 92–95.
71. Meier, M., Schindler, C., Gilles, S. *et al.* (2009) A Nonvolatile memory with resistively switching methyl-silsesquioxane. *IEEE Electron Device Letters*, **30**, 8–10.
72. Wei, Z., Takagi, T., Kanzawa, Y. *et al.* (2012) Retention model for high-density ReRAM, Proceedings of the 4th IEEE International Memory Workshop (IMW).
73. Kozicki, M.N., Park, M., and Mitkova, M. (2005) Nanoscale memory elements based on solid-state electrolytes. *IEEE Transactions on Nanotechnology*, **4**, 331–338.
74. Symanczyk, R., Dittrich, R., Keller, J. *et al.* (2007) Conductive bridging memory development from single cells to 2 MBit memory arrays. *Proceedings of the Non-Volatile Memory Technology Symposium (NVMTS)*, pp. 71–75.
75. Kügeler, C., Meier, M., Rosezin, R. *et al.* (2009) High density 3D memory architecture based on the resistive switching effect. *Solid State Electronics*, **53**, 1287–1292.

9

Electronic Effect Resistive Switching Memories

An Chen
GLOBALFOUNDRIES Inc., USA

9.1 Introduction

Mainstream memory technologies, including SRAM, DRAM, and Flash memories, are all CMOS-based and face scaling challenges. Recently, many novel memory device concepts have emerged as candidates for the next-generation nonvolatile memory technologies, including phase change memory (PCM), spin transfer torque RAM (STT-RAM), resistive RAM (RRAM), and so on. Although many emerging memories store information in switchable resistance levels, the name RRAM usually refers to the two-terminal metal–insulator–metal (MIM) resistive switching devices based on metal oxides. These devices can be electrically switched between a high resistance state (HRS, or off-state) and a low resistance state (LRS, or on-state). Multi-level cell (MLC) can be achieved with appropriate switching control (e.g., current compliance, switching voltage, etc.) or material stack engineering (e.g., multi-layer oxides). The HRS-to-LRS switching is known as "set" (or "program") and the LRS-to-HRS switching as "reset" (or "erase"). The set and reset switching may occur in the same voltage polarity direction ("unipolar switching") or in the opposite directions ("bipolar switching").

Resistive switching in metal oxides is not a new phenomenon [1–3]. Hysteretic current-voltage (I-V) characteristics or negative differential resistance (NDR) was reported on numerous metal oxides, for example, TiO_x [4,5], NiO_x [6], AlO_x [7], NbO_x [8], and so on. Some switching models were also proposed to explain these phenomena. One of them is based on the charge trapping that modifies band structure and transport properties, that is, Simmons-Verderber model [9]. Another model is based on the formation and annihilation of filaments [1], similar to the current understanding of resistive switching mechanisms. However, these early studies did not develop into more serious efforts to build memory devices based on these switching phenomena.

Emerging Nanoelectronic Devices, First Edition. An Chen, James Hutchby, Victor Zhirnov and George Bourianoff.
© 2015 John Wiley & Sons, Ltd. Published 2015 by John Wiley & Sons, Ltd.

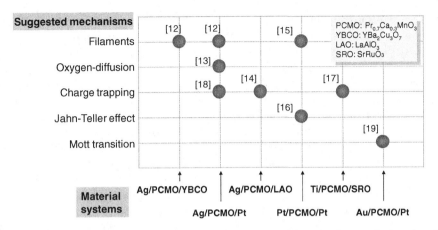

Figure 9.1 Various proposed resistive switching mechanisms for $Pr_{0.7}Ca_{0.3}MnO_3$-based resistive switching devices [12–19]

In the last 10 years, RRAM has received great attention and significant progress has been made in both the understanding of the switching mechanisms and the improvement of the switching properties through device engineering. Bipolar resistive switching in perovskite materials was reported in 2000 and suggested for memory applications [10]. Nonvolatile memories based on binary metal oxides were reported in 2004, with comprehensive characterization of memory performance [11]. These promising results stimulated strong interest in metal oxide-based RRAM for nonvolatile memory applications beyond Flash.

In this fast-growing field, it is inevitable to encounter contradictory results and interpretations even on the same target materials. For the purpose of illustration, Figure 9.1 summaries various switching models proposed for $Pr_{0.7}Ca_{0.3}MnO_3$-based resistive switching devices [12–19]. Several factors contribute to the contradiction and confusion in this field. First, multiple mechanisms may contribute to a switching process and different reports may focus on certain aspects of these mechanisms most related to the researchers' expertise. Second, the property of transitional metal oxides is highly sensitive to their composition and processing conditions; therefore, fabricated oxides with the same nominal composition may still have different properties due to process variations. Third, device structure (especially the electrodes and interface) could significantly affect the switching property.

Since the early exploration of the switching mechanisms of metal oxides, electronic effects have been considered as possible explanations. The dominant RRAM switching models today is based on the migration and reaction of ions or vacancies (e.g., oxygen ions or vacancies) that generates conductive paths within insulating oxides. However, several electronic effects or mechanisms may still contribute to the switching process in some RRAM materials. To categorize different RRAM switching mechanisms, the 2009 ITRS Emerging Research Device chapter classified the switching mechanisms into *thermal effect*, *ionic effect*, and *electronic effect* [20]. A real switching process may involve multiple effects and this classification is based on the dominant contributions. Thermal effect is related to power induced Joule heating and often involves the formation and disruption of localized conduction paths in an insulating material. Thermal effect resistive switching is usually unipolar. Ionic effect resistive switching involves the transport and electrochemical reactions of cations (e.g., Ag^+, Cu^+) and/or anions (e.g., O^{2-}) and is usually bipolar switching. Cation-based ionic switching mechanism is well understood and the physical process of switching induced by ionic migration can be confirmed by microscopic imaging. However, anion-based switching is more complicated and the exact switching physics is under extensive research. This early version of RRAM categories from ERD has evolved into new forms and the so-called electronic effective RRAM is no longer a separate category in the ERD chapter. However, some electronic effects may still be involved in resistive switching process and this chapter provides a brief review of them.

The "electronic effect" covers several different mechanisms. Charge injection and trapping could modify transport characteristics and induce abrupt resistance change. Mott materials may experience several orders of magnitude of resistance change at the triggering of metal–insulator transition. Ferroelectric polarization reversal in ferroelectric oxides may cause oxide resistance change, based on which a device concept of ferroelectric tunnel junction (FTJ) has been proposed. This chapter will review these three electronic switching mechanisms and assess their potentials for memory applications. The following three sections will discuss the three electronic effects in sequence: charge injection and trapping, Mott metal–insulator transition, and ferroelectric polarization reversal.

9.2   Charge Injection and Trapping

9.2.1   Charge Trapping Induced Resistive Switching

The basic model of charge trapping resistive switching process can be described as:

1. Injected charge carriers trapped inside of materials modify the band structure or interface properties, which may change transport properties and cause resistive switching.
2. The release of trapped charges recovers the material to its original states.
3. Retention of charge-trapping states depends on the release time of trapping charges.

The Simmons–Verderber model based on charge injection and trapping/detrapping well explained the measured switching I-V characteristics of an Au/SiO/Al structure [9]. Switching is as fast as a few nanoseconds, which was cited as evidence to relate the observed switching phenomena to electronic effects (rather than ionic or other effects involving atomic motions). It is suggested that during a "forming" process, Au ions are injected into SiO to introduce a broad band of localized impurity levels within the forbidden band of the insulator. Electrons can then move through the insulator by tunneling between adjacent sites within the impurity band. The memory effect is caused by charge (electron) storage within the insulator, which changes the band structure of the insulator and the field at the interface. The release of the trapped charge recovers the device back to its original state.

Recent publications claiming charge-trapping switching mechanisms are often based on two types of observations: (1) hysteretic I-V characteristics of a rectifying junction that can be attributed to interfacial modification by charge trapping; (2) I-V characteristics dominated by space charge limited conduction (SCLC) that is related to empty and filled trap states.

A $Ti/Pr_{0.7}Ca_{0.3}MnO_3$ junction is found to be rectifying with hysteretic I-V characteristics [17]. The reversible bipolar resistive switching can be explained by the change of the width and/or height of a Schottky-like barrier, induced by trapped charge carriers in the interface states. A forming process is required before stable switching can be achieved, which may be related to the generation of oxygen defect induced states at the Ti/PCMO interface. In addition to metal/oxide junctions, hysteretic I-V characteristics are also observed in a rectifying epitaxial heterojunction $SrRuO_3/SrTi_{0.99}Nb_{0.01}O_3$ between two oxides, which can also be explained by charge trapping model [21].

An experiment was carefully designed to prove the interfacial switching effect by inserting a thin layer of $Sm_{0.7}Ca_{0.3}MnO_3$ (SCMO) between Ti electrode and $La_{0.7}Sr_{0.3}MnO_3$ (LSMO) [22]. While the Ti/LSMO interface shows no resistive switching, the insertion of SCMO of even one unit cell (uc) thickness induces resistive switching. As the thickness of SCMO increases, both hysteresis and rectification develop. Switching time also decreases from ~10 µs for 1 uc SCMO to ~100 ns for 5 uc SCMO. The effect of SCMO saturates after 5 uc, indicating that active layer thickness for resistive switching is several uc of SCMO adjacent to the Ti/SCMO interface.

In an $Ag/Pr_{0.7}Ca_{0.3}MnO_3/Pt$ structure, the hysteretic I-V characteristics can be accurately described by SCLC model; therefore, the resistive switching is linked to the transition from trap-unfilled SCLC to trap-filled SCLC [18]. Based on similar observations of SCLC-dominated I-V characteristics, charge-trapping model is adopted to explain the resistive switching behavior in Cu_2O [23,24], TiO_x [25], $(Ba_{0.5}Sr_{0.5})(Zr_{0.2}Ti_{0.8})O_3$ [26,27], $SrTiO_{3-\delta}/Nb:SrTiO_3$ heterojunction [28], ZrO_2 [29], HfO_2 [30], Gd-doped TiO_2 [31], and so on. However, it should be noticed that the interpretation of the switching mechanisms in these devices may evolve over time and change with new observations.

By combining scanning electron microscopy (SEM) and electron beam induced current (EBIC) imaging with transport measurement on thin-film $SrZrO_3$ doped with 0.2% Cr, it was shown that current conduction and switching are confined in areas localized at defects [32]. The memory effects were explained by the storage and release of charge carriers within the insulator. Similar switching is also observed in 10 µm Cr-doped single-crystal $SrTiO_3$ [33]. Stable switching has been shown for more than 6000 write/erase cycles, with no change of signal after more than 10^5 readouts. Because of the existence of high initial resistance up to 100 V, it is suggested that bulk (not the interface) determines current flow across the insulator and switching originates from a change in bulk properties.

The reported switching parameters of some charge-trapping resistive switching materials reviewed in this section are summarized in Table 9.1 in Section 9.5 [9,17,18,21,22,25–27, 29–32,34,35].

9.2.2 Charge-Trapping Resistive Switching Memory Performance

Scaling: Since charge-trapping mechanism involves defects or charge trapping centers, fluctuation or inhomogeneity of the defect density may limit the scalability of these devices. A defect modification model suggests a statistical size limit on the order of 10–100 nm [36]. Localized conduction and switching are also observed in some charge-trapping resistive switching devices [32], which are often considered evidence of promising scalability. Fundamentally, the size limit of these localized conduction and switching region is still subject to the density fluctuation of the trapping sites.

Speed: The write/erase speed depends on how fast the charge trapping/detrapping processes take place. Reported speed in literature varies from tens of µs to several ns. Slow measured speed could be attributed to parasitic RC delay in large device structures. An electronic effect switching process may occur in tens of nanosecond or shorter time [21]. However, the feasibility of sub-nanosecond operation in large arrays is unclear. Tradeoff between the on/off ratio and switching speed is observed in Ti/PCMO [17] and Ag/PCMO [18]. The study on the switching in SiO_x shows a "dead time" (from tens of milliseconds up to several seconds) that prevents immediate repeating of cycles [9]. If such a dead time exists, memory access

Table 9.1 Summary of switching characteristics of charge trapping resistive memory devices

Devices	d_1 (nm)	V_{sw} (V)	I_{sw} (A)	On/off ratio	Speed (s)	Retention	Cycling	References
Charge injection and trapping								
Au/SiO/Al	20–300	4–8	$\sim10^{-2}$	~100	10^{-9}–10^{-7}	—	—	[9]
Ti/Pr$_{0.7}$Ca$_{0.3}$MnO$_3$/SrTiO$_3$	100	2–5	~0.1	1–4	$\sim10^{-3}$	—	—	[17]
SrRuO$_3$/SrTi$_{0.99}$Nb$_{0.01}$O$_3$/SrRuO$_3$	100	2–4	$\sim10^{-2}$	1–100	$\sim10^{-4}$	—	—	[21]
Ti/Sm$_{0.7}$Ca$_{0.3}$MnO$_3$/La$_{0.7}$Sr$_{0.3}$MnO$_3$	1–100	2–5	10^{-4}–0.1	>500	$\sim10^{-4}$	—	—	[22]
Ag/Pr$_{0.7}$Ca$_{0.3}$MnO$_3$/Pt	150	1–2	0.1–0.01	10–10^4	$\sim10^{-7}$	—	—	[18]
Al/TiO$_x$/Ti	10	3	$\sim10^{-3}$	$\sim10^3$	—	—	>100	[25]
Pt/SrTiO$_x$/Pt	167	2–5	$\sim10^{-4}$	$\sim10^2$	P: 5×10^{-8} E: 5×10^{-4}	10^6 at 125°C	10^5	[34]
Pt/(Ba$_{0.5}$Sr$_{0.5}$)(Zr$_{0.2}$Ti$_{0.8}$)O$_3$/Pt	200	1–3	10^{-7}–10^{-5}	~230	—	—	—	[26]
Au/Cr/Zr$^+$-implanted-ZrO$_2$/n$^+$-Si	70	3–4	$\sim10^{-2}$	$\sim10^4$	—	>2500s	—	[29]
TiN/AlCu/HfO$_2$/TiN	20	0.5–1.5	10^{-2}	~4	$\sim5\times10^{-8}$	$>10^4$ at 85°C	$>10^5$	[30]
(W or Al)/Gd:TiO$_2$/Pt	40	0.5–2	$\sim10^{-2}$	$\sim10^3$	—	—	>10	[31]
Pt/SrZrO$_3$(0.2% Cr)/SrRuO$_3$	35	—	$\sim10^{-4}$	45–90	—	$\sim5\times10^4$ s	>10	[32]
Al/SiO$_x$/p-Si/Al	5–7	~2	—	$\sim10^2$	—	>2000s	$>10^3$	[35]
Mott transition								
Au/Pt$_{0.7}$Ca$_{0.3}$MnO$_3$/Pt	360	~0.8	10^{-2}–10^{-1}	~10	—	—	—	[19]
Ag/CeO$_2$/La$_{0.67}$Ca$_{0.33}$MnO$_3$	80	2–4	$\sim10^{-2}$	10^2–10^4	•	—	—	[46]
Pt/TiO$_2$/TiN/Pt	2.5	~2	—	$>10^2$	2×10^{-8}	—	>10	[48,49]
Au/VO$_2$ (sol-gel)/Au	200	>50	6×10^{-4}	$\sim10^3$	—	—	$>10^8$	[52]
Au/VO$_2$ (anodic oxidation)/Au	200	1–10	10^{-5}–10^{-4}	—	10^{-9}–10^{-4}	—	—	[51]
Au/Cr/VO$_2$ (epitaxy)/Cr/Au	~90	10–20	$\sim10^{-2}$	—	—	—	—	[53]
Ferroelectrical polarization reversal								
Au/PbTiO$_3$/La$_{0.5}$Sr$_{0.5}$CoO$_3$	200	2–4	10^{-6}–10^{-4}	$\sim10^2$	$<10^{-2}$	—	—	[65]
Pt/Pb(Zr$_{0.52}$Ti$_{0.48}$)O$_3$/SrRuO$_3$	4–6	0.5–1	$\sim10^{-4}$	~4	—	—	—	[67]
Pt/SrTiO$_3$/Pt	4.5	~0.8	$\sim10^{-2}$	—	—	Unspecified extended period	—	[68]
BaTiO$_3$ (probed by PFM)	1–50	~3	—	2–10^3	—	—	—	[70]
Au/Co/BaTiO$_3$/La$_{0.67}$Sr$_{0.33}$MnO$_3$	500	~2	$\sim10^{-5}$	10^2	10^{-8}	—	>900	[71]
P(VDF:70%-TrFE:30%)	2 mono-layer	<1	—	~6	—	—	—	[73]

algorithms need to be designed accordingly to minimize delay. The reading speed of a charge-trapping resistive switching device is determined by peripheral circuitry. With a few orders of magnitude on/off ratio in these devices, reading is unlikely to be a concern, as long as reading voltage is not high enough to disturb device states.

On/off ratio: On/off ratio is determined by the resistance difference between the charge-trapped state and the charge-released state. The ratio could vary in a large range depending on trap-controlled conduction mechanisms. Ratios from<10 up to 10^5 have been reported. The on/off ratio in some devices was found to decrease with shorter switching pulse width, which presents a performance trade-off. A high on/off ratio helps to expedite reading and reduce errors.

Retention: Retention depends on the release time of trapped charges [18], which may depend exponentially on the trap depth or some activation energies [23]. Some estimation has shown that 10-year retention at room temperature is achievable for activation energy of 1.0–1.1 eV. Charge-trapping resistive switching materials may be engineered to achieve 10-year retention, but tradeoffs may exist between retention and other switching parameters. Retention of 1600 minutes at 85 °C has been measured on PCMO [12].

Endurance: The endurance of charge-trapping devices is limited by the stability of defects and trapping sites. A microscopic imaging study of localized switching process shows variation of the conduction/switching sites that indicates defect generations during switching [33]. Cycling fatigue due to defect generation is also observed in Pt/PCMO/Pt devices after 10^4 cycles [37].

Switching energy: In principle the minimum switching energy is associated with the trap depth; however, in reality much higher energy is required to switch these devices. The switching energy (E) can be estimated from the measured switching voltage (V), switching current (I) and switching time (t), that is, $E = V \cdot I \cdot t$. As an example, the switching parameters for AlCu/HfO$_2$/TiN are $V \approx 1.5$ V, $I \approx 0.3$ mA, and $t \approx 50$ ns, which results in $E \approx 2 \times 10^{-11}$ J/bit [31]. This demonstrated switching energy is in a typical range for resistive switching materials. It has been shown that switching voltage, current, and time can be further reduced from the values chosen above by material engineering and device optimization. Switching energy smaller than 10^{-12} J/bit appears feasible for charge-trapping resistive switching memories.

CMOS compatibility: Some resistive switching memories, including these based on charge-trapping mechanisms have been integrated in standard CMOS structure [11,23,38]. Some novel materials used in charge-trapping resistive switching devices have been shown to be compatible with Si CMOS process. The simple two-terminal structure of resistive switching devices makes it easy for integration in CMOS without area penalty (e.g., at via locations). In the 1-transistor-1-resistor (1T1R) memory structure, it is the selection transistors that limit memory scalability.

Analysis has shown that for charge-trapping switching mechanism there exists a tradeoff between switching speed and retention [39]. Sufficiently high trap depth or barrier height is required to suppress the escape of the trapped charges to achieve long retention; however, high barriers constrain the magnitude of current density required for fast reading and switching.

9.3 Mott Transition

9.3.1 Resistive Switching Induced by Mott Transition

Charge injection into some materials may induce a transition from strongly correlated to weakly correlated electrons, leading to a Mott metal–insulator transition [40–44]. Resistive switching may occur as a result of this transition.

An observation on the electrical trigger of resistive switching in magnetoresistive manganite PCMO at low temperature suggests the possibility of dielectric breakdown of the charge-ordered state in Mott insulators [45]. In the bipolar resistive switching observed in Au/$Pr_{0.7}Ca_{0.3}MnO_3$/Pt structure at room temperature, the mixed valence state Mn^{4+}/Mn^{3+} at the Au/PCMO interface is found critical for switching [19]. No resistive switching is observed at Au/PMO interface with only Mn^{3+} valence state or at Au/CMO interface with only Mn^{4+} valence state. An increase of Mn^{4+}/Mn^{3+} ratio caused by oxygen annealing results in increasing HRS/LRS resistance ratio. It is postulated that the mixed Mn^{4+}/Mn^{3+} valence states at the metal/PCMO interface can regulate current through metal–insulator transition induced by a critical carrier density modulated by electrostatic doping, which leads to resistive switching.

Similarly, bipolar resistive switching observed in an Ag/CeO_2/$L_{0.67}Ca_{0.33}MnO_3$ structure is also explained by mixed valence states of Ce^{4+}/Ce^{3+} and Mott transition [46]. The on/off ratio reaches 10^5. Electrical field initiates the process of oxygen migration and the formation of oxygen vacancies facilitates the shift in the valence state of Ce cations. Such a valence shift of cerium oxide may propagate and self-assemble under the applied electric field through the insulator until a conductive percolation path of neighboring Ce^{3+} sites is created when some critical density of valence shifted sites is reached. The use of conductive oxide LCMO as the bottom electrode ensures free oxygen migration through the CeO_2/LCMO interface.

In a 0.2 mol% Cr-doped $SrTiO_3$ resistive switching material, x-ray absorption near-edge structure (XANES) spectra reveals significant valence change of Cr from Cr^{3+} to Cr^{4+} during electrical forming process [47]. It is suggested that the resistance change in Cr:$SrTiO_3$ is caused by Mott transition associated with internal doping of the Ti 3d band due to a change of Cr-dopant valence.

Mott transition is also proposed as the mechanism for the bipolar resistive switching observed in a 2.5 nm anatase TiO_2 layer grown by oxidation of a TiN diffusion barrier [48,49]. Mott transition may be triggered by the increase in donor concentration caused by the electron injection from the top electrode (Pt), which is accompanied by oxygen vacancy formation and O^{2-} migration in the anatase TiO_2 layer. Devices can be switched within 20–30 ns with ~2 V voltage pulses. The on/off ratio reaches almost 400.

Vanadium oxide (VO_2) is a well-known Mott material with potentials for high-speed device applications [50–57]. Sharp resistance change can be triggered by temperature [50,54,55] or current/voltage [51–53,56]. The metal–insulator transition is often accompanied by thermal structural phase transition (SPT), which complicates the switching mechanism and degrades device performance [53]. However, study has shown that thermal effect alone cannot describe the transition and electron correlation effects also contribute to the metal–insulator transition in VO_2 [51,57]. With careful design of device geometry, material parameters and thermal condition, it may be possible to induce electronic transition without triggering thermal SPT [53,57].

The transition time of VO_2 is shown to be dependent on voltage and can be as short as 1.5 ns [51]. The fundamental limit of the transition time may be on the order of ps [51]. On/off ratio is usually on the order of 10^3–10^5. Endurance as long as 10^8 cycles has been observed on sol-gel VO_2 film [52], and on sputtered VO_2 thin-film ~100 switching cycles is also measured [54]. Some studies also show chaotic switching behaviors inside the hysteresis region near the onset of the transition [55]. Although material properties at nano-scale may differ drastically from those of bulk materials, switching characteristics of devices as small as 200 nm in diameter are found to be similar to that of larger devices [56].

The reported values of switching parameters of the Mott transition devices discussed above are summarized in Table 9.1 [19,46,48,49,51–53].

Mott transition is incorporated in a phenomenological model that describes an insulating medium as a nonpercolative structure of bulk domains and electrode/insulator interface as smaller domains. Mott transition due to strong electron correlation at the interface plays a key role for switching [58,59]. Another model for resistive switching (colossal electro-resistance) is developed based on interface Mott transition using the density matrix renormalization group (DMRG) method [60].

In addition to the usual two-terminal structure, Mott transition based nonvolatile memory may also be built in a three-terminal transistor structure, where a Mott transition layer is included in the gate stack to allow or block tunneling current into a charge trapping layer [61]. In another proposed transistor concept based on Mott transition, additional carriers are drawn into the channel electrostatically from source/drain by gate voltage to induce Mott transition and create a thin conducting layer at the interface [62,63]. It is suggested that Mott transition field-effect transistors can overcome some intrinsic limitations of Si MOSFET, such as lower carrier/dopant fluctuations and lower gate leakage.

9.3.2 Mott Transition Resistive Switching Memory Performance

Scaling: Properties of Mott transition materials in nano-scale may differ drastically from the properties of bulk materials [64]. Although switching devices on the scale of 200 nm has been shown to have the characteristics similar to these observed in larger devices, this size is much larger than competitive sizes of state of the art memory technologies. More research on the size effect of the Mott transition properties is needed to address the fundamental scaling limit of this type of devices.

Speed: Switching speed as short as 1.5 ns has been observed on VO_2 [51], and TiO_2-based devices can be switched with 20–30 ns pulses [48,49]. Since the transition is driven by electron–electron correlations, the fundamental speed limit is expected to be on the order of $\varepsilon \cdot \rho$ where ε is the high-frequency dielectric constant and ρ is the specific resistivity of the material [51]. This value is on the order of 10^{-12} s, equivalent to a frequency limit of 1 THz.

On/off ratio: Experiments have shown on/off ratio in the range of 10^3–10^5 on Mott transition devices [46,54]. This is comparable with charge-trapping and ferroelectric resistive switching devices.

Retention: Although retention has been shown in some materials associated with Mott transition, the retention mechanism needs further study. The electrical conditions that triggered the Mott transition have to be continuously present to sustain the transition states for retention, for example, by ferroelectric field or charge trapping [47]. Notice that the volatile switching due to Mott transition may also be utilized as memory select devices.

Endurance: Endurance as long as 10^8 cycles has been shown on VO_2-based switching devices [51]. Without structural change, electrically triggered Mott transition is expected to have long endurance.

Switching energy: A rough estimation of switching energy can be made using the reported data on $Pt/TiO_2/TiN$ [48,49]. Reset switching can be done with 2.2 V, 30 ns pulses and LRS has the resistance of 0.2 kΩ. The switching energy is estimated to be $E = V^2/R \cdot t \approx 7 \times 10^{-10}$ J/bit, a relatively high value due to large current. Lower switching energy may be achievable with improved materials and devices.

9.4 Ferroelectric Resistive Switching

9.4.1 Ferroelectric Resistive Switching Mechanism

If the "insulator" layer in the MIM structure is made of ferroelectric materials, ferroelectric polarization can modify the carrier transport characteristics and induce bistable resistive switching. It should be noted that unlike Ferroelectric RAM (FeRAM) where ferroelectric materials are used as capacitors, ferroelectric resistive switching memory operates based on the transport property across a ferroelectric layer which can be electrically modulated due to ferroelectric polarization reversal.

In an $Au/PbTiO_3/La_{0.5}Sr_{0.5}CoO_3$ Schottky diode with ferroelectric $PbTiO_3$ layer, the interface charge and the band diagram can be continuously and permanently modified by an external electric field [65]. Bistable I-V behaviors are observed during voltage sweep, where the threshold voltage for set switching from HRS to LRS is related to the coercive field. The thickness of $PbTiO_3$ layer is 0.2 μm, and HRS conduction is found to be either bulk controlled (SCLC) or electrode controlled (Schottky emission), depending on the direction and history of the applied electric field. The changing conduction mechanisms are related to the polarization dependence of the band bending and depletion region width, which is caused by the local field-dependent permittivity of the ferroelectric material. The Schottky barrier width is reduced (increased) when the polarization in the space–charge region of the diode is parallel (anti-parallel) to the internal electric field. Therefore, the resistive switching can be explained by the change of the Schottky barrier width caused by abrupt change of the polarization at coercive field.

If the ferroelectric layer is made thin enough to allow carrier tunneling, the tunneling property of the ferroelectric layer may also be modified electrically, leading to a new device concept named "ferroelectric tunnel junction" (FTJ) [66]. The electric field induced polarization reversal of a ferroelectric barrier may have pronounced effect on the conductance of FTJs. The polarization reversal changes the sign of the polarization charge at the barrier/electrode interface, the positions of ions in ferroelectric unit cells, and lattice strains inside the barrier [66].

In a resistive switching heteroepitaxial junction $Pt/Pb(Zr_{0.52}Ti_{0.48})O_3/SrRuO_3$, the switching voltage matches well with the coercive electric field, suggesting the polarization reversal in the ferroelectric barrier as the origin of the resistive switching phenomena [67]. However, recent experiments suggest the possibility of ionic contributions in the switching process in these devices. Resistive switching observed in a junction between Pt electrode and ultrathin $SrTiO_3$ (4.5 nm) is also attributed to ferroelectric polarization [68]. Although bulk, undoped $SrTiO_3$ is not ferroelectric, epitaxial $SrTiO_3$ thin film may show strain-induced ferroelectricity [69].

Robust ferroelectricity and resistive switching in highly strained ultra thin $BaTiO_3$ (down to 1 nm thickness) are demonstrated by piezoresponse force microscopy (PFM) and conductive tip atomic force microscopy (CTAFM) [70]. The measured tunnel electro-resistance (TER, i.e., on/off ratio) increases from ~200% at the thickness of 1 nm to ~75 000% at 3 nm. The observed thickness dependence of the TER suggests that the ferroelectric resistive switching is caused by the polarization induced changes of the tunnel barrier characteristics (height, thickness, tunneling mass, etc.) rather than the changes in the density of states (DOS) at the barrier–electrode interface. Scalability of ferroelectric resistive switching in $BaTiO_3$ is demonstrated down to device as small as 70 nm. A FTJ device with $Au//Co/BaTiO_3/La_{0.67}Sr_{0.33}MnO_3$ structure and diameter of 500 nm is fabricated and characterized [71,72].

It demonstrates on/off ratio as high as 100, low switching current of $\sim 1 \times 10^4$ A/cm^2, endurance over 900 cycles, and switching speed faster than 10 ns.

Resistive switching induced by polarization reversal is also observed on ferroelectric polymers [73,74]. The tunneling current measured by scanning tunneling spectroscopy (STS) on a two monolayer thick ferroelectric copolymer of VDF:70%-TrFE:30% (vinylidene fluoride with trifluoroethylene) can change more than six times when flipping the polarity of probing voltage [73]. This local "switching" of vertical conductance can be explained by rotation of the top layer polarization direction.

Some reported key switching parameters of the ferroelectric resistive switching devices discussed here are summarized in Table 9.1 [65,67,68,70,71,73].

The resistive switching characteristics can be quantitatively described by theoretical models [75–77]. The change of potential profile at different polarization directions in a FTJ structure could lead to significant resistance change and calculation shows that electroresistance (i.e., on/off ratio) of FTJs can be as high as 10^3 [75]. Another FTJ model is developed based on strain-induced changes of the barrier thickness, electron effective mass, and conduction band edge at the coercive field [76]. The model predicts that asymmetric FTJs with dissimilar top and bottom electrodes have larger on/off ratio than symmetric FTJs.

More novel device concepts may be developed by combining ferroelectrics and ferromagnetism. It is found that resistance of a ferroelectric/ferromagnetic LiNbO$_3$/La$_{0.69}$Ca$_{0.31}$MnO$_3$ junction can be tuned by both magnetic field and electric field [78]. The voltage pulse induced resistive switching is attributed to ferroelectric polarization at the junction interface. Combining ferromagnetic electrodes with a ferroelectric tunnel barrier may produce multiferroic tunnel junctions (MFTJ). A four-state resistive switching device concept built on magnetic tunnel junctions with ferroelectric barriers is predicted based on first-principles calculations [79]. A Fe/BaTiO$_3$/La$_{0.67}$Sr$_{0.33}$MnO$_3$ structure was used to experimentally demonstrate local, large, and nonvolatile control of spin polarization by ferroelectric polarization reversal [80].

A critical requirement for functional FTJs is the existence of ferroelectricity in nanometer thick barrier materials [66]. Ferroelectricity is a cooperative phenomenon; therefore, it is generally expected that there is a critical thickness (t_c) below which ferroelectricity collapses due to the depolarization field or finite size effects [81–84]. Recent theoretical and experimental studies have shown that either t_c could be really small (a few unit cells) or there is essentially no intrinsic ferroelectric size limit for practical devices [83–96]. Figure 9.2 summaries the reported t_c on

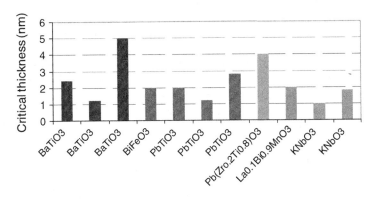

Figure 9.2 Reported critical thickness of some ferroelectric materials [83–96]

some ferroelectric materials. These results indicate that FTJ devices with ultrathin ferroelectric layer (down to a couple of nm thickness) are feasible.

Major challenges still exist before FTJ or MFTJ can be implemented in device applications. For example, parasitic effects such as local conductivity and transport via localized states must be eliminated. Better understanding is needed regarding the mechanisms of domain formation, nucleation and switching in nanoscale ferroelectrics, and tunneling transport across polar thin film dielectrics [66].

9.4.2 Ferroelectric Resistive Switching Memory Performance

Resistive switching memory based on FTJ is an emerging area with many open questions, especially ferroelectric properties in nano-scale. Interestingly, much of the nano-ferroelectric domain behavior can be derived from bulk classical physics [96]. In addition, extensive research results are available for FeRAM and may provide useful references for the research on ferroelectric resistive switching memory [97,98].

Scaling: Numerous studies on finite size effects in ferroelectrics point to a critical particle size or film thickness below which ferroelectricity disappears [99]. This critical size may impose a scaling limit on ferroelectric resistive switching devices. It has been shown that ferroelectricity may persist in films as thin as a few nanometers. Achievable memory density is determined by lateral size limit. A study on self-assembly $PbTiO_3$ and $Pb(Zr_xTi_{1-x})O_3$ nano-grains shows that ferroelectricity is preserved down to grain size of 20 nm [100]. Furthermore, $BaTiO_3$ nanowires as small as 10 nm in diameter are shown to retain ferroelectricity [101]. Therefore, it seems feasible to scale ferroelectric resistive switching devices below 10 nm.

Speed: Switching speed of ferroelectric resistive switching memory is determined by the ferroelectric polarization reversal time. The reversal switch typically takes place by the generation of new reverse domains at particularly sites [98], whose rate in submicron ferroelectrics is limited by nucleation rather than by domain wall motion and can be less than 1 ns [96]. Switching speed shorter than 4 ns has been measured in $LiTaO_3$ nano-domains using scanning nonlinear dielectric microscopy (SNDM) [102]. In FeRAM technology, access time as small as 280 ps has been tested in laboratory [98].

On/off ratio: On/off ratio of $\sim 10^2$ has been demonstrated on Schottky diode made on thicker $PbTiO_3$ film [65] and the ratio could be even higher in ultrathin $BaTiO_3$ [70]. Although it has been shown that the on/off ratio decreases with the thickness of the ferroelectric tunnel barriers, a ratio of nearly 10^3 can be achieved with a 3 nm tunnel barrier [70,75].

Retention: Destabilization of the ferroelectric polarization state may cause the loss of information recorded in ferroelectric resistive switching memory. For example, oxygen migration under the depolarization field or other built-in field may cause retention degradation [98,103]. Nonvolatility of nano-scale ferroelectric thin film has been proven by experiments, for example, ferroelectric state in a 4 nm $Pb(Zr_{0.2}Ti_{0.2})O_3$ film is stable for over 140 h [93]. The retention time of induced polarization in $BaTiO_3$ nanowires exceeds 5 days [98]. Some ferroelectric field effect devices have shown retention of 1–2 weeks or even over 10 months at room temperature [104–106]. Under optimal conditions, retention of ferroelectric memories may exceed 10 years at room temperature [97].

Cycling endurance: Ferroelectric memories have shown excellence endurance of 10^{11}–10^{12} cycles [107–109]. Data projection predicts endurance longer than 10^{16} cycles [108]. The cycling limit may be imposed by polarization fatigue, that is, the reduction of switchable ferroelectric polarization by repetitive electrical cycling [98]. Further study is needed to understand the endurance failure in nano-scale ferroelectric films.

Switching energy: Switching voltage of ferroelectric resistive switching memory is related to the coercive field that has been shown to decrease with the film thickness into the nanometer range [67,96]. This is desirable for low-voltage, low-power memory applications. Switching energy <10 fJ/bit is estimated based on measured characteristics of Au//Co/BaTiO$_3$/La$_{0.67}$Sr$_{0.33}$MnO$_3$ FTJ devices [71].

9.5 Perspectives

Resistive switching memories have presented promising opportunities and also great challenges. In the category of electronic effect resistive switching, there are wide range of materials with different switching mechanisms and characteristics. The exact switching processes and device physics are still not clearly understood. Some of the claimed mechanisms for the reported devices discussed in this chapter may be found to be incomplete or even incorrect with further investigations.

Table 9.1 summarizes reported parameters of the three types of electronic effect resistive switching memories. Notice that some parameters (e.g., speed) may be significantly below the best achievable values due to parasitic effects. Charge injection and trapping are well known processes in semiconductor device physics and are adopted to explain many resistive switching phenomena. Fewer examples of switching devices based on Mott transition and ferroelectric polarization reversal have been reported, partially due to the difficulty of conclusively attributing a switching phenomenon to these mechanisms. Some intriguing research on these switching mechanisms and device application is based on theoretical modeling. For example, resistive switching is predicted in a two-dimensional electron gas (2DEG) induced by ferroelectric reversal in oxide LaAlO$_3$/SrTiO$_3$ heterostructures, which may allow device scaling down to several nanometers [110]. Figure 9.3 summarizes the measured switching current *vs.* voltage of these devices, with different colors for different types of

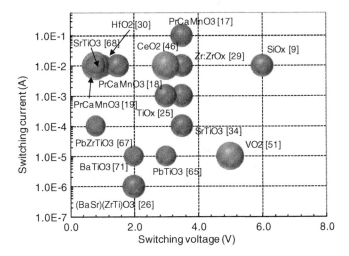

Figure 9.3 Reported switching current versus voltage of electronic-effect resistive switching devices. Blue: charge trapping; orange: Mott transition; pink: ferroelectric reversal. Sizes of the symbols have no physical meaning and are varied to differentiate the three mechanisms (Color in electronic version)

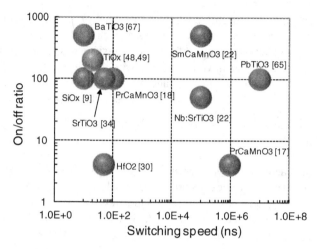

Figure 9.4   Reported on/off ratio versus switching speed of electronic-effect resistive switching devices. Blue: charge trapping; orange: Mott transition; pink: ferroelectric reversal (Color in electronic version)

electronic effects: charge trapping (blue), Mott transition (orange), and ferroelectric reversal (pink). Figure 9.4 plots the measured on/off ratio versus switching speed of these devices.

In any binary switching devices for information representation and storage, some physical changes have to occur in the material to enable the transition between two distinguishable states. Intuitively one can expect that the switching involving only electronic changes (e.g., electron spin flipping, charge motion, injection, and trapping, etc.) may consume lower power, switch faster, and have longer endurance than changes involving atomic/ionic motions (e.g., mass transport, ionic migration, electrochemical reactions, etc.) In this sense, electronic effect resistive switching memory may have certain advantages over the switching devices based on thermally induced physical changes (e.g., phase change memory) or ionic motions (e.g., ionic effect memory).

An important requirement for emerging nonvolatile memories is the scalability superior to Flash memories. As discussed earlier, the scaling of charge-trapping memories is limited by the fluctuation or inhomogeneity of the defect density. Ferroelectricity is a collective phenomenon; therefore, it is expected that the size of switching devices due to ferroelectric reversal has to be large enough for the material to remain ferroelectric. However, this ferroelectric size limit has been shown to be as small as a few nanometers. The scaling limit of Mott transition still requires more study, but there does not appear to be a fundamental size limit. Overall, it appears that electronic effect RRAM devices can be scaled to several nanometers.

With the current memory technologies already at <45 nm technology nodes, increasing memory density by device scaling alone offers limited potential. Multi-level cell (MLC) design and 3D stackable structures provide more growth space for high-density memories. For devices with high on/off ratios and cumulative switching mechanisms (e.g., charge trapping), it may be possible to program devices to multiple levels by accurately controlling the resistance levels. FTJ-based devices are not a natural choice for MLC, because of the binary nature of ferroelectric polarization. However, the MFTJ discussed in Section 9.4.1 may enable MLC by combining ferroelectrics and ferromagnetism in one device. 3D stackable memories still face key challenges in memory select device.

Table 9.2 Summary of three electronic resistive switching mechanisms

Types	Charge trapping	Mott transition	Ferroelectric
Data representation	Resistance change due to different transport properties of trap filled and trap empty states	Resistance change due to different transport properties of metallic and insulating states of Mott insulators	Resistance change due to different transport properties of ferroelectric tunnel layer at opposite polarization
Switching mechanism	Field-controlled trap filling and detrapping	Carrier injection induced Mott transition	Field-controlled polarization reversal
Scaling limit	Statistic fluctuation of defects and traps	Highly scalable, for example, interface 2DEG	Ferroelectric size effect; scalable to <10 nm
Multi-level cell capability	Possible; require accurate control of trap filling process	Probably difficult because of high sensitivity to small parameter change	Unlikely for ferroelectric tunnel junction (FTJ); possible with multiferroic tunnel junction (MFTJ)
CMOS compatibility	Compatible	Compatible	Compatible
Cross-bar compatibility	Prefer rectifying and nonlinear I-V, high on/off ratio; may require selection devices	Require high on/off ratio and asymmetric I-V; may require selection devices	Require high on/off ratio; may require selection devices

Table 9.2 qualitatively summarizes the three electronic effect resistive switching memories and their potentials for the next-generation nonvolatile memories. Table 9.3 summarizes reported values of some key switching parameters for the three mechanisms reviewed in the paper.

As a final note, a unique advantage of electronic effect resistive switching memory may lie in the large variety of materials and mechanisms. The combination of different materials and switching physics may provide innovative approaches to engineer the switching process to achieve better device performance. For example, the insertion of ferroelectric PZT layer at the

Table 9.3 Electronic effect resistive switching memory performance metrics

	Charge trapping	Mott transition	Ferroelectric
Cell element	1 T-1R, 1D-1R	1 T-1R, 1D-1R	1 T-1R, 1D-1R
Scaling limit (nm)	10–100	<200	≤ 10
On/off ratio	$\geq 10^2$	$\geq 10^3$	$\geq 10^3$
W/E time (ns)	<50	≤ 1 (1.5 ns demonstrated)	≤ 4
Retention time	10 yr (1600 min at 85 °C demonstrated)	—	10 yr (140 h demonstrated)
Cycling endurance	$>10^4$ cycles	$>10^8$ cycles	$>10^{12}$ cycles
W/E voltage (V)	1–5	≤ 1	≤ 1
Read voltage (V)	<1	≤ 1	<1
Write energy (J/bit)	$<1 \times 10^{-11}$	$<10^{-10}$	$<1 \times 10^{-11}$
Challenges	Defect fluctuation; material stability	Possible structural change; thermal stability	Ferroelectric fatigue

interface of an Ag/PCMO junction enhances the switching effects and improves device characteristics [111]. The combination of a ferroelectric layer and a ferromagnetic layer in the $LiNbO_3/La_{0.69}Ca_{0.31}MnO_3$ junction may create multiferroic functionalities [78]. The prerequisite is clear understanding of these different mechanisms and the ability to differentiate and control these different processes.

9.6 Summary

This chapter reviews three electronic effects that may contribute to resistive switching mechanisms in oxide-based RRAM devices: charge injection and trapping, Mott metal–insulator transition, and ferroelectric reversal. Recent research on RRAM switching mechanisms has converged to ionic reaction and migration that induce the formation and rupture of filamentary conductive paths. Pure electronic effects have been considered less likely to dominate the switching process. However, various electronic effects may still contribute partially to the switching process. For Mott materials and ferroelectric oxides, metal–insulator transition and ferroelectric reversal are possible mechanisms that trigger switching processes. Although the classification of the electronic effect mechanisms is somewhat subjective, evidence of electronic effect switching process is continuously reported in literature. Exploring these potential electronic effect mechanisms is still valuable for the understanding of the complex resistive switching process in RRAM devices.

References

1. Pagnia, H. and Sotnik, N. (1988) Bistable switching in electroformed metal-insulator-metal devices. *Physica Status Solidi (a)*, **108**, 11–65.
2. Dearnaley, G., Stoneham, A.M., and Morgan, D.V. (1970) Electrical phenomena in amorphous oxide films. *Reports on Progress in Physics*, **33**, 1129–1191.
3. Oxley, D.P. (1977) Electroforming, switching and memory effects in oxide thin films. *Electrocomponent Science and Technology*, **3**, 217–224.
4. Ansari, A.A. and Qadeer, A. (1985) Memory switching in thermally grown titanium oxide films. *Journal of Physics D: Applied Physics*, **18**, 911–917.
5. Argall, F. (1968) Switching phenomena in titanium oxide thin films. *Solid-State Electronics*, **11**, 535–541.
6. Bruyere, J.C. and Chakraverty, B.K. (1970) Switching and negative resistance in thin films of nickel oxide. *Applied Physics Letters*, **16**(1), 40–43.
7. Hickmott, T.W. (1965) Electron emission, electroluminescence, and voltage-controlled negative resistance in Al-Al$_2$O$_3$-Au diodes. *Journal of Applied Physics*, **36**(6), 1885–1896.
8. Hiatt, W.R. and Hickmott, T.W. (1965) Bistable switching in niobium oxide diodes. *Applied Physics Letters*, **6**(6), 106–108.
9. Simmons, J.G. and Verderber, R.R. (1967) New conduction and reversible memory phenomena in thin insulating films. *Proceedings of the Royal Society A*, **301**, 77–102.
10. Beck, A., Bednorz, J.G., Gerber, Ch. *et al.* (2000) Reproducible switching effect in thin oxide films for memory applications. *Applied Physics Letters*, **77**(1), 319–321.
11. Baek, I.G., Lee, M.S., Seo, S. *et al.* (Dec. 2004) Highly scalable non-volatile resistive memory using simple binary oxide driven by asymmetric unipolar voltage pulses, IEDM Tech. Dig., pp. 587–560.
12. Liu, S.Q., Wu, N.J., and Ignatiev, A. (2000) Electric-pulse-induced reversible resistance change effect in magnetoresistive films. *Applied Physics Letters*, **76**(19), 2479–2481.
13. Nian, Y.B., Strozier, J., Wu, N.J. *et al.* (2007) Evidence for an oxygen diffusion model for the electric pulse induced resistance change effect in transition-metal oxides. *Physical Review Letters*, **98**(14), 146403–146406.
14. Tsui, S., Baikalov, A., Cmaidalka, J. *et al.* (2004) Field-induced resistive switching in metal-oxide interfaces. *Applied Physics Letters*, **85**(2), 317–319.
15. Fujimoto, M., Koyama, H., Kobayashi, S. *et al.* (2006) Resistivity and resistive switching properties of Pr0.7Ca0.3MnO3 thin films. *Applied Physics Letters*, **89**(24), 2435041-3.

16. Hsu, S.T., Li, T., and Awaya, N. (2007) Resistance random access memory switching mechanism. *Journal of Applied Physics*, **101**(2), 024517-1-8.
17. Sawa, A., Fujii, T., Kawasaki, M., and Tokura, Y. (2004) Hysteretic current–voltage characteristics and resistance switching at a rectifying Ti/Pr$_{0.7}$Ca$_{0.3}$MnO$_3$ interface. *Applied Physics Letters*, **85**(18), 4073–4075.
18. Odagawa, A., Sato, H., Inoue, I.H. *et al.* (2004) Colossal electroresistance of a Pr$_{0.7}$Ca$_{0.3}$MnO$_3$ thin film at room temperature. *Physical Review B-Condensed Matter*, **70**(22), 224403-1-4.
19. Kim, D.S., Kim, Y.H., Lee, C.E., and Kim, Y.T. (2006) Colossal electroresistance mechanism in a Au/Pr$_{0.7}$Ca$_{0.3}$MnO$_3$/Pt sandwich structure: Evidence for a Mott transition. *Physical Review B-Condensed Matter*, **74**(17), 174430-1-6.
20. International Roadmap Committee (2009) Emerging Device Research Chapter, in the International Technology Roadmap of Semiconductors.
21. Fujii, T., Kawasaki, M., Sawa, A. *et al.* (2005) Hysteretic current–voltage characteristics and resistance switching at an epitaxial oxide Schottky junction SrRuO$_3$/SrTi$_{0.99}$Nb$_{0.01}$O$_3$. *Applied Physics Letters*, **86**(1), 012107-1-3.
22. Sawa, A., Fujii, T., Kawasaki, M., and Tokura, Y. (2006) Interface resistance switching at a few nanometer thick perovskite manganite active layers. *Applied Physics Letters*, **88**(23), 232112-1-3.
23. Chen, A., Haddad, S., Wu, Y.C. *et al.* (Dec. 2005) Non-volatile resistive wwitching for advanced memory applications, IEDM Tech. Dig., pp. 765–768.
24. Chen, A., Haddad, S., Wu, Y.C. *et al.* (2007) Switching characteristics of Cu$_2$O metal-insulator-metal resistive memory. *Applied Physics Letters*, **91**(12), 123517-1-3.
25. Yu, L., Kim, S., Ryu, M. *et al.* (2008) Structure effects on resistive switching of Al/TiO$_x$/Al devices for RRAM applications. *IEEE Electron Device Letters*, **29**(4), 331–333.
26. Xia, Y., He, W., Chen, L. *et al.* (2007) Field-induced resistive switching based on space-charge-limited current. *Applied Physics Letters*, **90**(2), 022907-1-3.
27. Xia, Y., Liu, Z., Wang, Y. *et al.* (2007) Conduction behavior change responsible for the resistive switching as investigated by complex impedance spectroscopy. *Applied Physics Letters*, **91**(10), 102904-1-3.
28. Ni, M.C., Guo, S.M., Tian, H.F. *et al.* (2007) Resistive switching effect in SrTiO$_{3-\delta}$/Nb-doped SrTiO$_3$ heterojunction. *Applied Physics Letters*, **91**(18), 183502-1-3.
29. Liu, Q., Guan, W., Long, S. *et al.* (2008) Resistive switching memory effect of ZrO$_2$ films with Zr$^+$ implanted. *Applied Physics Letters*, **92**(1), 012117-1-3.
30. Lee, H., Chen, P., Wu, T. *et al.* (Apr. 2008) HfO$_2$ bipolar resistive memory device with robust endurance using AlCu as electrode, VLSI-TSA, pp. 146–147.
31. Liu, L.F., Tang, H., Wang, Y. *et al.* (Oct. 2006) Reversible resistive switching of Gd-doped TiO$_2$ thin films for nonvolatile memory applications, Inter. Conf. Sol. State and Inter. Circ. Tech. (ICSICT), pp. 833–835.
32. Rossel, C., Meijer, G.I., Brémaud, D., and Widmer, D. (2001) Electrical current distribution across a metal-insulator-metal structure during bistable switching. *Journal of Applied Physics*, **90**(6), 2892–2894.
33. Watanabe, Y., Bednorz, J.G., Bietsch, A. *et al.* (2001) Current-driven insulator–conductor transition and nonvolatile memory in chromium-doped SrTiO$_3$ single crystals. *Applied Physics Letters*, **78**(23), 3738–3740.
34. Choi, D., Lee, D., Sim, H. *et al.* (2006) Reversible resistive switching of SrTiO$_x$ thin films for nonvolatile memory applications. *Applied Physics Letters*, **88**(8), 082904-1-3.
35. Tsai, C., Chang, T., Liu, P. *et al.* (2008) Low temperature improvement on silicon oxide grown by electron-gun evaporation for resistance memory applications. *Applied Physics Letters*, **93**(5), 052903-1-3.
36. Tsui, S., Wang, Y.Q., Xue, Y.Y., and Chu, C.W. (2006) Mechanism and scalability in resistive switching of metal-Pr$_{0.7}$Ca$_{0.3}$MnO$_3$ interface. *Applied Physics Letters*, **89**(12), 123502-1-3.
37. Papagianni, C., Nian, Y.B., Wang, Y.Q. *et al.* (Nov. 2004) Impedance study of reproducible switching memory effect, Non-Volatile Memory Tech. Workshop, pp. 125–128.
38. Zhuang, W.W., Pan, W., Ulrich, B.D. *et al.* (Dec. 2002) Novell colossal magnetoresistive thin film nonvolatile resistance random access memory (RRAM), IEDM Tech. Dig., pp. 193–196.
39. Schroeder, H., Zhirnov, V.V., Cavin, R.K., and Waser, R. (2010) Voltage-time dilemma of pure electronic mechanisms in resistive switching memory cells. *Journal of Applied Physics*, **107**(5), 054517-1-8.
40. Mott, N.F. (1968) Metal-insulator transition. *Reviews of Modern Physics*, **40**, 677.
41. Mott, N.F. (1982) Review lecture: metal-insulator transitions. *Proceedings of the Royal Society A*, **382**, 1–24.
42. Imada, M., Fujimori, A., and Tokura, Y. (1998) Metal-insulator transitions. *Reviews of Modern Physics*, **70**(4), 1039–1263.
43. Edwards, P.P., Johnston, R.L., Rao, C.N.R. *et al.* (1998) The metal-insulator transition: a perspective. *Philosophical Transactions of the Royal Society of London A*, **356**(1735), 5–22.

44. Ahn, C.H., Bhattacharya, A., Ventra, M.D. *et al.* (2006.) Electrostatic modification of novel materials. *Reviews of Modern Physics*, **78**(4), 1185–1212.
45. Asamitsu, A., Tomioka, Y., Kuwahara, H., and Tokura, Y. (1997) Current switching of resistive states in magnetoresistive manganites. *Nature*, **388**, 50–52.
46. Fors, R., Khartsev, S.I., and Grishin, A.M. (2005) Giant resistance switching in metal-insulator-manganite junctions: Evidence for Mott transition. *Physical Review B-Condensed Matter*, **71**, 045305-1-10.
47. Meijer, G.I., Staub, U., Janousch, M. *et al.* (2005) Valence states of Cr and the insulator-to-metal transition in Cr-doped $SrTiO_3$. *Physical Review B-Condensed Matter*, **72**(15), 155102-1-5.
48. Fujimotoa, M., Koyama, H., Konagai, M. *et al.* (2006) TiO_2 anatase nanolayer on TiN thin film exhibiting high-speed bipolar resistive switching. *Applied Physics Letters*, **89**(22), 223509-1-3.
49. Fujimotoa, M., Koyama, H., Hosoi, Y. *et al.* (2006) High-speed resistive switching of TiO_2/TiN nano-crystalline thin film. *Japanese Journal of Applied Physics Letters*, **45**, L310–L312.
50. Morin, F.J. (1959) Oxides which show a metal-to-insulator transition at the Neel temperature. *Physical Review Letters*, **3**(1), 34–36.
51. Stefanovich, G., Pergament, A., and Stefanovich, D. (2000) Electrical switching and Mott transition in VO_2. *Journal of Physics: Condensed Matter*, **12**(41), 8837–8845.
52. Guzman, G., Beteille, F., Morineau, R., and Livage, J. (1996) Electrical switching in VO_2 sol-gel films. *Journal of Materials Chemistry*, **6**(3), 505–506.
53. Kim, H.T., Chae, B.G., Youn, D.H. *et al.* (2005) Raman study of electric-field-induced first-order metal-insulator transition in VO_2-based devices. *Applied Physics Letters*, **86**(24), 242101-1-3.
54. Ko, C. and Ramanathan, S. (2008) Stability of electrical switching properties in vanadium dioxide thin films under multiple thermal cycles across the phase transition boundary. *Journal of Applied Physics*, **104**(8), 086105-1-3.
55. Almeida, L.A.L. de, Deep, G.S., Lima, A.M.N., and Neff, H. (2000) Thermal dynamics of VO_2 films within the metal–insulator transition: evidence for chaos near percolation threshold. *Applied Physics Letters*, **77**(26), 4365–4367.
56. Ruzmetov, D., Gopalakrishnan, G., Deng, J. *et al.* (2009) Electrical triggering of metal-insulator transition in nanoscale vanadium oxide junctions. *Journal of Applied Physics*, **106**(8), 083702-1-5.
57. Gopalakrishnan, G., Ruzmetov, D., and Ramanathan, S. (2009) On the triggering mechanism for the metal–insulator transition in thin film VO_2 devices: electric field versus thermal effects. *Journal of Materials Science*, **44**(19), 5345–5353.
58. Rozenberg, M.J., Inoue, I.H., and Sánchez, M.J. (2005) A model for non-volatile electronic memory devices with strongly correlated materials. *Thin Solid Films*, **486**(1–2), 24–27.
59. Rozenberg, M.J., Inoue, I.H., and Sánchez, M.J. (2006) Strong electron correlation effects in nonvolatile electronic memory devices. *Applied Physics Letters*, **88**(3), 033510-1-3.
60. Oka, T. and Nagaosa, N. (2005) Interfaces of correlated electron systems: proposed mechanism for colossal electroresistance. *Physical Review Letters*, **95**(26), 266403-1-4.
61. Park, W.J., Lee, J.W., Jeon, S.H., and Kim, C.W. (2008) "Non-volatile memory device including metal-insulator transition material", US Patent Application 0157186.
62. Zhou, C., Newns, D.M., Misewich, J.A., and Pattnaik, P.C. (1997) A field effect transistor based on the Mott transition in a molecular layer. *Applied Physics Letters*, **70**(5), 598–600.
63. Newns, D.M., Misewich, J.A., Tsuei, C.C. *et al.* (1998) Mott transition field effect transistor. *Applied Physics Letters*, **73**(6), 780–782.
64. Sharoni, A., Ramírez, J.G., and Schuller, I.K. (2008) Multiple avalanches across the metal-insulator transition of vanadium oxide nanoscaled junctions. *Physical Review Letters*, **101**(2), 026404-1-4.
65. Blom, P.W.M., Wolf, R.M., Cillessen, J.F.M., and Krijn, M.P.C.M. (1994) Ferroelectric Schottky diode. *Physical Review Letters*, **73**(15), 2107–2110.
66. Tsymbal, E.Y. and Kohlstedt, H. (2006) Tunneling across a ferroelectric. *Science*, **313**(5784), 181–183.
67. Rodríguez Contreras, J., Kohlstedt, H., Poppe, U. *et al.* (2003) Resistive switching in metal–ferroelectric–metal junctions. *Applied Physics Letters*, **83**(22), 4595–4597.
68. Son, J., Cagnon, J., and Stemmer, S. (2009) Electrical properties of epitaxial $SrTiO_3$ tunnel barriers on (001) Pt/$SrTiO_3$ substrates. *Applied Physics Letters*, **94**(6), 062903-1-3.
69. Pertsev, N.A., Tagantsev, A.K., and Setter, N. (2000) Phase transitions and strain-induced ferroelectricity in $SrTiO_3$ epitaxial thin films. *Physical Review B-Condensed Matter*, **61**(2), R825–R829.
70. Garcia, V., Fusil, S., Bouzehouane, K. *et al.* (2009) Giant tunnel electroresistance for non-destructive readout of ferroelectric states. *Nature*, **460**, 81–84.

71. Chanthbouala, A. *et al.* (2012) Solid-state memories based on ferroelectric tunnel junctions. *Nature Nanotechnology*, **7**, 101–104.

72. Chanthbouala, A. *et al.* (2012) A ferroelectric memristor. *Nature Materials*, **11**, 860–864.

73. Qu, H., Yao, W., Garcia, T. *et al.* (2003) Nanoscale polarization manipulation and conductance switching in ultrathin films of a ferroelectric copolymer. *Applied Physics Letters*, **82**(24), 4322–4324.

74. Bune, A.V., Fridkin, V.M., Ducharme, S. *et al.* (1998) Two-dimensional ferroelectric films. *Nature*, **391**, 874–877.

75. Zhuravlev, M.Y., Sabirianov, R.F., Jaswal, S.S., and Tsymbal, E.Y. (2005) Giant electroresistance in ferroelectric tunnel junctions. *Physical Review Letters*, **94**(24), 246802-1-4.

76. Kohlstedt, H., Pertsev, N.A., Contreras, J.Rodriguez., and Waser, R. (2005) Theoretical current-voltage characteristics of ferroelectric tunnel junctions. *Physical Review B-Condensed Matter*, **72**(12), 125341-1-10.

77. Velev, J.P., Duan, C.G., Belashchenko, K.D. *et al.* (2007) Effect of ferroelectricity on electron transport in Pt/BaTiO$_3$/Pt tunnel junctions. *Physical Review Letters*, **98**(13), 137201-1-4.

78. Guo, S.M., Zhao, Y.G., Xiong, C.M. *et al.* (2007) Current-voltage characteristics of LiNbO$_3$/La$_{0.69}$Ca$_{0.31}$MnO$_3$ heterojunction and its tunability. *Applied Physics Letters*, **91**(14), 143509-1-3.

79. Velev, J.P., Duan, C.G., Burton, J.D. *et al.* (2009) Magnetic tunnel junctions with ferroelectric barriers: prediction of four resistance states from first principles. *Nano Letters*, **9**(1), 427–432.

80. Garcia, V. *et al.* (2010) Ferroelectric control of spin polarization. *Science*, **327**, 1106–1110.

81. Batra, I.P. and Silverman, B.D. (1972) Thermodynamic stability of thin ferroelectric films. *Solid State Communications*, **11**(1), 291–294.

82. Batra, I.P., Wurfel, P., and Silverman, B.D. (1973) New type of first-order phase transition in ferroelectric thin films. *Physical Review Letters*, **30**(9), 384–387.

83. Kohlstedt, H., Pertsev, N.A., and Waser, R. (2002) Size effects on polarization in epitaxial ferroelectric films and the concept of ferroelectric tunnel junctions including first results. *Materials Research Society Symposium Proceedings*, **688**, C6.5.

84. Junquera, J. and Ghosez, P. (2003) Critical thickness for ferroelectricity in perovskite ultrathin films. *Nature*, **422**, 506–509.

85. Gerra, G., Tagantsev, A.K., Setter, N., and Parlinski, K. (2006) Ionic polarizability of conductive metal oxides and critical thickness for ferroelectricity in BaTiO$_3$. *Physical Review Letters*, **96**(10), 107603-1-4.

86. Kim, Y.S., Kim, D.H., Kim, J.D. *et al.* (2005) Critical thickness of ultrathin ferroelectric BaTiO$_3$ films. *Applied Physics Letters*, **86**(10), 102907-1-3.

87. Petraru, A., Kohlstedt, H., Poppe, U. *et al.* (2008) Wedgelike ultrathin epitaxial BaTiO$_3$ films for studies of scaling effects in ferroelectrics. *Applied Physics Letters*, **93**(7), 072902-1-3.

88. Bea, H., Fusil, S., Bouzehouane, K. *et al.* (2006) Ferroelectricity down to at least 2 nm in multiferroic BiFeO$_3$ epitaxial thin films. *Japanese Journal of Applied Physics*, **45**, L187–L189.

89. Chu, Y.H., Zhao, T., Cruz, M.P. *et al.* (2007) Ferroelectric size effects in multiferroic BiFeO$_3$ thin films. *Applied Physics Letters*, **90**(25), 252906-1-3.

90. Despont, L., Koitzsch, C., Clerc, F. *et al.* (2006) Direct evidence for ferroelectric polar distortion in ultrathin lead titanate perovskite films. *Physical Review B-Condensed Matter*, **73**(9), 094110-1-6.

91. Fong, D.D., Stephenson, G.B., Streiffer, S.K. *et al.* (2004) Ferroelectricity in ultrathin perovskite films. *Science*, **304**(5677), 1650–1653.

92. Lichtensteiger, C., Dawber, M., Stucki, N. *et al.* (2007) Monodomain to polydomain transition in ferroelectric PbTiO$_3$ thin films with La$_{0.67}$Sr$_{0.33}$MnO$_3$ electrodes. *Applied Physics Letters*, **90**(5), 052907-1-3.

93. Tybell, T., Ahn, C.H., and Triscone, J.M. (1999) Ferroelectricity in thin perovskite films. *Applied Physics Letters*, **75**(6), 856–858.

94. Gajek, M., Bibes, M., Fusil, S. *et al.* (2007) Tunnel junctions with multiferroic barriers. *Nature Materials*, **6**, 296–302.

95. Duan, C., Sabirianov, R.F., Mei, W. *et al.* (2006) Interface effect on ferroelectricity at the nanoscale. *Nano Letters*, **6**(3), 483–487.

96. Scott, J.F. (2007) Applications of modern ferroelectrics. *Science*, **315**(5814), 954–959.

97. Scott, J.F. and de Araujo, C.A.P. (1989) Ferroelectric memories. *Science*, **246**(4936), 1400–1405.

98. Dawber, M., Rabe, K.M., and Scott, J.F. (2005) Physics of thin-film ferroelectric oxides. *Reviews of Modern Physics*, **77**(4), 1083–1130.

99. Ahn, C.H., Rabe, K.M., and Triscone, J.M. (2004) Ferroelectricity at the nanoscale: local polarization in oxide thin films and heterostructures. *Science*, **303**(5657), 488–491.

100. Roelofs, A., Schneller, T., Szot, K., and Waser, R. (2003) Towards the limit of ferroelectric nanosized grains. *Nanotechnology*, **14**(2), 250–253.

101. Yun, W.S., Urban, J.J., Gu, Q., and Park, H. (2002) Ferroelectric properties of individual barium titanate nanowires investigated by scanned probe microscopy. *Nano Letters*, **2**(5), 447–450.

102. Fujimoto, K. and Cho, Y. (2003) High-speed switching of nanoscale ferroelectric domains in congruent single-crystal $LiTaO_3$. *Applied Physics Letters*, **83**(25), 5265–5267.

103. Boikov, Y.A., Goltsman, B.M., Yarmarkin, V.K., and Lemanov, V.V. (2001) Slow capacitance relaxation in $(BaSr)TiO_3$ thin films due to the oxygen vacancy redistribution. *Applied Physics Letters*, **78**(24), 3866–3868.

104. Watanabe, Y. (1995) Epitaxial all-perovskite ferroelectric field effect transistor with a memory retention. *Applied Physics Letters*, **66**(14), 1770–1772.

105. Watanabe, Y., Tanamura, M., and Matsumoto, Y. (1996) Memory retention and switching speed of ferroelectric field effect in $(Pb, La)(Ti, Zr)O_3/La_2CuO_4$:Sr heterostructure. *Japanese Journal of Applied Physics*, **35**, 1564–1568.

106. Tseng, T.Y. and Lee, S.Y. (2003) Improvement in retention time of metal–ferroelectric–metal–insulator–semi-conductor structures using MgO doped $Ba_{0.7}Sr_{0.3}TiO_3$ insulator layer. *Applied Physics Letters*, **83**(5), 981–983.

107. Chung, Y. (2002) Experimental 128-kbit ferroelectric memory with 10^{12} endurance and 10-year data retention. *IEE Proceedings - Circuits, Devices and Systems*, **149**(2), 136–142.

108. Lee, E., Jung, D., Kang, Y. *et al.* (2008) A characterization of endurance in 64 Mbit ferroelectric random access memory by analyzing the space charge concentration. *Japanese Journal of Applied Physics*, **47**, 2725–2727.

109. Yoon, S., Lee, N., Ryu, S. *et al.* (2005) Effect of ferroelectric switching time on fatigue behaviors of (117)- and (00*l*)-oriented $(Bi,La)_4Ti_3O_{12}$ thin films. *Thin Solid Films*, **484**(1–2), 374–378.

110. Niranjan, M.K., Wang, Y., Jaswal, S.S., and Tsymbal, E.Y. (2009) Prediction of a switchable two-dimensional electron gas at ferroelectric oxide interfaces. *Physical Review Letters*, **103**(1), 016804-1-4.

111. Xing, Z.W., Wu, N.J., and Ignatiev, A. (2007) Electric-pulse-induced resistive switching effect enhanced by a ferroelectric buffer on the $Pr_{0.7}Ca_{0.3}MnO_3$ thin film. *Applied Physics Letters*, **91**(5), 052106-1-3.

10

Macromolecular Memory

Benjamin F. Bory and Stefan C.J. Meskers
Eindhoven University of Technology, The Netherlands

10.1 Chapter Overview

Macromolecular materials, such as polymers, can be used to make electronic memory cells. This chapter focuses on data storage based on resistive switching in diodes incorporating macromolecules. The organization of the chapter is as follows. We start with a brief introduction into the chemical structure and properties of macromolecules. We then sketch the phenomenology of electronic memory effects involving macromolecules and review the current state of the art in polymer memory technology. The subsequent sections are devoted to the electrical and semi-(conducting) properties of polymers and macromolecules and physical–chemical mechanisms of the electronic memory effects. Finally we classify different types of macromolecular memory based on the physical–chemical characteristics of the material involved.

10.2 Macromolecules

10.2.1 Chemical Structure

Macromolecules can be defined as large molecules built up of more than, say, 10^2 atoms. The class of macromolecules includes synthetic and biological polymers, polyelectrolytes and also encompasses frameworks such as carbon nanotubes and graphene. The backbone of these macromolecules often consists mainly of carbon atoms, sometimes silicon. Two carbon atoms in the chain can be linked via either a single, double or triple bond depending on the number of electron pairs that are shared between the two atoms. Hydrogen and hetero atoms such as oxygen, nitrogen, sulfur can also be incorporated in the chain in various ways and this makes the possible variation in chemical structure almost endless. Compared to other solids, polymers can be characterized as "soft" condensed matter. The solid state structure of a polymer is usually disordered and has considerable internal free volume. The free volume leaves room for motion of the chains in the solid, which in turn allows for migration and diffusion of small molecules and ions and for rearrangements of intermolecular and intramolecular ordering.

Emerging Nanoelectronic Devices, First Edition. An Chen, James Hutchby, Victor Zhirnov and George Bourianoff.
© 2015 John Wiley & Sons, Ltd. Published 2015 by John Wiley & Sons, Ltd.

Furthermore, because of the large size of the molecules, relaxation times to reach thermodynamic equilibrium are often very long. As a consequence, macromolecular materials commonly occur in long-lived, metastable thermodynamic states and might not even reach thermodynamic equilibrium in a time span of years.

10.2.2 Memristive Effects in Macromolecular Materials

The free volume and internal motion allow macromolecules to adapt their structure and conformation in response to external stimuli and perturbations. This also extends to the electrical properties of macromolecules. In the past decade it has been shown that using a macromolecular material as insulator in a Metal–Insulator–Metal (MIM) structure, one can realize diodes whose electrical resistance can be modified by application of electrical bias voltage. Table 10.1 gives an overview of review articles that describe progress and achievements in this field.

The change in resistance may be transient and disappear when short-circuiting the diode (volatile memory) or could be semi-permanent under short-circuit conditions (nonvolatile memory). For the latter *memristive* devices, one can differentiate between write once read many (WORM, W) and rewriteable memory cells. The rewriteable cells can be further subdivided into cells that require write and erase voltages of opposite polarity (Bipolar, B) and cells that can be operated using write and erase voltages of the same polarity (Unipolar, U).

Attractive features of macromolecular memory include the possibility to use solution-based deposition methods such as printing and casting of the active layers. In addition, flexible, lightweight, and disposable substrates may be used. Furthermore because macromolecular memory is essentially a thin film technology and one can make economic use of raw materials (Figure 10.1).

10.2.3 Current State of Macromolecular Memory

In the last decade enormous progress has been made in the realization of macromolecular memory. Resistive switching has been shown to occur locally and downscaling to nm lengths is feasible. Projected retention times on the order of 10 years have been realized and switching is possible using short voltage pulses (submicrosecond). Integration into arrays has been

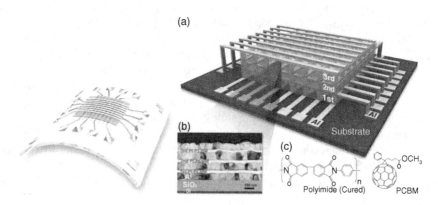

Figure 10.1 Macromolecular memory: flexible substrates (left, [60]) and 3-D integration (right, a–c, [26]. Copyright: Wiley VCH

Table 10.1 Reviews on diodes incorporating organic materials showing electronic memory effects in their electrical resistivity

Title of review	Reference details	Description
Electrical switching and bistability in organic/polymeric thin films and memory devices [1]	*Adv. Funct. Mater.* vol. 16, pp. 1001–1014, 2006.	Resistive switching in π-conjugated polymers
Nonvolatile memory elements based on organic materials [2]	*Adv. Mater.* vol. 19, pp. 1452–1463, 2007.	Classification of electrical characteristics and minimal performance specifications
Nanoionics-based resistive switching memories [3]	*Nat. Mater.* vol. 6, pp. 833–840, 2007.	Redox process involving metal ions; growth/dissolution of filaments
Polymer electronic memories: materials, devices and mechanisms [4]	*Prog. Polym. Sci.* vol. 33, pp. 917–978, 2008.	Mechanism in relation to chemical structure of polymers
Polymer and organic nonvolatile memory devices [5]	*Chem. Mater.* vol. 23, pp. 341–358, 2011.	Principles of operation and feasibility
Recent results on organic-based molecular memories [6]	*Curr. Appl. Phys.* vol. 11, pp. e49–e57, 2011.	Molecular aspects of organic memory
Progress in nonvolatile memory devices based on nanostructured materials and nanofabrication [7]	*J. Mater. Chem.* vol. 21, pp. 14 907–14 112, 2011.	Fabrication and integration of organic memory and device architecture
Donor–acceptor polymers for advanced memory device applications [8]	*Polym. Chem.* vol 2, pp. 2169–2174, 2011.	Polyimide polymers
Organic resistive memory devices: performance enhancement, integration, and advanced architectures [9]	*Adv. Funct. Mater.* vol 21, pp. 2806–2829, 2011.	Architecture, operation and integration of polymer memory; inventory of the different mechanisms and the integration into electronics
Organic resistive nonvolatile memory materials [10]	*MRS Bull.* vol. 37, pp. 144–149, 2012.	Performance and requirements; integrated circuit architectures
Advanced polyimide materials: syntheses, physical properties and applications [11]	*Prog. Polym. Sci.* vol. 37, pp. 907–974, 2012.	Synthesis and properties of polyimide materials; functionalized polyimids for resistive memory
Polyimide memory: a pithy guideline for future applications [12]	*Polym. Chem.* vol. 4, pp. 16–30, 2013.	Classification of memory effects in polyimides; electronic structures of polyimide materials in relation to memory

demonstrated. So far, a complete device technology seems not to have been disclosed. A main challenge for macromolecular memory is still to understand and control the resistive switching. As will become clear, macromolecular memory operates close to the threshold for electrical breakdown. Reproducibility of the switching characteristics is often an issue. In view of the very stringent requirements regarding uniformity, cycle endurance, and bit error rate for application as NAND flash, variability in the characteristic of cells may be an issue.

10.3 Elementary Physical Chemistry of Macromolecular Memory

10.3.1 Required Activation Barrier between Stable States

For a rewriteable, nonvolatile memory, one would like to have a molecular system that is bistable, having two distinct stable states ("ON" and "OFF") separated by a barrier in energy (see Figure 10.2). For electronic memory applications, the ON and OFF states should differ in charge transport related properties, namely delocalization length of charge carriers and energies of the electron and holes states on the polymer. The transition between the ON and OFF states is in essence a monomolecular chemical reaction. The transition state theory developed by Eyring describes the rate of chemical reactions and allows one to estimate the rate of interconversion between ON and OFF state, $k_{ON/OFF}$ in s^{-1}, as a function of the height of the energy barrier separating the two minima [13]. To warrant a lifetime of at least 10 years for the ON and OFF state, the barrier in between the two states should have a height of at least 1.25 eV of Gibbs free energy.

The barrier should have sufficient width to suppress quantum mechanical tunneling between the states. Application of a potential difference over the memory cell should lead to rapid conversion between the ON and OFF states. From the estimated barrier height it then follows that the minimal voltages needed to switch the cell will exceed 1.25 V.

Furthermore, memory cells relying on trapping of charges can most likely not be realized with semiconducting materials with a bandgap smaller than 1.25 eV.

One might argue that small molecules are inherently less stable and more prone to thermodynamical and quantum mechanical fluctuations than bulk phases in large crystals. So is there any experimental evidence for small bistable molecular systems with lifetime of the individual states exceeding 10 years? Amino acids are small molecules with fewer than 12

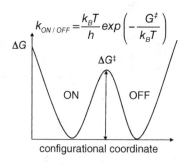

$$k_{ON/OFF} = \frac{k_B T}{h} exp\left(-\frac{G^{\ddagger}}{k_B T}\right)$$

Figure 10.2 Gibbs Free energy ΔG as function of a generalized configurational coordinate for a bistable molecule. The rate of spontaneous interconversion between ON and OFF states, $k_{ON/OFF}$, is determined by the height of the barrier ($\Delta G^{\ddagger}$) in between the two stable states and temperature. For a lifetime of ON and OFF states of 10 years at room temperature, a minimal barrier height of $\Delta G^{\ddagger} = 1.25$ eV is required according to molecular transition state theory

carbon atoms, that serve as monomeric units of proteins and polypeptides. Amino acids can exist in two mirror image related stable configurations labeled L and D. Even though the two forms have identical energies, in living plants and animals only the L form is produced and biochemically active. Analysis of fossils has shown that interconversion of the L into the D forms takes place on a time scale of 10^3 years [14]. This shows that bistable molecular systems with interconversion sufficiently slow for electronic memory applications indeed exist.

10.3.2 Electronic Structure of Macromolecules: Insulators, Semiconductors, Conductors

Depending on the type of connection between the carbon atoms, macromolecules behave electrically as either insulators, semiconductors or conductors. The origin of the diversity lies in the different possible types of coordination of substituents around the carbon atom. Two neighboring carbon atoms can form a chemical bond by sharing two valence electrons between them, with each atom donating one electron. The two electrons forming the bond must have antiparallel spin. Because carbon has four valence electrons it can bind to maximally four other atoms. Electrostatic repulsion between the electron pairs forming the bonds is minimized when the four carbon atoms binding to the central one are arranged in the form of a tetrahedron. Hybridization of the hydrogen like $2s$ and three $2p$ orbitals around the carbon nucleus into four equivalent sp^3 orbitals facilitates the formation of these four bonds. The resulting atomic arrangement is that of diamond, a material that can be characterized as a wide (5.4 eV) indirect bandgap semiconductor.

Within the diamond lattice, electrons and holes have high and almost equal mobility (2×10^3 cm^2/Vs).

In another allotrope of carbon, graphite, each carbon atom is bound to only three neighboring atoms and one valence electron per atom is delocalized over the honeycomb hexagonal sheet of carbon atoms. In this case, repulsion between electron pairs forming the bonds results in a planar triangular arrangement of the three substituents. Hybridization of the carbon $2s$ with *two* $2p$ orbitals to sp^2 orbitals can facilitate this bonding. This leaves one $2p$ orbital unhybridized containing one electron. Overlap between unhybridized $2p$ orbitals on neighboring sp^2 hybridized carbon atoms allows for quantum mechanical delocalization of these electrons and gives rise to (semi-)conducting properties of the molecular material. The delocalization of the electrons in the $2p$ atomic orbitals contributes to the bonding between two neighboring carbon atoms (now sharing a *double* bond) and the two atoms are said to be π-conjugated. A single layer of graphite, graphene, contains delocalized electrons and may be classified as a zero bandgap semiconductor. Electron and hole mobilities are exceptionally high (estimated $> 2 \times 10^4$ cm^2/Vs [15]) and relate, from a chemical point of view, to the stability of the two-dimensional hexagonal lattice against distortion.

Replacing two of the coordinating carbon atoms by hydrogen, a polymeric structure may be obtained (polyacetylene, see Figure 10.3). In this macromolecule, or π-conjugated polymer (named after the double or π-bond) one electron per carbon atom is still delocalized. Due to pairing of neighboring carbon atoms in the chain (Peierls distortion), a band gap opens (1.5 eV) and the resulting chain is alternating longer, singly bonded carbon atom pairs and shorter doubly bonded pairs. Because of the presence of double or π-bonds, polyacetylene is a π-conjugated polymer. The mobilities of charge carriers along the chain are reduced because of polaron formation and bulk mobilities are severely limited by poor intermolecular transport of carriers and trapping on impurities and defects. Carrier mobilities in purified molecular

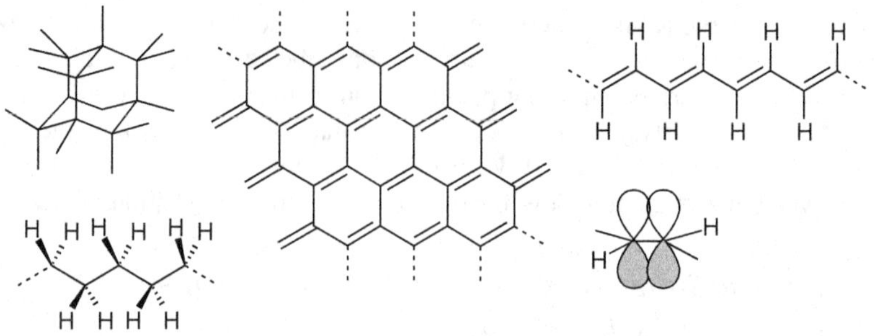

Figure 10.3 Left: Diamond lattice and polyethene with sp^3 hybridized carbon atoms and saturated bonds. Middle: Graphene sheet. Right: Polyacetylene polymer with two overlapping atomic $2p_z$ orbitals on two neighboring sp^2 hybridized carbon atoms forming a double bond

materials rarely exceed 1 cm^2/Vs and for π-conjugated polymers usually fall in the range 10^{-3}–10^{-7} cm^2/Vs. π-Conjugated polymers can be made conductive by oxidative or reductive doping. In the case of polyacetylene, doping with molecular iodine (I_2) results in a hole conducting material with conductivity approaching that of metals. By introducing dopants which themselves are macromolecules, diffusion, sublimation, and electromigration of the dopants can be suppressed and stable molecular conductors can be obtained. A well known example is poly(3,4-ethylenedioxythiophene): poly(styrenesulfonate), or PEDOT:PSS, where positive charge carriers on the π-conjugated EDOT polymer are stabilized by the deprotonated and negatively charged polystyrene sulfonate polymer (PSS).

10.3.3 Inducing Changes by Application of Bias Voltage: Electrochemistry

As mentioned above, bias voltages ≥ 1.25 V will be needed for switching between bistable states of macromolecular memory. Currently in NAND flash technology, voltages up to 20 V are used to program the floating gate transistor. The question arises: which chemical changes might one expect to be induced by applying voltages in the range 1–10^1 volt? From physical chemistry, the energies needed to break a chemical bond are well known. For a carbon–carbon single bond, 3.6 eV is required to break the bond. To split the strongest homodimeric bond known in organic chemistry, the nitrogen–nitrogen triple bond N≡N, 9.8 eV is required. Taking into account that macromolecular memory is a thin film technology and that the bias voltage applied often results in a potential drop localized over only a narrow spatial region in the organic materials, programming voltages up to 10 V could well lead to charge carriers with up to 10 eV of kinetic energy. In principle, the 10 eV energy of the electron suffices to break any chemical bond in a molecule. We remark that, in electrochemistry, energies needed for a chemical reaction can be measured quantitatively, using polar solvents and adding salt ions (electrolyte) to localize the electric field strength in a small region close to an electrode. From electrochemistry, it is well known that within a voltage range up to a few volts, almost any chemical conversion can be driven, including oxidation and reduction of all metals. This simple line of argument illustrates the tremendous difficulties that one may face when trying to unravel mechanistic aspects of memristive effects in organic molecular materials: with programming voltages typically in the range 2–8 V a multitude of chemical processes is likely to be induced.

10.3.4 Filamentation

As mentioned above, by applying potential differences on the order of 10 V over a thin organic layer, one can drive a molecular material very far out of thermodynamic equilibrium. Note that the low mobility of charge carriers contributes to localizing the electric field strength to narrow regions. In soft condensed matter, the behavior of molecular systems far away from equilibrium has been studied in considerable detail in the last half century. One of the striking outcomes is that when a system is driven far out of equilibrium, spontaneous self-organization into patterns and structures is possible. In contrast, under near equilibrium conditions, such self-organization processes would not be favorable because of the reduction of entropy associated with the emergence of organization, patterns and structures [16]. One of the self-organization processes studied in detail is *filamentation*, which occurs for instance in the transmission of light when the strength of the optical electric field becomes so high that the refractive index of the material changes. The light can then focus itself into small filaments [17]. Other examples of filamentation are the formation of plasma arcs [18]. Filamentation of electrical current in planar organic memory diodes is often observed and should most likely be considered as an integral part of the response of the materials to the large thermodynamic driving force applied.

10.4 Classes of Macromolecular Memory Materials and Their Performance

In this section we give an overview of recent reports on macromolecular memory (see Table 10.2). Because the mechanism of switching is often unclear we classify the memory diodes according to the type of macromolecular material used.

10.4.1 Saturated Macromolecules

Polymers that do not contain any double bonds are insulators and charge carrier injection via a metal contact into these materials is practically impossible. However these polymers may still be used as host materials in memory applications, when combined with additives that can take care of carrier transport. For instance, polyethylene oxide (PEO) is a polymer that can solvate ions. PEO combined with the silver salt $AgClO_4$ can be used to realize memory cells. In $AgClO_4$-PEO cells, electrochemical conversion of silver ions into metallic silver gives rise to the formation and dissolution of metal filaments, leading to resistive switching [19]. A disadvantage of these electrochemical cells is that the amount of charge needed to switch the diode depends on the history of the diode. The amount of current needed to switch the diode is only minimal close to the percolation threshold via the metallic filament. Some saturated polymers containing fluor substituents are ferroelectric and can be used to make a capacitive memory. When blended with semiconducting polymer resistive switching diodes have been accomplished [33]. For further information, we refer the reader to the discussion of ferroelectric memories in Chapters 6 and 9.

10.4.2 Macromolecules with π-Conjugation

10.4.2.1 Dynamic Doping

Organic molecules containing double bonds (π-conjugation) usually have semiconducting properties. The presence of charged dopants in these materials can change barriers for charge injection and the charge carrier density. Migration of charged dopants under the influence of

Table 10.2 Memory performance characteristics for macromolecular memory

Material	MIM architecture	Type	V_{on}	V_{off}	On/Off	Retention (s)	Endurance
PEO-AgClO$_4$ [19]	Pt/Ag	B	1	−1	10^5	10^5	10
π-Conjugated polyelectrolyte [20]	Ag/p$^+$-Si	B	3	−2	10^4	10^5	nd
Donor–acceptor polyimide [21]	ITO//Al	B	2.7	−1.5	10^4	10^4	nd
Donor–acceptor polyimide	ITO//Al	U	2.2	1	10^4	10^4	nd
Donor–acceptor polyimide	ITO//Al	W	2				
PEDOT:PSS [22]	Al//Al	B	−2.5	1.2	10^5	10^5	10
PEDOT:PSS [23]	Au//Au	U	4	1	10^2	10^4	nd
PEDOT:PSS [24]	Al/Cu	B	1	−0.8	10^4	10^4	nd
Fullerene [25]	ITO//Al	U	4	2	10^5	10^4	10^1
Fullerene [26]	Al//Al	U	5	10	10^3	10^7	10^2
Carbon nanotube [27]	Al//Al	B	−2	2	10^2	10^5	10^2
Diamond-like carbon [28]	W//Pt	U	1.2	0.4	300	10^3	10^3
Diamond-like carbon [29]	Ti/Pt	U	1.4	−1.5	10	10^4	nd
Amorphous Carbon [30]	Ag//Ag	B	6	−7	30	10^6	30
Graphite [31]		U	3	6	10^6	10^6	10^6
Graphene [32]	ITO/PMMA/PMMA/Al	B	3.5	−4	10^6	10^5	10^5

B: bipolar; U: unipolar; W: WORM; nd: not determined.

applied bias leads to memristive effects [34]. The memory functionality of the macromolecular material can be enhanced further by doping with metal ions that can also be reduced electrochemically to metal filaments.

10.4.2.2 Donor–Acceptor Polymers

A large research effort has focused on polyimide polymers functionalized with π-conjugated moieties that have pronounced electron donating and electron accepting properties [8,11,12]. These materials often show volatile resistive switching, but nonvolatile behavior has also been observed, including both WORM and rewritable characteristics. The volatile behavior has been related to the presence of an ionized electronic state of the donor acceptor polymer (charge transfer state) in which the electron donating unit has transferred an electron to the electron accepting moieties.

10.4.2.3 Conducting Polymers

Conducting polymers such as PEDOT:PSS have been used as WORM, relying on destructive processes induced at high current densities [35,36]. However also rewritable memory cells have been realized with PEDOT:PSS and a number of studies indicate the importance of the interface between polymer and metal contact in the switching and indicate an active role of native oxides [37–39].

10.4.2.4 Carbon Allotropes

Fullerenes are spheroidal π-conjugated molecules that in comparison with π-conjugated polymers show high electron and hole mobilities. When mixed with a host polymer, a solution processable material is obtained that shows rewriteable memory effects after optimization of the mixing ratio characteristic [40–42]. Very promising memory characteristics have been reported and integration of cells into 2-D [43,44] and 3-D [26] has been achieved. A number of studies have indicated the importance of an interfacial oxide layer in the resistive switching process [26,43–46]. Carbon nanotubes can also be applied in memories and also here indications for the involvement of an interfacial oxide layer in the switching have been obtained (Al_2O_3 [27,47], SiO_2 [48]). Graphene can give rise to bipolar switching [49] and sometimes WORM [50] memory effects. Graphene oxide functionalized with polymer chains gives a solution processable macromolecular material, displaying memory effects in combination with ITO and Al electrodes [51–53]. Finally, amorphous carbon and diamond-like carbon are amorphous carbon allotropes that show electronic memory effects in MIM devices. Amorphous carbon contains both sp^3 and sp^2 hybridized carbon atoms. Variation in the ratio and distribution of the semi-conducting (sp^2) and insulating (sp^3) varieties induced by the application of electrical stress has been suggested to lead to memory effects [54,55].

10.4.2.5 Hybrid Diodes Involving a Dedicated Inorganic Switching Layer

A number of studies have been devoted to MIM structures incorporating a (macro-)molecular layer in combination with a dedicated inorganic switching layer consisting of metal oxide [56,57] or metal halide [58]. Similar to metal oxide memory cells, these hybrid devices require an electroforming step before resistive memory effects become apparent. The electroforming process is induced by bias voltage stress and can be subdivided into different stages.

The early stages involve charge trapping at the interface between the macromolecular semiconductor and the metal oxide [59]. The final step of electroforming involves soft breakdown of the inorganic insulator and is related to hole injection [58].

10.5 Perspectives

Memory cells using macromolecular materials that rely on resistive switching show excellent memory characteristics when operated as isolated cells under single write/read/erase events. Large-scale integration does not seem to have been realized, and therefore performance as integrated memory is currently difficult to estimate. The occurrence of a so-called dead time in repeated switching could be a major hurdle. Key opportunities of macromolecular memory include the low cost of raw materials and deposition methods, for example, printing and coating. Furthermore, memory cells incorporating environmentally friendly materials can be realized. This provides applications in low-end, disposable electronics on, for example, flexible substrates such as paper or plastic.

Summary of macromolecular memory parameters:

Cell size	Demonstrated	130 nm
	Projected	10 nm
Density	Demonstrated	Not known
	Projected	Comparable to ferroelectric and phase change
Device speed (switching speed for logic devices;	Demonstrated	15 ns
reading/writing speed for memory devices	Projected	<10 ns
Circuit speed	Demonstrated	Not known
	Projected	Depends on "dead-time"
Operation voltages	Demonstrated	<6 V
	Projected	Not known
Switching energy	Demonstrated	5E-11 Joule/bit
	Projected	Not known
Endurance	Demonstrated	>1E5
	Projected	Not known
Retention (for memory)	Demonstrated	~1 yr
	Projected	
Binary throughput	Demonstrated	Not known
	Projected	Not known

10.6 Summary

Macromolecular materials can give rise to a wide variety of electronic memory effects. Although impressive memory characteristics have been obtained, lack of understanding and control over the switching mechanisms is still a major issue.

Acknowledgments

The work of BBF forms part of research programme of the Dutch Polymer Institute (DPI 704). We thank D.M. De Leeuw and H.L. Gomes for stimulating discussion.

References

1. Yang, Y., Ouyang, J., Ma, L.P. *et al.* (2006) Electrical switching and bistability in organic/polymeric thin films and memory devices. *Advanced Functional Materials*, **16**(8), 1001–1014.
2. Scott, J.C. and Bozano, L.D. (2007) Nonvolatile memory elements based on organic materials. *Advanced Materials*, **19**(11), 1452–1463.
3. Waser, R. and Aono, M. (2007) Nanoionics-based resistive switching memories. *Nature Materials*, **6**(11), 833–840.
4. Ling, Q.-D., Liaw, D.-J., Zhu, C. *et al.* (2008) Polymer electronic memories: materials, devices and mechanisms. *Progress in Polymer Science*, **33**(10), 917–978.
5. Heremans, P., Gelinck, G.H., Muller, R. *et al.* (2011) Polymer and organic nonvolatile memory devices. *Chemistry of Materials*, **23**(3), 341–358.
6. Salvo, B.De., Buckley, J., and Vuillaume, D. (2011) Recent results on organic-based molecular memories. *Current Applied Physics*, **11**(2), E49–E57.
7. Lee, J.-S. (2011) Progress in non-volatile memory devices based on nanostructured materials and nanofabrication. *Journal of Materials Chemistry*, **21**(37), 14097–14112.
8. Liu, C.-L. and Chen, W.-C. (2011) Donor–acceptor polymers for advanced memory device applications. *Polymer Chemistry*, **2**(10), 2169–2174.
9. Cho, B., Song, S., Ji, Y. *et al.* (2011) Organic resistive memory devices: Performance enhancement, integration, and advanced architectures. *Advanced Functional Materials*, **21**(15), 2806–2829.
10. Lee, T. and Chen, Y. (2012) Organic resistive nonvolatile memory materials. *MRS Bulletin*, **37**(2), 144–149.
11. Liaw, D.-J., Wang, K.-L., Huang, Y.-C. *et al.* (2012) Advanced polyimide materials: syntheses, physical properties and applications. *Progress in Polymer Science*, **37**(7), 907–974.
12. Kurosawa, T., Higashihara, T., and Ueda, M. (2013) Polyimide memory: a pithy guideline for future applications. *Polymer Chemistry*, **4**(1), 16–30.
13. Atkins, P.W. and Depaula, J. (2009) *Physical Chemistry*, Oxford University Press, Oxford.
14. Poinar, H.N., Höss, M., Bada, J.L., and Pääbo, S. (1996) Amino acid racemization and the preservation of ancient DNA. *Science*, **272**(5263), 864–866.
15. Geim, A.K. and Novoselov, K.S. (2007) The rise of graphene. *Nature Materials*, **6**(3), 183–191.
16. Cross, M.C. and Hohenberg, P.C. (1993) Pattern formation outside of equilibrium. *Reviews of Modern Physics*, **65**(3), 851–1112.
17. Couairon, A. and Mysyrowicz, A. (2007) Femtosecond filamentation in transparent media. *Physics Reports-Review Section of Physics Letters*, **441**(2–4), 47–189.
18. Purwins, H.-G. and Berkemeier, J. (2011) Self-organized patterns in planar low-temperature DC gas discharge. *IEEE Transactions on Plasma Sciences*, **39**(11), 2116–2117.
19. Wu, S., Tsuruoka, T., Terabe, K. *et al.* (2011) A polymer-electrolyte-based atomic switch. *Advanced Functional Materials*, **21**(1), 93–99.
20. Cho, B., Yun, J.M., Song, S. *et al.* (2011) Direct observation of Ag filamentary paths in organic resistive memory devices. *Advanced Functional Materials*, **21**(20), 3976–3981.
21. Li, Y., Fang, R., Ding, S., and Shen, Y. (2011) Rewritable and non-volatile memory effects based on polyimides containing pendant carbazole and triphenylamine groups. *Macromolecular Chemistry and Physics*, **212**(21), 2360–2370.
22. Yang, J., Zeng, F., Wang, Z.S. *et al.* (2011) Modulating resistive switching by diluted additive of Poly(vinyl-pyrrolidone) in Poly(3,4-ethylenedioxythiophene):Poly(styrenesulfonate). *Journal of Applied Physics*, **110**(11), 114518.
23. Liu, X., Ji, Z., Tu, D. *et al.* (2009) Organic nonpolar nonvolatile resistive switching in Poly(3,4-ethylene-dioxythiophene): Polystyrenesulfonate thin film. *Organic Electronics*, **10**(6), 1191–1194.
24. Wang, Z., Zeng, F., Yang, J. *et al.* (2012) Resistive switching induced by metallic filaments formation through Poly (3,4-ethylene-dioxythiophene): Poly(styrenesulfonate). *ACS Applied Materials & Interfaces*, **4**(1), 447–453.
25. Hahm, S.G., Kang, N.-G., Kwon, W. *et al.* (2012) Programmable bipolar and unipolar nonvolatile memory devices based on Poly(2-(N-carbazolyl)ethyl methacrylate) end-capped with fullerene. *Advanced Materials*, **24**(8), 1062–1066.
26. Song, S., Cho, B., Kim, T.-W. *et al.* (2010) Three-dimensional integration of organic resistive memory devices. *Advanced Materials*, **22**(44), 5048–5053.
27. Hwang, S.K., Lee, J.M., Kim, S. *et al.* (2012) Flexible multilevel resistive memory with controlled charge trap B- and N-Doped Carbon Nanotubes. *Nano Letters*, **12**(5), 2217–2221.

28. Fu, D., Xie, D., Feng, T. *et al.* (2011) Unipolar resistive switching properties of diamondlike carbon-based RRAM devices. *IEEE Electron Device Letters*, **32**(6), 803–805.

29. Peng, P., Xie, D., Yang, Y. *et al.* (2012) Resistive switching behavior in diamond-like carbon films grown by pulsed laser deposition for resistance switching random access memory application. *Journal of Applied Physics*, **111**(8), 084501.

30. Chai, Y., Wu, Y., Takei, K. *et al.* (2011) Nanoscale bipolar and complementary resistive switching memory based on amorphous carbon. *IEEE Transactions on Electron Devices*, **58**(11), 3933–3939.

31. Li, Y., Sinitskii, A., and Tour, J.M. (2008) Electronic two-terminal bistable graphitic memories. *Nature Materials*, **7**(12), 966–971.

32. Son, D.I., Kim, T.W., Shim, J.H. *et al.* (2010) Flexible organic bistable devices based on graphene embedded in an insulating Poly(methyl methacrylate) polymer layer. *Nano Letters*, **10**(7), 2441–2447.

33. Asadi, K., De Leeuw, D.M., De Boer, B., and Blom, P.W.M. (2008) Organic Non-volatile memories from ferroelectric phase-separated blends. *Nature Materials*, **7**(7), 547–550.

34. Sim, R., Chan, M.Y., Wong, A.S.W., and Lee, P.S. (2011) Alternative resistive switching mechanism based on migration of charged counter-ions within conductive polymers. *Organic Electronics*, **12**(1), 185–189.

35. D. M., DeLeeuw., Geuns, T.C.T., and De Brito, B.C. (2011) "Switching element used as memory element e.g. write-once-read-many memory element" WO patent WO2008087566-A1.

36. Wang, J., Cheng, X., Caironi, M. *et al.* (2011) Entirely solution-processed write-once-read-many-times memory devices and their operation mechanism. *Organic Electronics*, **12**(7), 1271–1274.

37. Kim, J.Y., Jeong, H.Y., Kim, J.W. *et al.* (2011) Critical role of top interface layer on the bipolar resistive switching of Al/PEDOT:PSS/Al Memory Device. *Current Applied Physics*, **11**(2), e35–e39

38. Ha, H. and Kim, O. (2010) Electrode-material-dependent switching characteristics of organic nonvolatile memory devices based on Poly(3,4-ethylene dioxythiophene): poly(styrenesulfonate) Film. *IEEE Electron Device Letters*, **31**(4), 368–370.

39. Jeong, H.Y., Kim, J.Y., Yoon, T.H., and Choi, S.-Y. (2010) Bipolar resistive switching characteristics of Poly(3,4-ethylene-dioxythiophene): Poly(styrenesulfonate) thin film. *Current Applied Physics*, **10**(1), e46–e49

40. Chen, J.-C., Liu, C.-L., Sun, Y.-S. *et al.* (2012) Tunable electrical memory characteristics by the morphology of self-assembled block copolymers: PCBM nanocomposite films. *Soft Matter*, **8**(2), 526–535.

41. Lian, S.-L., Liu, C.-L., and Chen, W.-C. (2011) Conjugated fluorene based rod-coil block copolymers and their PCBM composites for resistive memory switching devices. *ACS Applied Materials & Interfaces*, **3**(11), 4504–4511.

42. Hsu, J.-C., Liu, C.-L., Chen, W.-C. *et al.* (2011) A Supramolecular approach on using Poly(fluorenylstyrene)-block-poly(2-vinylpyridine):PCBM composite thin films for non-volatile memory device applications. *Macromolecular Rapid Communications*, **32**(6), 528–533.

43. Kim, J.J., Cho, B., Kim, K.S. *et al.* (2011) Electrical characterization of unipolar organic resistive memory devices scaled down by a direct metal-transfer method. *Advanced Materials*, **23**(18), 2104–2108.

44. Cho, B., Song, S., Ji, Y., and Lee, T. (2010) Electrical characterization of organic resistive memory with interfacial oxide layers formed by O2 plasma treatment. *Applied Physics Letters*, **97**(6), 063305.

45. Ko, S.H., Yoo, C.H., and Kim, T.W. (2012) Electrical bistabilities and memory stabilities of organic bistable devices utilizing C_{60} molecules embedded in a polymethyl methacylate matrix with an Al_2O_3 blocking layer. *Journal of the Electrochemical Society*, **159**(8), G93–G96.

46. Siebeneicher, P., Kleemann, H., Leo, K., and Lüssem, B. (2012) Non-volatile organic memory devices comprising SiO_2 and C_{60} showing 10^4 switching cycles. *Applied Physics Letters*, **100**(19), 193301.

47. Ávila-Niño, J.A., Machado, W.S., Sustaita, A.O. *et al.* (2012) Organic low voltage rewritable memory device based on PEDOT:PSS/f-MWCNTs Thin Film. *Organic Electronics*, **13**(11), 2582–2588.

48. Yao, J., Jin, Z., Zhong, L. *et al.* (2009) Two-terminal nonvolatile memories based on single-walled carbon nanotubes. *ACS Nano*, **3**(12), 4122–4126.

49. Wu, C., Li, F., Zhang, Y. *et al.* (2011) Highly reproducible memory effect of organic multilevel resistive-switch device utilizing graphene oxide sheets/polyimide hybrid nanocomposite. *Applied Physics Letters*, **99**(4), 042108.

50. Ji, Y., Choe, M., Cho, B. *et al.* (2012) Organic nonvolatile memory devices with charge trapping multilayer graphene film. *Nanotechnology*, **23**(10), 105202.

51. Liu, G., Zhuang, X., Chen, Y. *et al.* (2009) Bistable electrical switching and electronic memory effect in a solution-processable graphene oxide-donor polymer complex. *Applied Physics Letters*, **95**(25), 253301.

52. Zhang, B., Liu, Y.-L., Chen, Y. *et al.* (2011) Nonvolatile rewritable memory effects in graphene oxide functionalized by conjugated polymer containing fluorene and carbazole units. *Chemistry - A European Journal*, **17**(37), 10304–10311.

53. Yu, A.-D., Liu, C.-L., and Chen, W.-C. (2012) Supramolecular block copolymers: graphene oxide composites for memory device applications. *Chemical Communications*, **48**, 383–385.

54. Ufert, K. (2011) "Memory element using reversible switching between SP2 and SP3 hybridized carbon" US patent US8030637-B2.

55. Sebastian, A., Pauza, A., Rossel, C. *et al.* (2011) Resistance switching at the nanometre scale in amorphous carbon. *New Journal of Physics*, **13**, 013020.

56. Muller, Ch., Deleruyelle, D., Müller, R. *et al.* (2011) Resistance change in memory structures integrating CuTCNQ nanowires grown on dedicated HfO_2 switching layer. *Solid-State Electronics*, **56**, 168–174.

57. Verbakel, F., Meskers, S.C.J., Janssen, R.A.J. *et al.* (2007) Reproducible resistive switching in nonvolatile organic memories. *Applied Physics Letters*, **91**(19), 192103.

58. Bory, B.F., Gomes, H.L., Janssen, R.A.J. *et al.* (2012) Role of hole injection in electroforming of lif-polymer memory diodes. *Physical Chemistry C*, **116**(23), 12443–12447.

59. Chen, Q., Bory, B.F., Kiazadeh, A. *et al.* (2011) Opto-electronic characterization of electron traps upon forming polymer oxide memory diodes. *Applied Physics Letters*, **99**(8), 083305.

60. Li, T., Hu, W., and Zhu, D. (2010) A review of experimental testbeds of molecular electronic devices. *Advanced Materials*, **22**, 286.

11

Molecular Transistors

Mark A. Reed[1], Hyunwook Song[2], and Takhee Lee[3]

[1]*Departments of Electrical Engineering and Applied Physics, Yale University, USA*
[2]*Department of Applied Physics, Kyung Hee University, Korea*
[3]*Department of Physics, Seoul National University, Korea*

11.1 Introduction

Since Aviram and Ratner initially proposed a molecular rectifier in 1974 to predict the feasibility of constructing a functional molecular device using single molecules as the active elements [1], the field of molecular electronics has attracted significant interest over the past few decades [2–9]. The concept of making a functional device based on the properties inherent in a single molecule offers fascinating possibilities due to the potentially diverse electronic functions of the component molecules that can be tailored by chemical design and synthesis. Within the past few years, a wide range of characteristic functions illustrated by single molecules has been reported, including diodes [10–12], transistors [13–16], switches [17–21], and memory [21–23]. Single molecules provide ideal systems to investigate charge transport on the molecular scale, which is a subject of intense current interest for both practical applications and achieving a fundamental understanding of novel physical phenomena that take place at this length scale. This chapter focuses primarily on experimental aspects of devices that consist of one or very few molecules contacted between external electrodes. In particular, we concentrate on the characterization and manipulation of charge transport in this regime.

11.2 Experimental Approaches

11.2.1 Fabrication Methods

The fabrication of single-molecule electronic devices is a very challenging task. Conventional lithography is still unable to deliver resolution at the molecular scale, and it is beyond the capability of traditional microfabrication technologies. Nevertheless, a broad range of groups have devised a number of sophisticated experimental techniques. For an extended discussion, we refer the interested reader to the excellent reviews on various experimental testbeds of

Emerging Nanoelectronic Devices, First Edition. An Chen, James Hutchby, Victor Zhirnov and George Bourianoff.
© 2015 John Wiley & Sons, Ltd. Published 2015 by John Wiley & Sons, Ltd.

molecular electronic devices by Chen *et al.* [24], Akkerman *et al.* [6], McCreery *et al.* [8], and Li *et al.* [25] The common concept in all of these methods is the ability to form nanometer-sized gap (nanogap) electrodes. Individual molecules can occasionally bridge a gap between electrodes, thus creating reliable molecular junctions that allow charge transport measurements through constituent single molecules.

A common approach adopted to create molecular-scale electrode gaps is called a "break junction." Break junctions can be categorized into two types: mechanically controllable break junctions and electromigrated break junctions. Mechanically controllable break junctions (MCBJs) were introduced by Moreland *et al.* [26] and Muller *et al.* [27] This technique consists of a lithographically defined, metallic free-suspended bridge or a notched wire above a gap etched in an insulating (polymer or oxide) layer, fixed on the top of a bendable substrate [28–39]. The bendable substrate is most often made from a phosphor–bronze sheet owing to its superior mechanical deformation properties. This substrate is put in a three-point bending geometry, where it can be bent by moving a piezo-controlled pushing rod, as illustrated in Figure 11.1. As the substrate is bent, the metallic wire is elongated until finally the metallic constriction breaks and two fresh electrode surfaces are created. The molecules can be assembled between the separate gap electrodes by different methods. For example, one can break the electrodes while molecules are present either in solution [32] or in the gas phase [37] or by adding a solution with the desired molecules after the breakage of the metallic wire [38,39]. The first example of MCBJs to make molecular junctions was illustrated by Reed *et al.* (see Figure 11.1b and c) in 1997 [32]. In this study, a gold wire was covered with a self-assembled monolayer (SAM) of 1,4-benzenedithiol (BDT), which is able to bind to two gold electrodes through thiol groups. The gold wire was subsequently elongated in the molecular solution until breakage. Once the wire was broken, the solvent was evaporated and the wires were brought together until the onset of a conductance value. With the proper control experiments (which were performed identically but without the molecules), the measured conductance value could be ascribed to a small number (ideally one) of BDT molecules bridging the gap. One of the main advantages of MCBJs is that the contact size can be continuously adjusted under the precise control of a piezoelectric component without polluting the junction. Furthermore, the ability to repeat back and forth bending of the flexible substrates allows statistics to be obtained using a large number of measurements of the target molecule [30,31,37]. Although the exact local configuration of the junctions is unknown, it is evident from theoretical studies that the exact shape, configuration, and mechanical stress of the metal-molecule contacts are very important in influencing the result of experiments on single molecules [40–43].

Electromigrated break junctions (EBJs) were first developed by Park *et al.* in 1999 [44]. The controlled passage of a large density current or the application of a large direct current voltage to a continuous thin metal wire predefined by electron-beam lithography causes the electro-migration of metal atoms and the eventual breakage of the metal wire (Figure 11.2a). If performed properly, a separate electrode pair with distances of approximately 1–2 nm can be created so that the target molecule can subsequently bridge the gap between the broken electrodes. To incorporate the molecules into EBJs, two different approaches can be taken. One approach is to either deposit the molecules onto the electrode surface, after which the breaking process proceeds, or to first break and then assemble the molecules into the separate electrodes. Because a gate electrode can be readily fabricated on the substrate before the breaking process is performed by electromigration, the EBJs are especially advantageous in making three-terminal device configurations (see Figure 11.2a) [14,16,45]. In contrast to MCBJs, the

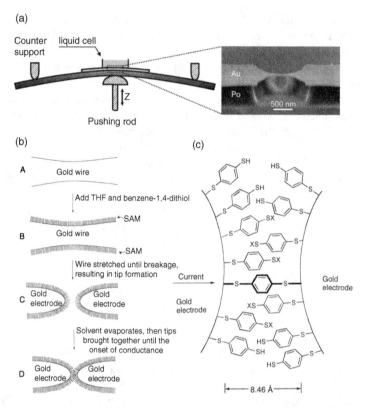

Figure 11.1 (a) Schematics of the MCBJ principle with a liquid cell and a SEM image of the central part of the microfabricated Au junction. Po is a polymer insulating layer. Reproduced with permission from [28]. Copyright 2008 American Chemical Society. (b) Schematic of the measurement process. Step A: The gold wire of the break junction before breaking and tip formation. Step B: After addition of 1,4-benzenedithiol, SAMs form on the gold wire surfaces. Step C: Mechanical breakage of the wire in solution produces two opposing gold contacts that are SAM-covered. Step D: After the solvent is evaporated, the gold contacts are slowly moved together until the onset of conductance is achieved. Steps C and D (without solution) can be repeated numerous times to test for reproducibility. Reproduced with permission from [32]. Copyright 1997 Science. (c) Schematic of a 1,4-benzenedithiol SAM between proximal gold electrodes formed in an MCBJ. The thiolate is normally H-terminated after deposition; end groups denoted as X can be either H or Au, with the Au potentially arising from a previous contact/ retraction event. These molecules remain nearly perpendicular to the Au surface, making other molecular orientations unlikely. Reproduced with permission from [32]. Copyright 1997 Science

nanogap junctions formed by electromigration cannot make a large repetitive collection of measurements with the same junction. Thus, a large number of devices must be fabricated to examine the statistical behavior of the electromigration breaking process [44,46,47]. Moreover, the technique must be used with care. The local heating of the junction during electromigration can increase the temperature, resulting in large gaps, the destruction of the molecules, and the formation of gold islands inside the gap [48]. Unintentional metal debris in the gap interferes with the insertion of the molecules of interest and can mask the intrinsic molecular signals [49–52]. Careful correlation of spectroscopies can be used to eliminate the presence of metal

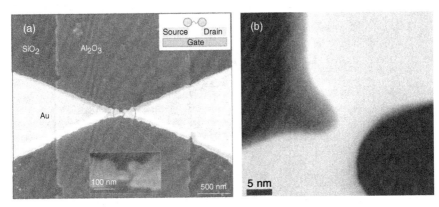

Figure 11.2 (a) SEM image of the metallic electrodes fabricated by electron beam lithography and the electromigrated break junction technique. The image shows two gold electrodes separated by 1 nm above an aluminum pad, which is covered with an 3-nm layer of aluminum oxide. The whole structure is defined on a silicon wafer. The central regions correspond to a gold bridge with a thickness of 15 nm and a minimum lateral size of 100 nm. The outer regions represent portions of the gold electrodes with a thickness of 100 nm. Reproduced with permission from [45]. Copyright 2002 Nature Publishing Group. (b) TEM images of a typical electromigrated nanogap on SiNx membrane. Reproduced with permission from [53]. Copyright American Chemical Society

islands. Recently, a few groups prepared electromigrated nanogaps on free-standing transparent SiN_x membranes to permit the use of transmission electron microscopy (TEM) to image the nanogap formation *in situ* (Figure 11.2b) [53,54].

Scanning tunneling microscopy (STM) and conducting-probe atomic force microscopy (CP-AFM) has also been widely used to measure the charge transport properties of a very small number of molecules (from several tens of molecules to a single molecule). The strength of STM lies in its combination of high-resolution imaging and spatially resolved electrical spectroscopy (so-called scanning tunneling spectroscopy, STS), providing the local density of states with atomic spatial resolution [55–57]. In general, the electrical contact is accomplished through the air gap (or vacuum tunneling gap, in ultrahigh vacuum STM) between the molecule or the molecular monolayer and the STM tip, which leads to considerable difficulty in evaluating the true conductance of single molecules. A significant improvement was demonstrated by Xu *et al.* [58], who measured the conductance of a single molecule by repeatedly forming several thousands of metal–molecule–metal junctions. This technique is referred to as a STM-controlled break junction (STM-BJ). In STM-BJs, molecular junctions are repeatedly and quickly formed by moving the STM tip into and out of contact with a metal electrode surface in a solution containing the molecules of interest. Single or a few molecules, bearing two anchoring groups at their ends, can bridge the gap formed when moving the tip back from the surface (Figure 11.3a). Because of the large number of measurements possible, this technique provides robust statistical analysis of the conductance data, and histograms of the conductance evolution during breaking show evidence of the formation of molecular junctions [58–68].

In CP-AFM [69–76], the metal-coated tip, acting as the top electrode, is gently brought into direct contact with the molecules on a conducting substrate, acting as the bottom electrode (this process is monitored by the feedback loop of the AFM apparatus) while an

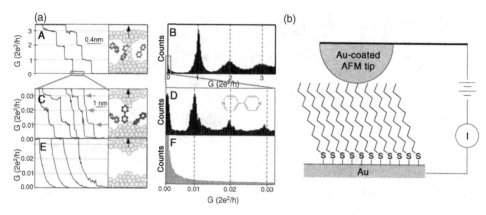

Figure 11.3 (a) Graph A: Conductance of a gold contact formed between a gold STM tip and a gold substrate decreases in quantum steps near multiples of G_0 ($= 2e^2/h$) as the tip is pulled away from the substrate. Graph B: Corresponding conductance histogram constructed from 1000 conductance curves as shown in graph A shows well-defined peaks near 1 G_0, 2 G_0, and 3 G_0 due to conductance quantization. Graph C: When the contact shown in graph A is completely broken, corresponding to the collapse of the last quantum step, a new series of conductance steps appears if molecules such as 4,4′ bipyridine are present in the solution. These steps are due to the formation of the stable molecular junction between the tip and the substrate electrodes. Graph D: Conductance histogram obtained from 1000 measurements as shown in graph C shows peaks near $1 \times$, $2 \times$, and $3 \times 0.01\ G_0$ that are ascribed to one, two, and three molecules, respectively. Graphs E and F: In the absence of molecules, no such steps or peaks are observed within the same conductance range. Reproduced with permission from [58]. Copyright 2003 Science. (b) Formation of a molecular junction by contacting an alkanethiol self-assembled monolayer with an Au-coated AFM tip. Reproduced with permission from [69]. Copyright 2000 American Chemical Society

external circuit is used to measure the current–voltage characteristics (Figure 11.3b). This procedure eliminates the current reduction caused by the extra tunneling gap in the STM setup [67,74–76]. However, the conducting probe tip of the CP-AFM coated with a metallic layer is significantly larger than an atomically sharp STM tip [69,75]. This difference produces a higher uncertainty in the number of molecules measured. Furthermore, one needs to consider the roughness and morphology of the bottom electrode substrate to estimate the number of molecules under investigation. The critical requirement for CP-AFM measurements is the very sensitive control of the tip-loading force to avoid applying excessive pressure to the molecules [77]. Excessive pressure may modify the molecular conformation and thus its electronic properties. On the other hand, the ability to apply a controlled mechanical pressure to a molecule to change its conformation can be a powerful tool to investigate the relationship between conformation and charge transport in molecular junctions [72,73,78].

A number of other approaches have been tried to create two electrodes with a molecular-sized gap for electronic transport experiments on molecular junctions. For example, Morpurgo *et al.* [79] proposed electrochemical deposition in which the inter-electrode distance can be tuned on the atomic scale in an aqueous solution by depositing (or removing) atoms at a low rate. Another method to control the inter-electrode distance on the molecular scale was reported by Kubatkin *et al.* [15] Using a shadow mask technique and evaporation at variable angles in ultrahigh vacuum (UHV) conditions, they obtained well-defined molecular devices under clean conditions and at low temperatures. An alternative method to overcome the mismatch between the resolution of lithographic methods and the molecular size was described by Dadosh *et al.* [80] using gold nanoparticles with a typical diameter of 10 nm. The molecules can be attached to the gold particles by thiol bonds such that they form particle–molecule–particle dumbbells from solution. Recently, single-walled carbon nanotubes (SWNT) have been used as electrodes separated by a nanogap (less than 10 nm) [81]. The nanogap electrodes can be obtained by a precise oxidation cutting of the SWNT, and the two facing SWNT ends that are terminated by carboxylic acids are covalently bridged by the molecules of interest functionalized with amine groups at both ends. These functionalized contacts can be used to fabricate devices with a variety of molecules, acting as pH sensors [81], photogated switches [82], and DNA hybridization sensors [83]. Very recently, Bjørnholm *et al.* [84] showed a new method for the direct synthesis and growth of end to end linked gold nanorods using gold nanoparticle seeds with a dithiol-functionalized poly(ethylene glycol) (SH-PEG-SH) linker. This method results in a nanogap with a size of 1–2 nm between two gold rods, which suggests the possibility of fabricating nanogap electrodes incorporating a single molecule or several molecules by bottom-up chemical assembly.

11.2.2 Electronic Transport Fundamentals

A full understanding of the transport properties of a molecular junction represents a key step towards the realization of single-molecule electronic devices and requires detailed microscopic characterization of the active region of the junction. Indeed, a hurdle in most single-molecule electronic devices is the unambiguous demonstration that the charge transport occurs only through a single molecule of interest. For these reasons, the analysis of the transport properties of the molecular junction attracts much attention in the field, and a variety of experimental techniques have been established in recent years.

The charge transport mechanism of a molecular junction can be revealed by the characteristic temperature[85–87] and length dependences [87–89]. Therefore, measurements of temperature- and length-variable transport for the molecular junction are necessary to examine the charge transport mechanism. In particular, two distinct transport mechanisms have been extensively discussed in the literature [2,5,6,8,85–90]: coherent transport via tunneling or superexchange and incoherent thermally activated hopping. Coherent tunneling or superexchange dominates through relatively short molecules and the conductance value (G) decreases exponentially as the molecular length increases, according to Equation 11.1:

$$G \propto \exp(-\beta d) \qquad (11.1)$$

where d is the molecular length, and β is the tunneling decay coefficient (varying between 0.7–0.9 Å^{-1} for alkyl chains and 0.2–0.5 Å^{-1} for π-conjugated molecules). In addition to the exponential decay of the conductance with molecular length, this coherent tunneling process is characterized by temperature-independent transport. On the other hand, incoherent hopping is known to be responsible for charge transport along long-conjugated molecular wires, and the conductance follows an Arrhenius relation given by:

$$G \propto \exp\left(\frac{-E_a}{k_B T}\right) \qquad (11.2)$$

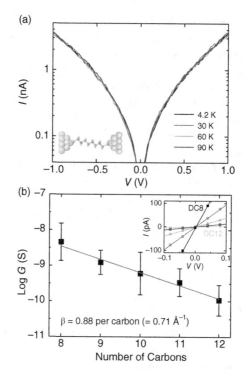

Figure 11.4 (a) Semilog plot of temperature-variable I(V) characteristics for Au–1,8-octanedithiol–Au junctions at selected temperatures (4.2, 30, 60, and 90 K). (b) Semilog plot of the conductance versus the number of carbon atoms for five different length alkanedithiol nanogap junctions. The decay coefficient (β) can be determined from the linear fit (the solid line), yielding a β value of 0.88 ($= 0.71 \, \text{Å}^{-1}$) per carbon atom. Inset shows length-dependent I(V) curves in the low-bias linear regime, where a conductance value is obtained from linear fits to the data. Reproduced with permission from [87]. Copyright American Chemical Society

where k_B is the Boltzmann constant, T is the temperature, and E_a represents the hopping activation energy. The incoherent charge hopping is also characterized by a weak length-dependent transport that results in conductance that scales linearly with the inverse of the molecular length.

From Reference [87], Figure 11.4a shows a representative temperature-variable current–voltage [I(V)] of 1,8-octanedithiol bridging the Au nanogap electrodes broken by electro-migration as described in the previous section. The I(V) curves were measured between 4.2 and 90 K, and no temperature dependence was observed [87]. The temperature-independent I(V) characteristic is a clear manifestation of coherent tunneling transport and eliminates many other potential mechanisms. In this study [87], the conductance of five different alkanedithiols having between eight (DC8) and 12 (DC12) carbon atoms was also measured to examine the length-dependent conductance (Figure 11.4b). In accordance with Equation 11.1, a semilog plot of the conductance versus the molecular length was linear. From the linear fit (the solid line across data points) in Figure 11.4b, the β value was found to be 0.88 ($= 0.71 \, \text{Å}^{-1}$) per carbon atom, assuming a through-bond tunneling [87]. This β value is in good agreement with the previously reported values of alkyl chains in literature [70,86].

Thus far, a consistent picture has emerged for the coherent tunneling mechanism of saturated alkyl chains and short-length conjugated molecules [86,91,92]. The coherent tunneling transport can be reasonably expected when the Fermi energy of the electrode lies within the large energy gap between the highest occupied molecular orbital (HOMO) and the lowest unoccupied molecular orbital (LUMO) of the short molecules. Collectively, the correct exponential decrease of conductance upon a molecular length increase, the temperature-independent I(V) characteristics, and the agreement with decay coefficients all point to the formation of a valid molecular junction.

Conjugated molecules made of repeating units with modulated molecular length are ideal for understanding charge transport mechanisms because these molecular systems permit the investigation of not only coherent tunneling and incoherent hopping but also the transition between two distinct transport mechanisms by systematically changing the molecular length [88,90]. This transition from tunneling to hopping was observed by Frisbie and colleagues [88], who synthesized oligophenyleneimine (OPI) molecules of various lengths (ranging from 1.5 to 7.3 nm) bonded to Au through a thiolate linkage. The OPI molecular wires were grown on the Au substrate by stepwise imination with alternating addition of benzene-1,4-dicarboxaldehyde and benzene-1,4-diamine, as shown in Figure 11.5a. The transport characteristics of the OPI wires were then measured using CP-AFM. In the semilog plot of resistance versus molecular length (Figure 11.5b), a clear transition of the length dependence of resistance was observed near 4 nm (OPI 5), indicating that the transport mechanism is different in short (OPI 1 to 4) and long (OPI 6 to 10) wires. In the short wires, the linear fit in Figure 11.5b indicates that the data are well-described by Equation 11.1 for coherent nonresonant tunneling. The β value is found to be 3.0 nm^{-1}, which is within the range of β values of typical conjugated molecules [93]. For long OPI wires, a much flatter resistance versus molecular length relation ($\beta \sim 0.9$ nm^{-1}) was shown. The extremely small β value suggests that the principal transport mechanism is incoherent hopping [88]. A plot of resistance versus molecular length for long OPI wires is linear (see Figure 11.5b, inset), which is consistent with hopping as described above and indicates that Equation 11.1 does not apply for the long wires [88]. The change in transport mechanism apparent in the length-dependent measurements was verified by the temperature dependence. Figure 11.5c shows that the resistance for OPI 4 is independent of temperature from 246 to 333 K, as expected for tunneling. On the other hand, both OPI 6 and OPI 10 display the strongly thermally activated transport that is characteristic of hopping [88]. The activation energies determined from the slopes of the data were identical at 0.28 eV for both OPI 6 and OPI 10 [93].

Hines *et al.* [90] also reported results that support a transition between tunneling and hopping in series of conjugated single-molecule junctions by carrying out both length- and temperature-dependent measurements of conductance using a STM-BJ method. These results provide experimental support for a theoretically predicted transition from tunneling for short molecules to thermally activated hopping for longer molecules [94,95].

Poot *et al.* [96] showed the temperature dependence of gated three-terminal molecular junctions containing sulfur end-functionalized tercyclohexylidenes. At low temperatures, they found temperature-independent transport, and at temperatures above 150 K, the current increased exponentially with increasing temperature (Figure 11.6). Over the entire temperature range (10–300 K) and for different gate voltages, a simple model [97] of transport through a single level well-described the experimental results, which indicates that the temperature dependence arises from the Fermi distribution in the leads [96].

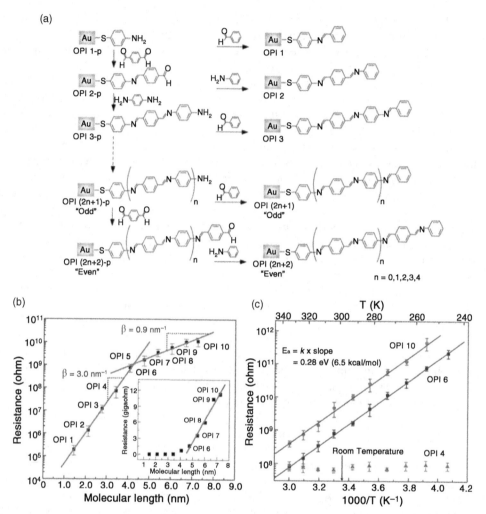

Figure 11.5   (a) Molecular structure and synthetic route to oligophenyleneimine wire precursors (OPI-p) and OPI monolayers on gold substrates. (b) Measurements of molecular wire resistance with CP-AFM. A gold-coated tip was brought into contact with an OPI monolayer on a gold substrate. The I-V traces were obtained over ± 1.5 V for OPI 3 to 10 and ± 1.0 V for OPI 1 and 2 at a load of 2 nN on the tip contact. Semilog plot of resistance versus molecular length for the gold/wire/gold junctions. Each data point is the average differential resistance obtained from 10 I(V) traces in the range -0.3 to $+0.3$ V. Error bars, 1 SD. Straight lines are linear fits to the data according to Eq. Eq. (11.1). (Inset) A linear plot of resistance versus molecular length, demonstrating linear scaling of resistance with length for the long OPI wires. (c) Arrhenius plot for OPI 4, OPI 6, and OPI 10. Each data point is the average differential resistance obtained at six different locations on samples in the range -0.2 to $+0.2$ V. Error bars, 1 SD. Straight lines are linear fits to the data. Reproduced with permission from [88]. Science

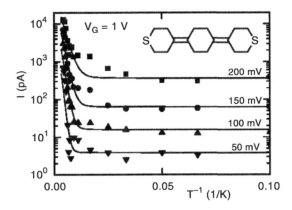

Figure 11.6 Current as a function of inverse temperature of the sulfur end-functionalized tercyclohexylidene molecule (see the inset) for four different source–drain voltages (50, 100, 150, and 200 mV), plotted for gate voltage of 1.0 V. Reproduced with permission from [96]. Copyright 2000 American Chemical Society

11.2.3 Molecular Junction Spectroscopies

A critical question is the identity of the molecules that inhabit the active region of the junction. Inelastic electron tunneling spectroscopy (IETS), an all-electronic spectroscopy due to localized molecular vibrational modes, was discovered in 1966 by Jaklevic and Lambe [98]. This pioneering work clearly showed the ability to detect the vibrational features of molecules buried in the interface of a metal–insulator–metal (MIM) device. To explain the principles of IETS (see also Reference [99]), Figure 11.7 shows the energy-band diagrams of a tunnel junction and the corresponding I(V), dI/dV, and d^2I/dV^2 plots. When a negative bias (small with respect to the tunnel barrier) is applied to the left metal electrode, the left Fermi level is lifted. An electron from an occupied state on the left side tunnels into an empty state on the right side, and its energy is conserved (process a). This process is elastic tunneling. During this process, the current increases linearly with the applied small bias (less than the vibrational energy; Figure 11.7b). However, if there is a vibrational mode with a frequency of ω localized inside this barrier, then the electron can lose a quantum of energy, $\hbar\omega$, to excite the vibrational mode and tunnel into another empty state when the applied bias is large enough such that $eV \geq \hbar\omega$ (process b) [100,101]. This process opens an inelastic tunneling channel for the electron, and its overall tunneling probability is increased. Thus, the total tunneling current has a kink that is a function of the applied bias (Figure 11.7b). This kink becomes a step in the differential conductance (dI/dV) plot and a peak in the d^2I/dV^2 plot. Typically, only a very small fraction of electrons tunnel inelastically (the cross-section for such an excitation is very small because the electron traversal time is much smaller than the oscillator period), and thus the IETS conductance step is often too small to be conveniently detected. In practice, investigators use a phase-sensitive ("lock-in") detection technique to directly measure the peaks of the second derivative of I(V). The IETS signal, which is proportional to the second derivative of I(V), is usually measured by an AC modulation method. Theoretically, the signal can also be determined by a mathematical differential approach that computes the numerical derivatives of the directly measured I(V) characteristics [102]. However, this method is

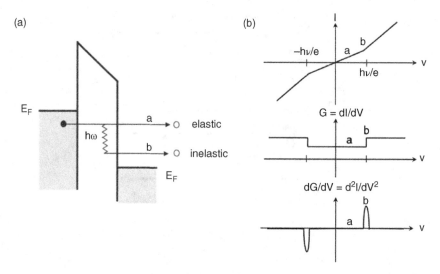

Figure 11.7 (a) Energy band diagram of a tunnel junction with a vibrational mode of frequency ω localized inside: "a" is the elastic tunneling process; "b" is the inelastic tunneling process. (b) Corresponding I(V), dI/dV, and d^2I/dV^2 characteristics. Reproduced with permission from [99]. Copyright 2008 Elsevier

generally not feasible in practice due to insufficient signal to noise ratios or bit resolutions of the instrumentation used to acquire the data.

IETS recently became a primary characterization technique to identify the component molecules present in molecular junctions (not an adlayer or impurity, but molecules forming the active region of a junction) [103–106], analogous to infrared and Raman spectroscopy for macroscopic samples, for the unambiguous determination of the molecular species in the junction. An example of experimental IETS measurements is shown in Figure 11.8 [106], which shows the I(V) curve, the differential conductance (dI/dV), and the IETS (d^2I/dV^2) spectrum of Au-octanedithiol (ODT)-Au and Au-benzenedithiol (BDT)-Au junctions measured at 4.2 K using an electromigrated break junction. Although the I(V) characteristics seem to be linear over the bias range measured, the plots of dI/dV and d^2I/dV^2 exhibit significant features corresponding to vibrational modes of the molecules under investigation. Standard AC modulation techniques with a lock-in amplifier are used to directly obtain the first and second harmonic signals proportional to dI/dV and d^2I/dV^2, respectively [107]. As explained above, a molecular vibration coupled to tunneling charge carriers gives rise to an increase in slope of the dI/dV curve owing to an inelastic tunneling process, which then appears as a step and peak in the first (dI/dV) and second (d^2I/dV^2) derivatives, respectively. The plot of d^2I/dV^2 versus V is referred to as the IETS spectrum. The observed spectral features were assigned to specific molecular vibrations by comparison with previously reported infrared, Raman, and IETS measurements and by density functional theory calculations. For the ODT junction (Figure 11.8c), peaks were reproducibly observed at 92, 119, 143, 161, 181, and 355 mV, which correspond to ν(C-S) stretching, $δ_r$(CH₂) rocking, ν(C-C) stretching, $γ_w$(CH₂) wagging, $δ_s$(CH₂) scissoring, and ν(C-H) stretching modes, respectively. The absence of a prominent peak corresponding to the ν(S-H) stretching mode at 319 mV ($2575\,cm^{-1}$) suggests that the thiol (−SH) anchoring group reacts with the Au electrode pairs broken during the electromigration. In the IETS spectrum of the BDT junction (Figure 11.8f), three prominent peaks

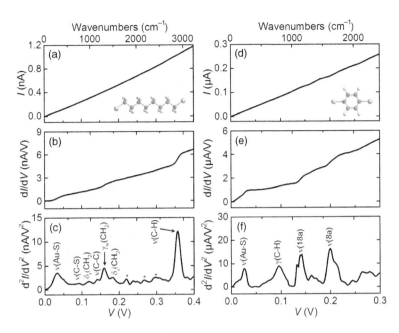

Figure 11.8 Transport properties of gold junctions measured at 4.2 K: (a)–(c) Au-ODT-Au and (d)–(f) Au-BDT-Au. (a) and (d) I(V) characteristics. The insets display the chemical structure of each molecule. (b) and (e) Differential conductance (dI/dV) obtained from lock-in first harmonic signal. (c) and (f) IETS spectrum (d^2I/dV^2) obtained from lock-in second harmonic signal. The peaks are labeled with their assigned vibrational modes. Reproduced with permission from [106]. Copyright 2009 American Institute of Physics

reproducibly appeared at 96, 142, and 201 mV, corresponding to γ(C−H) aryl out of plane bending, ν(18a) stretching, and ν(8a) stretching modes, respectively. These modes originate from vibrations of the phenyl ring. A theoretical study predicted that the ν(18a) and ν(8a) ring modes should have strong vibronic coupling in phenylene molecules [108] and is consistent with these results. The dominance of aromatic ring modes in IETS spectra has also been experimentally observed for various conjugated molecules [109,110]. The fully assigned IETS spectrum provides unambiguous experimental evidence of the existence of the desired molecules in the region of the junction and, correlating with the other characteristics of the junction transport, only leaves the IETS-identified molecule as the only element in the junction through which tunneling is occurring.

The first observation of IETS in a single molecule was obtained in STM [111]. The possibility of performing IETS studies using STM was discussed soon after its invention [112]. However, due to difficulties in achieving the extreme mechanical stability that is necessary to observe small changes in tunneling conductance, this technique has only recently been realized [111]. In the STM implementation of IETS, the MIM tunnel junction is replaced by a STM junction consisting of a sharp metallic tip, a vacuum gap, and a surface with the adsorbed molecules. Using STM-IETS, elegant imaging and probing can be performed at the same time, and vibrational spectroscopy studies on a single molecule can be achieved [113]. One of the most fruitful techniques for IETS of molecular structures arose from the pioneering work of Gregory [114] in 1990, in which a junction between two crossed wires was delicately made by a deflecting Lorentz force. Kushmerick *et al.* [115] demonstrated that reproducible molecular junctions could be formed with sufficient stability and robustness for clear IETS signatures [109]. The ease of electrode and molecular exchange has allowed elegant and thorough investigations of structure–function relationships. The technique has also enabled investigations into such areas as selection rules [116] and pathways [117], illustrating the power of IETS in characterizing and understanding nanoscale junctions. Wang *et al.* also reported an IETS study of an alkanedithiol self-assembled monolayer (SAM) using a nanometer-scale device (nanopore technique) [103]. Remarkably, the authors were able to verify that the observed spectra were indeed valid IETS data by examining the peak width as a function of temperature and AC modulation voltage (refer to the following paragraph) [103]. Recently, Hihath *et al.* [118] reported the IETS spectra of a single 1,3-propanedithiol molecule using an STM break junction at cryogenic temperatures. In particular, these authors were able to observe simultaneous changes in the conductance and vibrational modes of a single molecule as the junction was stretched. This ability allowed them to correlate the changes in the conductance with the changes in the configuration of a single-molecule junction. Moreover, the authors were also able to conduct a statistical analysis of the phonon spectra to identify the most relevant modes. These vibrational modes matched the IR and Raman spectra well and have been described by a simple one-dimensional model [118]. Another useful example of IETS for studying molecular junctions was reported by Long *et al.* [110] This study provides insight into changing transport characteristics resulting from exposure to air. IETS spectra have shown that molecular conduction could be significantly affected by rapid hydration at the gold–sulfur contacts. The detrimental effects of hydration on molecular conduction are important for understanding charge transport through gold–thiol molecular junctions exposed to atmospheric conditions.

An important tool to verify that the obtained spectra are indeed valid IETS data is to examine the vibrational peak width broadening as a function of temperature and applied modulation voltage. The width of a spectral peak includes a natural intrinsic linewidth, W_I, and two broadening effects: thermal broadening ($5.4k_BT/e$, where k_B is Boltzmann's constant and T denotes temperature) due to the breadth of the Fermi level and modulation broadening ($1.7V_m$, where V_m is the AC modulation voltage) due to the dynamic detection technique used to obtain the second harmonic signals [119]. The full width at half maximum (FWHM) of the d^2I/dV^2 vibrational peak in IETS is given by [98,120]:

$$W = [(1.7V_m)^2 + (5.4k_BT/e)^2 + (W_I)^2]^{1/2}. \tag{11.3}$$

Figure 11.9 illustrates a study of the linewidth broadening of a vibrational peak in IETS measurements. Figure 11.9a shows the modulation broadening of a representative IETS feature {from Reference [16, the ν(C−H) stretching mode of ODT molecules in the electromigrated break junction} at a constant temperature of 4.2 K. The data points show the FWHM of the experimental peak. Considering the known thermal broadening and modulation broadening, the intrinsic linewidth, W_I, can be determined from a fit to the modulation broadening data (Figure 11.9a, solid line), giving $W_I = 4.94 \pm 0.89$ meV (following Equation 11.3). Figure 11.9b shows the thermal broadening of the same ν(C−H) peak at a fixed modulation, demonstrating excellent agreement between the experimental FWHM values (circles) and theoretical values (squares).

Over the last few years, IETS has evolved into an essential tool in the field of molecular electronics. Although IETS requires cryogenic temperatures, it is the only available method that provides both structural and electronic information about a single-molecule electronic

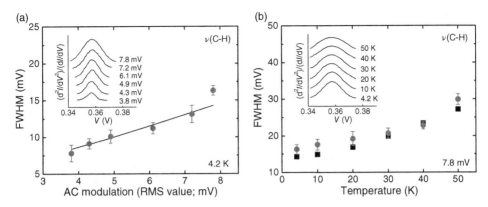

Figure 11.9 Full width at half maximum (FWHM) of the peak corresponding to the ν(C−H) stretching mode (∼357 mV) as a function of AC modulation voltage (a) and temperature (b). The circles indicate experimental data, and the solid line (a) and squares (b) show theoretical values. The error bars are determined by the Gaussian fitting. Insets show successive IET spectroscopy scans for the ν(C−H) mode under increasing AC modulation voltage (a) and increasing temperature (b), as indicated. r.m.s., root mean squared. Reproduced with permission from [16]. Copyright 2009 Nature Publishing Group

device for a particular conformation and contact geometry of the molecular junction at low temperature [8]. From sophisticated comparisons between experiments and theoretical computations, IETS can be more useful for characterizing numerous aspects of molecular junctions, such as identification of the molecule, information on the nature of the interfaces, orientation of the molecule, and even electronic pathways [2].

Recently an additional tool, transition voltage spectroscopy (TVS), became an increasingly popular spectroscopic tool for molecular junctions [16,121–125] and other diverse nano-electronic systems [126]. Specifically, TVS is used to give insight into the energy offset between the contact Fermi level and the nearest molecular level responsible for charge transport in molecular junctions by measuring the transition voltage (V_{trans}) required to generate the inflection behavior of a Fowler–Nordheim (F-N) plot, that is, the corresponding analysis of ln (I/V^2) against $1/V$ for I(V) characteristics [121,122].

By combining TVS with ultraviolet photoelectron spectroscopy (UPS), Beebe *et al.* [121] correlated the charge transport properties of π-conjugated molecules with their effective band lineup. In this study, CP-AFM and crossed-wire tunnel junction measurements on molecular junctions revealed a characteristic minimum in the F-N plot at a bias voltage, V_{trans} (Figure 11.10a), which scaled linearly with the HOMO energy (which is the nearest molecular level for the measured molecules) obtained from UPS (Figure 11.10b) [121]. These results show that the magnitude of V_{trans} is molecule-specific (as a form of spectroscopy) and depends directly on the manner in which the conjugation path is extended. In general, the HOMO–LUMO gap of π-conjugated molecules decreases with an increase in conjugation length [127]. It is thus reasonable to expect longer conjugated molecules to exhibit a smaller value of V_{trans} than shorter conjugated molecules within a given molecular series [122]. Recently, TVS also facilitated the calibration of orbital energy positions in molecular transistors [16,128].

TVS was initially interpreted by a simple barrier picture for charge tunneling in a junction [121]. Within this interpretation, the transition behavior in the F-N plots corresponds to a change in the tunneling mechanism from direct tunneling through a trapezoidal barrier to

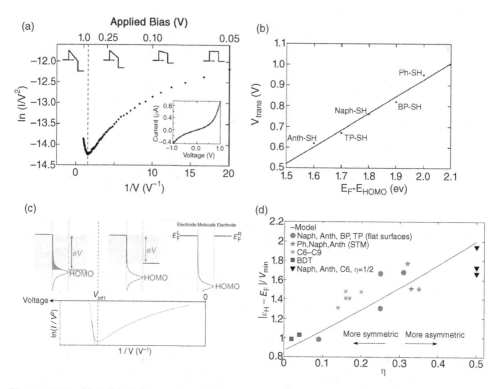

Figure 11.10 (a) Solid circles represent the average of 100 I–V curves for an Au-anthracenethiol-Au junction measured by CP-AFM. The dashed line corresponds to the voltage at which the tunneling barrier transitions from trapezoidal to triangular (V_{trans}). Also shown are representations of the barrier shape at various values of applied bias. The inset shows current-voltage data on standard axes. (b) V_{trans} (CP-AFM) versus EF-EHOMO energy difference (UPS). Reproduced with permission from [121]. Copyright 2006 American Physical Society. (c) Schematic of the theoretical model [129] to qualitatively explain the inflection of F-N curve. Also shown are representations of the barrier shape at various values of applied bias. The inset shows current–voltage data on standard axes. Reproduced with permission from [129]. Copyright 2010 American Physical Society. (d) Ratio between the HOMO energy (at zero bias) and the transition voltage, (denoted as V_{min} in the figure), versus asymmetry parameter, η. The solid line is obtained from a Lorentzian transmission function and symbols are results of *ab initio* finite bias calculations (see Reference [130] for details). Reproduced with permission from [130]. Copyright 2010 American Physical Society

Fowler–Nordheim tunneling (or field emission) through a triangular barrier (see the barrier shapes in Figure 11.12a) [121]. The transition voltage equals the barrier height, which is interpreted as the energy gap from the metal electrode's Fermi level to the nearest molecular level. However, as pointed out by Huisman *et al.* [124], the naïve tunnel barrier model is inconsistent with experimental data. On the other hand, the TVS experiments on molecular junctions are more appropriately described by the coherent Landauer approach with a single transport level [124]. Charge transport through such a junction is described by a transmission function, which is assumed to have a Lorentzian shape. Within the coherent Landauer transport picture, V_{trans} can be directly scaled with the barrier height (Φ_B) in molecular junctions, thus giving valid information on molecular energy levels. Araidai *et al.* [129] theoretically

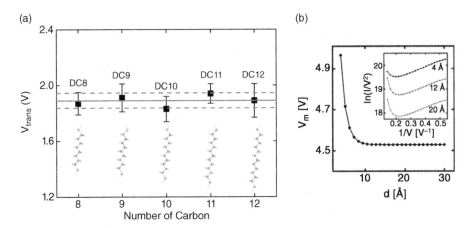

Figure 11.11 (a) V_{trans} as a function of molecular length for a series of alkanedithiols from DC8 to DC12. The solid line represents the mean value of V_{trans} for five different length alkanedithiols, and two dashed lines show the standard deviation for averaging. Error bars on each data point also denote the standard deviation across individual measurements for different devices. Chemical structures for each molecule are displayed in the inset. Reproduced with permission from [87]. Copyright 2010 American Chemical Society. (b) V_{trans} (denoted as V_m in the figure) versus molecular length d. V_{trans} becomes length independent for $d > 8$ Å. Reproduced with permission from [124]. Copyright 2010 American Chemical Society

investigated the origin of an inflection behavior appearing in the F-N plot of I(V) characteristics for molecular junctions. The results show that the inflection does not necessarily indicate the transition between the two regimes of direct tunneling and Fowler–Nordheim tunneling. Their close examination of the relation between the behavior of the F-N curve and the transmission function showed that the inflection takes place when the molecular level responsible for the charge transport approaches the edge of the electrode-bias window (Figure 11.10c) [129]. Although the origin of the inflection behavior drastically differs from the conventional model, the F-N plots obtained from their calculations show very similar behavior to those from recent experiments [129]. Recently, Chen *et al.* [130] reported extensive *ab initio* calculations to simulate TVS for a broad class of molecular junctions. The numerical data closely follow the trend expected from an analytical model with a Lorentzian-shaped transmission function. Interestingly, the ratio of V_{trans} to the HOMO level position was found to vary between 0.8 and 2.0 depending on the junction asymmetry (as shown in Figure 11.10d), which means that it is necessary to consider the asymmetry of the molecular junction to use TVS as a quantitative spectroscopic tool to probe the molecular levels [130].

The interpretation of TVS, fulfilled by comparison of the Simmons model and the coherent Landauer approach, suggests that length-dependent TVS measurements for saturated alkyl chains can provide a critical test to distinguish true molecular junctions from a vacuum tunnel junction with no molecules [124]. The saturated alkyl molecular system constitutes an important control series in molecular transport experiments to corroborate valid molecular junctions because molecular energy levels remain nearly unchanged with molecular length, and the transport mechanism has been extensively established. To this end, TVS on a series of alkanedithiol molecules having different lengths (ranging from DC8 to DC12) was performed, employing electromigrated break junctions [87]. V_{trans} for a series of alkanedithiol molecules is summarized graphically in Figure 11.11a. The average value of V_{trans}, represented by the solid line in Figure 11.11a, falls within the standard deviation (within two dashed lines) of measured values for each of the molecules, thereby illustrating that V_{trans} is invariant with molecular length for alkanedithiols. This result is consistent with the fact that the HOMO–LUMO gap of these molecules is virtually length independent. Thus, this result agrees with the data of Beebe *et al.* [122]. Similarly, ultraviolet photoelectron spectroscopy measurements have shown that the energy offset between the HOMO and the electrode's Fermi level for different length alkyl chains is constant [131,132]. It should be noted that such length constancy in V_{trans} for the alkanedithiol junctions is fully in agreement with the results expected from a coherent molecular transport model based on the Landauer formula, that is, V_{trans} is independent of molecular length (longer than $\sim$8 Å) for constant barrier heights (Φ_B; see Figure 11.11b) [124]. These findings on length-dependent TVS measurements provide additional verification of the formation of a true molecular junction [87].

An additional technique, thermopower (also called the Seebeck coefficient, S), has been utilized to investigate single molecule junctions. Paulsson and Data stressed in a theoretical paper that thermoelectric measurements can provide new insights into charge transport in molecular junctions [133]. Analogous to the hot point probe measurements commonly used to establish the p- or n-type character of semiconductors [134], the thermoelectric voltage yields valuable information regarding the location of the Fermi energy. Previous proposals have suggested that the location of the Fermi energy can be deduced from the asymmetry of the I(V) curve caused by asymmetric contacts [135]. However, this measurement is performed far from equilibrium and requires detailed knowledge of the contacts [133]. In contrast, the thermoelectric voltage in molecular junctions is large enough to be measured but is rather insensitive to the detailed coupling to the contacts, and it gives valuable information about the position of the Fermi energy relative to the molecular levels [133]. Interestingly, these thermoelectric measurements suggest that a molecular junction could be the basis for not only molecular electronics but also thermoelectric energy conversion devices [136].

The first experiment on the thermoelectric measurement in single-molecule junctions was reported by Reddy *et al.* in 2007 [136]. These authors used STM break junctions to trap molecules between two gold electrodes with a temperature differential and statistically measured the thermoelectric voltage of 1,4-benzenedithiol (BDT), 4,4-dibenzenedithiol (DBDT), and 4,4-tribenzenedithiol (TBDT) in contact with gold at room temperature. These data were used to construct histograms for each temperature differential, which were used to estimate the average and the variation in the junction thermopower (or Seebeck coefficient), $S_{junction}$.

The relation between $S_{junction}$ of the Au-molecule-Au junction and the measured thermoelectric voltage, ΔV, is given by [136]:

$$S_{junction} = S_{Au} - \frac{\Delta V}{\Delta T}, \qquad (11.4)$$

where S_{Au} is the Seebeck coefficient of bulk Au, which is 1.94 μV/K at 300 K [137]. ΔV_{peak} is plotted as a function of ΔT, where ΔV_{peak} corresponds to ΔV at the peak of the distribution in the histogram. From the slope $\Delta V_{peak}/\Delta T$ and Equation 11.4, it was found that $S_{Au\text{-}BDT\text{-}Au} = +8.7 \pm 2.1$ μV/K, where the error is the FWHM [136]. Similar experiments were also performed with DBDT and TBDT, and statistical analysis revealed that $S_{Au\text{-}DBDT\text{-}Au} = +12.9 \pm 2.2$ μV/K and $S_{Au\text{-}TBDT\text{-}Au} = +14.2 \pm 3.2$ μV/K [136]. There seems to be a linear dependence of the thermopower with molecular length, which is in contrast to the exponential dependence of electrical resistance that is generally attributed to tunneling across the molecule. The relative position of the HOMO and LUMO levels with respect to the E_F of the metal electrodes can be related to the measured value of $S_{junction}$ [133]. The Landauer formula [138] is used to relate $S_{junction}$ to the transmission function, $\tau(E)$. It is shown that $S_{junction}$ can be obtained as [136]:

$$S_{junction} = -\frac{\pi^2 k_B^2 T}{3e} \frac{\partial \ln[\tau(E)]}{\partial E} |E = E_F|, \qquad (11.5)$$

where k_B is the Boltzmann constant. The transmission function for the case of the Au-BDT-Au junction, which was derived using the nonequilibrium Green's function formalism in conjunction with the extended Huckel theory [133]. It is clear that $\tau(E) \sim 1$ when E_F aligns with either the HOMO or the LUMO levels and decreases rapidly to below 0.01 in between [136]. Using this transmission function in Equation 11.5, it was shown that, depending on the position of the Fermi energy with respect to molecular levels, the thermopower can be either positive or negative. If E_F is closer to the HOMO, the sign is positive, indicative of hole-dominated transport. If, on the contrary, the LUMO is closer to E_F, then the thermopower is negative, indicative of electron-dominated transport. $S_{Au\text{-}BDT\text{-}Au}$ is positive (p-type), and thus, E_F is closer to the HOMO level (hole-dominated transport) [136]. Using the measured value of $S_{Au\text{-}BDT\text{-}Au} = +8.7 \pm 2.1\,\mu V/K$ from Figure 11.14c, E_F was also estimated to be 1.2 eV from the HOMO level [136].

Baheti *et al.* [139] also performed thermopower measurements to elucidate the effects of chemical structure on the electronic structure and charge transport in molecular junctions. Again, the authors used a STM break-junction technique to measure the Seebeck coefficient of several benzene derivatives, where 1,4-benzenedithiol (BDT) was modified by the addition of electron-withdrawing or electron-donating groups such as fluorine, chlorine, and methyl on the benzene ring, and the thiol end groups on BDT were replaced by cyanide end groups [139]. For the substituted BDT molecules, it was observed that the thermopower of the molecular junction decreases for electron-withdrawing substituents (fluorine and chlorine) and increases for electron-donating substituents (methyl) [139]. In fact, this change in the measured thermopower is reasonably predictable. Electron-withdrawing groups remove electron density from the σ-orbital of the benzene ring, allowing the ring's high-energy π system to stabilize. Because the HOMO has a largely π character, its energy is therefore decreased, shifting it further away from E_F. As discussed above, such a shift results in a decrease of the thermopower. Alternatively, the addition of electron-donating groups increases the σ-orbital electron density in the benzene ring, leading to an increase in the energy of the π-system and thereby shifting the HOMO closer to E_F. This shift causes the enhancement of the thermopower in this case. Moreover, the sign of the thermopower for cyanide end groups were found to be negative, which indicates that charge transport in 1,4-benzenedicyanide is dominated by the LUMO [139]. Thus, these measurements show that it is possible to tune the thermoelectric properties of molecular junctions in a controllable way by the addition of substituents.

Combined optical and transport experiments on molecular junctions can reveal a wealth of additional information beyond that available from purely electronic measurements [140]. Exciting recent works have demonstrated that simultaneous single-molecule optical spectroscopy and transport measurement is possible [140–142]. These experiments are based on the fact that the metallic nanogap electrodes used to create molecular junctions are able to act as

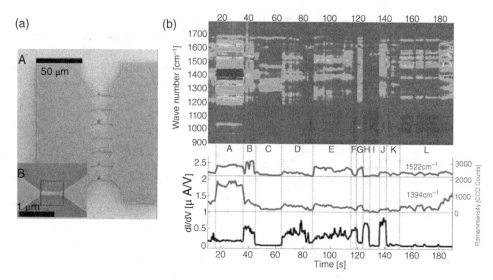

Figure 11.12 (a) Micrograph A: Full multibowtie structure, with seven nanoconstrictions. Inset B: Close-up of an individual constriction after electromigration. Note that the resulting nanoscale gap (<5 nm at closest separation, as inferred from closer images) is toward the right edge of the indicated red square. Reproduced with permission from [141]. Copyright 2010 American Chemical Society. (b) Waterfall plot of Raman spectrum (1 s integrations) and conduction measurements for a pMA sample. The device experiences periods of correlation (regions B, D, E) and anticorrelation (region L) between Raman intensity and conduction. Distinct changes in conduction are observed with every significant change in the Raman spectrum and are indicated by vertical lines. The modes near 1394 and 1522 cm^{-1} show similar intensity fluctuations except at region B and the end of region L. This results in the saturation of the signal at region A which would otherwise resolve into well-defined peaks. The 1522 cm^{-1} mode has been shifted upward on the lower graph for clarity. Reproduced with permission from [142]. Copyright 2010 American Chemical Society

tremendously effective plasmonic antennas, leading to dramatic surface-enhanced Raman scattering in the junctions [141,142], which thus makes it possible to perform surface-enhanced Raman spectroscopy (SERS) of a target molecule placed on the nanogaps (for a review of SERS, see Reference [143]).

Ward *et al.* [140–142,144] performed a series of optical experiments on Au nanogap structures prepared with electromigration using a confocal Raman microscope. The initial experiments examined nanogaps as a potential SERS substrate [141], with para-mercaptoaniline (pMA) as the molecule of interest. Nanoconstrictions were placed in parallel to allow the simultaneous electromigration of seven nanogaps at one time (Figure 11.12a). Samples were characterized using a Raman microscope via spatial maps and time spectra of the SERS response. Prior to electromigration, no significant SERS response was detected anywhere on the devices. Following electromigration, the authors observed a SERS response that was strongly localized to the resulting gaps. Successive spectra measured directly over the SERS hotspot revealed "blinking" and spectral diffusion, phenomena often associated with single- or few-molecule Raman sensitivity [141]. Blinking occurs when the Raman spectrum rapidly changes on the second timescale, with the amplitudes of different modes changing independently of one another. Spectral shifts as large as $\pm 20\,\mathrm{cm}^{-1}$ were observed, making it difficult to directly compare SERS spectra with other published results. Blinking and spectral shifts are attributed to the movement or rearrangement of the molecule relative to the metallic substrate. It is unlikely that an ensemble of molecules would experience the same rearrangements synchronously, and thus the observation of blinking and wandering is expected only in situations where a few molecules are probed [141].

In a subsequent experiment [142], the same group performed simultaneous SERS and transport measurements, including Raman microscope observations over the center of nanogap devices during electromigration. Molecules of interest, pMA or a fluorinated oligomer (FOPE), were assembled on the Au surface prior to electromigration. Once the device resistance exceeded approximately 1 kΩ, SERS could be seen. This result indicates that localized plasmon modes responsible for the large SERS enhancements may now be excited. As the gap migrates further, the SERS response scaled logarithmically with the device resistance until the resistance reached approximately 1 MΩ. In most samples, the Raman response and conduction of the nanogap become decoupled at this point, with the conduction typically changing little while uncorrelated Raman blinking occurs. In about 11% of 190 devices, however, the Raman response and conduction showed very strong temporal correlations [142]. A typical correlated SERS time spectrum and conductance measurement for a FOPE device are presented in Figure 11.12b. Because the conduction in nanogaps is dominated by approximately a single molecular volume, the observed correlations between conductance and Raman measurements strongly indicate that the nanogaps have single-molecule Raman sensitivity. It is then possible to confirm that electronic transport is taking place through the molecule of interest from the characteristic Raman spectrum. Data sets such as those shown in Figure 11.12b implicitly contain an enormous amount of information about the configuration of the molecule in the junction [140].

Tian *et al.* [36] reported a combined SERS and mechanically controllable break-junction method to measure the SERS signals of molecules located inside the nanogap between two electrodes on a Si chip [874]. They showed that the SERS signal depends critically on the separation of the electrodes and the incident light polarization. In particular, when the incident laser polarization was along the two electrodes, the field in the nanogap was the strongest because of the coupling to the localized surface plasmon resonance of the two gold electrodes [145]. Moskovits and colleagues carried out SERS measurements on Rhodamine (R6G) adsorbed in nanogaps produced in single Ag nanowire by electromigration [146]. For gaps that divide the nanowire uniformly across its width, the SERS intensity was maximum when the electric vector was oriented parallel to the long axis of the nanowire (i.e., across the gap). In the experiment of Ioffe *et al.* [147], SERS was also used as a tool to spectroscopically monitor heating (and cooling) processes in conducting molecular junctions, which involved measuring both the Stokes and anti-Stokes components of the Raman scattering.

11.3 Molecular Transistors

11.3.1 Orbital Gated Transport

Theoretical proposals have indicated that the field-effect gating of a molecular junction is possible in a fashion similar to a conventional field-effect transistor (FET) [148–152]. Indeed, the experimental demonstration of a true molecular transistor, one that depends on the external modulation of molecular orbitals, has been the outstanding challenge of the field of molecular electronics since soon after its inception [1].

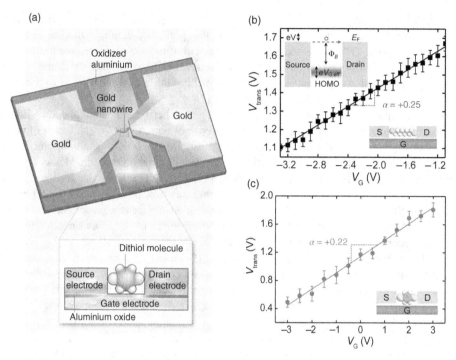

Figure 11.13 (a) Schematic illustration of a molecular transistor. Diagram A: Each device consists of a fractured gold nanowire overlaid on a strip of oxidized aluminum. Inset B: Side-on, close-up view of a device. The broken ends of the nanowire form the source and drain electrodes of the transistor, and the oxidized aluminum forms the gate electrode. Aluminum oxide on the surface of the gate electrode provides a necessary layer of insulating material known as the gate dielectric. A single molecule (here, an aromatic dithiol) connects the source and drain electrodes. The components of the device are not drawn to scale. Reproduced with permission from [144]. Copyright 2010 Nature Publishing Group. (b) Linear scaling of V_{trans} in terms of V_G for 1,8-octanedithiol. Inset: Schematic of the energy band for HOMO-mediated hole tunneling, where $eV_{G,eff}$ describes the actual amount of molecular orbital shift produced by gating. (c) Linear scaling of V_{trans} in terms of V_G for 1,4-benzenedithiol. Reproduced with permission from [16]. Copyright 2010 Nature Publishing Group

We recently presented the construction and characterization of molecular transistors [16], where the transport current is controlled by directly modulating the energy of the molecular orbitals of a single molecule (based purely on electrostatic gate control). As illustrated in Figure 11.13a, individual molecules are connected to source and drain electrodes with a bottom-gate control electrode in a FET configuration. In such devices, the energies of the molecular orbitals with respect to the Fermi level of the electrodes can be directly tuned by adjusting the gate voltage, V_G. We made such devices using the electromigration technique of fracturing a continuous gold wire (coated with the desired molecules in vacuum at 4.2 K) that is placed over an oxidized aluminum gate electrode [14,45,153]. This arrangement produces source and drain electrodes with a nanometer-scale gap that are often bridged by single or very few molecules.

We verified that the charge transport properties were through the inserted molecules, using a combination of transport techniques that gives a self-consistent characterization of the

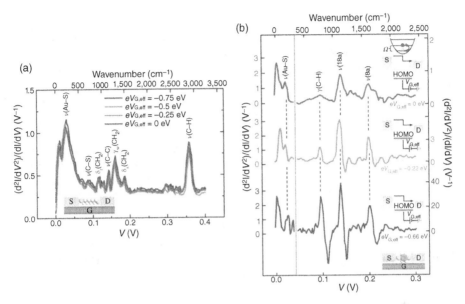

Figure 11.14 (a) IETS spectra for a Au-ODT-Au junction measured at 4.2 K for different values of $eV_{G,\text{eff}}$, with vibration modes assigned. (b) IETS spectra for a Au-BDT-Au junction measured at 4.2 K for different values of $eV_{G,\text{eff}}$, with vibration modes assigned. The left-hand y axis corresponds to the left shaded region of the spectra, and the various right-hand y axes (with different scales) correspond to the related spectra in the right region. The vertical dotted line corresponds to $V = 45$ mV (363 cm^{-1}). Significant modification in the spectral intensity and line shape for the benzene ring modes, g(C−H), n(18a) and n(8a), was observed for different values of $eV_{G,\text{eff}}$, as indicated. Insets: Energy diagrams illustrating inelastic tunneling as the position of the HOMO resonance shifts as a result of gating. Reproduced with permission from [16]. Copyright 2010 Nature Publishing Group

molecular junction [16]. Inelastic electron tunneling spectroscopy (IETS) was used to measure the interactions between the tunneling charge carriers and the vibrational modes of the molecules in the devices. This technique provides definitive proof that the measured currents actually pass through the molecules in molecular transistors. We tested two types of transistor, each with a different molecule in the junction: either an alkanedithiol (containing two SH groups connected by a saturated hydrocarbon chain) or an aromatic dithiol (containing two SH groups connected by a benzene ring). Because each dithiol has its own vibrational fingerprint (see Figure 11.14), the IETS spectra of the devices provide unambiguous identification of the component molecules in the junctions.

Transition voltage spectroscopy (TVS), which measures the transition voltage (V_{trans}) required to generate inflection behavior in a Fowler–Nordheim plot was also used. It has previously been shown [121,122] that V_{trans} is proportional to the difference in energy between the gating orbital of the molecular junction (the orbital that modulates charge transport) and the Fermi levels of the source and drain electrodes, where the Fermi level is the highest possible energy for a conducting electron in an electrode. By measuring V_{trans} using TVS at different applied gate voltages, we demonstrated that a linear relationship exists between the gate voltages and molecular orbital energy in their devices (Figure 11.13b and c), as expected for molecular transistors [128]. The slope in the linear relationship, $\alpha = \Delta V_{\text{trans}}/\Delta V_G$, is the gate efficiency factor, which describes the effectiveness of molecular orbital gating. The actual amount of molecular orbital shift produced by the applied gate voltage can also be determined in terms of an effective molecular orbital gating energy, $eV_{G,eff} = |\alpha| eV_G$. In a three-terminal device, a negative or positive gate voltage would respectively raise or lower the orbital energies in the molecules relative to E_F [149,150]. Therefore, a positive value of α indicates HOMO-mediated hole tunneling (p-type-like; Figure 11.13b, inset). Conversely, α is negative for LUMO-mediated electron tunneling (n-type-like). By extrapolating the y-intercept from the linear fit in Figure 11.13b, Song *et al.* obtained the zero-gate transition voltage, $V_{trans,0} = 1.93 \pm 0.06$ V for Au-octanedithiol (ODT)-Au junctions, which provides an estimate of the original position (at $V_G = 0$ V) of the HOMO level relative to E_F. For Au-benzenedithiol (BDT)-Au junctions, the positive sign of α explicitly indicates that HOMO-mediated tunneling is the dominant transport channel (Figure 11.13c). It was found to be $V_{trans,0} = 1.14 \pm 0.04$ V for the BDT junction, which is much less than the value for the ODT junction owing to the π-conjugated BDT molecule having a smaller HOMO–LUMO gap.

We further examined the dependence of the IETS spectra on molecular orbital gating [16]. The IETS spectra of the transistors that incorporate alkanedithiol (ODT) were essentially unaffected by the gate voltage (Figure 11.14a). This finding indicates that charge transport through the device is nonresonant; that is, there is a large energy difference between the dithiol's HOMO and the electrode's Fermi level. Conversely, we observed that the applied gate voltage strongly modulates the IETS spectra of transistors that incorporate an aromatic dithiol (BDT; Figure 11.14b) [16]. Specifically, when a negative gate voltage is applied (which brings the energy of the molecular junction's HOMO closer to that of the electrode's Fermi level), the signal intensities of the spectra increase greatly and the shapes of the vibrational peaks change. The change in peak shape is a clear indication of increased coupling between the tunneling charge carriers and the molecular vibrations, owing to a near resonance between the HOMO and the Fermi level [154,155].

Collectively, these results demonstrate direct gate modulation of molecular orbitals in molecular transistors. The IETS spectra reveal which orbitals are resonantly enhanced, and dramatic differences are seen in the comparison between near resonant and far from resonant systems. These observations validate the concept of orbital-modulated carrier transport and elucidate both charge transport mechanisms and the electronic structure of molecular junctions.

Subsequently, Cao *et al.* [156] reported electrode structures with a controllable molecular-scale gap between source and drain electrodes and a third terminal of a buried gate using photolithography and molecular lithography with self-assembled mono/multiple molecule layer(s) as a resist. The synthesized thiolated phthalocyanine-derivative molecules were assembled between the tailored molecular gap of the fabricated FET electrode structures in solution via Au-S bonding, forming stable contacts between the electrodes and the molecules [156]. The electrical measurements at room temperature show that the device has transport characteristics of a typical p-type FET device with large gate modulation [156]. In addition, the transistor effect has been also observed at a monolayer level, for example, for alkanethiol [157] and various conjugated molecules [158–160].

11.3.2 Coulomb Blockade and Kondo Regimes

Solid-state molecular transistors have been used to study two additional mechanisms in single molecule quantum dots [13–15,45–47,153]: Coulomb blockades, in which the flow of electrons is controlled by the sequential charging of a molecule due to electron–electron

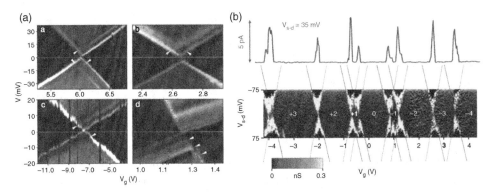

Figure 11.15 (a) Different conductance plots as a function of the bias voltage and the gate voltage obtained from four different devices. The dark triangular regions correspond to the conductance gap, and the bright lines represent peaks in the differential conductance. The arrows mark the point where the conductance lines intercept the conductance gap. Reproduced with permission from [13]. Copyright 2000 Nature Publishing Group. (b) Measurements of the differential conductance for OPV5 as a function of the bias voltage and the gate voltage. The full solid line at the top of the figure shows a representative I(V) trace. Reproduced with permission from [15]. Copyright 2003 Nature Publishing Group

Coulomb repulsion, and the Kondo effect, in which conducting electrons interact with the local spin (intrinsic angular momentum) of the quantum dot in a molecular junction, leading to an increase of the conductance at low bias.

For example, a Coulomb blockade was observed for molecules such as fullerene (C_{60}) [13] and p-phenylenevinylene oligomers [15], which have five benzene rings connected through four double bonds (OPV5) weakly coupled to the source-drain electrodes. Park *et al.* [13] prepared a single-C_{60} transistor by depositing a dilute toluene solution of C_{60} onto a ~1 nm gap between gold electrodes created by electromigration. The entire structure was defined on a SiO_2 insulating layer on top of a degenerately doped silicon wafer, which served as a gate electrode that modulated the electrostatic potential of the C_{60} molecule. The devices exhibited strongly suppressed conductance near zero bias voltage followed by steplike current jumps at higher voltages. These transport features clearly indicate that the conduction in this device is dominated by the Coulomb blockade effect. From analysis of the stability diagram in Figure 11.15a (conductance plots as a function of the bias voltage and the gate voltage), Park *et al.* [13] showed that the charging energy of the C_{60} molecule can exceed 150 meV. This value is much larger than in semiconductor quantum dots. Notice that in the stability diagrams of Figure 11.15a, there are running lines (marked by arrows) that intersect the main diamonds or conductance gap regions, indicating the presence of internal excitations of the C_{60} molecules. The energies of these excitations (of a few meV) are too small to correspond to electronic excitations. The authors of Reference [13 suggested that these lines could correspond to the vibrational excitation of the center of mass oscillation of C_{60} within the confinement potential that binds it to the gold surface. In the case of OPV5, Kubatkin *et al.* [15] observed up to eight successive charge states of the molecule, as shown in Figure 11.15b. This result suggests that the transport experiment had access to many different charge or redox states. While the spectroscopic HOMO–LUMO gap for this molecule is of the order of 2.5 eV, the extracted gap from the stability diagram in Figure 11.15b is one order of magnitude smaller (0.2 eV). The authors argued that this discrepancy is due to the fact that the intrinsic electronic levels of the molecules are significantly altered in the metallic contacts. In particular, they suggested that the image charges generated in the source and drain electrodes by the charges on the molecule are probably the origin of this effect. With organometallic molecules bearing a transition metal, such as the cobalt terpiridynil complex and divanadium complex, Kondo resonance has been observed in addition to Coulomb blockades [14,45]. Kondo resonance is observed when increasing the coupling between the molecule and the electrodes (e.g., by changing the length of the insulating tethers between the metal ion and the electrodes) [14]. van der Zant *et al.* and Natelson *et al.* have achieved much in this field, such as transport measurements through single molecular magnets [161], the observation of the Kondo effect in gold break junctions with the presence of magnetic impurities [162], inelastic tunneling features via molecular vibrations in the Kondo regime [163], and fundamental scaling laws that govern the nonequilibrium standard spin–1/2 Kondo effect [164]. We recommend the following reviews on this subject: References [46,47,165–167].

Several groups have also demonstrated the FET-like behavior of redox molecules using an electrochemical gate [168–170]. This method has the advantage that the potential at the molecule is well-defined because the potential drop at each electrode is maintained by a double layer established with respect to a reference electrode [5]. Because the gate voltage falls across the double layers at the electrode–electrolyte interfaces, the effective gate thickness is in the order of a few solvated ions, which results in a large gate field. This large field allows the reversible switching of redox molecules between oxidized and reduced states. For example, a study of perylene tetracarboxylic diimide (PTCDI), a redox molecule, found that the current through the molecule can be reversibly varied over nearly three orders of magnitude [169]. In terms of the energy diagram, decreasing the gate voltage shifts the LUMO of the PTCDI redox molecule towards the Fermi levels of the electrodes, causing a large increase in the current through the molecules (i.e., n-type FET-like behavior). Recently, Diez-Perez *et al.* [171] synthesized two coronene derivatives, each consisting of 13 aromatic rings arranged into a well-defined honeycomb structure. Using the STM break–junction method, they studied charge transport in single coronene molecules as a function of the electrochemical gate voltage and observed pronounced n-type gating effects in these single coronene devices [171].

11.4 Molecular Design

It is clear that the charge transport through a molecular junction crucially depends on the position of the frontier orbitals (HOMO or LUMO) of the molecule with respect to the electrode's Fermi energy and on its character (degree of delocalization). In principle, side groups or substituents can control the conformation of a molecule, which in turn determines the degree of conjugation (i.e., delocalization of the molecular orbitals), and they can adjust the position of the frontier orbitals (i.e., the intrinsic electronic structure of a molecule). All of these effects have a strong influence on the conductance of the molecular junction. Thus, the transport characteristics can be chemically tuned and designed, to a certain extent, with the inclusion of appropriate side groups or substituents.

Venkataraman *et al.* [172] measured the low-bias conductance values of a series of very short conjugated molecules (substituted 1,4-diaminobenzenes) using a STM-based break–junction technique in a solution of the molecules. Transport through these substituted benzenes was confirmed using coherent nonresonant tunneling or superexchange, with the molecular junction conductance depending on the alignment of the metal Fermi level to the nearest molecular level. Electron-donating substituents, which increase the energy of the occupied molecular orbitals, increase the junction conductance, while electron-withdrawing substituents have the opposite effect. In detail, the highest occupied molecular orbital (HOMO) in 1,4-diaminobenzene is best described as a combination of the lone pairs on each of the N atoms and some component of p–π density on each of the two C atoms to which the N atoms are bonded. As electron-donating substituents replace H atoms on the ring, the energy of the HOMO increases. When a H atom is replaced by a methoxy group (OCH_3), the $O(2p\pi)$ lone pair delocalizes into the benzene π space, thereby raising the HOMO energy. On the contrary, when electron-withdrawing substituents replace H atoms, the energy of the HOMO is lowered. When H is replaced by Cl, the more electronegative Cl removes electron density from the σ space of the benzene, thereby deshielding the π space and lowering the HOMO energy. These shifts in the isolated molecule are measured as changes in the ionization potential (IP): electron-donating substituents decrease the IP, while electron-withdrawing substituents increase the IP. Thus, for the measured molecular series of the substituted 1,4-diaminobenzenes, the conductance values vary inversely with the calculated ionization potential of the molecules. These results reveal that the HOMO is closest to the gold Fermi energy, which is consistent with hole transport (i.e., transport dominated by the HOMO of the molecules) [172]. Interestingly, this study is consistent with the results reported by Baheti *et al.* [139], in which the thermopower of molecular junctions based on several 1,4-benzenedithiol (BDT) derivatives was investigated. The thermopower of the measured molecules was modified in the same way by the addition of electron-withdrawing or -donating groups such as fluorine, chlorine, and methyl on the benzene ring (see Section 11.3.4).

As shown in the example above, chemical modification via side groups or substituents can be used to control the alignment between the molecular levels and the Fermi energy of the metallic electrodes. In other words, one can use chemical substituents to "dope" molecular junctions. Another example of modulating conductance via chemical substituents was described in Reference [173], in which the authors examined the low-bias conductance of 1,4-bis-(6-thiahexyl)-benzene derivatives to study the correlation between the low-bias conductance and the position of the frontier orbitals. For this purpose, these authors theoretically determined the position of these orbitals using density functional theory (DFT) calculations. It was shown that the more electron-rich benzene rings (with a higher HOMO) have higher conductances, which is also consistent with hole transport (i.e., via the benzene HOMO). These results stress that the actual position of the frontier orbitals of the molecule in the junction are the true determinants of the conductance, which in principle may differ from the corresponding frontier orbitals in the gas phase. Thus, it would be highly desirable to obtain information about the molecular-level alignment with the electrode Fermi energy in molecular junctions. It is still rather challenging to show this fundamental effect in a systematic manner.

The fact that the molecular conformation must have a major impact on the conduction through a molecular junction has been reasonably predicted [174]. Some experiments have verified this effect [175], but the most illustrative example has been reported by Venkataraman *et al.* in Reference [91]. In the simple case of a biphenyl, a molecule with two phenyl rings linked by a single C–C bond, the conductance is expected to change with the relative twist angle between the two rings, with the planar conformation having the highest conductance. The authors used amine linking groups to form single-molecule junctions

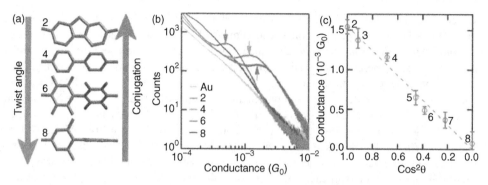

Figure 11.16 Biphenyl junction conductance as a function of molecular twist angle. (a) Structures of a subset of the biphenyl series studied, shown in order of increasing twist angle or decreasing conjugation. (b) Conductance histograms obtained from measurements using molecule 2 (constructed from 15 000 traces and scaled by 1/15), molecule 4 (constructed from 7000 traces and scaled by 1/7), molecule 6 (constructed from 11 000 traces and scaled by 1/11) and molecule 8 (constructed from 5000 traces and scaled by 1/5). Also shown is the control histogram obtained from measurements without molecules between the contacts (constructed from 6000 traces and scaled by 1/6). Arrows point to the peak conductance values obtained from Lorentzian fits. All data were taken at a bias voltage of 25 mV. (c) Position of the peaks for all the molecules studied plotted against cos2θ, where θ is the calculated twist angle for each molecule. Error bars are determined from the standard deviation of the peak locations determined from the fits to histograms of 1000 traces. Reproduced with permission from [91]. Copyright 2006 Nature Publishing Group

with a series of biphenyl molecules with different ring substitutions that alter the twist angle of the molecules. The two benzene rings of a biphenyl can rotate relative to each other. For such a molecule, as the twist angle (θ) between the two rings increases and the degree of π conjugation between them simultaneously decreases, the junction conductance is expected to decrease because molecular electron transfer rates scale as the square of the π overlap [174,176]. When neglecting the contribution of tunneling transport through the σ orbitals, the theory predicts a $\cos^2 \theta$ relation [177]. The authors of Reference [91] illustrated that the conductance for the biphenyl series decreases with increasing twist angle, consistent with the cosine-squared relation predicted for transport through π-conjugated biphenyl systems [177]. Specifically, Figure 11.16b shows the measured conductance histograms for the biphenyl molecules in Figure 11.16a. For the flat molecule (2), the histogram peak yields an average junction conductance value of $1.5 \times 10^{-3} G_0$ ($G_0 = 2e^2/h$, where e is the charge on an electron and h is Planck's constant). For 4,4′-diaminobiphenyl (4) with an equal molecular length but a twist angle θ of 34°, the conductance histogram shows a peak occurring at a lower conductance of $1.1 \times 10^{-3} G_0$. As the angle between the two aromatic rings is increased to 52° in 4,4′-diaminooctafluorobiphenyl (6), the conductance value also drops further, to $4.9 \times 10^{-4} G_0$. When all four hydrogen atoms on the proximal carbons of this molecule are replaced with methyl groups to give 2,6,2′,6′-tetramethyl-4,4′-diaminobiphenyl (8) with a twist angle θ of 88°, most counts in the histogram occur for low junction conductance values. Figure 11.16c shows a plot of the peak conductance value measured for the seven molecules against $\cos^2 \theta$, and the data are fitted with a $\cos^2 \theta$ curve (dashed line in Figure 11.16c). The good quality of the fit indicates that the electronic effects of the substituents do not significantly alter the simple picture in which junction conductance may be adjusted by simply decreasing the π overlap between phenyl rings within the bridging molecule. However, the role of the geometric and electronic effects of the substituents is still being actively discussed [178,179].

Recently, Wandlowski and co-workers investigated the conductance of a family of biphenyl-dithiol derivatives with conformationally fixed torsion angles using the STM break–junction method [180]. They found that the measured conductance value depends on the torsion angle θ between two phenyl rings; twisting the biphenyl system from flat ($\theta = 0°$) to perpendicular ($\theta = 90°$) decreased the conductance by a factor of 30. Detailed calculations of transport based on DFT and a two-level model supported the experimentally obtained $\cos^2 \theta$ correlation between the junction conductance and the torsion angle.

In fact, the role of conjugation in the conduction through molecular systems has been clearly illustrated by the comparison of the conductance through alkanes to that through conjugated molecules with extended π electron states (e.g., oligophenyleneethynylene, OPE and oligophenylenevinylene, OPV). The π-conjugated molecules show substantially higher conductance than the alkanes, consistent with a rational dependence on the HOMO–LUMO gap [92,181].

The strength of the metal–molecule coupling can be also chemically tuned using appropriate anchoring groups to bind a molecule to metallic electrodes. Tao and coworkers have systematically studied the effect of anchoring groups on the conductance of single molecules using alkanes terminated with dithiol, diamine, and dicarboxylic acid groups as a model system [182]. The conductance values of these molecules were found to be independent of temperature, indicating coherent tunneling. For each anchoring group, the authors reported an exponential decay of the conductance with the molecular length, given by $G = A\exp(-\beta N)$, which also suggests the tunneling mechanism. They observed differences in β between different anchoring groups with the trend β (dithiol) $> \beta$ (diamine) $\geq \beta$ (dicarboxyl acid). The β values, ranging between 0.8 and 1.0 per –CH$_2$ unit or between 0.6 and 0.8 Å^{-1}, were close to the values reported in the literature [93]. The differences in β are small for the three anchoring groups, but this small difference can significantly affect long-distance electron transport in the molecules due to the exponential dependence of the conductance on length. Both experimental evidence [183] and theoretical calculations [184,185] indicate that β depends on the alignment of the molecular energy levels relative to the Fermi energy level of the electrodes. Based on these considerations, the observed differences in β may be attributed to the differences in the energy alignments [182]. The prefactor of the exponential function, A, which is a measure of contact resistance, is highly sensitive to the type of the anchoring group, which varies in the order of Au-S > Au-NH$_2$ > Au-COOH. This large dependence was attributed to the different coupling strengths provided by the different anchoring groups between the alkane and the electrodes [182].

Other important factors are the stability of the contact and the variability of the bonding between anchoring groups and metallic electrodes, which can play an important role in the reproducibility of the experimental results. For example, Venkataraman and co-workers in Reference [64] suggested the use of amine (NH$_2$) groups to obtain well-defined values of the conductance of molecular junctions. In this work, the conductance of amine-terminated molecules was measured by breaking Au point contacts in a molecular solution at room temperature. It was found that the variability of the observed conductance for the diamine molecule-Au junctions is much less than the variability for diisonitrile- and dithiol-Au junctions. This narrow distribution in the measured conductance values enables unambiguous conductance measurements of single molecules. For an alkane diamine series with 2–8 carbon atoms in the hydrocarbon chain, Venkataraman et $al.$'s results showed a systematic trend in the conductance from which a tunneling decay constant of 0.91 ± 0.03 per methylene group was found [64]. The authors hypothesized that the diamine link binds preferentially to under-coordinated Au atoms in the junction. This theory is supported by DFT theory calculations that show the amine binding to a gold adatom with sufficient angular flexibility for easy junction formation and well-defined electronic coupling of the N lone pair to the Au. Therefore, the amine linkage leads to well-defined conductance measurements of a single molecule junction in a statistical study.

On the other hand, with respect to the spread of the peaks in the conductance histograms, there were no significant differences between thiols and amines in the work of Tao and co-workers [182]. Similar findings were also reported by Martin *et al.* [186] Using the MCBJ method, these authors measured the conductance histogram for benzenediamine. In contrast to Reference [91], they did not find a pronounced peak structure. According to these authors, this difference may be due to the absence of a solvent in their experiment and to the fast rupture of the metal–molecule bond that must have reduced the probability of forming stable molecular junctions.

In addition, Park *et al.* [187] compared the low bias conductance of a series of alkanes terminated on various ends with dimethyl phosphines, methyl sulfides, and amines and found that junctions formed with dimethyl phosphine-terminated alkanes have the highest conductance. In this work, they observed a clear conductance signature with these linker groups, indicating that the binding is well-defined and electronically selective. Martin *et al.* [188] designed and synthesized a linear and rigid C_{60}-capped molecule, 1,4-bis(fullero[*c*]pyrrolidin-1-yl) benzene (BDC$_{60}$), and compared its electrical characteristics to those of 1,4-benzenedi-amine (BDA) and 1,4-benzenedithiol (BDT) using lithographic MCBJs. The main conclusion of this work is the suitability of fullerene anchoring for single-molecule electronic measurements. In particular, the fullerene-anchoring leads to a considerably lower spread in low-bias conductance compared to thiols due to its higher junction stability, which minimizes fluctuations due to atomic details at the anchoring site.

Finally, the exact position of anchoring group is also critical. Mayor *et al.* [189] showed that the conductance of a thiol-terminated rod-like conjugated molecule crucially depends on the position of the thiol group. They showed that by placing the thiol group in the *meta* position of the last phenyl ring, the conjugation is partially interrupted and the current decreases significantly compared to the case in which the thiol group is in the *para* position.

11.5 Perspectives

It seems unlikely that molecular-based electronics will ever replace the silicon-based electronics of today. However, there are convincing reasons to believe that they can complement silicon devices by providing, for example, novel functionalities beyond the scope of conventional solid-state devices because the small dimensions of the molecule with a variety of electronic, optical, mechanical, and thermoelectric properties are able to lead to valuable new physical phenomena. As described in this review, we have learned much about the fundamental charge transport in single molecules, which plays a key role in both basic science and technological applications. Although it is still difficult to fabricate reliable electronic devices made with single molecules, and many of the remaining challenges are still formidable, some concepts and techniques are now well-established and discussed in this review. It is certain that this emerging field has advanced considerably over recent years, and such research efforts will help us to incorporate molecular components into traditional microelectronic devices.

Acknowledgments

T.L. acknowledges financial support from the National Creative Research Laboratory program (Grant No. 2012026372) by the National Research Foundation of Korea (NRF) grant funded by the Korea government (MSIP). M.A.R. acknowledges financial support from the US Army Research Office (W911NF-08-1-0365), the Canadian Institute for Advanced Research (CIfAR), and the National Science Foundation under Grant No. 1309898.

References

1. Aviram, A. and Ratner, M.A. (1974) *Chemical Physics Letters*, **29**, 277.
2. Cuevas, J.C. and Scheer, E. (2009) *Molecular Electronics: An Introduction to Theory and Experiment*, World Scientific, Singapore.
3. Cuniberti, G., Fagas, G., and Richter, K. (eds) (2005) *Introducing Molecular Electronics*, Springer-Verlag, Berlin, Heidelberg.
4. Reed, M.A. and Lee, T. (2003) *Molecular Nanoelectronics*, American Scientific, Stevenson Ranch.
5. Tao, N.J. (2006) *Nature Nanotechnology*, **1**, 173.
6. Akkerman, H.B. and de Boer, B. (2008) *Journal of Physics: Condensed Matter*, **20**, 013001.
7. Heath, J.R. (2009) *Annual Review of Materials Research*, **39**, 1.
8. McCreery, R.L. and Bergren, A.J. (2009) *Advanced Materials*, **21**, 4303–4322.
9. Ulgut, B. and Abruna, H.D. (2008) *Chemical Reviews*, **108**, 2721–2736.
10. Díez-Pérez, I., Hihath, J., Lee, Y. *et al.* (2009) *Nature Chemistry*, **1**, 635.
11. Metzger, R.M., Chen, B., Hopfner, U. *et al.* (1997) *Journal of the American Chemical Society*, **119**, 10455.
12. Ebling, M., Ochs, R., Koentopp, M. *et al.* (2005) *Proceedings of the National Academy of Sciences of the United States of America*, **102**, 8815–8820.
13. Park, H., Park, J., Lim, A.K.L. *et al.* (2000) *Nature*, **407**, 57.
14. Park, J., Pasupathy, A.N., Goldsmith, J.I. *et al.* (2002) *Nature*, **417**, 722.
15. Kubatkin, S., Danilov, A., Hjort, M. *et al.* (2003) *Nature*, **425**, 698.
16. Song, H., Kim, Y., Jang, Y.H. *et al.* (2009) *Nature*, **462**, 1039.
17. van der Molen, S.J. and Liljeroth, P. (2010) *Journal of Physics: Condensed Matter*, **22**, 133001.
18. Quek, S.Y., Kamenetska, M., Steigerwald, M.L. *et al.* (2009) *Nature Nanotechnology*, **4**, 230.
19. Blum, A.S., Kushmerick, J.G., Long, D.P. *et al.* (2005) *Nature Materials*, **4**, 167.
20. Choi, B.Y., Kahng, S.J., Kim, S. *et al.* (2006) *Physical Review Letters*, **96**, 156106.
21. Lörtscher, E., Ciszek, J.W., Tour, J., and Riel, H. (2006) *Small*, **2**, 973.
22. Green, J.E., Choi, J.W., Boukai, A. *et al.* (2007) *Nature*, **445**, 414–417.
23. Lee, J., Chang, H., Kim, S. *et al.* (2009) *Angewandte Chemie*, **121**, 8653.
24. Chen, F., Hihath, J., Huang, Z. *et al.* (2007) *Annual Review of Physical Chemistry*, **58**, 535.
25. Li, T., Hu, W., and Zhu, D. (2010) *Advanced Materials*, **22**, 286.
26. Moreland, J. and Ekin, J.W. (1985) *Journal of Applied Physiology*, **58**, 3888.
27. Muller, C.J., van Ruitenbeek, J.M., and de Jongh, L.J. (1992) *Physica C*, **191**, 485.
28. Huber, R., Gonzalez, M.T., Wu, S. *et al.* (2008) *Journal of the American Chemical Society*, **130**, 1080.
29. Wu, S., González, M.T., Huber, R. *et al.* (2008) *Nature Nanotechnology*, **3**, 569–574.
30. González, M.T., Wu, S., Huber, R. *et al.* (2006) *Nano Letters*, **6**, 2238.
31. Lörtscher, E., Weber, H., and Riel, H. (2007) *Physical Review Letters*, **98**, 176807.
32. Reed, M.A., Zhou, C., Muller, C.J. *et al.* (1997) *Science*, **278**, 252.
33. van Ruitenbeek, J., Scheer, E., and Weber, H.B. (2005) *Lecture Notes in Physics: Introducing Molecular Electronics*, vol. **680** (eds G. Cuniberti, G. Fagas, and K. Richter), Springer, Heidelberg, pp. 253–274.
34. Reichert, J., Ochs, R., Beckmann, D. *et al.* (2002) *Physical Review Letters*, **88**, 176 804.
35. Djukic, D. and van Ruitenbeek, J.M. (2006) *Nano Letters*, **6**, 789.
36. Tian, J.H., Liu, B., Li, X.L. *et al.* (2006) *Journal of the American Chemical Society*, **128**, 14 748.
37. Smit, R.H.M., Noat, Y., Untiedt, C. *et al.* (2002) *Nature*, **419**, 906.
38. Kergueris, C., Bourgoin, J.P., Palacin, S. *et al.* (1999) *Physical Review B-Condensed Matter*, **59**, 19.
39. Dulic, D., van der Molen, S.J., Kudernac, T. *et al.* (2003) *Physical Review Letters*, **91**, 207402.
40. Lee, M.H., Speyer, G., and Sankey, O.F. (2006) *Physica Status Solidi B-Basic Research*, **243**, 2021.

41. Grigoriev, A., Sköldberg, J., Wendin, G., and Crljen, Z. (2006) *Physical Review B-Condensed Matter*, **74**, 045401.
42. Basch, H., Cohen, R., and Ratner, M.A. (2005) *Nano Letters*, **5**, 1668.
43. Romaner, L., Heimel, G., Gruber, M. *et al.* (2006) *Small*, **2**, 1468.
44. Park, H., Lim, A.K.L., Alivisatos, A.P. *et al.* (1999) *Applied Physics Letters*, **75**, 301.
45. Liang, W.J., Shores, M.P., Bockrath, M. *et al.* (2002) *Nature*, **417**, 725.
46. van der Zant, H.S.J., Kervennic, Y.-V., Poot, M. *et al.* (2006) *Faraday Discuss*, **131**, 347.
47. Natelson, D., Yu, L.H., Ciszek, J.W. *et al.* (2006) *Chemical Physics*, **324**, 267–275.
48. Trouwborst, M.L., van der Molen, S.J., and van Wees, B. (2006) *Journal of Applied Physiology*, **99**, 114316.
49. Heersche, H.B., de Groot, Z., Folk, J.A. *et al.* (2006) *Physical Review Letters*, **96**, 017 205.
50. Houck, A.A., Labaziewicz, J., Chan, E.K. *et al.* (2005) *Nano Letters*, **5**, 1685.
51. Taychatanapat, T., Bolotin, K.I., Kuemmeth, F., and Ralph, D.C. (2007) *Nano Letters*, **7**, 652.
52. Luo, K., Chae, D.H., and Yao, Z. (2007) *Nanotechnology*, **18**, 465203.
53. Strachan, D.R., Smith, D.E., Fischbein, M.D. *et al.* (2006) *Nano Letters*, **6**, 441.
54. Heersche, H.B., Lientschnig, G., O'Neill, K. *et al.* (2007) *Applied Physics Letters*, **91**, 072107.
55. Crommie, M.F., Lutz, C.P., and Eigler, D.M. (1993) *Nature*, **363**, 524.
56. Repp, J., Meyer, G., Stojkovic, S.M. *et al.* (2005) *Physical Review Letters*, **94**, 026803.
57. Manoharan, H.C., Lutz, C.P., and Eigler, D. (2000) *Nature*, **403**, 512.
58. Xu, B. and Tao, N.J. (2003) *Science*, **301**, 1221–1223.
59. Grüter, L., González, M.T., Huber, R. *et al.* (2005) *Small*, **1**, 1067.
60. Huisman, E.H., Trouwborst, M.L., Bakker, F.L. *et al.* (2008) *Nano Letters*, **8**, 3381.
61. Tsutsui, M., Shoji, K., Taniguchi, M., and Kawai, T. (2008) *Nano Letters*, **8**, 345.
62. Xu, B.Q., Zhang, P.M., Li, X.L., and Tao, N.J. (2004) *Nano Letters*, **4**, 1105.
63. Jang, S.-Y., Reddy, P., Majumdar, A., and Segalman, R.A. (2006) *Nano Letters*, **6**, 2362.
64. Venkataraman, L., Klare, J.E., Tam, I.W. *et al.* (2006) *Nano Letters*, **6**, 458.
65. Xu, B., Xiao, X., and Tao, N.J. (2003) *Journal of the American Chemical Society*, **125**, 16164.
66. Xia, J.L., Diez-Perez, I., and Tao, N.J. (2008) *Nano Letters*, **8**, 1960.
67. Li, X., He, J., Hihath, J. *et al.* (2006) *Journal of the American Chemical Society*, **128**, 2135.
68. Kamenetska, M., Koentopp, M., Whalley, A.C. *et al.* (2009) *Physical Review Letters*, **102**, 126803.
69. Wold, D.J. and Frisbie, C.D. (2000) *Journal of the American Chemical Society*, **122**, 2970–2971.
70. Wold, D.J. and Frisbie, C.D. (2001) *Journal of the American Chemical Society*, **123**, 5549–5556.
71. Kim, B.S., Beebe, J.M., Olivier, C. *et al.* (2007) *The Journal of Physical Chemistry C*, **111**, 7521.
72. Song, H., Lee, H., and Lee, T. (2007) *Journal of the American Chemical Society*, **129**, 3806.
73. Wang, G., Kim, T.W., Jo, G., and Lee, T. (2009) *Journal of the American Chemical Society*, **131**, 5980.
74. Engelkes, V.B., Beebe, J.M., and Frisbie, C.D. (2004) *Journal of the American Chemical Society*, **126**, 14287.
75. Beebe, J.M., Engelkes, V.B., Miller, L.L., and Frisbie, C.D. (2002) *Journal of the American Chemical Society*, **124**, 11268.
76. Sakaguchi, H., Hirai, A., Iwata, F. *et al.* (2001) *Applied Physics Letters*, **79**, 3708.
77. Son, K.-A., Kim, H.I., and Houston, J.E. (2001) *Physical Review Letters*, **86**, 5357–5360.
78. Park, J.Y., Maier, S., Hendriksen, B., and Salmeron, M. (2010) *Materials Today*, **13**, 38–45.
79. Morpurgo, A.F., Marcus, C.M., and Robinson, D.B. (1999) *Applied Physics Letters*, **74**, 2084.
80. Dadosh, T., Gordin, Y., Krahne, R. *et al.* (2005) *Nature*, **436**, 677.
81. Guo, X.F., Small, J.P., Klare, J.E. *et al.* (2006) *Science*, **311**, 356.
82. Whalley, A.C., Steigerwald, M.L., Guo, X.F., and Nuckolls, C. (2007) *Journal of the American Chemical Society*, **129**, 590.
83. Guo, X.F., Gorodetsky, A.A., Hone, J. *et al.* (2008) *Nature Nanotechnology*, **3**, 163.
84. Jain, T., Westerlund, F., Johnson, E. *et al.* (2009) *ACS Nano*, **3**, 828.
85. Galperin, M., Ratner, M.A., Nitzan, A., and Troisi, A. (2005) *Science*, **319**, 1056.
86. Wang, W., Lee, T., and Reed, M.A. (2003) *Physical Review B-Condensed Matter*, **68**, 035416.
87. Song, H., Kim, Y., Jeong, H. *et al.* (2005) *The Journal of Physical Chemistry C*. Article ASAP doi: 10.1021/jp104760b
88. Choi, H., Kim, B., and Frisbie, C.D. (2008) *Science*, **320**, 1482.
89. Lafferentz, L., Ample, F., Yu, H. *et al.* (2009) *Science*, **323**, 1193–1197.
90. Hines, T., Diez-Perez, I., Hihath, J. *et al.* (2010) *Journal of the American Chemical Society*, **132**, 11658–11664.
91. Venkataraman, L., Klare, J.E., Nuckolls, C. *et al.* (2006) Dependence of single molecule junction conductance on molecular conformation. *Nature*, **442**, 904–907.

92. Wold, D.J., Haag, R., Rampi, M.A., and Frisbie, C.D. (2002) *The Journal of Physical Chemistry B*, **106**, 2813.
93. Salomon, A., Cahen, D., Lindsay, S. *et al.* (2003) *Advanced Materials*, **15**, 1881.
94. Segal, D. and Nitzan, A. (2001) *Chemical Physics*, **268**, 315.
95. Segal, D. and Nitzan, A. (2002) *Chemical Physics*, **281**, 235.
96. Poot, M., Osorio, E., O'Neill, K. *et al.* (2006) *Nano Letters*, **6**, 1031.
97. Datta, S. (2004) *Nanotechnology*, **15**, S433.
98. Lambe, J. and Jaklevic, R.C. (1968) *Physical Review*, **165**, 821.
99. Reed, M.A. (2008) *Materials Today*, **11**, 46.
100. Hansma, P.K. (1977) *Physical Letters C Physical Report*, **30**, 145.
101. Adkins, C.J. and Phillips, W.A. (1985) *Journal of Physics C*, **18**, 1313.
102. Horiuchi, T. *et al.* (1989) *Review of Scientific Instruments*, **60**, 994.
103. Wang, W., Lee, T., Kretzschmar, I., and Reed, M.A. (2004) *Nano Letters*, **4**, 643.
104. Kushmerick, J.G., Lazorcik, J., Patterson, C.H. *et al.* (2004) *Nano Letters*, **4**, 639.
105. Galperin, M., Ratner, M.A., Nitzan, A., and Troisi, A. (2008) *Science*, **319**, 1056.
106. Song, H., Kim, Y., Ku, J. *et al.* (2009) *Applied Physics Letters*, **94**, 103110.
107. Jaklevic, R.C. and Lambe, J. (1966) *Physical Review Letters*, **17**, 1139.
108. Troisi, A., Ratner, M.A., and Nitzan, A. (2003) *Journal of Chemical Physics*, **118**, 6072.
109. Kushmerick, J.G., Lazorcik, J., Patterson, C.H. *et al.* (2004) *Nano Letters*, **4**, 639.
110. Long, D.P., Lazorcik, J.L., Mantooth, B.A. *et al.* (2006) *Nature Materials*, **5**, 901.
111. Stipe, B.C., Rezaei, M.A., and Ho, W. (1998) *Science*, **280**, 1732.
112. Binnig, G., Garcia, N., and Rohrer, H. (1985) *Physical Review*, **B32**, 1336.
113. Ho, W. (2002) *Journal of Chemical Physics*, **117**, 11033.
114. Gregory, S. (1990) *Physical Review Letters*, **64**, 689.
115. Kushmerick, J.G., Holt, D.B., Yang, J.C. *et al.* (2002) *Physical Review Letters*, **89**, 086802.
116. Beebe, J.M., Moore, H.J., Lee, T.R., and Kushmerick, J.G. (2007) *Nano Letters*, **7**, 1364.
117. Troisi, A., Beebe, J.M., Picraux, L.B. *et al.* (2007) *Proceedings of the National Academy of Sciences of the United States of America*, **104**, 14255.
118. Hihath, J., Arroyo, C.R., Rubio-Bollinger, G. *et al.* (2008) *Nano Letters*, **8**, 1673.
119. Hansma, P.K. (1977) *Physical Letters C Physical Report*, **30**, 145.
120. Lauhon, L.J. and Ho, W. (2001) *Review of Scientific Instruments*, **72**, 216.
121. Beebe, J.M., Kim, B., Gadzuk, J.W. *et al.* (2006) *Physical Review Letters*, **97**, 026801.
122. Beebe, J.M., Kim, B., Frisbie, C.D., and Kushmerick, J.G. (2008) *ACS Nano*, **2**, 827.
123. Yu, L.H., Gergel-Hackett, N., Zangmeister, C.D. *et al.* (2008) *Journal of Physics: Condensed Matter*, **20**, 374114.
124. Huisman, E.H., Guédon, C.M., van Wees, B.J., and van der Molen, S.J. (2009) *Nano Letters*, **9**, 3909.
125. Liu, K., Wang, X., and Wang, F. (2008) *ACS Nano*, **2**, 2315.
126. Chiu, P.W. and Roth, S. (2008) *Applied Physics Letters*, **92**, 042107.
127. O'Neill, L. and Byrne, H.J. (2005) *The Journal of Physical Chemistry B*, **109**, 12685.
128. Baldea, I. (2014) *Chemical Physics*. doi: 10.1016/j.chemphys.2010.08.009
129. Araidai, M. and Tsukada, M. (2010) *Physical Review B-Condensed Matter*, **81**, 235114.
130. Chen, J., Markussen, T., and Thygesen, K.S. (2010) *Physical Review B-Condensed Matter*, **82**, 121412(R).
131. Alloway, D.M., Hofmann, M., Smith, D.L. *et al.* (2003) *The Journal of Physical Chemistry B*, **107**, 11690.
132. Duwez, A.S., Pfister-Guillouzo, G., Delhalle, J., and Riga, J. (2000) *The Journal of Physical Chemistry B*, **104**, 9029.
133. Paulsson, M. and Datta, S. (2003) *Physical Review B-Condensed Matter*, **67**, 241403.
134. Pierret, R.F. (1996) *Semiconductor Fundamentals*, Addison-Wesley, Reading, MA, Chap. 3.2.2.
135. Krzeminski, C., Delerue, C., Allan, G. *et al.* (2001) *Physical Review B-Condensed Matter*, **64**, 085405.
136. Reddy, P., Jang, S.Y., Segalman, R.A., and Majumdar, A. (2007) *Science*, **315**, 1568.
137. Blatt, F.J. (1976) *Thermoelectric Power of Metals*, Plenum Press, New York, pp. xv, 264.
138. Buttiker, M., Imry, Y., Landauer, R., and Pinhas, S. (1985) *Physical Review B-Condensed Matter*, **31**, 6207.
139. Baheti, K., Malen, J.A., Doak, P. *et al.* (2008) *Nano Letters*, **8**, 715.
140. Ward, D.R., Scott, G.D., Keane, Z.K. *et al.* (2008) *Journal of Physics: Condensed Matter*, **20**, 374118.
141. Ward, D.R., Grady, N.K., Levin, C.S. *et al.* (2007) *Nano Letters*, **7**, 1396.
142. Ward, D.R., Halas, N.J., Ciszek, J.W. *et al.* (2008) *Nano Letters*, **8**, 919.
143. Moskovits, M. (2005) Surface-enhanced Raman spectroscopy. *Journal of Raman Spectroscopy*, **36**, 485.
144. Ward, D.R., Hueser, F., Pauly, F. *et al.* (2010) *Nature Nanotechnology*, **5**, 732–736.
145. MÄuhlschlegel, P., Eisler, H.-J., Martin, O.J.F. *et al.* (2005) *Science*, **308**, 1607.

146. Baik, J.M., Lee, S.J., and Moskovits, M. (2009) *Nano Letters*, **9**, 672.
147. Ioffe, Z., Shamai, T., Ophir, A. *et al.* (2008) *Nature Nanotechnology*, **3**, 727.
148. Di Ventra, M., Pantelides, S.T., and Lang, N.D. (2000) *Applied Physics Letters*, **76**, 3448.
149. Damle, P., Rakshit, T., Paulsson, M., and Datta, S. (2002) *IEEE Transactions on Nanotechnology*, **1**, 145–1153.
150. Ghosh, A.W., Rakshit, T., and Datta, S. (2004) *Nano Letters*, **4**, 565–568.
151. Lang, N.D. and Solomon, P.M. (2005) Charge control in a model biphenyl molecular transistor. *Nano Letters*, **5**, 921–924.
152. Solomon, P.M. and Lang, N.D. (2008) The biphenyl molecule as a model transistor. *ACS Nano*, **2**, 435–440.
153. Osorio, E.A., O'Neill, K., Wegewijs, M. *et al.* (2007) *Nano Letters*, **7**, 3336.
154. Galperin, M., Ratner, M.A., and Nitzan, A. (2004) *Journal of Chemical Physics*, **121**, 11965–11979.
155. Persson, B.N.J. and Baratoff, A. (1987) *Physical Review Letters*, **59**, 339–342.
156. Cao, L., Chen, S., Wei, D. *et al.* (2010) *Journal of Materials Chemistry*, **20**, 2305–2309.
157. Hwang, G.J., Jeng, R.P., Lien, C. *et al.* (2006) Field effects on electron conduction through self-assembled monolayers. *Applied Physics Letters*, **89**, 133120.
158. Mottaghi, M., Lang, P., Rodriguez, F. *et al.* (2007) *Advanced Functional Materials*, **17**, 597.
159. Smits, E.C.P., Mathijssen, S.G.J., van Hal, P.A. *et al.* (2008) *Nature*, **455**, 956.
160. Tulevski, G.S., Miao, Q., Fukuto, M. *et al.* (2004) *Journal of the American Chemical Society*, **126**, 15048.
161. Heersche, H.B., de Groot, Z., Folk, J.A. *et al.* (2006) *Physical Review Letters*, **96**, 206801.
162. Heersche, H.B., de Groot, Z., Folk, J.A. *et al.* (2006) *Physical Review Letters*, **96**, 017 205.
163. Yu, L.H., Keane, Z.K., Ciszek, J.W. *et al.* (2004) *Physical Review Letters*, **93**, 266802.
164. Scott, D., Keane, Z.K., Ciszek, J.W. *et al.* (2009) *Physical Review B-Condensed Matter*, **79**, 165413.
165. Osorio, E.A., Bjørnholm, T., Lehn, J.-M. *et al.* (2006) *Journal of Physics: Condensed Matter*, **20**, 374121.
166. Moth-Poulsen, K. and Bjørnholm, T. (2009) *Nature Nanotechnology*, **4**, 551.
167. Scott, G.D. and Natelson, D. (2010) *ACS Nano*, **4**, 3560–3579.
168. Haiss, W., van Zalinge, H., Nichols, R.J. *et al.* (2003) *Journal of the American Chemical Society*, **125**, 15294–15295.
169. Xu, B., Xiao, X., Yang, X. *et al.* (2005) Large gate modulation in the current of a room temperature single molecule transistor. *Journal of the American Chemical Society*, **127**, 2386–2387.
170. Albrecht, T., Guckian, A., Ulstrup, J., and Vos, J.G. (2005) Transistor-like behavior of transition metal complexes. *Nano Letters*, **5**, 1451–1455.
171. Diez-Perez, I., Li, Z., Hihath, J. *et al.* (2010) *Nature Communications*, **1**, 31. doi: 10.1038/ncomms1029
172. Venkataraman, L., Park, Y.S., Whalley, A.C. *et al.* (2007) *Nano Letters*, **7**, 502.
173. Leary, E., Higgins, S.J., van Zalinge, H. *et al.* (2007) *Chemical Communications*, 3939.
174. Mujica, V., Nitzan, A., Mao, Y. *et al.* (1999) *Advances in Chemical Physics*, **107**, 403.
175. Moresco, F., Meyer, G., Rieder, K.H. *et al.* (2001) *Physical Review Letters*, **86**, 672.
176. Nitzan, A. (2000) Electron transmission through molecules and molecular interfaces. *Annual Review of Physical Chemistry*, **52**, 681–750.
177. Woitellier, S., Launay, J.P., and Joachim, C. (1989) The possibility of molecular switching: Theoretical study of $[(NH3)5Ru-4,4'-Bipy-Ru(NH3)5]^{5+}$. *Chemical Physics*, **131**, 481–488.
178. Pauly, F., Viljas, J.K., and Cuevas, J.C. (2008) Schn, G *Physical Review B-Condensed Matter*, **77**, 155312.
179. Finch, C.M., Sirichantaropass, S., Bailey, S.W. *et al.* (2008) *Journal of Physics: Condensed Matter*, **20**, 2022203.
180. Mishchenko, A., Vonlanthen, D., Meded, V. *et al.* (2010) *Nano Letters*, **10**, 156.
181. Huber, B., González, M.T., Langer, M. *et al.* (2008) *Journal of the American Chemical Society*, **130**, 1080–1084.
182. Chen, F., Li, X., hihath, J. *et al.* (2006) *Journal of the American Chemical Society*, **128**, 15874.
183. Kim, B., Beebe, J.M., Jun, Y. *et al.* (2006) *Journal of the American Chemical Society*, **128**, 4970–4971.
184. Yaliraki, S.N., Kemp, M., and Ratner, M.A. (1999) *Journal of the American Chemical Society*, **121**, 3428–3434.
185. Xue, Y.Q., Datta, S., and Ratner, M.A. (2001) *Journal of Chemical Physics*, **115**, 4292–4299.
186. Martin, C.A., Ding, D., van der Zant, H.S.J., and van Ruitenbeek, J.M. (2008) *New Journal of Physics*, **10**, 065008.
187. Park, Y.S., Whalley, A.C., Kamenetska, M. *et al.* (2007) *Journal of the American Chemical Society*, **129**, 15768.
188. Martin, C.A., Ding, D., Sørensen, J.K. *et al.* (2008) *Journal of the American Chemical Society*, **130**, 13198–13199.
189. Mayor, M., Weber, H.B., Reichert, J. *et al.* (2003) *Angewandte Chemie*, **115**, 6014.

12

Memory Select Devices

An Chen
GLOBALFOUNDRIES Inc., USA

12.1 Introduction

With Flash memories approaching the scaling limit, some novel memory devices have emerged as potential candidates for Flash memory replacement, as shown in the taxonomy in Figure 12.1 [1]. Some promising examples include phase change memory (PCM), resistive random access memory (RRAM), spin transfer torque random access memory (STTRAM), and so on.

Many emerging nonvolatile memories (NVMs) have a simple two-terminal structure compatible with crossbar array architecture. Crossbar array layout enables device footprint as small as $4F^2$ (*F*: critical dimension of a technology node). By stacking up multiple layers of crossbar arrays [2,3] or building vertical 3D memory structures [4,5], even smaller footprint and higher memory density can be achieved. On the other hand, crossbar memory arrays also face challenges in device selection and isolation, which needs to be addressed by functional memory select devices.

12.2 Crossbar Array and Memory Select Devices

In crossbar arrays, memory devices are built at the junction between horizontal wordlines (WLs) and vertical bitlines (BLs), as shown in Figure 12.2. A device (e.g., R_j) is selected by applying voltages to its access lines. The unselected devices form large number of sneak paths in parallel with the selected path, which degrades the accessibility to the selected device. One example of these sneak paths is shown in Figure 12.2. The impact of the sneak paths on the reading and writing operations of crossbar arrays has been analyzed in numerous papers [6–14].

To reduce the impact of sneak paths, select devices need to be inserted at every junction in series with the memory elements. Most test chips of emerging memories use transistors as select devices in a 1-transistor-1-resistor (1T1R) configuration; however, the three-terminal structure and large footprint of transistors compromise the scaling advantages of crossbar arrays. Instead, two-terminal select devices are more compatible with crossbar arrays and 3D memories.

Emerging Nanoelectronic Devices, First Edition. An Chen, James Hutchby, Victor Zhirnov and George Bourianoff.
© 2015 John Wiley & Sons, Ltd. Published 2015 by John Wiley & Sons, Ltd.

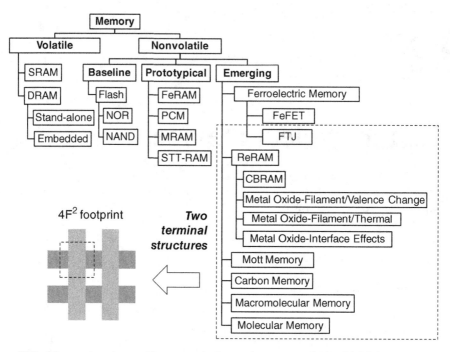

Figure 12.1 Memory taxonomy with two-terminal emerging memory devices highlighted. Inset shows a crossbar array layout with $4F^2$ device footprint

Figure 12.3 illustrates all possible sneak paths in parallel with a selected device at junction (i,j) in a $m \times n$ array. Notice that leakage current through any sneak paths flows from WL to BL, BL to WL and WL to BL, that is, along any sneak paths there always exists a segment with reverse direction for current flow. Therefore, an asymmetrical device with significantly higher reverse resistance can help to reduce sneak current. Figure 12.3 also shows that multiple unselected devices share the voltage applied on a selected device, that is, unselected devices usually have

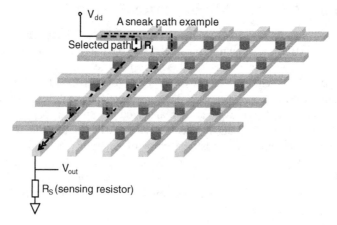

Figure 12.2 Illustration of a crossbar array with the selected current path (the dashed line) and an example of sneak paths (the dotted dashed line) highlighted

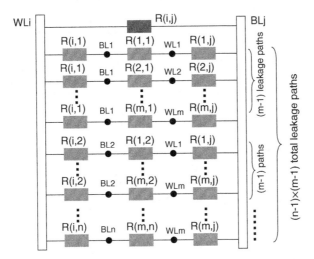

Figure 12.3 An illustration of all possible sneak paths in parallel with a selected device R(i,j) in a $m \times n$ array

lower bias than that of the selected device. This can also be ensured by proper bias of unselected WLs/BLs. So a nonlinear device with higher resistance at lower voltage can make sneak paths effectively more resistive and reduce sneak current. Therefore, both asymmetry and nonlinearity in device characteristics can help to reduce the impact of sneak paths in crossbar arrays.

Functional select devices should provide either asymmetry or nonlinearity. Asymmetry is readily available in rectifying diodes. Nonlinearity can be introduced by non-ohmic transport mechanisms (e.g., tunneling) or some volatile switching phenomena (e.g., threshold switching). Figure 12.4 categorizes these memory select device options.

To work for a given array size, select devices need to have sufficient difference between conductive state and blocking state, measured by resistance. This can be described by an *ON/OFF*

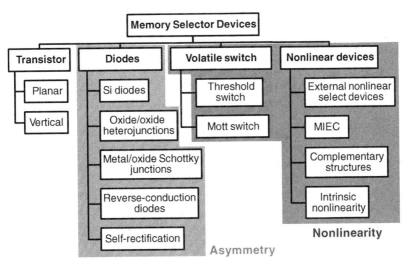

Figure 12.4 Taxonomy of memory select devices

Table 12.1 Parameters of select devices

Parameter	Explanations
ON/OFF ratio	Resistance ratio between ON and OFF state; affects array size
Maximum ON current	Needs to be sufficiently high for the switching of memory elements
Scalability	Needs to be as scalable as the memory elements.
Threshold voltage	The minimal voltage required to turn on the select devices.
Operation polarity	ON and OFF states in the same (unipolar) or opposite (bipolar) directions.
Switching speed	How fast select devices turn on and off affects the speed of memory.
Endurance	Select devices are turned on and off during both writing and reading operations and are therefore subject to stronger requirements on endurance.
Manufacturability	Compatibility with memory integration process, material stability, thermal budget, process variability, and so on.

ratio, which is the *rectification ratio* for diodes and the *nonlinearity ratio* for nonlinear devices. Select devices also need to provide sufficiently high current for the switching of the memory elements, so the *maximum ON current* is another key parameter. It is desirable that select devices are at least as scalable as the memory element to avoid compromising their scaling advantage. The key parameters of memory select devices are summarized in Table 12.1.

Some emerging memories (e.g., RRAM) require current control during switching to achieve stable performance. Transistors as select devices can provide active switching control via gate voltage (V_g). However, two-terminal select devices generally lack the capability to control switching current. Peripheral transistors outside the crossbar array may be used to control switching current; however, the effectiveness may be weakened by parasitic components.

Variability of select devices needs to be carefully controlled. For diode-type select devices, at least one side of a junction needs to have low or intermediate doping; otherwise, breakdown voltage would be too low. For a volume of $10\,\mathrm{nm}^3$, a doping concentration as high as $10^{18}\,\mathrm{cm}^{-3}$ means just a single atom. In such highly scaled devices, random dopant fluctuation may cause large variation and affect diode functions. Diode scaling is also limited by the finite size of the depletion region [15]. It is estimated that there is a minimum length of $\sim$8 nm for Si diodes, due to the lateral depletion. Contact resistance constrains maximum ON current of all select devices and should be minimized. It is estimated that contact resistivity $<10^{-7}\,\Omega\cdot\mathrm{cm}^2$ is needed for functional select devices.

12.3 Memory Select Device Options

12.3.1 Transistors as Memory Select Devices

Transistors are probably the best select devices in terms of performance. Flash memory is a combination of memory element (charge storage in the floating gate) and select device (transistor) in one device. Although transistors have sufficient ON/OFF ratio and ON current, their three-terminal structure and large footprint are unsuitable for crossbar arrays and 3D memories. Transistor V_g controls the current flowing through the memory element during switching, which affects the switching properties of some memories, for example, RRAM. Transistor characteristics and scaling trend are well understood; therefore, their compatibility for future-generation memories can be better projected than other less mature select devices. Gate all around (GAA) FET structures are considered a promising solution to address some scaling challenges of transistors. The GAA structure can also be applied on vertical transistors and tunnel FET (TFET).

The major disadvantages of transistors as select devices are the $>4F^2$ footprint of planar transistors and the high processing temperature that limits transistors in 3D stackable memories. Vertical transistors with $4F^2$ footprint have been explored but their feasibility is not yet proven [16,17]. In addition, the half-pitch dimension of vertical transistors is related to the thickness of the channel, gate dielectrics, and gate contact, which affects the scaling of vertical select transistors. $4F^2$ BJT-access RRAM was demonstrated in [18].

In 1T1R configuration transistor channel resistance needs to be significantly lower than the resistance of the memory element; otherwise, the effective memory window will be dominated by the transistor rather than the memory element. However, transistor resistance has increased steadily while half-pitch scales down. Therefore, at highly scaled nodes memory elements also need to have relatively high resistance to be functional in 1T1R configuration.

Due to these challenges, 1T1R does not appear competitive against 3D NAND. An alternative function of transistors could be access devices in the peripheral for a block of small crossbar array [3]. The peripheral transistors will control the access to the block and limit the voltage/current during switching. The memory elements inside of the array are connected to two-terminal select devices. Since peripheral transistors do not need to scale as aggressively as the memory element, transistor resistance can be reduced. However, the effectiveness of switching control by peripheral transistors depends on parasitic components.

12.3.2 Diodes as Memory Select Devices

Diode select devices have the advantage in scalability but still face the challenges in ON/OFF ratio and maximum ON current [2,19,20]. Silicon diodes require high processing temperature and therefore low-temperature oxide-based diodes have been actively explored as alternatives. Regular diodes only work for unipolar switching devices (i.e., set and reset in the same voltage polarity direction). For bipolar switching memories (i.e., set and reset in the opposite directions), reverse-conduction diodes (e.g., Zener diode, BARITT diode, etc.) are in principle possible solutions; however little progress has been made in reverse-conduction diodes for memory select devices.

12.3.2.1 Silicon Diodes

Both single-crystal Si [21] and poly-Si diodes [22–24] have been developed as a select device for PCM arrays. To operate PCM with diode selectors, diodes need to provide ON-current density above 8 MA/cm^2 and OFF-current density below 100 A/cm^2. To provide such a high ON-current density, the contact resistivity needs to be reduced to $<10^{-7}\,\Omega\cdot\text{cm}^2$. By engineering the metal electrodes and electrode-Si interface, total contact resistivity of top and bottom electrodes was reduced to $3.5\times10^{-8}\,\Omega\cdot\text{cm}^2$ [22]. To minimize the OFF-current density, impurity concentration at junction needs to be reduced, which is achieved with an undoped Si layer between p- and n-layers [22]. A short-time annealing technique was developed to further reduce the OFF-current density and enlarge the ON/OFF ratio. Poly-Si technology can achieve ON-current density of 10^7 A/cm^2 (at $\sim$1.8 V), OFF-current density of 10^{-1} A/cm^2 (at $\sim$−1.5 V), that is, ON/OFF ratio of 10^8. The ideality factor is 2.2. It is believed that Si diode can be scaled beyond 20 or 10 nm. Poly-Si diode select devices have been integrated in PCM crossbar arrays [22], 3D vertical chain-cell type PCM [23], and a 1 Gb PCM test chip [24]. A major challenge of Si diodes is the high processing temperature (above 1000 °C) required to crystallize Si to reduce contact resistivity and OFF-current.

12.3.2.2 Oxide Heterojunction Diodes

Heterojunction diodes can be built using semiconducting oxides deposited at relatively low temperature, for example, 200–300 °C [25]. The rectifying oxide heterojunctions can be utilized as select devices, especially for oxide-based RRAM memory elements [25–28]. A p-NiO$_x$/n-TiO$_x$ diode has demonstrated a rectification ratio of 10^5 at ± 3 V, ON current density of 5×10^3 A/cm^2 (at ~2.5 V), and ideality factor of 4.3 [25]. It was integrated with NO$_x$-based RRAM elements to form 1D1R memory device. A p-CuO$_x$/n-InZnO$_x$ diode achieved higher ON current density of 10^4 A/cm^2 (at ~1.3 V). It was integrated with NiO$_x$ RRAM in a two-layer 8×8 crossbar array [26,27] and with Al$_2$O$_3$ antifuse in a one-time programmable (OTP) memory [28]. Oxide p-n heterojunction was also demonstrated in p-ZnO·Rh$_2$O$_3$/n-InGaZnO$_4$ and achieved on/off ratio of 10^3 (at ± 5 V), ideality factor of ~2.3, and threshold voltage of 2.1 V [29]. Si substrates can be used as a part of heterojunction diodes as demonstrated in n-ZnO/p-Si [30] and n-Ge-nanowire/p-Si [31]. The ON current of oxide-based heterojunction diodes is often limited by both contact resistance and density of states of the oxide materials.

12.3.2.3 Metal/Oxide Schottky Diodes

Metal/oxide Schottky diodes can also work as memory select devices [32–37]. In a TiO$_x$-based diode with (In,Sn)$_2$O$_3$ (ITO) and Pt as electrodes, temperature-dependent current–voltage (I-V) characteristics confirms that Schottky barrier of ~0.55 eV forms at the TiO$_x$/Pt interface [32]. The rectification ratio is ~1.6×10^4 at ± 1 V. ON current of ~100 mA (at ~1.5 V) is provided through a diode area of 7500 μm^2, which corresponds to a low ON current density of ~13 A/cm^2. A Pt/TiO$_2$/Ti diode with Pt as the Schottky contact and Ti the ohmic contact achieved rectification ratio of 10^7–10^9 at ± 1 V, although ON current remains low at <10 A/cm^2 (for an area of 60 000 μm^2) [33]. Another demonstration of Pt/TiO$_2$/Ti Schottky diodes showed rectification ratio of ~2.4×10^6 at ± 2 V and ON current density of ~3×10^5 A/cm^2 at 2 V on a 4 μm^2 area [34]. Measurement showed that diode current density increases with decreasing area, indicating that current is not uniform across the diode area, which may be explained by edge leakage. Consequently, higher ON current density values can be obtained at smaller diode sizes. An Ag/n-ZnO Schottky diode with non-alloyed Ti/Au ohmic contact demonstrated a rectification ratio of 10^5 and forward current density over 10^4 A/cm^2 at 2 V [35]. A rectifying Ti/TiO$_2$/Pt diode was found to transition into a resistive switching device after a strong forming process, which is attributed to the destruction of the interface Schottky barrier by conductive filaments [36]. Similarly, a Pt/TiO$_x$/Pt was also shown to behave as a switchable rectifier [37,38]. The coexistence of rectifying and switching behaviors in oxide diodes manifests the complexity of these material systems, which may cause reliability concerns. In addition to oxide Schottky diodes, Si Schottky diodes are also utilized as select devices, for example, Al/p-Si [39].

Figure 12.5 compares the forward versus reverse current density of some of the diodes discussed above [22,25,26,29,32,33,36,38]. Most diode select devices still cannot reach the ON current density target of 1 MA/cm^2 marked in the figure. The diagonal direction dashed lines represent rectification ratio (e.g., 1, 10^4, and 10^8). Many diode devices can achieve rectification ratio above 10^4 at ± 1 V. Rectification ratio is clearly lower at lower voltage (± 0.5 V) because forward current decreases from 1.0 to 0.5 V and reverse current only changes slightly. Notice that the forward current density of some diodes changes much less than the exponential dependence between 1.0 and 0.5 V. This is because the forward current of these

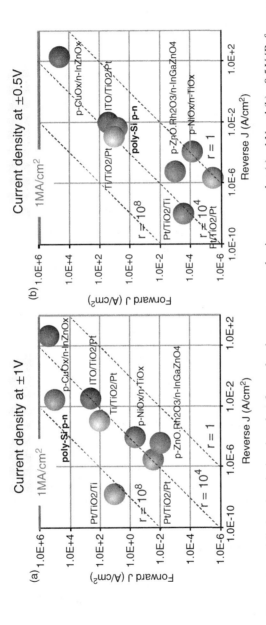

Figure 12.5 Comparison of diode select devices based on forward and reverse current density measured at (a) ±1 V and (b) ±0.5 V [References: 22,25,26,29, 32,33,36,38]

Table 12.2   Summary of the processing temperature and ideality factor of some diode select devices

Diodes	Processing temperature (°C)	Ideality factor, n	Ref.
p-NiO$_x$/n-TiO$_x$	<300	4.3	25
p-CuO$_x$/n-InZnO$_x$	Room temperature	—	27
p-ZnO·Rh$_2$O$_3$/n-InGaZnO$_4$	Room temperature	2.3	29
Pt/TiO$_2$/Ti	250	1.02	33
ITO/TiO$_2$/Pt	250	—	32
Poly-Si p-n	>1000	2.2	22

diodes is already limited by contact resistance at 0.5–1.0 V and deviates from the exponential diode I-V characteristics.

Table 12.2 summarizes the processing temperature and ideality factors of some diode select devices discussed above. Low processing temperature is one of their major advantages.

12.3.2.4   Self-rectifying Resistive Switching Memories

Some RRAM devices are found to be self-rectifying, which may enable crossbar memory arrays without external select devices. In a RRAM device made from ZrO$_2$ doped with Au nano-crystals, the low resistance state (LRS) exhibits asymmetrical characteristics with a rectification ratio of 10^2–10^3 measured at 0.5 V [40]. The self-rectification property is attributed to the Au/ZrO$_2$ interface, while the resistive switching is explained by the formation of conductive filaments inside of ZrO$_2$. A RRAM device with Al/PCMO interface was shown to be rectifying in LRS with a rectification ratio of $\sim10^2$ and symmetrical in high resistance state (HRS) [41]. The device was scaled down to 50 nm diameter and achieved endurance of 10^5 cycles for reading (0.5 V, 10 µs) and 10^4 cycles for writing (±7 V, 10/1 µs). A unipolar NiSi/HfO$_x$/TiN RRAM device was also found to be rectifying in LRS with a rectification ratio of $\sim10^3$ at ±1 V due to the Schottky barrier at the NiSi/HfO$_x$ interface [42]. When heavily doped Si was used as one electrode for HfO$_x$ RRAM, self-rectifying unipolar resistive switching characteristics were observed on n$^+$-Si but not on p$^+$-Si [43,44]. Similar self-rectifying unipolar resistive switching characteristics were also reported on Ni/AlO$_x$/n$^+$-Si [39]. A bipolar Ni/TiO$_2$/HfO$_2$/Ni resistive switch exhibits rectifying characteristics in LRS, which was explained by the asymmetric tunnel barrier at the TiO$_2$/HfO$_2$ interface [45]. A Cu/α-Si/ WO$_3$/Pt device shows rectifying behavior in LRS with rectification ratio of 10^2 at ±0.75 V, which was attributed to the interface between the conductive filament in WO$_3$ and the α-Si layer [46]. In a conductive-bridge RAM (CBRAM) with Cu/SiO$_2$/n$^+$-Si structure, it was found that the deposition of Cu on the n$^+$-Si electrode results in the formation of a Schottky interface with a potential barrier of 0.65–0.75 eV [47]. Similar behavior was also reported in an Ag/α-Si-based CBRAM [48].

Most of the self-rectification is due to the Schottky contact between a metal electrode and the oxides/filaments. Self-rectifying memory devices could simplify the design and fabrication of crossbar arrays, especially in 3D vertical crossbar arrays. However, the combination of rectification and switching characteristics in one structure may also complicate the performance optimization of both the memory and select devices.

Some key parameters (e.g., rectification ratio, ON-state current) of reported rectifying diode select devices are summarized in Table 12.3.

Table 12.3 Key parameters of some reported rectifying diode select devices

Device types	Materials	Size^a	Rectification ratio	V_{on}^{b}	I_{on} (J_{on})	Ref.
Rectifying diodes for select devices (only compatible with unipolar memory devices)						
pn diodes	Single-crystal Si	90 nm	10^7–10^8 (at ±1 V)	1.8 V	1.8 mA; 2.8×10^7 A/cm^2	21
	Poly-crystal Si	80 nm	$>10^6$ (at ±2 V)	2.0 V	0.5 mA; 8×10^6 A/cm^2	22
Schottky diodes	Ag/n-ZnO	—	$\sim 3 \times 10^5$ (at ±2 V)	2.0 V	—; 1×10^4 A/cm^2	35
	Pt/TiO$_2$	6×10^4 μm^2	10^9 (at ±1 V)	1.0 V	6 mA; 1×10^1 A/cm^2	33
	ITO/TiO$_2$	7.5×10^3 μm^2	3×10^4 (at ±1 V)	1.0 V	30 mA; 4×10^2 A/cm^2	32
	Pt/TiO$_2$	2×2 μm^2	2.4×10^6 (at ±2 V)	2.0 V	10 mA; 3×10^5 A/cm^2	34
	Pt/TiO$_2$	—	5×10^5 (at ±2 V)	2.0 V	—; 5×10^2 A/cm^2	36
Heterojunction diodes	n-ZnO/p-Si	100 μm	10^3	3.0 V	25 mA; 2.5×10^2 A/cm^2	30
	p-CuO$_x$/n-InZnO$_x$	0.5×0.5 μm^2	3×10^4 (at ±1 V)	1.0 V	0.15 mA; 6×10^4 A/cm^2	28
	p-NiO$_x$/n-TiO$_x$	30×30 μm^2	5×10^5 (at ±2 V)	2.0 V	1.8 mA; 2×10^2 A/cm^2	25–27
	p-ZnO·Rh$_2$O$_3$/n-InGaZnO$_4$	500×500 μm^2	2.5×10^2 (at ±1 V)	1.0 V	18 μA; 7.2×10^{-3} A/cm^2	29

(continued)

Table 12.3 (*Continued*)

Device types	Materials	Size[a]	Rectification ratio	V_{on}[b]	I_{on} (J_{on})	Ref.
		Rectifying diodes for select devices (only compatible with unipolar memory devices)				
Self-rectifying memory devices	Au/Au-nc:ZrO_2	600×600 μm²	1×10^3 (at ±1 V)	1.0 V	0.6 mA	40
	Al/$Pr_{0.7}Ca_{0.3}MnO_3$	—	~80 (at ±3 V)	3.0 V	1.7×10^{-1} A/cm² ~0.8 mA	41
	Ni/HfO_x/n^+-Si	7850 μm²	7×10^3 (at ±0.8 V)	0.8 V	0.2 mA 2.6 A/cm²	43
	Ni/AlO_x/n^+-Si	—	4×10^3 (at ±1 V)	1.0 V	~1.5 mA	44
	α-Si/WO_3	—	10^2 (at ±2 V)	2.0 V	~0.15 mA	46
	Ag/α-Si	100×100 nm²	10^6 (at ±0.5 V)	0.5 V	4 nA 4×10^1 A/cm²	48

[a] Size is diameter (d) or area (A). If not specified in the paper, device is assumed to be circular, that is, $A = \pi(d/2)^2$.

[b] V_{on} is where the forward current I_{on} is measured.

12.3.3 Volatile Switches as Memory Select Devices

Volatile resistive switching devices can also be utilized as memory select devices. They provide access to a selected memory element in their ON state and block sneak paths in OFF state. The volatile switching characteristics allow them to be switched quickly between ON and OFF states.

12.3.3.1 Threshold Switches

In chalcogenide materials, a volatile threshold switching process occurs before stable non-volatile memory switching is triggered. It was reported that chalcogenide-based threshold switches could be used as access devices in PCM arrays [49]. Niobium oxide is found to possess both memory switching and threshold switching properties at different compositions. $Pt/NbO_{2-x}/Pt$ threshold switching device becomes conductive above a threshold voltage (V_{th}) and falls back to high-resistance state below a hold voltage (V_{hold}) [50]. The threshold switching property remains stable up to 160 °C. By combining memory switching property of Nb_2O_{5-x} and threshold switching of NbO_{2-x}, a hybrid memory ($W/bi-layer-NbO_x/Pt$) was demonstrated in a 1 kb array [50]. In Si-As-Te ternary alloy, the composition (controlled by the sputtering power during deposition) determines the emergence of threshold switching [51]. Both V_{th} and V_{hold} vary with composition, which may provide a method to optimize the selector operation window. Functional 1-selector-1-resistor (1S1R) memory device with SiAsTe-based threshold switch selector and TiO_x-based RRAM was demonstrated experimentally. Another threshold switch device based on AsTeGeSiN was shown to be scalable to 30 nm with current density exceeding 10 MA/cm^2 and endurance over 10^8 cycles [52]. It was demonstrated as a functional select device for TaO_x-based RRAM devices. NiO_x was found to change from nonvolatile switching to threshold switching at higher oxygen concentration [53], which could be utilized to build all-NiO_x memory devices with the select device and memory element both made from NiO_x with different compositions [54].

12.3.3.2 Mott Transition Switches

Metal–insulator transition (MIT or Mott transition) induces several orders of magnitudes change of resistance in Mott materials [55], which could be utilized in memory select devices. A well-known Mott material is vanadium oxide (VO_x) with typical transition temperature around 67 °C. It is believed that the intrinsic switching speed of VO_x is below tens of ns or even sub-ns. The feasibility of VO_2-based select devices is demonstrated by integrating a $Pt/VO_2/Pt$ device with NiO_x-based RRAM [56]. Resistance of the VO_2 select device can be switched over three orders of magnitudes with V_{th} of 0.6 V and V_{hold} of 0.4 V. The 1S1R memory device can be switched within tens of ns. Another demonstration of $Pt/VO_2/Pt$ select device achieved switching speed <20 ns, ON current of ~1 MA/cm^2, and V_{th} of 0.35 V, although the ON/OFF ratio is relative low (~50) [57]. A functional 1S1R memory structure is demonstrated by integrating the VO_2-switch with ZrO_x/HfO_x bipolar RRAM devices. Both threshold switching and memory switching have been observed in VO_x, depending on O content [58]. A key challenge of VO_2-based select devices is the low operation temperature. Mott materials with higher transition temperatures need to be found to provide feasible select device options [59].

12.3.4 Nonlinearity for Device Selection

As discussed earlier, nonlinearity improves the accessibility to a selected device by making unselected devices effectively more resistive via the voltage-dependent resistance.

12.3.4.1 Nonlinear Memory Select Devices

Nonlinearity can be easily achieved with non-ohmic transport mechanisms, for example, tunneling [3]. A Ni/TiO$_2$/Ni device shows nearly symmetrical exponential I-V characteristics with six orders of magnitude of current change for voltage swing from 0 V to ± 2 V [60]. It is integrated with HfO$_2$-RRAM to demonstrate a 1S1R memory structure. Another TiO$_2$-based nonlinear selector uses Pt/TiO$_2$/TiN structure and is integrated with bilayer Pt/TiO$_{2-x}$/TiO$_2$/W RRAM for a functional memory device [61]. A so-called "varistor" select device uses a 4 nm TiO$_2$ layer sandwiched between two 10 nm TaO$_x$ layers and achieves 10^4 resistance difference between 1.5 and 0.75 V [62]. It was found that the substitution of Ti^{4+} in TiO$_2$ by Ta^{5+} ions increases the conductivity of the initially insulating TiO$_2$ layer. The ON current of nonlinear select devices can be modulated by oxide thickness and oxidation conditions. Some papers attributed the exponential I-V dependence to Schottky barriers at the metal/oxide interface; however, it did not explain why a back to back Schottky diodes can provide high ON current [60,61]. A more plausible explanation is the tunneling-based transport mechanisms in these oxide materials [62]. Temperature-dependent transport characterization would help to identify the actual mechanisms. Another possible mechanism for symmetrical exponential I-V characteristics in a back to back diode structure is provided in a n$^+$/p/n$^+$ poly-Si device model, where the middle p-layer is fully depleted and a drain induced barrier lowering (DIBL) effect causes exponential current increase with applied voltage [63].

12.3.4.2 Mixed Ionic Electronic Conduction Devices

A so-called mixed ionic electronic conduction (MIEC) device is developed as select devices for PCM [64–67]. The device is made from Cu-containing MIEC materials sandwiched between an inert top electrode (TE; e.g., TiN, W) and a bottom electrode (BE). Negative voltage applied on TE pulls Cu$^+$ in MIEC away from the BE and create vacancies near BE. The hole and vacancy concentrations depend exponentially on the applied voltage. Symmetrical diode-like I-V characteristics can be achieved with two inert electrodes. Large fraction of mobile Cu$^+$ enables high current density exceeding tens of MA/cm^2 [64]. For example, a 40 nm MIEC device is able to provide a current >200 μA, that is, a current density >15 MA/cm^2. The device selection capability of MIEC for large array is characterized by the voltage margin (V_m) between the positive and negative voltages measured at certain current level (e.g., 10 nA). It was shown that V_m increases with decreasing device size [64]. Endurance above 10^8 cycles has been demonstrated on MIEC devices in small arrays [65]. The failure mechanism was attributed to the accumulation of Cu near the TE, which was also found curable by thermal annealing. Endurance of MIEC increases exponentially with decreasing current and is independent of device sizes. The MIEC select devices were also integrated with PCM in a 512 kb testing array using 180 nm CMOS process and achieved 100% yield [66]. Electrode optimization plays critical roles in the yield improvement. It was demonstrated that MIEC devices tolerate processing temperature up to 500 °C and can be fabricated with manufacturing-level single-target sputter deposition. The scalability of MIEC select devices was tested to below 30 nm in diameter and below 12 nm in thickness [67]. An integrated MIEC-PCM memory device can be switched at the speed of 15 ns and read within 1 μs at typical reading current (~5 μA).

12.3.4.3 Complementary Resistive Switch

A complementary resistive switch (CRS) provides a self-selecting memory by connecting two bipolar RRAM devices anti-serially [68]. It may be considered a "constructed nonlinearity" with re-defined logic states and operations. Both states "0" and "1" have high resistance in CSR, which helps to minimize leakage through sneak paths. In either state, one of the two RRAMs is in LRS and the other in HRS. When reading a "1" state, the HRS device is switched to LRS and both devices end up in LRS. When reading a "0" state, no switching occurs and CSR remains in HRS. Notice that the reading operation is destructive, which may degrade the overall endurance of CRS. A non-destructive readout method based on capacitive voltage divider was also proposed [69]. CRS has been demonstrated in different resistive switching devices, for example, $Cu/SiO_2/Pt$ bipolar resistive switches [70], amorphous carbon-based RRAM [71], TaO_x–based RRAM [72–74], multi-layer TiO_x device [75], HfO_x RRAM [76], ZrO_x/HfO_x bi-layer RRAM [77], Cu/TaO_2 atomic switch [78], Nb_2O_{5-x}/NbO_y RRAM [79], and so on.

An important issue with CRS is the "ON window." For CRS built from devices with asymmetric I-V curve, there will be no stable ON state if the asymmetry is so large that $V_{reset} < V_{set}/2$. A potential solution is to use series resistors to make the I-V curves symmetric. For RRAM based on valence change mechanism (VCM), CSR on-window may also be engineered by changing oxygen content and oxide thickness [75]. It was also suggested that hetero-devices (two different RRAM devices connected anti-serially) can be used to control the operation voltages of CSR [80]. In spite of these promising results of CRS, some critical challenges still remain for CRS. In-depth understanding of the HRS/LRS $\rightarrow$ LRS/LRS $\rightarrow$ LRS/HRS transitions for different materials is needed. Destructive reading method requires RRAM devices with good write endurance, since every read operation of CRS also involves a write operation. The operation voltage of CRS is also larger than that of single devices. The feasibility of CRS for crossbar arrays depends on solutions to these challenges.

12.3.4.4 Intrinsic Nonlinearity in Memory Device Characteristics

Nonlinear I-V characteristics may also be an intrinsic property of some resistive switching devices [81]. A tunneling oxide layer can be intentionally built in a RRAM device to provide nonlinearity I-V dependence. A TiO_x-RRAM device is combined with different tunnel oxides, HfO_x, ZrO_x, or AlO_x [82]. Appropriately chosen tunneling oxide layer not only provide built-in nonlinearity but also assist resistive switching process. In a TiO_x/Ta_2O_5 bi-layer RRAM device, the nonlinearity and resistive switching characteristics can be modulated by processing conditions, which may enable a self-selecting crossbar array [83]. It was reported that a 64 Mb multi-layer RRAM array was fabricated with 0.13 µm CMOS process based on an oxide stack with an insulating oxide layer on top of a conductive oxide layer [84]. The insulating oxide provides variable tunnel barrier height and nonlinear characteristics.

Similar to the self-rectifying characteristics, intrinsic nonlinearity in resistive switching memories could simplify processing and enable really compact memory design. However, the interaction between the resistive switching mechanism and tunneling mechanism may also complicate the optimization of both properties.

The key parameters of some reported volatile switch and nonlinear select devices are summarized in Table 12.4.

Table 12.4 Key parameters of some reported volatile switch and nonlinear select devices

Device types	Materials	Size^a	On/off ratio	V_{th} ^{b}	I_{on} ^{c} (J_{on})	V_{hold} ^{d}	I_{off} (J_{off})	Ref.
Volatile switch select devices (compatible with both bipolar and unipolar memory devices)								
Threshold switch	Pt/NbO$_{2-x}$/Pt	250 nm	>10 (at ±1.5 V)	~1.6 V	>1 mA; >4×10^6 A/cm^2	~1.3 V	~0.1 mA (at 1.4 V); 4×10^5 A/cm^2	50
	Si-As-Te^e	—	>10^2 (at ±1.0 V)	1–2 V	>5 mA; —	0.5–1 V	~50 µA (at 1.0 V); -	51
	Ti/AsTeGeSiN/Ti	30 nm	>25 (at ±1.5 V)	~1.75 V	>0.1 mA; >1×10^7 A/cm^2	~1.4–1.5 V	~4 µA (at 1.5 V); 4×10^5 A/cm^2	52
	Pt/NiO/Pt^f	30×30 µm^2	>10 (at ±1.75 V)	~1.8 V	>50 mA; >6×10^3 A/cm^2	~1.6 V	~5 mA (at 1.75 V); 6×10^2 A/cm^2	54
Mott transition	Pt/VO$_2$/Pt	—	~300 (at 0.5 V)	~0.7 V	~10 mA	~0.25 V	~30 µA (at 0.5 V)	56
	Pt/VO$_2$/Pt	250 nm	>75 (at ±0.3 V)	0.35–0.4 V	>3 mA; >6×10^6 A/cm^2	~0.2 V	~40 µA (at 0.3 V); ~8×10^4 A/cm^2	57

Device types	Materials	Size^a	Nonlinearity ratio^g	Full V	$I(V)$ $J(V)$	Half V	$I(V/2)$ $J(V/2)$	Ref.
Nonlinear select devices (compatible with both bipolar and unipolar memory devices)								
Nonlinear MIM select devices	Ni/TiO$_2$/Ni	10^4 µm^2	>10^2 (at 2/1 V)	2 V	(4–40)×10^{-6} A; (4–40)×10^{-2} A/cm^2	1 V	(4–8)×10^{-8} A; (4–8)×10^{-4} A/cm^2	60
	Pt/TiO$_2$/TiN	0.049 µm^2	~10^2 (at 2/1 V)	2 V	(1–15) µA; (2–30)×10^3 A/cm^2	1 V	(2.5–10)×10^{-8} A; (5–20)×10^1 A/cm^2	61
	Pt/TaO$_x$/TiO$_2$/TaO$_x$/Pt	250 nm	~10^3 (at 2/1 V)	2 V	~1×10^{-2} A; ~2×10^7 A/cm^2	1 V	~1×10^{-5} A; ~2×10^4 A/cm^2	62
MIEC^h	Cu-containing material	<30 nm	>10^4 (at 1.0/0.5 V)	1 V	>10^{-6} A; >1.4×10^5 A/cm^2	0.5 V	~5×10^{-11} A; ~7 A/cm^2	67

^{a} Size is diameter (d) or area (A). If not specified in the paper, device is assumed to be circular, that is, $A = \pi(d/2)^2$.

^{b} Threshold voltage: the voltage above which devices switch on.

^{c} In some measurements current is capped at certain limit, so some values here are the limiting current.

^{d} Hold voltage: the voltage below which devices recover to the off state.

^{e} Parameters depend on composition. Values here are for composition 1 in the paper.

^{f} NiO$_x$ exhibits both memory (at low O partial pressure) and threshold (at high O partial pressure) switching.

^{g} The "nonlinearity ratio" definition here uses the current ratio at chosen full voltage and half voltage.

^{h} MIEC = "mixed ionic electronic conduction."

12.4   Challenges of Memory Select Devices

Functional memory select devices have to meet both array design specifics (e.g., array size) and memory device switching requirements (e.g., switching current). High ON/OFF ratio and maximum ON current are two critical parameters. In addition, good scalability, low processing temperature, fast speed, and high endurance are also important factors to determine the feasibility of select devices.

Oxide-based diodes are explored mainly because of their low processing temperature and material compatibility with oxide-based RRAMs. As shown in Figure 12.5, many reported oxide-based diodes still cannot meet the ON current requirement. Oxide diode current is determined mainly by density of states and carrier mobility, both of which are not better in oxides than in Si. Therefore, it is unlikely that oxide diodes will outperform Si diodes in terms of ON/OFF ratio and ON current. They may be more suitable for medium-size crossbar arrays in 3D stackable structures where low processing temperature is critical. An important limitation of rectifying diodes is one-way conduction that only works with unipolar RRAMs. It is generally believed that bipolar RRAMs are more stable and reliable than unipolar RRAMs. In terms of scalability, simulation has shown that in extremely scaled diodes barrier thickness may decrease with lateral diode size, which increases tunneling and degrades rectification ratio [85]. In vertical 3D memory arrays, the depletion width of select diodes becomes part of lateral dimension and a limiting factor for lateral scaling.

Volatile switching devices are characterized with threshold voltage (V_{th}) where devices turn on and hold voltage (V_{hold}) where devices turn off. These voltages and the ON/OFF resistance of volatile switches need to be balanced with memory parameters. Figure 12.6 uses load line analysis to illustrate a working combination of a volatile switch selector and a memory element in (a) and a dysfunctional combination in (b). Assume that the memory element is in LRS with resistance (R_{LRS}). Initially the switch-based selector has high resistance (R_H). The voltage distribution based on R_{LRS}-R_H combination at point "1" results in a voltage above V_{th} applied on the selector, which turns it on (R_L). Once the selector turns on, voltage redistributes based on R_{LRS}-R_L combination and selector voltage decreases. With a wide enough window between V_{hold} and V_{th} in (a), the select device voltage stays above V_{hold}, which keeps the select device ON and allows access to the memory element. However, the narrow V_{hold}-V_{th} window in (b) will shut off the selector as soon as it turns on and disable the access to the memory. This simplified analysis illustrates the importance of balanced parameters between volatile switch select devices and memory elements.

Nonlinearity has become an increasingly important select device options, because nonlinear I-V characteristics are relatively easy to implement (e.g., tunneling transport mechanisms). Nonlinear select device parameters can be modulated by different processing conditions.

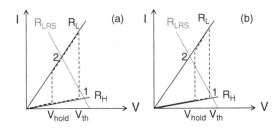

Figure 12.6   (a) Balanced 1S1R structure. (b) Nonfunctional 1S1R structure

Nonlinearity can also be constructed, for example, using complementary structures. Tradeoffs between ON current and nonlinearity ratio need to be carefully addressed. Since nonlinearity is voltage-dependent, performance of nonlinear select devices also depends on voltage distribution and data patterns in the array. As any other select devices, variability of nonlinear select devices needs to be minimized to control memory parameter distribution in a large array.

Self-selecting capability is attractive for crossbar arrays and may be enabled by self-rectification or intrinsic nonlinearity in memory elements. In either approach, there are often tradeoffs between selector parameters and memory performance. Optimization of the overall memory performance requires comprehensive solutions combining material choices, processing control, and device design. Connecting a memory element to an external select device often requires the middle metal electrode to decouple these two devices, which increases aspect ratio in planar arrays and enlarges the footprint in 3D vertical structure. Self-selecting solutions may help to eliminate this problem.

Most RRAM devices still require a forming process and switching current control. Select devices for RRAM need to provide sufficient current and voltage for forming. It's unlikely that any two-terminal select devices will be able to provide switching control like transistors. External line transistor or serial resistors may provide some current control, but their effectiveness depends strongly on parasitic components in the array. There has been limited research on the switching speed and transient effects of select devices, which could affect the dynamic behavior of memory devices.

Line resistance is neglected in most analysis of select device requirements and crossbar array performance. Line resistance degrades the voltage and current delivered to devices further away from voltage sources in the array, which affect the functionality of both the memory element and select devices. In highly scaled nodes, line resistance may become a critical limiting factor in crossbar array sizes [8]. Contact resistance presents a major challenge for all memory select devices, which becomes increasingly difficult with device scaling.

12.5 Summary

This chapter reviews device options and R&D status of memory select devices targeted for the applications in high-density crossbar arrays and 3D memories. Diodes, volatile switches, and nonlinear devices have been demonstrated as possible solutions for two-terminal select devices. Performance of these potential select devices has been significantly improved recently. Some array-level demonstration has also been reported. Although there are still many challenges to be overcome, it looks promising that some of these extensively explored select device options may provide feasible solutions to enable high-density and low-cost memory and storage technologies in the near future.

References

1. International Technology Roadmap of Semiconductors (ITRS) (2011) Emerging Research Devices.
2. Crowley, M. *et al.* (Feb. 2003) 512 Mb PROM with 8 Layers of antifuse/diode cells. ISSCC Tech. Dig. 16.4.
3. Kawahara, A. *et al.* (2013) An 8 Mb multi-layered cross-point ReRAM macro with 443MB/s write throughput. *IEEE Journal of Solid-State Circuits*, **48**(1), 178–185.
4. Tanaka, H. *et al.* (Jun. 2007) Bit Cost Scalable Technology with Punch and Plug Process for Ultra High Density Flash Memory. Symposium VLSI Tech., pp. 14–15.
5. Yoon, H.S. *et al.* (Jun. 2009) Vertical cross-point resistance change memory for ultra-high density non-volatile memory applications. Symposium VLSI Tech., pp. 26–27.

6. Flocke, A. and Noll, T.G. (Feb. 2007) Fundamental analysis of resistive nanocrossbars for the use in hybrid Nano/CMOS-memory. Proc. 33rd Eur. Solid-State Circuits Conf., pp. 328–331.

7. Flocke, A., Noll, T.G., Kugeler, C. et al. (Aug. 2008) A fundamental analysis of nano-crossbars with non-linear switching materials and its impact on TiO_2 as a resistive layer. Proc. 8th IEEE Conf. Nanotech., pp. 319–322.

8. Chen, A. (2013) Comprehensive crossbar array model with solutions for line resistance and nonlinear device characteristics. IEEE Transactions on Electron Devices, 60(4), 1316–1326.

9. Liang, J., Yeh, S., Wong, S.S., and Wong, H.-S.P. (May 2012) Scaling challenges for the cross-point resistive memory array to sub-10 nm node–an interconnect perspective. Proc. 4th IEEE Intern. Memory Workshop, pp. 1–4.

10. Amsinck, C.J., Spigna, N.H.D., Nackashi, D.P., and Franzon, P.D. (2005) Scaling constraints in nanoelectronic random-access memories. Nanotechnology, 16(10), 2251–2260.

11. Kim, G.H. et al. (2010) A theoretical model for Schottky diodes for excluding the sneak current in cross bar array resistive memory. Nanotechnology, 21(38), 385202-1–385202-7.

12. Ziegler, M.M. and Stan, M.R. (Aug. 2002) Design and analysis of crossbar circuits for molecular nanoelectronics. Proc. 2nd IEEE Conf. Nanotechnol., pp. 323–327.

13. Rose, G.S. et al. (Apr.–May 2006) Design approaches for hybrid CMOS/molecular memory based on experimental device data. Proc. 16th ACM Great Lakes Symp. VLSI, pp. 2–7.

14. Stan, M.R., Franzon, P.D., Goldstein, S.C. et al. (2003) Molecular electronics: From devices and interconnect to circuits and architecture. Proceedings of the IEEE, 91(11), 1940–1957.

15. Zhirnov, V.V., Meade, R., Cavin, R.K., and Sandhu, G. (2011) Scaling limits of resistive memories. Nanotechnology, 22(25), 254027-1-21.

16. Yang, B. et al. (2008) Vertical silicon-nanowire formation and gate-all-around MOSFET. IEEE Electron Device Letters, 29(7), 791–793.

17. Ghandi, R. et al. (2011) Vertical Si-nanowire n-type tunneling FETs with low subthreshold swing ($\leq$50mV/decade) at Room Temperature. IEEE Electron Device Letters, 32(4), 437–439.

18. Wang, C-H. et al. (Dec. 2010) Three-dimensional $4F^2$ ReRAM cell with CMOS logic compatible process. IEDM Tech. Dig., pp. 664–667.

19. Lee, K.J. et al. (Feb. 2007) A 90 nm 1.8 V 512 Mb Diode-Switch PRAM with 266MB/s Read Throughput. ISSCC Tech. Dig., 26.1.

20. Liu, T.Y. et al. (Feb. 2013) A 130.7 mm^2 2-layer 32 Gb ReRAM memory device in 24 nm technology. ISSCC Tech. Dig., 12.1.

21. Oh, J.H. et al. (Dec. 2006) Full integration of highly manufacturable 512 Mb PRAM based on 90 nm technology. IEDM Tech. Dig., pp. 515–518.

22. Sasago, Y. et al. (Jun. 2009) Cross-point phase change memory with $4F^2$ cell size driven by low-contact-resistivity poly-Si diode. Symposium VLSI Tech., pp. 24–25.

23. Kinoshita, M. et al. (Jun. 2012) Scalable 3-D vertical chain-cell-type phase-change memory with $4F^2$ poly-Si diodes. Symposium VLSI Tech., pp. 35–36.

24. Lee, S.H. et al. (Dec. 2011) Highly Productive PCRAM Technology Platform and Full Chip Operation: Based on $4F^2$ (84 nm Pitch) Cell Scheme for 1Gb and Beyond. IEDM Tech. Dig., pp. 47–50.

25. Lee, M.J. et al. (2007) A low-temperature-grown oxide diode as a new switch element for high-density, nonvolatile memories. Advanced Materials, 19(1), 73–76.

26. Lee, M.J. et al. (Dec. 2007) 2-stack 1D-1R cross-point structure with oxide diodes as switch elements for high density resistance RAM applications. IEDM Tech. Dig., pp. 771–774.

27. Lee, M.J. et al. (Dec. 2008) Stack friendly all-oxide 3D RRAM using GaInZnO peripheral TFT realized over glass substrates. IEDM Tech. Dig.

28. Ahn, S.E. et al. (2009) Stackable all-oxide-based nonvolatile memory with Al_2O_3 antifuse and p-CuO_x/n-$InZnO_x$ diode. IEEE Electron Device Letters, 30(5), 550–552.

29. Narushima, S. et al. (2003) A p-type amorphous oxide semiconductor and room temperature fabrication of amorphous oxide p-n heterojunction diodes. Advanced Materials, 15(17), 1409–1413.

30. Choi, Y. et al. (2010) High current fast switching n-ZnO/p-Si diode. Journal of Physics D-Applied Physics, 43, 345101-1-4.

31. Kim, S., Zhang, Y., McVittie, J.P. et al. (2008) Integrating phase-change memory cell with Ge nanowire diode for crosspoint memory–experimental demonstration and analysis. IEEE Transactions on Electron Devices, 55(9), 2307–2313.

32. Shin, Y.C. et al. (2008) (In,Sn)2O3/TiO2/Pt Schottky-type diode switch for the TiO_2 resistive switching memory array. Applied Physics Letters, 92(16), 162904–1-3.

33. Park, W.Y. *et al.* (2010) A Pt/TiO2/Ti Schottky-type selection diode for alleviating the sneak current in resistance switching memory arrays. *Nanotechnology*, **21**(19), 195201-1-4.
34. Kim, G.H. *et al.* (2012) Schottky diode with excellent performance for large integration density of crossbar resistive memory. *Applied Physics Letters*, **100**(21), 213508-1-3.
35. Huby, N. *et al.* (2008) New selector based on zinc oxide grown by low temperature atomic layer deposition for vertically stacked non-volatile memory devices. *Microelectronic Engineering*, **85**(12), 2442–2444.
36. Huang, J.J., Kuo, C.W., Chang, W.C., and Hou, T.H. (2010) Transition of stable rectification to resistive-switching in Ti/TiO$_2$/Pt oxide diode. *Applied Physics Letters*, **96**(26), 262901-1-3.
37. Shima, H. *et al.* (2008) Control of resistance switching voltages in rectifying Pt/TiOx/Pt trilayer. *Applied Physics Letters*, **92**(4), 043510–1-3.
38. Shima, H., Zhong, N., and Akinaga, H. (2009) Switchable rectifier built with Pt/TiOx/Pt trilayer. *Applied Physics Letters*, **94**(8), 082905–1-3.
39. Cho, B. *et al.* (2010) Rewritable switching of one diode–one resistor nonvolatile organic memory devices. *Advanced Materials*, **22**(11), 1228–1232.
40. Zuo, Q. *et al.* (2009) Self-rectifying effect in gold nanocrystal-embedded zirconium oxide resistive memory. *Journal of Applied Physiology*, **106**(7), 073724–1-5.
41. Jo, M. *et al.* (Jun. 2010) Novel Cross-point Resistive Switching Memory with Self-formed Schottky Barrier. Symposium VLSI Tech., pp. 53–54.
42. Tran, X.A. *et al.* (2012) A self-rectifying HfO$_x$-based unipolar RRAM with NiSi electrode. *IEEE Electron Device Letters*, **33**(4), 585–587.
43. Tran, X.A. *et al.* (2013) Self-selection unipolar HfO$_x$-Based RRAM. *IEEE Transactions on Electron Devices*, **60**(1), 391–395.
44. Tran, X.A. *et al.* (2012) A self-rectifying AlO$_y$ bipolar RRAM with sub-50-μA set/reset current for cross-bar architecture. *IEEE Electron Device Letters*, **33**(10), 1402–1404.
45. Hsu, C.W., Hou, T.H., Chen, M.C. *et al.* (2013) Bipolar Ni/TiO$_2$/HfO$_2$/Ni RRAM with multilevel states and self-rectifying characteristics. *IEEE Electron Device Letters*, **34**(7), 885–887.
46. Lv, H., Li, Y., Liu, Q. *et al.* (2013) Self-rectifying resistive-switching device with a-Si/WO$_3$ bilayer. *IEEE Electron Device Letters*, **34**(2), 229–231.
47. Puthentheradam, S.C., Schroder, D.K., and Kozicki, M.N. (2011) Inherent diode isolation in programmable metallization cell resistive memory elements. *Applied Physics A*, **102**(4), 817–826.
48. Kim, K.H., Jo, S.H., Gaba, S., and Lu, W. (2010) Nanoscale resistive memory with intrinsic diode characteristics and long endurance. *Applied Physics Letters*, **96**, 053106-1-3.
49. Kau, D. *et al.* (Dec. 2009) A stackable cross point phase change memory. IEDM Tech. Dig., pp. 617–620.
50. Kim, S. *et al.* (Jun. 2012) Ultrathin (<10 nm) Nb$_2$O$_5$/NbO$_2$ hybrid memory with both memory and selector characteristics for high density 3D vertically stackable RRAM applications. Symposium VLSI Tech., pp. 155–156.
51. Lee, J.H. *et al.* (2012) Threshold switching in Si-As-Te thin film for the selector device of crossbar resistive memory. *Applied Physics Letters*, **100**(12), 123505-1-4.
52. Lee, J.H. *et al.* (Dec. 2012) Highly-Scalable Threshold Switching Select Device based on Chaclogenide Glasses for 3D Nanoscaled Memory Arrays. IEDM Tech. Dig., pp. 33–35.
53. Seo, S. *et al.* (2004) Reproducible resistance switching in polycrystalline NiO films. *Applied Physics Letters*, **85**(23), 5655–5657.
54. Lee, M.-J. *et al.* (2011) A simple device unit consisting of all NiO storage and switch elements for multilevel terabit nonvolatile random access memory. *ACS Applied Materials & Interfaces*, **3**(11), 4475–4479.
55. Mott, N.F. (1968) Metal-insulator transition. *Reviews of Modern Physics*, **40**(4), 677–683.
56. Lee, M.-J. *et al.* (2007) Two series oxide resistors applicable to high speed and high density nonvolatile memory. *Advanced Materials*, **19**(22), 3919–3923.
57. Son, M. *et al.* (2011) Excellent selector characteristics of nanoscale VO$_2$ for high-density bipolar ReRAM applications. *IEEE Electron Device Letters*, **32**(11), 1579–1581.
58. Son, M. *et al.* (2012) Self-selective characteristics of nanoscale VO$_x$ devices for high-density ReRAM applications. *IEEE Electron Device Letters*, **33**(5), 718–720.
59. Ha, S.D., Aydogdu, G.H., and Ramanathan, S. (2011) Metal-insulator transition and electrically driven memristive characteristics of SmNiO$_3$ thin films. *Applied Physics Letters*, **98**(1), 012105-1-3.
60. Huang, J.J., Tseng, Y.M., Hsu, C.W., and Hou, T.H. (2011) Bipolar nonlinear Ni/TiO$_2$/Ni selector for 1S1R crossbar array applications. *IEEE Electron Device Letters*, **32**(10), 1427–1429.
61. Shin, J. *et al.* (2011) TiO$_2$-based metal-insulator-metal selection device for bipolar resistive random access memory cross-point application. *Journal of Applied Physiology*, **109**(3), 033712-1-4.

62. Lee, W. *et al.* (Jun. 2012) Varistor-type bidirectional switch ($J_{MAX} > 10^7$ A/cm^2, selectivity $\sim 10^4$) for 3D bipolar resistive memory arrays. Symposium VLSI Tech., pp. 37–38.

63. Song, Y.H., Park, S.Y., Lee, J.M. *et al.* (2011) Bidirectional two-terminal switching device for crossbar array architecture. *IEEE Electron Device Letters*, **32**(8), 1023–1025.

64. Gopalakrishnan, K. *et al.* (Jun. 2010) Highly-scalable novel access device based on mixed ionic electronic conduction (MIEC) materials for high density phase change memory (PCM) arrays. Symposium VLSI Tech., 19.4.

65. Shenoy, R.S. *et al.* (Jun. 2011) Endurance and scaling trends of novel access-devices for multi-layer crosspoint memory based on mixed ionic electronic conduction (MIEC) materials. Symposium VLSI Tech., T5B-1.

66. Burr, G.W. *et al.* (Jun. 2012) Large-scale (512 kbit) integration of multilayer-ready access-devices based on mixed-ionic-electronic-conduction (MIEC) at 100% yield. Symposium VLSI Tech., T5.4.

67. Virwani, K. *et al.* (Dec. 2012) Sub-30nm scaling and high-speed operation of fully-confined access-devices for 3D crosspoint memory based on mixed-ionic-electronic-conduction (MIEC) materials. IEDM Tech. Dig., pp. 36–39.

68. Linn, E., Rosezin, R., Kuegeler, C., and Waser, R. (2010) Complementary resistive switches for passive nanocrossbar memories. *Nature Materials*, **9**, 403–406.

69. Tappertzhofen, S. *et al.* (2011) Capacity based nondestructive readout for complementary resistive switches. *Nanotechnology*, **22**(39), 395203 1-7.

70. Rosezin, R. *et al.* (2011) Integrated complementary resistive switches for passive high-density nanocrossbar arrays. *IEEE Electron Device Letters*, **32**(2), 191–193.

71. Chai, Y. *et al.* (2011) Nanoscale bipolar and complementary resistive switching memory based on amorphous carbon. *IEEE Transactions on Electron Devices*, **58**(11), 3933–3939.

72. Lee, M.J. *et al.* (2011) A fast, high-endurance and scalable non-volatile memory device made from asymmetric $Ta_2O_{5(x}/TaO_{2(x}$ bilayer structures. *Nature Materials*, **10**, 625–630.

73. Yang, Y., Sheridan, P., and Lu, W. (2012) Complementary resistive switching in tantalum oxide-based resistive memory devices. *Applied Physics Letters*, **100**(20), 203112-1-4.

74. Schmelzer, S., Linn, E., Bottger, U., and Waser, R. (2013) Uniform complementary resistive switching in tantalum oxide using current sweeps. *IEEE Electron Device Letters*, **34**(1), 114–116.

75. Bae, Y.C. *et al.* (2012) Oxygen ion drift-induced complementary resistive switching in homo $TiO_x/TiO_y/TiO_x$ and hetero $TiO_x/TiON/TiO_x$ triple multilayer frameworks. *Advanced Functional Materials*, **22**(4), 709–716.

76. Nardi, F., Balatti, S., Larentis, S., and Ielmini, D. (Dec. 2011) Complementary switching in metal oxides: toward diode-less crossbar RRAMs. IEDM Tech. Dig., pp. 709–712.

77. Lee, J. *et al.* (Dec. 2010) Diode-less nano-scale ZrO_x/HfO_x RRAM device with excellent switching uniformity and reliability for high-density cross-point memory applications. IEDM Tech. Dig., pp. 452–455.

78. Banno, N. *et al.* (Jun. 2012) Nonvolatile 32 × 32 crossbar atom switch block integrated on a 65-nm CMOS platform. Symposium VLSI Tech., pp. 39–40.

79. Liu, X. *et al.* (2013) Complementary resistive switching in niobium oxide-based resistive memory devices. *IEEE Electron Device Letters*, **34**(2), 235–237.

80. Lee, D. *et al.* (2012) Operation voltage control in complementary resistive switches using heterodevice. *IEEE Electron Device Letters*, **33**(4), 600–602.

81. Xie, Y.W., Sun, J.R., Wang, D.J. *et al.* (2006) Reversible electroresistance at the $Ag/La_{0.67}Sr_{0.33}MnO_3$ interface. *Journal of Applied Physiology*, **100**(3), 033704-1-5.

82. Choi, H. *et al.* (May 2011) The effect of tunnel barrier at resistive switching device for low power memory applications. Proc. 3rd IEEE Intern. Memory Workshop.

83. Lee, H.D. *et al.* (Jun. 2012) Integration of 4F2 selector-less crossbar array 2 Mb ReRAM based on transition metal oxides for high density memory applications. Symposium VLSI Tech., pp. 151–152.

84. Chevallier, C.J. *et al.* (Feb. 2010) A 0.13 μm 64 Mb multi-layered conductive metal-oxide memory. ISSCC Tech. Dig., 14.3.

85. Smit, G.D.J., Rogge, S., and Klapwijk, T.M. (2002) Scaling of nano-Schottky-diodes. *Applied Physics A*, **81**(20), 3852–3854.

13

Emerging Memory Devices: Assessment and Benchmarking

Matthew J. Marinella[1] and Victor V. Zhirnov[2]
[1]Sandia National Laboratories, USA
[2]Semiconductor Research Corporation, USA

13.1 Introduction

We are at an interesting juncture in memory technology. As of the time of this writing, production NAND flash has been scaled to a critical dimension of 16 nm [1] and 3D Vertical NAND has entered commercial production [2]. However, endurance and retention have become strongly degraded as flash tunnel oxides become thinner, leading to requirements for extensive error correction code (ECC) schemes and substantial redundant storage requirements. Storage Class Memory (SCM) has identified the significant latency gap between NAND-based solid-state disks (SSDs) and DRAM [3,4] (SCM concepts are covered in detail in Chapter 25). Hence, an emerging or prototypical memory technology may supplement, or even supplant NAND flash in the coming decade. Furthermore, continued DRAM scaling faces numerous challenges, and does not yet have known manufacturable solutions past the 20 nm node [5]. Impending limitations of standard memory technologies combined with massive increases in data quantities have even led to proposals of a radical shift toward datacentric-based architectures such as nanostores [6].

Prototypical memory technologies, in particular spin transfer torque RAM (STT-RAM) and phase change RAM (PCRAM), have made improvements in recent years and are being commercially produced for niche markets. These technologies are discussed in detail in Chapter 4 (PCRAM) and Chapter 5 (STT-RAM) of this book. Emerging memory technologies, especially resistance switching memories (RRAM) have made rapid advancements in the past decade and offer the possibility of greater scaling and performance than prototypical or baseline technologies. Redox RAM (ReRAM) technologies have advanced at a particularly high rate in recent years, both in the scientific understanding and technological development of working prototype chips. ReRAM has been commercialized on a small scale [7,8], and multi-layered NAND-scale prototypes have been demonstrated [9,10]. In response to this

Emerging Nanoelectronic Devices, First Edition. An Chen, James Hutchby, Victor Zhirnov and George Bourianoff.
© 2015 John Wiley & Sons, Ltd. Published 2015 by John Wiley & Sons, Ltd.

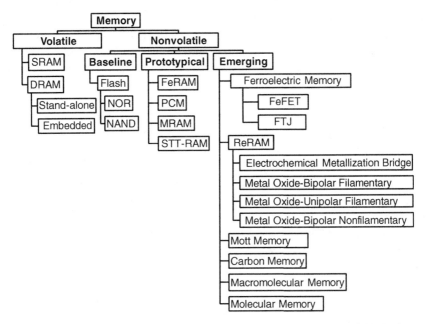

Figure 13.1 Taxonomy of memory technologies (adapted from the 2013 ITRS ERD chapter [5])

increase in ReRAM research and development, the 2013 ITRS Emerging Research Devices (ERD) chapter added an expanded coverage of this technology, which is also reflected in this chapter. Other emerging memory technologies are further from commercialization but if certain challenges were overcome, these could offer additional benefits in energy, speed, and reliability.

Figure 13.1 provides a visual categorization of modern solid state memory technologies.[1] A key distinguishing feature of these technologies is whether the technology requires an active power source to maintain data. If a technology loses data when the power source is removed, it is *volatile*, whereas if the information is retained without power it is *nonvolatile*. SRAM and DRAM are mature, volatile memory technologies whose key technological parameters, including critical dimensions, performance, and reliability have steadily advanced for several decades. NAND and NOR flash are mature nonvolatile memory (NVM) technologies that provide baseline performance, scaling, and reliability metrics against which prototypical and emerging memories are compared. Ferroelectric RAM (FeRAM or FRAM), PCM (or PCRAM), magnetic RAM (MRAM), and STT-RAM are prototypical technologies that have matured to the point that commercial products exist. When a technology reaches this stage, it is transferred from the ERD to the Process Integration, Devices, and Structures (PIDS) chapter of ITRS.

This chapter is written as a counterpart and expansion of the memory section of the ITRS Emerging Research Devices (ERD) chapter [5]. The intention is to provide a summary of the

---

[1] This chapter focuses on the devices considered by the ITRS Emerging Memory section of the ERD chapter, and considers memory technologies that can be integrated with CMOS. For this reason, traditional magnetic hard disk drives are not included in Figure 13.1.

emerging memory technologies that fit the ITRS criteria,[2] including an assessment of these technologies in the application space. Subsections 13.3 through 13.8 provides a brief overview of the physical principles governing the operation of each memory device, whereas more detailed information on physics can be found in the previous chapters of this book and external references. In addition, they summarize significant technological advancements of each technology in the past several years, as well as key research challenges. Finally, a comparative assessment of the collective group of emerging memory devices is made, with regards to the technological areas that emerging memories are most likely to compete in, namely NAND flash, SCM, and DRAM.

13.2 Common Emerging Memory Terminology and Metrics

All of the emerging memory technologies covered in this chapter are nonvolatile. Switching of a nonvolatile memory state is typically referred to as a either a SET or RESET operation. With the exception of the FeFET, all of the emerging memory technologies in Figure 13.1 are two terminal resistance switching devices (RRAM). For an RRAM cell, SET switching refers to the operation of changing from the high resistance state (HRS) to low resistance state (LRS). SET switching is also referred to as a "program" or "write" operation. The RESET operation (also referred to as "erase") switches the device from the LRS to the HRS. The HRS and LRS can be considered "0" and "1" bit states, respectively. The LRS state also might be referred to as a SET state, and the HRS as the RESET state.

Nonvolatile memories typically are assessed by the following key metrics:

- *Endurance:* the number of SET/RESET cycles a memory can survive before the difference between the two states cannot be adequately determined by the read circuitry. Historically, flash has a maximum endurance of 10^5–10^6 cycles. Current ultra-scaled cells have an endurance of 10^3 or less [5].
- *Retention:* the amount of time a device will retain a state, as sensed by the array circuitry, after the programming bias is removed. This can be evaluated on a virgin device or one that has been fully cycled. As a baseline, flash historically is expected to have state retention of 10 years at 85 °C after cycling to maximum endurance (i.e., 10^5 cycles).
- *Resistance ratio, R_{OFF}/R_{ON}:* The ratio of the HRS to LRS is an important metric, which needs to be large enough to avoid an overlap of the HRS and LRS distributions in an array. If these distributions overlap, it will lead to bit errors [11].
- *Switching voltage, current, and power:* SET and RESET switching voltages and currents are important metrics of a memory operation. Switching voltage and current are the primary factors determining the power, and hence heat generation in a densely packed memory array. Heating during switching can be the primary factor in determining how closely cells can be spaced. In addition, the switching current determines the size of drive transistor required. The maximum switching voltage determines the compatibility of BEOL technology with scaled CMOS. Most resistive memories switch with less than 3 V, whereas switching current ranges from nA to mA, and hence switching current is a common metric which is

---

[2] The baseline criterion that the Emerging Research Devices chapter of ITRS uses when deciding to include a new emerging memory or logic device in its tables are: (a) publication in peer reviewed journals or conferences proceedings by two or more groups, or (b) extensive publication by a single group. Extensive technical evaluations, discussions, and workshops among the ERD group members also play an important role in this decision.

used to compare. In ReRAM, typically the RESET currents are highest, which is why this is a particular metric of importance.

- *Switching time:* The time which it takes to transition from SET to RESET or RESET to SET for a memory cell. This is an important metric for system speed. In fact the "memory bottleneck" due to limited speed of memory access is one of the difficult problems of modern and future information processing systems.
- *Switching energy:* Switching energy is the product of switching time and power. A memory's switching energy is a key performance metric for applications ranging from high performance to mobile computing.

With the exception of the FeFET, all of the emerging memory technologies in Figure 13.1 are two terminal resistance switching devices. This makes them amenable to an efficient crossbar configuration, where perpendicular top electrode and bottom electrode lines intersect at each device. This is the most dense possible arrangement, as each individual device only requires an area of $4F^2$, where F is the minimum areal feature size of the device. However, in order to avoid so-called sneak paths interfering with the read and write of the cell, it is necessary to have a select device isolating each device. Select device concepts and candidate structures are covered in detail in Chapter 12.

13.3 Redox RAM

The emerging memory technology category of Redox RAM (ReRAM) has seen the greatest increase in research activity over the past two years, with roughly 593 publications in the literature [5]. The terminology redox RAM was suggested in 2007 in Reference [12] to describe the category of resistance switching resulting from an electrochemical reaction involving oxidation and reduction processes in a thin insulating film sandwiched between two electrodes. Redox memory can be split into four major categories based on the physical mechanisms governing their behavior, as illustrated in Figure 13.2. It is quite useful to categorize ReRAM in terms of physical mechanisms, as done in Reference [12]. However, it is possible that there is some combination of these mechanisms in a structure, especially when fabricated with a metal oxide switching layer. Therefore in Figure 3.2 the three metal oxide categories have been named by the electrical behavior they exhibit – namely distinguished by bipolar versus unipolar operation, as well as whether or not the current through a device is strongly dependent on area. These four categories of ReRAM are described in the following sections.

13.3.1 Electrochemical Metallization Bridge

Electrochemical metallization bridge (EMB) memory is also referred to as conducting bridge RAM (CBRAM) and programmable metallization cell (PMC) technology. A key attribute of EMB memory is that Switching is based on the motion of *cations*. An EMB type ReRAM consists of an asymmetric metal/insulator/metal structure with a reactive electrode composed of either Ag or Cu (see Figure 13.2). The inert electrode is chosen such that it will not strongly react with the insulator; typical choices are Pt, Ni, Au, and W. The insulator is often a solid electrolyte such as GeSe, GeS, a binary oxide, or a Cu-doped binary oxide (which are mixed conductors) [13]. A comprehensive table of EMB insulators is provided in Reference [14].

EMB switching is typically bipolar, and hence requires an electric field. Unipolar operation is also possible, but requires higher current and energy and is therefore uncommon [15,16]. Prior to the commencement of switching, an electroforming process is required to create an initial

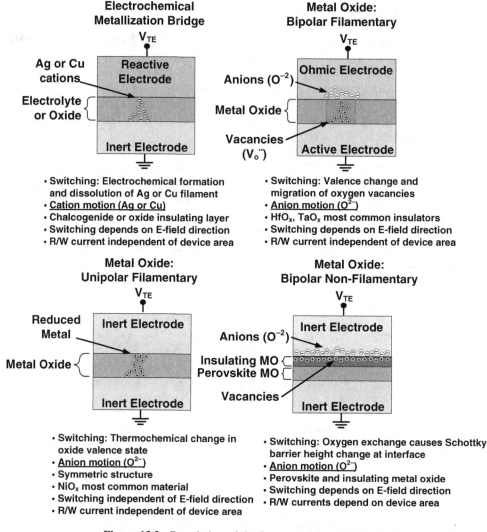

Figure 13.2 Description of the four categories of ReRAM

filament. Switching occurs based on the formation and dissolution of an Ag or Cu filament under an electric field. The filament oxides are formed based on the oxidation reaction (using Ag as an example) [13]:

$$Ag \Leftrightarrow Ag^+ + e^- \qquad (13.1)$$

The SET process occurs as the cations oxidize under the electric field and deposit on the inert electrode where they are reduced and form a filament. The RESET process involves dissolution of the filament initiated by high Joule heating at the thinnest portion of the filament, re-oxidizing the Ag. Then, electrochemical motion of the Ag^+ cations causes the filament to disperse into the insulator, returning the structure to a high resistance state. The specific details of the process are still the subject of some debate, and are comprehensively reviewed in References [13,14,17].

EMB ReRAM technology has achieved significant performance metrics. Endurance as high as 10^{10} with little degradation was demonstrated in GeSe cells (projected to well beyond 10^{11} cycles) [18]. High temperature retention due to random diffusion of Ag or Cu atoms in the insulator has historically been considered a moderate weakness for EMB ReRAM. However, recent results show excellent retention, with states holding after 1000 h at 200 °C [19]. Scalability to dimensions of 20 nm has been demonstrated for a GeSe EMB cell [20]. A recent experiment demonstrating quantized conductance corresponding to the conductivity of a filament formed from a single chain of Ag atoms showed that scaling to atomic dimensions is possible [21].

Significant progress has been made in the development of EMB ReRAM macros and low density commercial products – which are important to foster advancement of an emerging technology [11]. In 2012, a 1 Mbit integrated EEPROM replacement chip with 0.18 μm technology was released [15]. Recently, a very low power prototype chip which switches at 600 mV and achieves a 1 pJ SET and 8 pJ RESET energy was demonstrated [22]. Also, a $W/Al_2O_3/Ti/CuTe$-based EMB cell realizing subnanosecond SET and RESET times was reported [16].

A fundamental research challenge of EMB is the understanding of subtle details of the switching process. Recent research in this area has elucidated many of the details of the switching physics, but the scientific community lacks complete agreement on the exact nature of the filament formation and dissolution process details. Another key research challenge for EMB ReRAM (and other filament-based ReRAM technologies) is random telegraph noise (RTN) [11]. This effect manifests itself as a variation in read current, especially in the high resistance state where the fluctuation of small numbers of atoms in the filament may vary to cause conductivity changes [23]. EMB ReRAM cells often employ a solid electrolyte switching layer (e.g., GeSe), which are not easy to integrate with a CMOS back end of line (BEOL) process and may be damaged by high temperatures. For this reason, commercial EMB products are generally integrated during later steps of the BOEL process, whereas it may be desirable for maximum performance to integrate the cell before the first metallization layer.

13.3.2 Metal Oxide: Bipolar Filamentary

The metal oxide bipolar filamentary (MO-BF) category of ReRAM is an asymmetric metal/insulator/metal structure, in which switching occurs due to the change of oxygen *anion* concentration in a filament (see Figure 13.2). This type of memory is often referred to as Valence Change Memory (VCM) in the literature due to the change of valence states of oxygen atoms during switching [12]. The switching material is typically a transition metal oxide; recently the most common materials are HfO_x and TaO_x due to CMOS compatibility, scalability, and performance. Other common switching oxides include TiO_2, WO_x, AlO_x, as well as nitrides (AlN [24]), and oxynitrides (AlO_xN_y [25]). In addition, bilayer structures are often used with either two stoichiometries of the same oxide (e.g., Ta_2O_{5-x}/TaO_{2-x} [26]), or with a second layer composed of a different oxide intended to enhance nonlinearity and act as a select device [27]. The ohmic electrode is typically fabricated from a material with high oxygen reactivity, such as Ta, Hf, or Ti. This electrode may serve as an oxygen reservoir. The inert electrode (which may also be referred to as the active electrode), is typically formed from a material that will not react significantly with the metal oxide, such as Pt, Ir, TiN, or TaN. Switching is thought to occur in the metal oxide region directly adjacent to the interface with this inert electrode.

As with other filamentary ReRAM, prior to the resistance switching process, an electroforming step is required. In the case of MO-BF ReRAM, the electroforming step does permanent damage [17], and has similarities to a "soft breakdown" of an oxide [11,28]. "Forming free" processes have been reported [29], although it is suggested that a filament forming process still occurs at a very low current [30]. Switching most likely occurs due to a change in concentration and configuration of charged oxygen anions (O^{2-}) and oxygen vacancies ($V_O^{..}$) under an applied bias. The requirement of opposite polarities for SET and RESET switching are a clear indication that an electric field is involved. However, it is also probable that Joule heating plays a significant role in switching [31]. During SET switching, anions are removed from the switching filament near the inert electrode, creating a high concentration of $V_O^{..}$ and hence, the device enters a low resistance state. RESET switching involves the return of anions to this region, which lowers the $V_O^{..}$ concentration and raises the resistance. Details of the MO-BF switching process are not fully understood or agreed on in the scientific community. Chapter 8 and several recent reviews provide a thorough explanation of the most recent understanding [12,17,32,33].

MO-BF has achieved impressive performance at the single cell level. Endurances as high as 10^{12} have been reported by several groups using versions of a TaO_x-based cell [26,29,34]. Sub-nanosecond switching speeds have been reported at the cell level [35], as well as a SET switching energy of 115 fJ and RESET energy of 13 pJ (for a R_{OFF}/R_{ON} ratio of 2) [36]. At VLSI 2013, a $Ta/TaO_x/TiO_2/Ti$-based ReRAM array combining an endurance of 10^{12} cycles and 10 pJ write energy has been demonstrated [29]. A highly scaled HfO_x cell with 100 fJ SET energy has been demonstrated (RESET currents were not given, although it was stated that the energy was comparable to the SET energy) [37]. High endurance, speed, and low energy are important factors for a M-type SCM candidate as well as a potential future DRAM replacement.

A number of recent reports indicate potential for MO-BF to 10 nm and below. In 2011, a 10×10 nm HfO_x cell was demonstrated with an endurance of 5×10^7 cycles, 10 year retention at 100 °C (extrapolated from a 30 h bake at 200 °C), while maintaining an R_{OFF}/R_{ON} ratio >50 [37]. More recently, an ~8 nm HfO_x ReRAM structure was demonstrated with an endurance of 10^8 cycles, R_{OFF}/R_{ON} of 100, and 10 year retention (extrapolated from a 30 h, 150 °C bake) [38]. Using carbon nanotubes as electrodes, a 5×5 nm proof of concept cell AlO_x was operated, demonstrating the smallest operational ReRAM structure at the time of this writing [39].

Excellent retention has been demonstrated for MO-BF ReRAM. HfO_x cells have been reported to have a 10 year retention at 105 °C (where the temperature depends on the compliance current used during RESET), extrapolated from Arrhenius behavior evaluated from 150 to 250 °C [40]. A detailed retention study on an $Ir/TaO_x/TaN$ cell extrapolated 10 year retention at 85 °C from 1000 h at 150 °C [41].

In the past two years, there has also been very rapid progress in prototype ReRAM products integrated with CMOS. In 2012, an 8 Mb TaO_x-based ReRAM macro with a write throughput of 443 MB/s was reported [42]. In late 2013, the first commercial MO-BF ReRAM product was announced, which will be a TaO_x-based device integrated into an eight-bit microcontroller [7]. In early 2013, a bi-layer 32 Gb prototype chip was announced, which is nearing the capacity of modern NAND flash [9]. Device structure and material details were not given in this report.

As noted above, a key scientific research challenge continues to be the development of a complete physical understanding of electroforming, switching, and degradation mechanisms in metal oxides. The most pertinent technological challenges are variability and random telegraph noise. For example, Reference [11] points out that with a 1 MB MO-BF ReRAM-based test chip, the R_{OFF} varies over four orders of magnitude such that it overlaps with the R_{ON} spread, and hence causes bit errors in the range of 1000 ppm.

13.3.3 Metal Oxide: Unipolar Filamentary

The metal oxide unipolar filamentary (MO-UF) ReRAM consists of a metal/oxide/metal structure which typically employs a binary metal oxide switching layer. The electrodes are inert materials that are not highly reactive with the oxide, such as noble metals and TiN. MO-UF ReRAM was the first major transition metal oxide-based ReRAM to be demonstrated, and was originally named OxRRAM [43]. It is also often referred to as thermochemical memory, due to the thermal dominated switching mechanism [12]. Unlike other types of ReRAM, the MO-UF structure can be symmetric, although asymmetric structures may be used to tailor device properties. The earliest integrated MO-UF-based memory utilized a NiO cell [43]; subsequent cells have utilized HfO_x, TiO_x, and CuO.

As with other filamentary oxide ReRAM, MO-UF requires a soft breakdown of the oxide to form a switching filament [43]. This filament is thought to be formed of metal oxide of a reduced oxidation (i.e., more metallic) state than the surrounding oxide. SET switching uses a current compliance (cc), which causes a thermochemical reduction to occur. RESET switching results from the rupture of this filament from a higher current pulse (without cc). An in depth discussion of thermochemical switching mechanisms is provided in Reference [44].

Several demonstrations of SET/RESET endurance up to 10^6 cycles have been reported [43,45], as well as 10^{12} read cycles [43]. Notable results have been demonstrated starting in 2009 on a MO-BF variant known as "contact ReRAM" (CRRAM) [45]. CRRAM technology has been demonstrated scalable to 35 nm [46], 60 µA RESET current [46], and SET/RESET voltages of 2.0 and 1.5 V, respectively [47]. Endurance of 10^6 cycles and retention of 1000 h at 150 °C (400 h at 150 °C for a 35 nm cell) were demonstrated for CRRAM [45,46].

The most pertinent challenge for MO-UF ReRAM are high write currents and low relative endurances resulting from a thermochemical dominated switching process. Initial prototypes required RESET currents as high as 2 mA, although recent results are as low as 60 µA (for the unipolar W/TiO_xN_y CRRAM) [46]. This is a significant factor which limits the maximum integration density [11]. This has led research in recent years to trend toward a focus on bipolar technologies. MO-UF technology has the same variability and random telegraph noise problems associated with other filamentary ReRAMs.

13.3.4 Metal Oxide: Bipolar Nonfilamentary

Metal oxide–bipolar nonfilamentary (MO-BN) ReRAM is an asymmetric metal/oxide/metal structure, which generally utilizes a bilayer oxide switching film. The oxide layers are typically comprised of a thin insulating (often binary) metal oxide (IMO), and a conductive metal oxide (CMO), which are often perovskites [48]. A more recent MO-BN device variant using two binary oxide layers [49]. MO-BN ReRAM is unique in that conduction does not occur through a switching filament, but over the majority of the device area. This leads to a strong dependence of virgin, high, and low state resistances, and switching currents on device electrode area [48,49]. Due to the large area over which the current is spread out, thermal effects are not considered to play a major role in switching [48]. Unlike the more common filamentary ReRAM technologies, MO-BN typically does not require an electroforming step, which eliminates a source of device to device variation.

As with other types of ReRAM, the exact physics responsible for switching are not completely understood. However, as with oxide-based (anionic) ReRAM, the predominant theory of switching is based on modulation of O^{2-} concentrations under an applied bias. In the case of MO-BN ReRAM, O^{2-} concentrations are modulated through the majority of the TMO area, rather than in a localized switching filament. Modulation of the O^{2-} most likely causes a raising and lower of the Schottky barrier height at the interface. The current through the device is exponentially sensitive on this barrier height.

Data on a MO-BN device technology was presented in 2009 demonstrating typical SET/RESET voltages of ±3 V with a 1–$10\,\mu s$ write time and $\sim10^6$ cycle endurance [48]. In 2010, a $0.13\,\mu m$, 64 Mb test chip was presented. This was the largest ReRAM prototype at that time [10].

At IEDM 2013, a new version of MO-BN technology known as vacancy modulated conductive oxide resistive RAM (VMCO-RRAM) was presented [49]. This structure consists of two binary transition metal oxides (Al_2O_3 and TiO_2) and requires electroforming, unlike earlier MO-BN technology. The VMCO-RRAM has a predicted retention of 10 years extrapolated from measurements of $168\,h$ at $125\,°C$. Switching as fast as $10\,ns$ was demonstrated with SET/RESET voltages of ~3.5 V. A minimum SET/RESET of ±2.0 V was possible using longer program times. Scaling down to $40\,nm$ was demonstrated, and sub-nA switching current is predicted at sub-nm scales. Endurance has not yet been reported for VMCO-RRAM.

Between 2010 and late 2013, the significant progress was not reported for MO-BM technologies due to the single most important challenge for this ReRAM category: short retention. The VCMO-RRAM shows significant progress, with an estimated retention of greater than 10 years [49]. The other early research challenge for this technology is the need to improve CMOS processes compatibility, which has also been addressed by the VCMO-RRAM. Recent progress of MO-BN technology could make this a very attractive option.

13.4 Emerging Ferroelectric Memories

Ferroelectric RAM (FeRAM) was conceptually invented over half a century ago [50], and the first commercial prototype was demonstrated in 1988 [51]. FeRAM is now considered a prototypical memory technology, with several commercial parts available, and it is described in Chapter 3. There also two significant emerging memory technologies which rely on ferroelectric effects: the ferroelectric FET (FeFET) and the ferroelectric tunnel junction (FTJ). Both of these show great promise as a high endurance, scalable emerging memory due to recent advancements in ferroelectric materials [52].

Ferroelectric memories fundamentally rely on the same physical switching mechanism, although there are different methods of creating an electronic memory based on this effect. Switching is caused by a reversal of spontaneous electric polarization (P_r) of thin ferroelectric films under an electric field, as described in detail in Chapter 6.

Traditionally, the most common ferroelectric materials used in electronic memories are PbZrTi (PZT) and $SrBi_2Ta2O_9$ (SBT). However, the recent discover of ferroelectricity in the CMOS compatible material, (doped) HfO_2 [52,53], has provided great enthusiasm for the future prospects of a CMOS integrated ferroelectric memory.

13.4.1 Ferroelectric FET

The FeFET structure integrates a ferroelectric capacitor, similar to that in a FeRAM cell, into the gate of a MOSFET. A typical FeFET structure is shown in Figure 13.3 [54]. There are

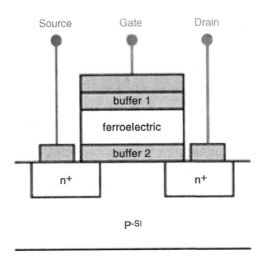

Figure 13.3 Typical ferroelectric FET structure. Reprinted with permission from Reference [54]

several different variations of this structure; it is possible to remove one or both buffer layers, or deposit the structure on a nonsilicon substrate. It was originally demonstrated in 1974 [55], although significant research progress was not made until roughly two decades later. Switching in the FeFET occurs due to the change of P_r in the gate ferroelectric, as described above. The change in polarization state is achieved by applying a positive or negative voltage pulse on the gate. When a positive pulse is applied, the resulting remnant polarization in the ferroelectric holds an electric field in the direction of the channel to the gate electrode, which raises the threshold voltage (V_T) of the MOSFET. This electric field can be thought of as subtracting a portion of the gate field, hence making it more difficult to turn the device on. Conversely, a negative gate pulse creates a remnant electric field which points from the gate to channel, and enhances the voltage applied by the gate. This field normally attracts electrons and creates an inversion layer in the n-FET, effectively creating a normally on device. For an in depth physical analysis and complete analytical model of the FeFET device physics, the reader is referred to Chapter 6 of this book and Reference [56].

The FeFET is the only FET-based memory described in this chapter, and hence it has a slightly different performance and reliability characterization methodology than other (e.g., resistive switching) nonvolatile memories. The retention is typically based on either the time dependent difference in V_T or difference in I_D of the SET and RESET states of the after a given time period. The slope from these V_T shifts following a few days to a few weeks are typically extrapolated to 10 years to provide a projected V_T window after this time.

Prior to around 2004, progress on developing a scaled FeFET with reasonable retention was relatively slow [57]. In 2004, a $10\,\mu m$ device was demonstrated with 10^{12} cycles of endurance which retained data for 12 days, maintaining an I_{ON}/I_{OFF} ratio of 10^6 (although no long term endurance projection was made) [58]. In 2011, the first submicron FeFET with 10 year retention projected [59]. These devices utilized a Pt/SBT/Hf-Al-O/Si gate stack and had a minimum gate length $0.26\,\mu m$, and gave an endurance of 10^8 cycles. In this work, utilizing $20\,\mu s$ writing pulses of $-4\,V$ and $+6\,V$, an initial voltage window of $500\,mV$ was obtained, and an estimated voltage window of $\sim 250\,mV$ is extracted from a one week retention measurement [59]. In 2012, the same group demonstrated a 64 kbit FeFET array based on the

Pt/SBT/Hf-Al-O/Si gate stack [60]. This array requires $10\,\mu s$ $\pm 7.5\,V$ SET/RESET pulses, and achieves 10^8 cycle endurance. An initial window of $350\,mV$ was demonstrated and a 10 year V_T window of $200\,mV$ was predicted (extrapolated from a 48 h measurement).

Recent demonstrations of ferroelectric silicon doped HfO_2 (Si: HfO_2) show some of best performance, reliability, and scaling data to date [53,61]. At IEDM 2011, HfSiO FeFETs were demonstrated which required 1 s programming with voltages of $-3\,V$ and $+4\,V$ and achieve an initial V_T window of $1000\,mV$, which is predicted to be $650\,mV$ after 10 years [53]. In 2012, a similar Si:HfO_2 FeFET with a gate length of 28 nm was demonstrated, by integrating the ferroelectric HfO_2 into the gate stack of a HKMG CMOS process [61]. This FeFET required 20 ns, $\pm 5\,V$ SET/RESET pulses and produced an initial V_T window of $1000\,mV$. It was estimated that these devices would maintain a $600\,mV$ window after 10 years (extrapolated from a 30 h measurement). Endurance on this highly scaled device was only 10^4 cycles due to significant charge injection.

Retention remains a key issue for the FeFET. All 10 year retention extrapolations are made from actual measurement lasting 12 days or less. Poor retention is due to two issues inherent in integrating a ferroelectric capacitor in a MOSFET gate stack: (a) the depolarization field and (b) charge trapping at the ferroelectric buffer- or ferroelectric-semiconductor interface [57]. The depolarization field occurs as a consequence of integrating a ferroelectric capacitor with a semiconductor. It is not possible for the ferroelectric to have the polarization charge compensated charge on the semiconductor side (even with a buffer layer) due to the capacitance of the semiconductor. The second issue, charge trapping at the interface of the ferroelectric and semiconductor, is caused by the remnant polarization field acting on this interface [53,57]. This charge eventually compensates the field created by the remnant polarization of the ferroelectric and effectively cancels the threshold voltage shift. Charge injection appears to be especially deleterious on highly scaled devices [61].

The FeFET is also difficult to integrate with CMOS, especially versions which rely on SBT or PZT. Doped HfO_2 offers an improved method of CMOS integration, but additional research is still needed to understand the origins of ferroelectricity in this material. In addition, the FeFET always requires an FET structure, and hence is not easy to integrate as a back end of line (BEOL) memory, as with metal/insulator/metal structures like ReRAM and the ferroelectric tunnel junction. However, if the extremely high endurance of the FeFET could be combined with high speed, low energy switching, it could serve as a strong M-class SCM or DRAM replacement candidate.

13.4.2   Ferroelectric Tunnel Junction

The ferroelectric tunnel junction (FTJ), also referred to as ferroelectric polarization reversal memory, is an asymmetric, metal/insulator/metal or metal/insulator/semiconductor resistive switching memory device. The insulator is a thin ferroelectric film which causes the resistance measured through the electrodes vary significantly based on the polarization direction due to the giant tunnel electroresistance (TER) effect. The theoretical concept of giant TER was first proposed as a polar switch by Leo Esaki in 1971, although experimental demonstration was not achieved until the early twenty first century [62]. Typical ferroelectric insulators are perovskites, such as $BaTiO_3$ (BTO) and $Pb(Zr,Ti)O_3$ (PZT). One electrode is often a metal such as Pt or Au. The other (e.g., bottom) electrode is typically a conducting perovskite, such as La_xSr_yMnO [63] or $SrRuO_3$ [64]. Recently, improved tunnel electroresistance was found by using a semiconducting bottom electrode, Nb:SrTiO3 [65].

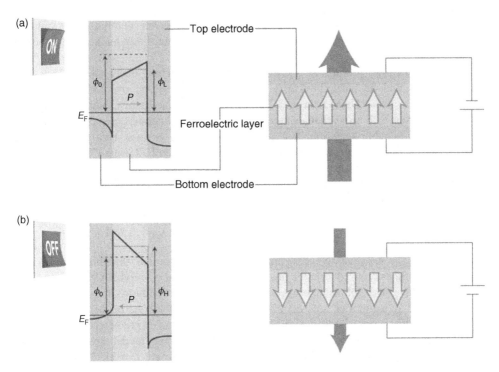

Figure 13.4 Physical depiction of a FTJ band diagram and device structure in: (a) low resistance (ON) state and (b) high resistance (OFF) state. The red arrows represent (a) high and (b) low current flow through the resistive element, based on the polarization direction, which influences the barrier height. Reprinted by permission from Macmillan Publishers Ltd: Nature, reference [66], copyright 2012

The SET, RESET, and read procedures are similar to bipolar ReRAM, but the physical mechanism is markedly different. The resistance switching in an FTJ can be explained with the aid of Figure 13.4. The device is set to a low resistance state (LRS) by applying a positive voltage pulse to the top electrode (2–5 V, and 10 ns to 100 µs are typical experimental values) [63,65]. This voltage pulse causes the electric field in the ferroelectric layer to polarize in the direction of the bottom to top electrode, as illustrated in Figure 13.4a. In this state, the electron tunnel barrier, ϕ, is low compared to the opposite polarization (the average barrier for the LRS is denoted ϕ_0 in Figure 13.4). A negative voltage pulse of roughly equal time and magnitude causes a reversal of the polarization, and the barrier rises to an average height of ϕ_L. The cell is read by measuring the current at a voltage low enough to avoid disturbing the polarization, typically on the order of 100 mV.

FTJ memory was realized for the first time in the past several years, and hence only a handful of proof of concept studies of memory cell characteristics have been performed. In 2012, a 50 nm FTJ memory was demonstrated [63], with an estimated potential to scale to 10 nm, based on the ferroelectric domain size [5]. This device had the very low write energy of 10 fJ/bit. The highest published endurance and retention values are given in Reference [67] for a $Co/BiFeO_3/Ca_{0.96}Ce_{0.04}MnO_3$ FTJ of 4×10^6 cycles and 10 years, respectively. Reference [65] elucidates a common, but important tradeoff between retention, endurance, and write voltage can be found. If a write voltage of 3.5 V is used, R_{OFF}/R_{ON} values of 10^4 are possible (which will increase retention), but endurance is limited to tens of cycles. In the opposite extreme, if a write voltage of 2.2 V is used the R_{OFF}/R_{ON} drops to about 40, which increases the endurance but reduces the retention times and decreases read margins in an array.

FTJ memory shows promise as a scalable, ultra-low switching energy technology, but is relatively new and has many research challenges that must be overcome before being considered as a viable NAND flash, SCM, or DRAM replacement candidate. One of the key research areas is to prove that a single FTJ can have an endurance greater than 10^6, while maintaining reasonable retention. Scaling down to 10 nm or less must also be proven. Furthermore, it will be technologically important to develop an FTJ using CMOS compatible materials; one possibility is to take advantage of ferroelectricity discovered recently in doped HfO [52]. Another important step will be to develop FTJ memory arrays and explore the cell to cell variation and yields.

13.5 Mott Memory

Mott memory is a resistive switching memory cell with a similar structure to metal oxide ReRAM, sometimes referred to in the literature as a correlated electron RAM (CeRAM) [68]. Specifically, this is a metal/insulator/metal structure utilizing a Mott insulator. Common Mott insulators used in memories are VO_x, $Pr_{1-x}Ca_xMnO_3$, and NbO_x. In these materials, the standard quantum theory of solids, such as the tight-binding model, predict that these materials should act as electrical conductors [69]. The standard tight-binding model uses the independent electron approximation, which does not account for strong electron-electron interactions, where a short screening length exists. In a Mott insulator, the high electron density under certain conditions creates a short enough electron screening length that electrons interact with each other such to create an energy gap and impede conduction. One can consider the simple physical picture: these extremely high densities of electrons strongly repulse each other and resist conduction. A Mott insulator transitions to metallic behavior past a particular material dependent temperature, known as the metal insulator transition (MIT) transition temperature. For example, this has been shown to occur at ~ 1070 K in NbO_2 [70].

Switching in a Mott insulator typically occurs due to a metal insulator transition induced by current injection in the device, and can be unipolar [71] or bipolar [68,72]. In fact, the electrical switching of a Mott memory shares similarities with oxide-based ReRAM (hence in some cases the Mott transition is considered an alternate explanation to ionic motion for resistance switching in certain metal oxides [71,72]). In a Mott memory, a voltage pulse forces the Mott transition (which naturally occurs at the MIT temperature). A voltage pulse of either opposite polarity [72], or different waveform [72] causes the device to return to its conducting state. The ability to use a current to cause a Mott insulator to reach the insulating state is expected, and accounts for the behavior of Mott oscillators [73].

Mott switching has the advantages of high speed and low energy. For example, switching at speeds below 3 ns, and endurance of 10^9 cycles, with energies around 100 fJ in NbO crossbar devices have been demonstrated [72]. However, these devices were not nonvolatile (nor were they intended to be – these were being explored as a resistive memory select device), and lost their state as soon as they cooled, typically in a few ns. Carbonyl doped NiO Mott memories with an endurance of 100 cycles were reported to retain data for 1 h at 300 °C [68].

Theoretical understanding and experimental demonstration of retention in Mott memory is an important research challenge. In devices that exhibit retention of conduction below MIT temperature, it will be necessary to demonstrate high endurance, scaling, and eventually functional arrays. If long retention cannot be demonstrated, it is also possible that Mott insulators can be used as a select device for resistive memory arrays [73] (see Chapter 12).

13.6 Macromolecular Memory

The category of macromolecular memory, also referred to as polymer, or organic memory, encompasses a wide variety of memory devices [74,75] (also see Chapter 10). The defining feature of this category is the incorporation of polymers in the switching layer. For brevity the following treatment focuses on the category of devices which exhibit resistance change due to this polymer layer. This category of macromolecular memory has demonstrated the best performance metrics [5]. A very thorough description of other types of polymer memories, including capacitor and FET structures is given in reference [74].

The resistive macromolecular memory structure consists of a metal/insulator/metal structure, where the insulating/switching layer is a polymer – which is often organic. There are numerous resistance switching mechanisms reported in macromolecular memories, which are not yet fully understood. The read and write operations in a resistive macromolecular memory are similar to those in ReRAM and it is possible define a similar set of categorization to the ReRAM scheme presented above. Bipolar voltage induced electrochemical formation and dissolution of an Ag filament was directly observed in an Ag/WPF-BT-FEO^3/Conducting-Si stack [76]. This is a strong indication that the mechanism is analogous to the electrochemical metallization bridge (EMB) ReRAM described in Section 13.3.1, except that the insulating layer has been replaced by an organic polymer. Anionic motion, similar to the unipolar and bipolar metal oxide ReRAM has been proposed as another switching mechanisms in several macromolecular memory devices [74]. A mechanism for bipolar switching reported by several groups is the creation and annihilation of a charge transfer (CT) complex due to an electric field [75,77,78]. This mechanism is unique to polymer memories. When formed, a charge transfer complex allows the flow of charge through the normally insulating polymer.

A number of promising macromolecular memory devices have been recently been reported. A 2×2 μm Al/parylene-C/parylene-C/W cell has been demonstrated to exhibit bilayer switching with a RESET current of 18 nA (a 150 nA SET current is required) at a power of 67 nW [79]. Assuming that the RESET switching is 15 ns, as reported for an earlier parylene-C memory [80], a switching energy of 1 fJ is possible. This device exhibited a retention of $\sim 10^4$ s (~ 3 h) at 200 °C and endurance of about 100 cycles [79]. The highest reported endurance in a macromolecular memory comes from an Al/polymide/ITO cell, using a polymide named "DAXIN-PI" [78]. This bipolar device exhibited 10^5 SET/RESET cycles (using 200 ns ± 2 V pulses) and with retention of ~ 3 h at 85 °C, following this endurance test. SET and RESET voltages were as low as 1.0 and -0.3 V, respectively, and a maximum R_{OFF}/R_{ON} ratio of 10^5 was demonstrated. As with other types of ReRAM, tradeoffs between switching voltage, switching time, endurance, and retention exist. A relatively long retention time[4] of over 10^6 (>32 days) was obtained by using a Al/PEDOT:PSS^5/Cu bipolar electrochemical metallization type cell [81].

A highly scaled 8×8 crossbar array of 100 nm Al/PI:PCBM/Au with unipolar resistive switching cells has been demonstrated [82]. For macromolecular memories which rely on

---

[3] The full polymer formula is: poly[(9,9-bis(6'-(N,N,N-trimethylammonium) hexyl)-2,7-fl uorene)-co-(9,9-bis(2-(2-methoxyethoxy)ethyl)-fluorene)-co-(2,1,3-benzothiadiazole)] dibromide.

[4] For a macromolecular memory cell.

[5] The full polymer formula is: poly(3,4-ethylene-dioxythiophene):poly(styrenesulfonate).

filamentary conduction, it is reasonable to assume that the ultimate scaling limits will follow that of EMB, or MO-BF type ReRAM due to small filament sizes. For example, in Reference [81], the size of the Cu filament is calculated to be about 4.1 nm (from resistivity calculations).

The 8×8 array work presented in reference [82] also provided baseline statistics for the 64 cells (statistical analysis was done for the array with $2 \times 2 \, \mu m^2$ devices). The yield was 50 out of the 64 cell array ($\sim$78%). Both the high resistance state had a spread of about two orders of magnitude, and the average R_{OFF}/R_{ON} was about 10^5, so the distributions did not overlap. A stacked three layer set of 8×8 crossbars of unipolar $200 \times 200 \, \mu m$ PI:PCBM-based cells, stacked between Al metal lines has been demonstrated [83]. In this case, the yield was 160/192 bits (83%). High resistance state values had a relatively large spread (about four orders of magnitude), but still did not overlap with the low resistance states. Large variation in the high resistance state of cells is a common attribute of metal oxide ReRAM [11]. Similar PI:PCBM 8×8 arrays have also been demonstrated on a flexible substrate with relatively similar performance [84], demonstrating the possibility of use as a NVM for flexible electronics.

Macromolecular memories are interesting from the fundamental material science perspective, and present interesting prospect as a future memory technology. However, several key challenges must be addressed to make macromolecular memory viable as a mainstream commercial product. An important, fundamental challenge of macromolecular memories is, as with ReRAM, to obtain an improved understanding and categorization of macromolecular resistive switching mechanisms [75]. At the cell level, endurance and retention in most reports are low (<200 cycles endurance and <3 h retention are typical [79,80,82]). Demonstration of proof of concept devices that combine low energy, high endurance, high retention, and excellent scalability in a single device is a key milestone. Furthermore, most polymers used in these memories are not considered CMOS compatible materials, and it is not clear that these memories will survive integration with a typical CMOS at typical BEOL temperatures.

13.7 Carbon-based Resistive Switching Memory

Restive switching in numerous forms of carbon has been reported in recent years. A straightforward method to categorize these emerging memory technologies is by the form of carbon that facilitates switching: carbon nanotubes, amorphous carbon (a-C), diamond-like carbon (DLC), graphene, and graphene oxide. The physical switching mechanisms reported vary widely, which include the change from the sp^2 to sp^3 state in carbon [85], redox mechanisms (as described above in Section 13.4 and in Chapter 8), charge trapping (described in Chapter 9), and nanoelectromechanical switching. In addition to reports of resistive switching in the carbon material, carbon nanotubes have been utilized as electrode materials for other types of memory including phase change [86] and metal oxide–bipolar filamentary ReRAM [39]. In the following, we briefly survey the state of the art of carbon-based resistive switching memories.

13.7.1 Amorphous Carbon and Diamond-like Carbon

An amorphous carbon (a-C) or diamond-like carbon (DLC)-based resistive switch consists of an electrode/x/electrode structure similar to other ReRAM cells, where x is an a-C or DLC layer. Resistive switching in amorphous carbon is reported to result from numerous different physical mechanisms. In several reports, unipolar switching in a-C [85,87] and DLC [88] is thought to occur due to the formation and rupture of a sp^3 filament in the material. This case is analogous to metal oxide-unipolar filamentary ReRAM, as the SET and RESET mechanism are both thermochemical.

In another case, when the structure includes a Cu or Ag electrode, and bipolar switching occurs, it is reasonable to propose that the electrochemical formation and dissolution of a metal filament cause switching. This has been observed in both Cu/a-C/electrode and Ag/a-C/ electrode structures [89]. In this case, the memory is an EMB ReRAM cell (as discussed above), where the insulating layer is composed of amorphous carbon rather than the more traditional EMB materials such as chalcogenides and metal oxides.

13.7.2 Graphene and Graphene Oxide

Graphene-based memories have been reported by multiple groups [90–93]. A resistive switching structure is typically comprised of a graphene strip contacted on both sides by electrodes (i.e., Pt [90] or Cr/Au [92]). One proposed mechanism of switching in graphene is the nanoelectromechanical breaking and reattaching of the strip [90]. Bipolar resistive switching in a graphene strip has also been explained by charging effects [90].

Bipolar switching has also been reported in metal/graphene oxide (GO)/metal structures. In the case where one of the electrodes is Cu, it is likely an electrochemical metallization bridging effect occurs, similar to the EMB ReRAM. A Cu/GO/Pt structure with bipolar switching at ~1 V, with a R_{OFF}/R_{ON} ratio of 20 and retention of 10^4 s has also been reported [90]. A flexible Al/GO/Al structure was demonstrated with bipolar switching at about 3.5 V, suggesting a valence change mechanism similar to that in MO-BF ReRAM.

13.7.3 Carbon Nanotubes

Carbon nanotube memory originally attracted significant attention in the year 2000, after the concept of a highly scaled memory array with each cell was defined by the cross-section of two carbon nanotubes was presented [94]. The proposed mechanism for switching in such an array was the electrostatic repulsion and attraction of the two crossing nanotubes. More recently, a full prototype memory chip IC was demonstrated based on lithographically defined carbon nanotube resistance switching layers [95]. The proposed switching mechanism is fundamentally the same electrostatic resistance switching mechanism reported in reference [94]. However, it was hypothesized in the lithographically defined nanotube layers that the connection and repulsion of many nanotubes in concert is responsible for the observed resistance change [95]. The 4 Mb test chip based on 22 nm carbon nanotube layer memory cells has a typical SET operation requiring a 5 V pulse for 500 ns, with a typical current of 1 μA. The RESET operation requires a 50 ns, 4.5 V pulse resulting in a typical current of 15 μA. Assuming both currents give are averages, the SET and RESET switching energies are 2.5 pJ and 3.4 pJ respectively. An endurance of 10^4 cycles and retention of 24 h at 125 °C are reported for this prototype. For this carbon nanotube layer-based NVM technology, few fundamental studies have been published, and hence it is difficult to verify the switching mechanism.

13.7.4 Research Challenges

Carbon memories present interesting features, such as high temperature operation, flexible arrays, and simple processes. However, carbon memory is a relatively young field with many research challenges. Foremost is the understanding of switching mechanisms in different materials. Resistive switching in amorphous carbon and graphene oxide appear to have similarities to redox memory; if this is verified it would be appropriate to consider them a member of the ReRAM family. More fundamental work is needed to understand and verify the proposed switching mechanisms in each category of emerging carbon memory.

13.8 Molecular Memory

Molecular memory is a resistive switching structure based on a single monolayer of molecules surrounded by electrode materials. The memory effect is related to the change in the charge state of a molecule such that conductivity is modulated. Molecular memory is intriguing from the scientific perspective, and offers the possibility of very high scalability and low energy.

One of the earliest demonstrations of molecular memory was made from self-assembled monolayers of four different benzene-based molecules between Au electrodes [96]. This memory was bipolar, requiring a +5 V SET and −5 V RESET process. The retention in this demonstration was only ∼15 min, which would categorize it as a volatile memory. Another significant effort focused on bipolar switching of a [2]rotaxane-based molecular switch tunnel junction (MSTJ) in crossbar configurations [97,98]. In 2007, a rotaxane-based 160 kb molecular memory array was demonstrated [99]. Switching was achieved with 200 ms pulses of +0.2 (SET) and −1.5 V (RESET). This array had a yield of about 25%, with an average retention of about 90 min and maximum retention of about 180 min.

Molecular memory has many challenges, and research activity has declined in recent years.[6] A major research topic related to molecular memory continues to be separation of the effects of the contacts from the molecular switching. Other key research challenges include very short retention times, low endurance, high variation, and low yield.

Recently, novel work using DNA as a storage medium was presented which offers an alternative promising molecular method of extremely dense information storage [100,101]. Arguably, the DNA molecule demonstrates storage density several orders of magnitude higher than other known storage technologies. The storage density of molecular DNA memory is ∼10^{19} bit/cm^3, and 1 kg of DNA could store a maximum theoretical capacity of ∼2 × 10^{18} Mbit, which exceeds the world's current total information storage capacity [102]. Recent progress in DNA synthesis and sequencing has made it possible to experimentally explore DNA storage beyond biological applications. A major breakthrough occurred in 2012–2013 when mainstream digital formats were demonstrated to be compatible with DNA storage, offering a 1000× improvement in density [100,101]. DNA volumetric memory density far exceeds (1000×) projected ultimate electronic memory densities [103]. Also, in the living cell, the memory read/write operations occur at high speed (<100 μs/bit) and require very low energy (∼10^{-11} W/GB) [103].

There are still many unknowns regarding both DNA operations in cells and with regard to the potential of DNA technology for massive storage applications. One of the research goals is to demonstrate miniaturized, on-chip integrated DNA storage. New technologies for DNA synthesis and sequencing described above are key components of this work. For example, there are recent promising demonstrations of micro-manufactured DNA devices, such as a "DNA transistor" for sequencing [104]. Based on the rapid and continuing progress in DNA technologies, it is reasonable to suggest that research toward the integration of DNA memory

---

[6] Research has declined as number of indexed publications recorded in the Research Activity section of the ITRS ERD chapter.

systems with semiconductor integrated circuits could provide an impetus for highly dense memory systems operating at very low power.

13.9 Assessment and Benchmarking

Key concepts, recent status, and challenges of 10 emerging memory device technologies have been presented. It is useful to compare and contrast the applicability of these technologies against the required benchmark of their prospective markets. Estimated quantitative and qualitative requirements for key markets are provided in Table 13.1. This table is populated from the 2013 ITRS ERD and PIDS chapters, as well as external sources where noted. The quantitative data provided are estimates, and can vary greatly depending on the specific implementation of the memory. For example, DRAM used in a supercomputer may have considerably higher performance requirements than that used in a mobile device, such as a tablet. The data presented is intended as a useful starting point against which emerging memory technologies can be benchmarked. It should also be noted that switching energy and read/write latency requirements are given at the system level. Numbers for emerging memories are typically given for the device or array level. Hence, system architecture and circuit-level considerations may determine switching energy and latency rather than performance of the memory itself. System versus device level energetics are discussed further for DRAM and flash memory in Chapter 3.

Table 13.1 is arranged in order of the approximate time emerging memory technologies would likely enter the market. Embedded EEPROM replacement involves the replacement of an EEPROM cell in an embedded microcontroller or field programmable gate array (FPGA). An example of this is provided in Reference [7], which describes a microcontroller product with a monolithically integrated ReRAM memory. As indicated by the requirements, the EEPROM market has the lowest barrier to entry and is a good niche starting point for emerging memories. Flash memory is currently a ~US$ 30 billion/year market, the majority of which is NAND. For this reason, displacing NAND flash will be an important near term goal of emerging memories. The most important attributes for this market are scaling to higher densities and at a lower cost than NAND flash. Storage Class Memory is a market opportunity for emerging and prototypical memories, based on the large latency gap between typical magnetic- or flash-based disk drives and the DRAM main memory. SCM would replace magnetic memory, solid state disks (SSDs), and possibly even DRAM with one or more emerging memory technologies (NAND flash is considered a low-end or early SCM technology). M-Class SCM has performance and endurance characteristics closer to those of DRAM, whereas S-Class is reminiscent of flash, with lower performance, but a high retention requirement. DRAM serves as the main working memory for a wide variety of systems, including PC and most mobile devices, such as tablets and smartphones. It has effectively unlimited endurance and a low read/write latency, but retention of less than 1 s. The final market in Table 13.1 does not currently exist, but is a straw man for a memory integrated with high performance logic; similar to that suggested by the "Nanostore" concept [6]. In the following, we assess quantitative characteristics of emerging technology, with regard to the requirements of these potential markets.

Tables 13.2 and 13.3 provide a compilation of quantitative performance parameters for emerging memory technologies, including best projected and demonstrated values of key metrics with references as given. The values are from the 2013 ITRS ERD tables and references therein [5].

Table 13.1 Approximate requirements for emerging memories categorized by available markets

	Embedded EEPROM replacement[a]	NAND flash replacement (e.g., SSD)[b]	S-type storage class memory[c]	M-type storage class memory[c]	Stand-alone DRAM (DIMM) replacement[d]	CMOS integrated DRAM/storage/main memory[e]
Time to implementation	Now	2–5 yr	2–5 yr	5–10 yr	5–10 yr	>10 yr
Quantitative requirements						
Minimum bit level endurance	10^6	10^3	10^6	10^9	10^{16}	10^{16}
Minimum bit level retention	10 yr	1 yr	10 yr	5 d	64 ms	10 yr
Maximum system level read/write latency	100 μs	100 μs	5 μs	200 ns	100 ns	10 ns
Maximum system level write energy (pJ)	10^4	100	25	100	100	1
Maximum feature size (nm)	180	12	20	20	20	10
Minimum 2D layer density (bit/cm)	10^9	10^{11}	10^{10}	10^{10}	10^9	10^{11}
Maximum cost (US$/GB)	30^f	2	4	10	10	10
Qualitative requirements						
Performance	Low	Low	Moderate	High	High	High
Reliability	High	Low/moderate	Moderate	Moderate	Moderate	High
CMOS compatibility	Required	Useful/not required	Useful/not required	Useful/not required	Useful/not required	Required
BEOL process	Required	Not required	Not required	Not required	Not required	Required
Layering capability	Not required	Required	Required	Useful/not required	Required	Required

[a] Based on common embedded microcontrollers with flash-based program/data memory.
[b] Based on modern NAND flash characteristics, considering a stand-alone module.
[c] Based on SCM info from 2013 ITRS ERD Tables.
[d] Based on modern DRAM characteristics.
[e] High performance logic CMOS integration based on estimated requirements for data-center level processor (e.g., a "nanostore" [6]). This could also be thought of as a "univeral memory" which does not require tradeoffs in performance or reliability.
[f] Based on the cost of a standalone external microcontroller memory; information on the cost per bit of flash integrated in a microcontroller is not available.

Table 13.2 Parameters for emerging memory technologies (excluding ReRAM). Numbers are from the 2013 ITRS ERD Table 4a and references therein [5]. Research activity is given in number of peer reviewed publications indexed from July 2011 through July 2013

		Emerging ferroelectric memory		Carbon memory	Mott memory	Macromolecular memory	Molecular memory
Subclass		FeFET	FE tunnel junction	NA	NA	NA	NA
Feature size F (nm)	Best projected	Same as CMOS transistor	<10	<5	5–10	5	5
	Demonstrated	28	50	22	110	100	30
Cell area	Best projected	$4F^2$	$4F^2$	$4F^2$	$4F^2$	$4F^2$	$4F^2$
	Demonstrated	$4F^2$	Not available	Not available	Not available	$4F^2$	Not available
Write/erase time	Best projected	<100 pS	<1 ns	Not available	<1 ns	<10 ns	<40 ns
	Demonstrated	20 ns, 10 ns	10 ns	10 ns	2 ns	15 ns	10 s, 0.2 s
Retention time	Best projected	10 yr	>10 y	Not available	Not available	> 1 yr	Not available
	Demonstrated	2.5×10^5 s (3 d)	3 d	168 h at 250 °C	Not available	10^5 s	2 mo
Write cycles	Best projected	$>10^{12}$	10^{14}	Not available	$>10^{16}$	Not available	$>10^{16}$
	Demonstrated	10^{12}	4×10^6	5×10^7	~100	10^5	2×10^3
Write operating voltage (V)	Best projected	Not available	1	Not available	Not available	~1	80 mV
	Demonstrated	±5	2–3	5–6	1.25/0.75	1.4	4.0~1.5
Read operating voltage (V)	Best projected	Not available	0.1 V	Not available	Not available	<0.1	0.3
	Demonstrated	0.5	0.1	1.5	0.2	0.2	0.5
Write energy per bit	Best projected	0.1 fJ	1 fJ	Not available	Not available	0.1 fJ	0.1 aJ
	Demonstrated	1 fJ	10 fJ	Not available	~1 fJ	10 fJ	Not available
Research activity		30	27	52	31	80	21

Table 13.3 Parameters for the four categories of ReRAM [5]. Numbers are from the 2013 ITRS ERD Table 4b and references therein. Research activity is given in number of peer reviewed publications indexed from July 2011 through July 2013

		Electrochemical metallization bridge	Metal oxide: bipolar filamentary	Metal oxide: unipolar filamentary	Metal oxide: bipolar nonfilamentary
Storage mechanism		Electrochemical filament formation	Valence change filament formation	Thermochemical effect filament formation	Change in tunneling characteristics near interface
Feature size F (nm)	Best projected	<5	<5	Not available	<10
	Demonstrated	20 (GeSe), 30 (CuS)	5 (AlO_x)	35	40
Cell area (2D)	Best projected	$4F^2$	$4F^2$	$4F^2$	$4F^2$
	Demonstrated	$4F^2$	$4F^2$	$4F^2$	$4F^2$
Write/erase time (ns)	Best projected	<1	<1	not available	10
	Demonstrated	<1	<1	10 (W), 5 (E)	<100
Retention time	Best projected (yr)	>10	>10	>10	>10
	Demonstrated	1000 h at 200 °C	3000 h at 150 °C	1000 h at 150 °C	4 h at 125 °C
Write cycles	Best projected	$>10^{11}$	$>10^{12}$	Not available	$>10^6$
	Demonstrated	10^{10}	10^{12}	10^6	10^6
Write operating voltage (V)	Best projected	<0.5	<1	Not available	Not available
	Demonstrated	0.6	1–3	1–3	2
Read operating voltage (V)	Best projected	<0.2	0.1	Not available	0.1
	Demonstrated	0.2	0.1–0.2	0.4	0.5
Write/erase energy (J/bit)	Best projected	Not available	0.1 fJ	Not available	Not available
	Demonstrated	1 pJ (W), 8 pJ (E)	115 fJ (W), <1 pJ (E)	Not available	1 pJ
Research activity	593 (includes all categories)				

13.9.1 Scaling

The scalability of an emerging memory technology is a key factor for it to be considered a strong candidate to replace traditional data storage technologies. Although NAND flash has been scaled to a 2D density of 16 nm, a density of roughly 9×10^{10} bits/cm^2 (or ~11 GB/cm^2), these ultra-scaled devices have suffered severely degraded retention, endurance, and yield [5]. DRAM also has "no known solutions" for scaling challenges below the 20 nm half pitch, occurring in 2017 [5]. The maximum density a memory technology can reach is defined by the minimum feature size F and cell area, in terms of F. The maximum density of a layer is plotted in Figure 13.5 for a given feature size using a cell area of 4, 6, and 8 F^2.

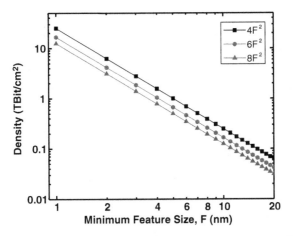

Figure 13.5 Density versus minimum feature size (F) for cell sizes of $4F^2$, $6F^2$, and $8F^2$

Scalability is one of the categories evaluated in both the qualitative and quantitative sections of the ERD chapter. Scalability below 10 nm is predicted for all emerging memories, although it has only been demonstrated for BF-MO ReRAM (see Table 13.3). This matches the qualitative opinion reflected in the expert survey that ReRAM has the highest potential scalability (Table 13.4). It is reasonable to conclude that other emerging memories that rely on filamentary switching, such as certain types of carbon and macromolecular memory, will scale to the same dimensions as standard filamentary ReRAM. The high scalability of ReRAM makes it especially attractive as a NAND flash replacement for S-type SCM technology.

In certain memory categories listed in Table 13.1, such as M-type SCM and DRAM replacement, scaling is less important than performance factors, such as endurance, write speed, and energy. For example, while it is possible that the FeFET may not scale as well as ReRAM, but the high endurance may still make it a strong DRAM replacement candidate.

13.9.2 Performance

Emerging memory performance is assessed quantitatively in Tables 13.2 and 13.3 by write/erase speed, energy, and voltage. Performance is assessed qualitatively from survey results in Table 13.4, as an assessment of speed, energy efficiency, and ON/OFF ratio. Demonstrated SET/RESET speeds are 20 ns or less for all technologies except molecular memory and MO-BN ReRAM. Where available, all technologies had demonstrated switching energies of less than 10 pJ, and in many cases less than 1 pJ. Interestingly, the qualitative assessments of speed and energy efficiency are lowest for macromolecular memory, even though there are very competitive values reported in the literature for this technology. Differences in performance records reported in the literature and the opinion conveyed by the survey of experts might reflect significant recent improvement in performance, or the anticipation of difficulty achieving record performances on production devices. As described in Chapter 3, array and architectural considerations will ultimately determine the system level speed and energy

Table 13.4 Qualitative assessment of emerging memory technologies sorted by overall score. Based on the expert survey presented in the 2013 ITRS ERD chapter [5]

	Overall	Scalability	Speed	Energy efficiency	ON/OFF "1"/"0" ratio	Operational reliability	Room temperature operation	CMOS technological compatibility	CMOS architectural compatibility
ReRAM	18.7	2.9	2.5	2.1	2.2	1.6	2.5	2.4	2.4
FeFET	17.4	2.0	2.4	2.3	2.1	1.7	2.4	2.3	2.4
FTJ	17.3	2.3	2.2	2.2	2.1	1.7	2.4	2.1	2.2
Carbon-based	17.0	2.2	2.2	2.0	2.3	1.7	2.4	2.0	2.2
Mott	16.6	2.1	2.4	2.1	2.2	1.7	1.9	2.0	2.2
Macromolecular	13.9	1.8	1.7	1.7	1.6	1.3	2.2	1.7	1.8
Molecular	13.9	2.6	1.7	2.0	1.3	1.1	2.0	1.6	1.8

requirements of each of these memories. For example, a DRAM cell only requires about 10 fJ at the cell level but at the system level requires ~50 pJ.

13.9.3 Reliability

Write cycles (endurance) and retention data are the main quantitative measures of reliability provided in Tables 13.2 and 13.3, although other factors, such as device to device variability and random telegraph noise (RTN) are also important reliability considerations in memory device reliability. Variability and RTN are somewhat difficult to assess in a quantitative comparison table, but are expected to influence the expert survey opinions of "operational reliability" provided in Table 13.4.

The minimum endurance required for an emerging memory to enter any of the application markets described in Table 13.1 is about 10^6 SET/RESET cycles. This endurance has been demonstrated in all categories of ReRAM, ferroelectric, and carbon memory (except graphene memories). If a memory technology has an endurance of less than 10^6, it is likely that workarounds such as error correction code (ECC) and garbage collection routines will be needed (as with modern NAND-flash memory). Mott, molecular, and macro-molecular memory have not yet met this requirement (although macromolecular has achieved 10^5 cycles), and hence achieving this endurance should be a key research in those technologies.

For M-type SCM and DRAM replacement applications, higher endurance is required. M-type SCM requires one billion cycles, whereas straightforward DRAM replacement would require 10^{16} SET/RESET cycles. The requirement of a billion cycles has been met by the FeFET, as well as EMB and MO-BF ReRAM. This high endurance, along with excellent performance makes these technologies potential M-type SCM and DRAM replacements. In the case of the FeFET, the low retention is not an impediment for these technologies.

The retention required for the traditional NVM applications of embedded EEPROM and NAND-flash, as well as S-type SCM is 10 years (typically retaining data at a temperature of 85 °C). Since it is difficult to make a measurement over 10 years, retention is typically projected from accelerated measurements [41], but there is not a standardized method of reporting these projections in an emerging memory device [105]. ReRAM (except for non-filamentary) are projected to meet this 10 year requirement from relatively detailed retention studies that derived from thorough arhennius studies. Ten year retention is also predicted for the FeFET and FTJ, although the longest demonstrated periods are about three days for each, and hence additional data is needed to verify this claim. For a DRAM replacement and M-type SCM memory retention is much less important (whereas endurance is key), and hence the FeFET and FTJ are still a strong contenders.

Qualitatively, all emerging memories scored low in the "operational reliability" category. This can be presumably attributed to several factors, especially variability in the SET and RESET states between devices and lack of long term reliability data. This illustrates the general fact that after basic functionality of an emerging memory technology is demonstrated, research and development efforts must prove reliability before being accepted in any of the markets listed in Table 13.1. Perhaps finding a niche, low consequence market which allows the memory product to demonstrate reliable operation of high quantities of parts is the best approach to gain widespread use of an emerging memory technology [11].

The scores for macromolecular and molecular memories were the lowest at 1.3 and 1.1, respectively. This indicates a strong need for these technologies to focus research efforts on proving that reliable operation is possible, otherwise commercial adaptation of these technologies is unlikely.

13.9.4 CMOS Compatibility and Cost

CMOS compatibility and cost are, perhaps, the most subjective requirements for emerging memory technology, although they are also often the greatest factors which determine commercial success in any market. All emerging memories discussed except for the FeFET employ back end of line (BEOL) processing, and hence their CMOS compatibility should be assessed with regard to BEOL materials and temperatures. ReRAM received the highest score on CMOS compatibility, which likely reflects that at least for certain ReRAM devices, there are no special materials required and that this technology can survive typical BEOL CMOS temperatures (which are typically a maximum of 400 °C). The lowest scores were given to macromolecular and molecular memories, and can be attributed to two problems. First, both of these technologies require special materials not typically found in fabrication, (such as polymers). Second, it is not clear that many of these polymers or molecules can survive BEOL temperatures and process conditions.

Another important factor in CMOS compatibility is the maximum voltage require in the SET or RESET operation. EMB ReRAM has the lowest demonstrated switching voltages (0.6 V has been demonstrated), which provides an advantage in the integration with advanced CMOS nodes without the need for a separate high voltage transistor.

13.9.5 Tradeoffs

One of the greatest shortcomings of Tables 13.2 and 13.3 is that they present record performance and reliability metrics for different devices of a particular technology column. For example, in the MO-BF ReRAM different devices were used to demonstrate scaling to 5×5 nm, sub-ns switching, 10^{12} cycle endurance, and 10 year retention at >100 °C. A single device with all of these characteristics has not been demonstrated. It should also be noted that these parameters come from array demonstrations when possible, but are often from single devices. More mature technologies like EMB ReRAM have metrics taken from sophisticated test chips which closely resemble future commercial products, whereas the least developed technologies generally have parameters available at the research device level. As discussed in Chapter 3, array parasitics and peripheral circuitry can significantly increase the write energy. Power considerations of passive resistive memory arrays have been modeled in reference [106]; and array level energetics of these arrays is an important area of future work.

Hence, it may not be possible to obtain a device or an array in a given emerging memory technology that has all of the characteristics of a column in Tables 13.2 and 13.3. This methodology is intended only to provide the most optimistic characteristics of emerging memory technology with a single data set.

It is therefore worthwhile to consider the tradeoffs that prevent the simultaneous achievement of every parameter for a column of Tables 13.2 or 13.3. One common example of this type of tradeoff is that between switching current (which is reflected in switching energy) and retention of a ReRAM cell. An example of this is provided in Reference [40], whereas the compliance current used during switching is the key factor which determines retention. When the compliance current of 100 μA is used to SET the device, a wider, more robust filament is formed compared to when 10 μA is used. Hence, the temperature at which 10 year retention is possible is 105 °C for a 100 μA cc but only 95 °C for a 10 μA cc. This demonstrates that it is possible to increase the retention of a ReRAM cell if one is willing to pay the price in switching current (and therefore switching power). Fundamental energy–space–time tradeoffs for ReRAM are discussed in detail in Reference [107].

13.10 Summary and Conclusions

Several emerging memory devices provide promising features to compete with dominant conventional memory technologies: floating gate (NAND/NOR flash and embedded EEPROM), and DRAM. They also may enable system improvements by serving as M- and S-type Storage Class Memory technologies. ReRAM has made especially rapid progress in the past two years, and may be nearing commercial viability for niche markets. Interest in FeFET and FTJ has been reinvigorated by the discovery of ferroelectricity in doped HfO_x, which may enable a surge of progress for emerging ferroelectric memories.

Reliability remains a key challenge for all emerging memory technologies – and should be a major focus of future research efforts. Although several emerging memory technologies have demonstrated excellent performance characteristics, they will not become commercially viable until reliable system-level operation in native environments is unquestionably established.

Acknowledgments

The authors would like to acknowledge R.J. Kaplar and P.R. Mickel (Sandia) and A. Chen (Globalfoundries), and J. Hutchby (SRC) for reviewing and making useful comments on this manuscript. One author (M.J.M.) would like to acknowledge funding from Sandia's Laboratory Directed Research and Development (LDRD) Program. Sandia National Laboratories is a multi-program laboratory managed and operated by Sandia Corporation, a wholly owned subsidiary of Lockheed Martin Corporation, for the US Department of Energy's National Nuclear Security Administration under contract DE-AC04-94AL85000.

References

1. Micron (2013) Micron Unveils 16-Nanometer Flash Memory Technology. Available: http://investors.micron .com/releasedetail.cfm?ReleaseID=777402 (accessed 16 July 2013).
2. Samsung (2013) Samsung Starts Mass Producing Industry's First 3D Vertical NAND Flash. Available: http:// www.samsung.com/global/business/semiconductor/news-events/press-releases/detail?newsId=12990 (accessed 16 July 2013).
3. Freitas, R.F. and Wilcke, W.W. (2008) Storage-class memory: The next storage system technology. *IBM Journal of Research and Development*, **52**, 439–447.
4. Burr, G.W., Kurdi, B.N., Scott, J.C. *et al.* (2008) Overview of candidate device technologies for storage-class memory. *IBM Journal of Research and Development*, **52**, 449–464.
5. (2013) The International Technology Roadmap for Semiconductors (ITRS). Available: www.itrs.net.
6. Ranganathan, P. (2011) From microprocessors to nanostores: Rethinking data-centric systems. *Computer*, **44**, 39–48.
7. Panasonic (2013) Panasonic Starts World's First Mass Production of ReRAM Mounted Microcomputers. Available: http://panasonic.co.jp/corp/news/official.data/data.dir/2013/07/en130730-2/en130730-2.html (accessed 16 July 2013).
8. Adesto (2013) Available: http://www.adestotech.com/ (accessed 16 September 2013).
9. Tz-Yi, L., Tian Hong, Y., Scheuerlein, R. *et al.* (2013) A 130.7 mm2 2-layer 32Gb ReRAM memory device in 24nm technology. ISSCC Dig. of Tech. Papers, pp. 210–211.
10. Chevallier, C.J., Chang Hua, S., Lim, S.F. *et al.* (2010) A 0.13 um 64 Mb multi-layered conductive metal-oxide memory. ISSCC Dig. of Tech. Papers, pp. 260–261.
11. Prall, K., Ramaswamy, N., Kinney, W. *et al.* (2012) An Update on Emerging Memory: Progress to 2X nm. 4th IEEE International Memory Workshop, pp. 1–5.

12. Waser, R., Dittmann, R., Staikov, G., and Szot, K. (2009) Redox-based resistive switching memories – nanoionic mechanisms, prospects, and challenges. *Advanced Materials*, **21**, 2632–2663.

13. Valov, I. and Kozicki, M.N. (2013) Cation-based resistance change memory. *Journal of Physics D-Applied Physics*, **46**, 074005.

14. Valov, I., Waser, R., Jameson, J.R., and Kozicki, M.N. (2011) Electrochemical metallization memories—fundamentals, applications, prospects. *Nanotechnology*, **22**, 254003.

15. Kozicki, M.N., Dandamudi, P., Barnaby, H.J., and Gonzalez-Velo, Y. (2013) (Invited) programmable metallization cells in memory and switching applications. *ECS Transactions*, **58**, 47–52.

16. Goux, L., Sankaran, K., Kar, G. *et al.* (2012) Field-driven ultrafast sub-ns programming in $W/Al_2O_3/Ti/CuTe$-based 1T1R CBRAM system. VLSI Technology Tech Dig., pp. 69–70.

17. Waser, R., Bruchhaus, R., and Menzel, S. (2013) Redox-based resistive switching memories, in *Nanoelectronics and Information Technology* (ed. R. Waser), Wiley-VCH, Weinheim, Germany.

18. Kozicki, M.N., Mira, P., and Mitkova, M. (2005) Nanoscale memory elements based on solid-state electrolytes. *IEEE Transactions on Nanotechnology*, **4**, 331–338.

19. Jameson, J., Blanchard, P., Cheng, C. *et al.* (2013) Conductive-Bridge Memory (CBRAM) with Excellent High-Temperature Retention, presented at the. International Electron Device Meeting (IEDM), Washington DC.

20. Kund, M., Beitel, G., Pinnow, C.U. *et al.* (2005) Conductive bridging RAM (CBRAM): an emerging non-volatile memory technology scalable to sub 20nm. IEDM Tech. Digest, pp. 754–757.

21. Jameson, J.R., Gilbert, N., Koushan, F. *et al.* (2012) Quantized conductance in $Ag/GeS2/W$ conductive-bridge memory cells. *IEEE Electron Device Letters*, **33**, 257–259.

22. Gilbert, N., Yanqing, Z., Dinh, J. *et al.* (2013) A 0.6 V 8 pJ/write non-volatile CBRAM macro embedded in a body sensor node for ultra low energy applications. 2013 VLSI Circuits, pp. C204–C205.

23. Soni, R., Meuffels, P., Petraru, A. *et al.* (2010) Probing Cu doped Ge0.3Se0.7 based resistance switching memory devices with random telegraph noise. *Journal of Applied Physics*, **107**, 024517.

24. Choi, B., Yang, J.J., Zhang, M.X. *et al.* (2012) Nitride memristors. *Applied Physics A*, **109**, 1–4.

25. Marinella, M.J., Stevens, J.E., Longoria, E.M., and Kotula, P.G. (2012) Resistive switching in aluminum nitride. Device Research Conference (DRC), 2012 70th Annual, pp. 89–90.

26. Lee, M.-J., Lee, C.B., Lee, D. *et al.* (2011) A fast, high-endurance and scalable non-volatile memory device made from asymmetric Ta_2O_{5-x}/TaO_{2-x} bilayer structures. *Nature Materials*, **10**, 625–630.

27. Yang, J.J., Zhang, M.-X., Pickett, M.D. *et al.* (2012) Engineering nonlinearity into memristors for passive crossbar applications. *Applied Physics Letters*, **100**, 113501.

28. Lohn, A.J., Mickel, P.R., and Marinella, M.J. (2013) Dynamics of percolative breakdown mechanism in tantalum oxide resistive switching. *Applied Physics Letters*, **103**, 173503.

29. Chung-Wei, H., Wang, I.T., Chun-Li, L. *et al.* (2013) Self-rectifying bipolar TaO_xTiO_2 RRAM with superior endurance over 10^{12} cycles for 3D high-density storage-class memory. VLSI Technology Tech. Dig., pp. T166–T167.

30. Lohn, A.J., Stevens, J.E., Mickel, P.R. *et al.* (2013) A CMOS Compatible, Forming Free TaOx Reram, presented at the. Electrochemical Society Meeting, San Francisco.

31. Larentis, S., Nardi, F., Balatti, S. *et al.* (2012) Resistive switching by voltage-driven ion migration in bipolar RRAM—Part II: Modeling. *IEEE Transactions on Electron Devices*, **59**, 2468–2475.

32. Doo Seok, J., Reji, T., Katiyar, R.S. *et al.* (2012) Emerging memories: Resistive switching mechanisms and current status. *Reports on Progress in Physics*, **75**, 076502.

33. Wong, H.S.P., Heng-Yuan, L., Shimeng, Y. *et al.* (2012) Metal oxide RRAM. *Proceedings of the IEEE*, **100**, 1951–1970.

34. Young-Bae, K., Seung Ryul, L., Dongsoo, L. *et al.* (2011) Bi-layered RRAM with unlimited endurance and extremely uniform switching. VLSI Technology Tech. Dig., pp. 52–53.

35. Torrezan, A.C., Strachan, J.P., Medeiros-Ribeiro, G., and Williams, R.S. (2011) Sub-nanosecond switching of a tantalum oxide memristor. *Nanotechnology*, **22**, 485203.

36. Strachan, J.P., Torrezan, A.C., Medeiros-Ribeiro, G., and Williams, R.S. (2011) Measuring the switching dynamics and energy efficiency of tantalum oxide memristors. *Nanotechnology*, **22**, 505402.

37. Govoreanu, B., Kar, G.S., Chen, Y. *et al.* (2011) 10 × 10 nm2 Hf/HfOx crossbar resistive RAM with excellent performance, reliability and low-energy operation. IEDM Tech. Dig., pp. 31.6.1–31.6.4.

38. Zhiping, Z., Yi, W., Wong, H.S.P., and Wong, S.S. (2013) Nanometer-scale HfOx RRAM. *Electron Device Letters*, **34**, 1005–1007.

39. Tsai, C.-L., Xiong, F., Pop, E., and Shim, M. (2013) Resistive random access memory enabled by carbon nanotube crossbar electrodes. *ACS Nano*, **7**, 5360–5366.

40. Yang Yin, C., Degraeve, R., Clima, S. *et al.* (2012) Understanding of the endurance failure in scaled HfO$_2$-based 1T1R RRAM through vacancy mobility degradation. IEDM Tech Dig, pp. 20.3.1–20.3.4.
41. Wei, Z., Takagi, T., Kanzawa, Y. *et al.* (2011) Demonstration of high-density ReRAM ensuring 10-year retention at 85C based on a newly developed reliability model. IEDM, pp. 31.4.1–31.4.4.
42. Kawahara, A., Azuma, R., Ikeda, Y. *et al.* (2012) An 8 Mb multi-layered cross-point ReRAM macro with 443MB/s write throughput. ISSCC Tech Dig, pp. 432–434.
43. Baek, I.G., Lee, M.S., Seo, S. *et al.* (2004) Highly scalable nonvolatile resistive memory using simple binary oxide driven by asymmetric unipolar voltage pulses. 2004 IEDM Technical Digest, pp. 587–590.
44. Ielmini, D., Bruchhaus, R., and Waser, R. (2011) Thermochemical resistive switching: Materials, mechanisms, and scaling projections. *Phase Transitions*, **84**, 570–602.
45. Tseng, Y.-H., Chia-En, H., Kuo, C.H. *et al.* (2009) High density and ultra small cell size of Contact ReRAM (CR-RAM) in 90nm CMOS logic technology and circuits. IEDM Tech Dig, pp. 1–4.
46. Shen, W.C., Mei, C.Y., Chih, Y.D. *et al.* (2012) High-K metal gate contact RRAM (CRRAM) in pure 28nm CMOS logic process. IEDM Tech. Dig., pp. 31.6.1–31.6.4.
47. Chang, M.-F., Wu, C.-W., Kuo, C.-C. *et al.* (2012) A 0.5V 4 Mb logic-process compatible embedded resistive RAM (ReRAM) in 65 nm CMOS using low-voltage current-mode sensing scheme with 45 ns random read time. ISSCC Tech Dig, pp. 434–436.
48. Meyer, R., Schloss, L., Brewer, J. *et al.* (2008) Oxide dual-layer memory element for scalable non-volatile cross-point memory technology. NVMTS, pp. 1–5.
49. Govoreanu, B., Redolfi, A., Zhang, L. *et al.* (2013) Vacancy-modulated conductive oxide resistive RAM (VMCO-RRAM): An area-scalable switching current, self-compliant, highly nonlinear and wide on/off-window resistive switching cell. IEDM Tech. Dig., pp. 10.2.1–10.2.4.
50. Buck, D.A. (1952) Ferroelectrics for Digital Information and Switching.
51. Eaton, S.S., Butler, D.B., Parris, M. *et al.* (1988) A Ferroelectric Nonvolatile Memory. Solid-State Circuits Conference, 1988. Digest of Technical Papers. ISSCC. 1988 IEEE International, pp. 130.
52. Boscke, T.S., Muller, J., Brauhaus, D. *et al.* (2011) Ferroelectricity in hafnium oxide thin films. *Applied Physics Letters*, **99**, 102903–3.
53. Boscke, T.S., Muller, J., Brauhaus, D. *et al.* (2011) Ferroelectricity in hafnium oxide: CMOS compatible ferroelectric field effect transistors. IEEE Tech. Dig., pp. 24.5.1–24.5.4.
54. Hoffman, J., Pan, X., Reiner, J.W. *et al.* (2010) Ferroelectric field effect transistors for memory applications. *Advanced Materials*, **22**, 2957–2961.
55. Wu, S.Y. (1974) A new ferroelectric memory device, metal-ferroelectric-semiconductor transistor. *IEEE Transactions on Electron Devices*, **21**, 499–504.
56. Miller, S.L. and McWhorter, P.J. (1992) Physics of the ferroelectric nonvolatile memory field effect transistor. *Journal of Applied Physics*, **72**, 5999–6010.
57. Ma, T.P. and Han, J.-P. (2002) Why is nonvolatile ferroelectric memory field-effect transistor still elusive? *Electron Device Letters, IEEE*, **23**, 386–388.
58. Sakai, S. and Ilangovan, R. (2004) Metal-ferroelectric-insulator-semiconductor memory FET with long retention and high endurance. *IEEE Electron Device Letters*, **25**, 369–371.
59. Hai, L.V., Takahashi, M., and Sakai, S. (2011) Downsizing of Ferroelectric-Gate Field-Effect-Transistors for Ferroelectric-NAND Flash Memory Cells. IEEE International Memory Workshop, pp. 1–4.
60. Zhang, X., Takahashi, M., Takeuchi, K., and Sakai, S. (2012) 64 kbit ferroelectric-gate-transistor-integrated NAND flash memory with 7.5V program and long data retention. *Japanese Journal of Applied Physics*, **51**, 04DD01.
61. Muller, J., Yurchuk, E., Schlosser, T. *et al.* (2012) Ferroelectricity in HfO$_2$ enables nonvolatile data storage in 28nm HKMG. VLSI Technology Tech. Dig., pp. 25–26.
62. Tsymbal, E.Y. and Kohlstedt, H. (2006) Tunneling across a ferroelectric. *Science*, **313**, 181–183.
63. Chanthbouala, A., Crassous, A., Garcia, V. *et al.* (2012) Solid-state memories based on ferroelectric tunnel junctions. *Nature Nanotechnology*, **7**, 101–104.
64. Gruverman, A., Wu, D., Lu, H. *et al.* (2009) Tunneling electroresistance effect in ferroelectric tunnel junctions at the nanoscale. *Nano Letters*, **9**, 3539–3543.
65. Wen, Z., Li, C., Wu, D. *et al.* (2013) Ferroelectric-field-effect-enhanced electroresistance in metal/ferroelectric/semiconductor tunnel junctions. *Nature Materials*, **12**, 617–621.
66. Ionescu, A.M. (2012) Nanoelectronics: Ferroelectric devices show potential. *Nature Nanotechnology*, **7**, 83–85.
67. Boyn, S., Girod, S., Garcia, V., Fusil, S., Xavier, S., Deranlot, C., Yamada, H., Carretero, C., Jacquet, E., Bibes, M., Barthelemy, A., and Grollier, J. (2013) High-performance ferroelectric memory based on fully patterned tunnel junctions. *Applied Physics Letters*, **104**, 052909.

68. McWilliams, C.R., Celinska, J., Paz de Araujo, C.A., and Xue, K.-H. (2011) Device characterization of correlated electron random access memories. *Journal of Applied Physics*, **109**, 091608.

69. Ashcroft, N.W. and Mermin, N.D. (1976) *Solid State Physics*, vol. 1976, Rinehart and Winston, New York.

70. Janninck, R.F. and Whitmore, D.H. (1966) Electrical conductivity and thermoelectric power of niobium dioxide. *Journal of Physics and Chemistry of Solids*, **27**, 1183–1187.

71. Xue, K.-H., Paz de Araujo, C.A., Celinska, J., and McWilliams, C. (2011) A non-filamentary model for unipolar switching transition metal oxide resistance random access memories. *Journal of Applied Physics*, **109**, 091602.

72. Rozenberg, M.J., Inoue, I.H., and Sanchez, M.J. (2006) Strong electron correlation effects in nonvolatile electronic memory devices. *Applied Physics Letters*, **88**, 033510–3.

73. Pickett, M.D. and Williams, R.S. (2012) Sub-100 fJ and sub-nanosecond thermally driven threshold switching in niobium oxide crosspoint nanodevices. *Nanotechnology*, **23**, 215202.

74. Ling, Q.-D., Liaw, D.-J., Zhu, C. *et al.* (2008) Polymer electronic memories: Materials, devices and mechanisms. *Progress in Polymer Science*, **33**, 917–978.

75. Lee, T. and Chen, Y. (2012) Organic resistive nonvolatile memory materials. *MRS Bulletin*, **37**, 144–149.

76. Cho, B., Yun, J.-M., Song, S. *et al.* (2011) Direct observation of Ag filamentary paths in organic resistive memory devices. *Advanced Functional Materials*, **21**, 3976–3981.

77. Ouyang, J., Chu, C.-W., Szmanda, C.R. *et al.* (2004) Programmable polymer thin film and non-volatile memory device. *Nature Materials*, **3**, 918–922.

78. Liu, S.-H., Wen-Luh, Y., Chi-Chang, W. *et al.* (2013) High-performance polyimide-based ReRAM for nonvolatile memory application. *IEEE Electron Device Letters*, **34**, 123–125.

79. Wenliang, B., Ru, H., Yimao, C. *et al.* (2013) Record low-power organic RRAM with sub-20-nA reset current. *Electron Device Letters, IEEE*, **34**, 223–225.

80. Kuang, Y., Ru, H., Yu, T. *et al.* (2010) Flexible single-component-polymer resistive memory for ultrafast and highly compatible nonvolatile memory applications. *IEEE Electron Device Letters*, **31**, 758–760.

81. Wang, Z., Zeng, F., Yang, J. *et al.* (2012) Resistive switching induced by metallic filaments formation through poly(3,4-ethylene-dioxythiophene):poly(styrenesulfonate). *ACS Applied Materials & Interfaces*, **4**, 447–453.

82. Kim, J.J., Cho, B., Kim, K.S. *et al.* (2011) Electrical characterization of unipolar organic resistive memory devices scaled down by a direct metal-transfer method. *Advanced Materials*, **23**, 2104–2107.

83. Song, S., Cho, B., Kim, T.-W. *et al.* (2010) Three-dimensional integration of organic resistive memory devices. *Advanced Materials*, **22**, 5048–5052.

84. Dong Ick, S., Jae Ho, S., Dong Hee, P. *et al.* (2011) Polymer–ultrathin graphite sheet–polymer composite structured flexible nonvolatile bistable organic memory devices. *Nanotechnology*, **22**, 295203.

85. Kreupl, F., Bruchhaus, R., Majewski, P. *et al.* (2008) Carbon-based resistive memory. Electron Devices Meeting, 2008. IEDM 2008. IEEE International, pp. 1–4.

86. Xiong, F., Liao, A.D., Estrada, D., and Pop, E. (2011) Low-power switching of phase-change materials with carbon nanotube electrodes. *Science*, **332**, 568–570.

87. Yang, C., Yi, W., Takei, K. *et al.* (2011) Nanoscale bipolar and complementary resistive switching memory based on amorphous carbon. *IEEE Transactions on Electron Devices*, **58**, 3933–3939.

88. Di, F., Dan, X., Tingting, F. *et al.* (2011) Unipolar resistive switching properties of diamondlike carbon-based RRAM devices. *IEEE Electron Device Letters*, **32**, 803–805.

89. Zhuge, F., Dai, W., He, C.L. *et al.* (2010) Nonvolatile resistive switching memory based on amorphous carbon. *Applied Physics Letters*, **96**, 163505.

90. Li, Y., Sinitskii, A., and Tour, J.M. (2008) Electronic two-terminal bistable graphitic memories. *Nature Materials*, **7**, 966–971.

91. Sinitskii, A. and Tour, J.M. (2009) Lithographic graphitic memories. *ACS Nano*, **3**, 2760–2766.

92. Shin, Y.J., Kwon, J.H., Kalon, G. *et al.* (2010) Ambipolar bistable switching effect of graphene. *Applied Physics Letters*, **97**, 262105.

93. Hong, A.J., Song, E.B., Yu, H.S. *et al.* (2011) Graphene flash memory. *ACS Nano*, **5**, 7812–7817.

94. Rueckes, T., Kim, K., Joselevich, E. *et al.* (2000) Carbon nanotube-based nonvolatile random access memory for molecular computing. *Science*, **289**, 94–97.

95. Kianian, S., Rosendale, G., Manning, M. *et al.* (2010) A 3D stackable Carbon Nanotube-based nonvolatile memory (NRAM). Proc. ESSDERC, pp. 404–407.

96. Reed, M.A., Chen, J., Rawlett, A.M. *et al.* (2001) Molecular random access memory cell. *Applied Physics Letters*, **78**, 3735–3737.

97. Luo, Y., Collier, C.P., Jeppesen, J.O. *et al.* (2002) Two-Dimensional Molecular Electronics Circuits. *Chem-PhysChem*, **3**, 519–525.

98. Chen, Y., Ohlberg, D.A.A., Li, X. *et al.* (2003) Nanoscale molecular-switch devices fabricated by imprint lithography. *Applied Physics Letters*, **82**, 1610–1612.
99. Green, J.E., Wook Choi, J., Boukai, A. *et al.* (2007) A 160-kilobit molecular electronic memory patterned at 1011 bits per square centimetre. *Nature*, **445**, 414–417.
100. Church, G.M., Gao, Y., and Kosuri, S. (2012) Next-generation digital information storage in DNA. *Science*, **337**, 1628–1628.
101. Goldman, N., Bertone, P., Chen, S. *et al.* (2013) Towards practical, high-capacity, low-maintenance information storage in synthesized DNA. *Nature*, **494**, 77–80.
102. Hilbert, M. and López, P. (2011) The world's technological capacity to store, communicate, and compute information. *Science*, **332**, 60–65.
103. Zhirnov, V.V. and Cavin, R.K. (2013) Future microsystems for information processing: Limits and lessons from the living systems. *IEEE Journal of the Electron Devices Society*, **1**, 29–47.
104. Luan, B., Stolovitzky, G., and Martyna, G. (2012) Slowing and controlling the translocation of DNA in a solid-state nanopore. *Nanoscale*, **4**, 1068–1077.
105. Chevallier, C.J. and Marinella, M.J. unpublished communication.
106. Chen, A. (2013) A comprehensive crossbar array model with solutions for line resistance and nonlinear device characteristics. *IEEE Transactions on Electron Devices*, **60**, 1318–1326.
107. Zhirnov, V.V., Cavin, R.K., Menzel, S. *et al.* (2010) Memory devices: Energy-space-time tradeoffs. *Proceedings of the IEEE*, **98**, 2185–2200.

Part Three

Nanoelectronic Logic and Information Processing

14

Re-Invention of FET

Toshiro Hiramoto
Institute of Industrial Science, The University of Tokyo, Japan

14.1 Introduction

The silicon metal oxide semiconductor field effect transistor (MOSFET) for very large scale integration (VLSI) has been scaled down for more than 40 years in order to attain higher speed, lower power, higher integration, and lower cost. The gate length has reached as small as 20 nm and more than a billion transistors are now integrated in a single chip. The silicon devices are certainly in both the nanometer regime and the giga-scale integration regime. It is predicted in the 2011 version of the International Technology Roadmap for Semiconductors (ITRS) [1] that FET will continue to miniaturize and its gate length will become less than 10 nm in 2022 in production. The semiconductor technologies will continue to be the basis of the contemporary information-oriented society.

However, there are a lot of technical barriers to realize such small and giga-scale devices with higher performance and lower power dissipation. It is now well recognized that simple scaling of conventional bulk MOSFETs will fail in the nanometer regime. Therefore, non-conventional MOSFETs with new transistor structures and/or new channel materials have been under development and even re-invention of new types of FETs has been pursued at research level.

In this chapter, the historical trend of MOSFETs in the past and the future trend predicted by ITRS are compared first in order to look into the technological barriers that should be targeted. Next, the present status of the development of nonconventional devices with new structures and new materials are briefly mentioned as near-term solutions. Then, completely new types of transistors that have been re-invented recently are described in detail.

14.2 Historical and Future Trend of MOSFETs

Figure 14.1 shows the historical and future trends of the speed and energy dissipation of MOSFETs. In this figure, CV/I is used as a figure of merit for evaluating the circuit speed that is a reciprocal of the circuit delay, where C is the gate capacitance (other load capacitances are

Emerging Nanoelectronic Devices, First Edition. An Chen, James Hutchby, Victor Zhirnov and George Bourianoff.
© 2015 John Wiley & Sons, Ltd. Published 2015 by John Wiley & Sons, Ltd.

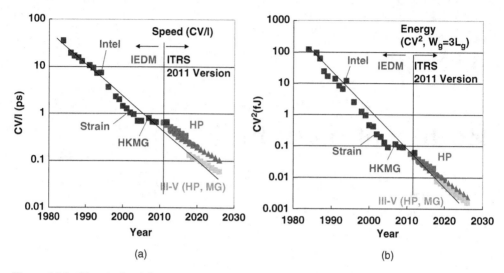

Figure 14.1 Historical and future trends of (a) the speed and (b) the energy dissipation of MOSFETs. The data up to 2011 were taken from typical device parameters presented in IEDM and the data after 2011 are based on ITRS prediction for HP and III-V MG devices

ignored for simplicity), V is the supply voltage, and I is the on-current of MOSFETs. Similarly, CV^2 is used as a figure of merit for evaluating energy dissipation. It is assumed that $W_g = 3L_g$, where W_g is the gate width and L_g is the gate length. Please note that CV^2 represents active energy dissipation only and does not include static energy dissipation. Data up to 2011 were taken from the typical device parameters presented at the International Electron Devices Meeting (IEDM), and data after 2011 are based on ITRS prediction for high performance (HP) devices and III-V multigate (MG) devices [1].

It is clearly seen that both circuit delay and energy decreased rapidly until the mid2000s. However, advancement has started saturating since then. In the ITRS prediction, advancement will continue, but the trends of delay and energy advancement will slow down. Speed and energy will be much better when MG devices are realized with III-V channel materials, which have higher mobility than silicon. Therefore, the development of high mobility channel transistors is certainly necessary.

In order to look into these trends in more detail, the historical and future trends of supply voltage (V_{dd}) and on-current (I_{on}) are summarized in Figure 14.2, and L_g and equivalent oxide thickness (EOT) are shown in Figure 14.3.

It is found that V_{dd}, I_{on}, L_g, and EOT show similar behavior: rapid advancement until mid 2000s but saturating since then, indicating that device technology has faced severe technical barriers. Figures 14.2 and 14.3 indicate the years when strain technology [2] and high k metal gate (HKMG) [3] were first introduced. These technologies certainly contributed to the device advancement (especially, HKMG contributed largely to the EOT reduction), but it is surprising to know that the advancements by these technologies were only temporary. In the prediction by ITRS, III-V devices will have much higher Ion while keeping EOT relatively thicker than silicon HP transistors, contributing largely to higher speed and lower energy.

The off current (I_{off}) is plotted in Figure 14.4a. Ioff exponentially increased until the mid2000s and completely stopped increasing since then. This means that I_{off} had reached the

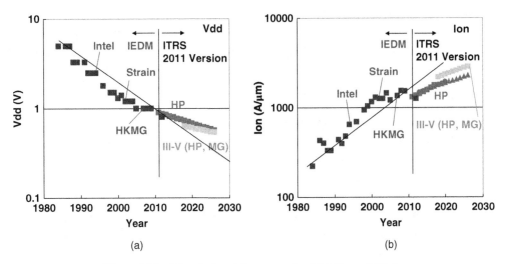

Figure 14.2   Historical and future trends of (a) V_{dd} and (b) I_{on}

highest limit for HP transistors, indicating that threshold voltage (V_{th}) has not been reduced since then. Apparently, one of the reasons for the saturation of transistor speed, energy, V_{dd}, and I_{on} after the mid2000s in Figures 14.1 and 14.2 is that V_{th} had to remain almost constant and overdrive voltage ($V_{dd} - V_{th}$) rapidly decreased as V_{dd} decreased. Therefore, new device technology that has high overdrive voltage at a constant I_{off} is strongly required.

On the other hand, the M1 pitch is plotted in Figure 14.4b. The M1 pitch has been continuously reduced unlike L_g and will be reduced in a similar rate in the prediction. The M1 pitch directly corresponds to the transistor integration level, which will not lose pace because the M1 pitch is purely determined by the lithography and is almost independent of the technological barriers for transistor speed and energy.

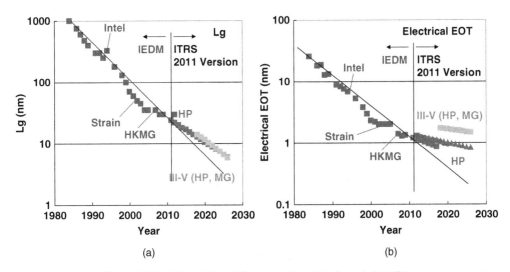

Figure 14.3   Historical and future trends of (a) L_g and (b) EOT

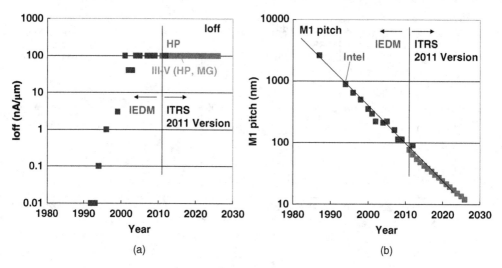

Figure 14.4 Historical and future trends of (a) I_{off} and (b) M1 pitch

14.3 Near-term Solutions

Considering the historical trend and future trend in Figures 14.1–14.4, a schematic of device characteristics that should be targeted is illustrated in Figure 14.5.

14.3.1 High Mobility Channel Transistor

First, in order to attain higher I_{on} at constant I_{off} at lower V_{dd}, the mobility improvement of channel material is certainly necessary. The characteristics of a high mobility channel transistor are schematically shown in Figure 14.5a, comparing with a conventional silicon channel bulk

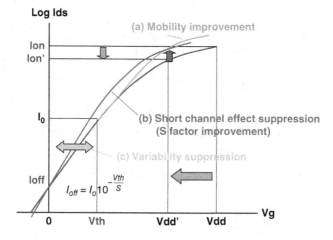

Figure 14.5 Schematic of device characteristics that should be targeted

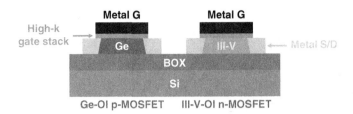

Figure 14.6 Schematic of the integration of III-V nFETs and Ge pFETs in a chip [4]

transistor. Thanks to high mobility, a higher I_{on} can be obtained with a given overdrive voltage, or the same I_{on} can be obtained with reduced overdrive voltage, than with a conventional device.

Generally, III-V compound semiconductors, including GaAs, InP, InAs, InSb, and their ternary or quaternary alloy semiconductors, have much higher electron mobility than silicon, and they are promising materials for n-type FETs. On the other hand, Ge and SiGe have higher hole mobility than silicon and they are attractive for p-type FETs. High mobility transistors using III-V or Ge have been widely studied, and the integration of III-V nFETs and Ge pFETs in a chip has been also achieved at research level [4], as shown in Figure 14.6. Moreover, a Ge nFET, which has higher electron mobility than silicon, and a III-V pFET, whose hole mobility is enhanced by strain, are also being studied.

14.3.2 Multi-gate Transistor

Second, in order to attain lower V_{th} at constant I_{off}, the steep subthreshold slope is necessary. The subthreshold properties are generally characterized by the subthreshold gate voltage swing to achieve 10× drain current change (Subthreshold swing, SS). It is well known that the minimum SS is kT/q as long as the operation principle is based on conventional FET, where k is the Boltzmann constant, T the temperature, and q the elementary charge. The minimum SS at room temperature is 60 mV/dec. In conventional bulk transistors, the SS is degraded due to the short channel effect and the typical value of SS is approximately 100 mV/dec. The ideal subthreshold characteristics with SS = 60 mV/dec are schematically shown in Figure 14.5b, compared with conventional single-gate bulk transistors. A higher I_{on} can be obtained with a given I_{off} and V_{dd}, or the same I_{on} can be obtained with a reduced V_{dd}, than with conventional devices.

The subthreshold characteristics close to the ideal SS can be obtained by better electrostatic control by gate electrode, which can be achieved by changing the transistor structure. Figure 14.7 shows the evolution of transistor structure from single-gate structure to three-dimensional (3D) multi-gate structures. The FinFET is one of the double-gate structures. In a tri-gate transistor, three surfaces of the channel are surrounded by the gate electrodes. The ideal structure in terms of the electrostatic control by gate is the gate all around (GAA) nanowire structure [5].

The tri-gate transistor has been in production, where the short channel effect is well suppressed and SS of approximately 70 mV/dec has been attained in transistors with L_g as small as 30 nm [6], as shown in Figure 14.8. The data for a tri-gate transistor in the year 2011 are also plotted in Figures 14.1–14.4. Thanks to better SS, V_{dd} can be reduced to 0.8 V, leading to an improvement of the energy, as shown in Figure 14.1b.

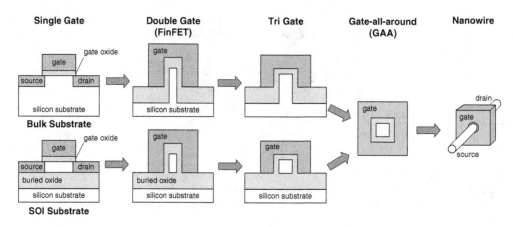

Figure 14.7 Evolution of transistor structure from single-gate to 3D MG structures

14.3.3 Intrinsic Channel Transistor

Third, in order to suppress the characteristic variability, intrinsic channel transistors should be realized. One of the main obstacles that prevent further device scaling and V_{dd} lowering is the random variability of transistor characteristics, which is not represented in Figures 14.1–14.4. It is known that the main origin of random variability in conventional bulk transistors is random dopant fluctuation (RDF) [7]. Since the number and the position of dopant atoms in the transistor channel are randomly distributed, V_{th} of each transistor also varies randomly. It is experimentally confirmed that V_{th} of bulk transistor follows a normal distribution up to $\pm 5\sigma$ [8].

The only way to suppress the variability due to RDF is to eliminate dopant atoms from the transistor channel, which is impossible in conventional bulk planar transistors because a heavily doped channel is necessary to suppress the short channel effect. The intrinsic channel or very lightly doped channel, whose channel is fully depleted (FD), can be realized using a

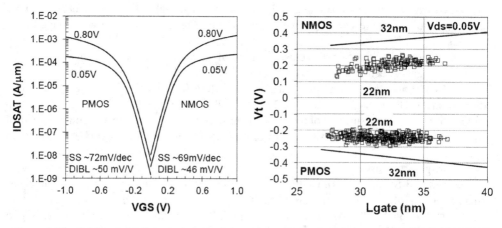

Figure 14.8 Subthreshold characteristics and short channel behavior of tri-gate FETs [6]. © 2012 IEEE

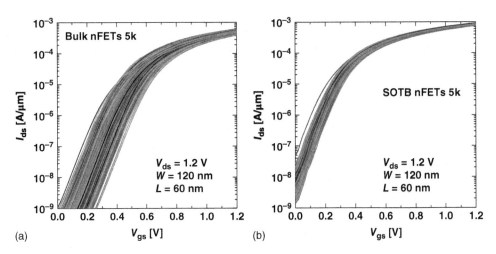

Figure 14.9   Characteristics of (a) 1 k bulk and (b) 1 k intrinsic channel FD SOI transistors [9]

silicon on insulator (SOI) transistor or a 3D multi-gate transistor, where the short channel effect is well suppressed not by a heavily doped channel but by better gate electrostatic control.

Figure 14.9 compares measured characteristics variability in 1 k bulk transistors and 1 k intrinsic channel FD SOI transistors fabricated using 65 nm technology [9]. Compared to bulk transistors, the intrinsic channel FD SOI transistors have much smaller random Vth variability. On-current (I_{on}) variability at $V_{gs} = V_{ds} = 1.2\,V$ is also suppressed in the intrinsic channel transistors. The intrinsic channel is a very effective way to suppress random variability.

However, intrinsic channel transistors still have random variability, which may be caused by other origins than RDF, including random variations of gate work function. As the transistor size shrinks and the integration level increases, the variability problem becomes more severe. Any possible origins of variability should be strictly suppressed in nanometer and giga-scale transistors.

The variability problem will also arise in any of the new types of devices that will be introduced in the following sections. The origins of variability in devices with new operation principles may be different from conventional FETs. The variability problem should be primarily considered in the exploration of new types of devices for VLSI applications.

14.4   Long-term Solutions

14.4.1   Energy Efficiency

Before discussing new types of devices as long-term solutions, the energy efficiency of CMOS logic circuits are reconsidered in this subsection. Figure 14.10 shows calculated delay, power, and energy of a CMOS logic circuit [10] as a function of V_{dd}, where parameters of the 65 nm technology generation are assumed, both active and static power dissipations are considered, and the energy is defined as the product of power and delay. As V_{dd} is reduced, the delay increases, while the power dissipation dramatically decreases. Here, we should pay attention to the energy, which has the optimum (minimum) value (E_{opt}) at a certain V_{dd}. The energy decreases as V_{dd} decreases due to the reduction of active energy, while the energy increases as V_{dd} is further reduced due to the increase in the static energy. V_{opt} is defined as V_{dd} at which the

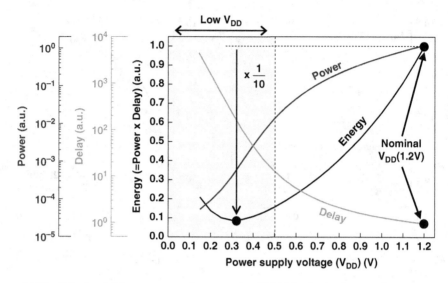

Figure 14.10   Calculated delay, power, and energy of a CMOS logic circuit assuming 65 nm CMOS process as a function of V_{dd} [10]

energy becomes optimum. Since the operation at V_{opt} can achieve the most energy efficiency, V_{opt} is the target supply voltage for ultra-low power and energy efficient applications [10].

Figure 14.11 shows energy versus V_{dd} assuming various subthreshold swings (SS). In the calculation, I_{off} at $V_{ds} = V_{dd}$ is fixed. It should be noted that the energy dissipation strongly depends on SS, and both E_{opt} and V_{opt} drastically decreases as SS decreases. The ideal ultra-low energy circuits cannot be achieved by conventional CMOS devices whose SS is more than 60 mV/dec. This is a strong motivation for the invention of new types of FETs that have steeper subthreshold swing less than 60 mV/dec.

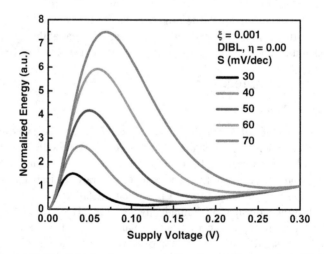

Figure 14.11   Calculated energy as a function of V_{dd} at various SS [11]

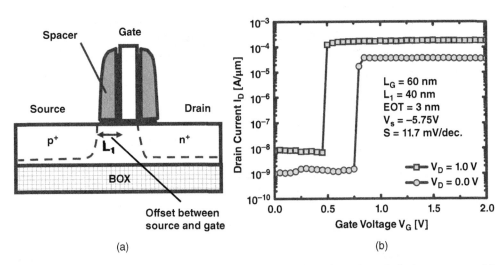

Figure 14.12 Impact-ionization MOS (I-MOS). (a) A schematic n-channel device structure. (b) Experimental result of silicon p-channel I-MOS showing very steep subthreshold swing [12]

14.4.2 Impact-ionization MOS

One of the early attempts to pursue steep subthreshold device was impact-ionization MOS (I-MOS) [12]. The I-MOS is in the form of a gated p-i-n structure and uses a gain mechanism caused by the modulation of impact-ionization related breakdown voltage. Since the impact-ionization is an abrupt function of the electric field, the device has a subthreshold slope much lower than kT/q.

Figure 14.12a shows a schematic n-channel device structure [12]. The main differences between I-MOS and a conventional n-channel MOSFET are an n-type source and an offset between source and gate in I-MOS. Figure 14.12b shows experimental I-V characteristics of silicon-based p-channel I-MOS [12]. Extremely steep subthreshold swing ($\sim$10 mV/dec) is achieved. One of the drawbacks of I-MOS is that the device inherently requires high drain voltage to give rise to the breakdown. The channel material should have a narrow bandgap to realize low voltage operation. Another concern of I-MOS is the delay associated with positive feedback that is intrinsic to impact ionization phenomena.

14.4.3 Tunnel FET

One of the most promising and most actively researched steep SS devices is the tunnel FET (TFET). TFETs make use of tunneling current instead of diffusion to avoid the limitation by kT/q in the subthreshold region. Figure 14.13 shows a schematic device structure and band diagram of an n-channel TFET that consists of p+ source, n+ drain, and gate [13]. When the gate voltage is 0 V, the channel is fully depleted and the device is in the OFF state (Figure 14.13b). When the positive gate voltage is applied, the device turns on by the Zener tunneling (Figure 14.13c). Since the Fermi tail is cutoff by the bandgap the subthreshold swing is not limited to kT/q and the off-current can be significantly below that of the conventional MOSFET [13]. Detailed operation principles of TFETs are found in review papers [13,14].

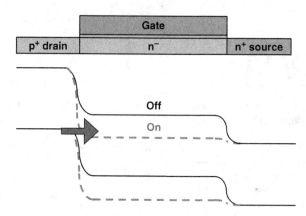

Figure 14.13 Schematic structure and band diagram of a n-channel tunnel FET showing the operation principle [13]. (a) A simple pn junction without the gate. (b) A tunnel FET with gate in OFF state. (c) A tunnel FET with gate in ON state

TFETs have been fabricated and a sub-60 mV/dec subthreshold slope has been observed in various channel materials including Si, Ge, and III-V semiconductors. Figure 14.14a shows an example of a silicon TFET [15]. The minimum SS of 36 mV/dec is obtained, although this small SS is achieved in only narrow Ids range (approximately one order) and at very small I_{ds} ($I_{ds} \sim 1$ pA/μm). The average SS in three orders of magnitude of I_{ds} is 81 mV/dec, that is larger than the ideal SS in conventional MOSFETs.

Figure 14.14b shows another example of a TFET composed of heterojunction InGaAs, compared with a thin body InGaAs MOSFET at matched $I_{off} = 200$ pA, $L_g = 150$ nm, EOT = 1.15 nm, and $V_{ds} = 0.3$ V [16]. SS of the InGaAs TFET is smaller than 60 mV/dec in the range from 3×10^{-10} to 3×10^{-9} A. Due to the steeper SS, the InGaAs TFET shows gain over the InGaAs MOSFET at low overdrive, but Ion is much smaller at $V_{gs} - V_{th} \sim 0.6$ V. The smaller I_{on} is a significant disadvantage of TFETs.

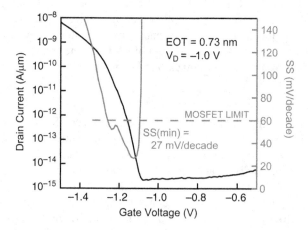

Figure 14.14 Examples of I-V characteristics of TFETs. (a) S silicon TFET [15]. SS of 36 mV/dec is observed only in very narrow I_{ds} range. (b) A heterojunction InGaAs TFET compared with a thin body InGaAs MOSFET [16]

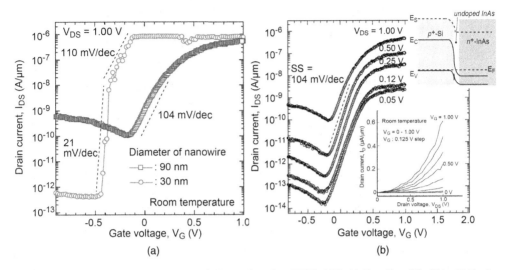

Figure 14.15 I-V characteristics of InAs/Si heterojunction TFETs [17]. (a) Excellent SS of 21 mV/dec is achieved. (b) Large V_{ds} dependence of V_{th} is observed

Figure 14.15a shows an example of sub-60 mV/dec characteristics of an InAs/Si heterojunction p-channel TFET [17]. Excellent SS of 21 mV/dec is achieved in a wide I_d range of four orders of magnitude. The on/off ratio reaches approximately 10^6. However, this InAs/Si heterojunction p-channel TFET has another disadvantage. Figure 14.15b shows another example of an InAs/Si heterojunction p-channel TFET showing I-V characteristics with various drain voltage (V_{ds}) dependence [17]. It is found that V_{th} decreases as V_{ds} increases. In conventional MOSFETs, this phenomenon of V_{ds} dependence of V_{th} is known as drain induced barrier lowering (DIBL) and characterized by $\eta = |\Delta V_{th}/\Delta V_{ds}|$. Although the mechanism may be different in TFETs, some TFETs show very high η, especially heterojunction TFETs. η reaches as large as 0.74 in Figure 14.15b [17].

In order to investigate the effect of η in steep subthreshold devices, the energy dissipation is calculated at fixed SS of 30 mV/dec at various η, as shown in Figure 14.16 [11]. I_{off} at $V_{ds} = V_{dd}$ is fixed in the calculation, and hence V_{th} increases as V_{ds} decreases in high η devices. As η increases, the energy as well as V_{opt} and E_{opt} increases. This is because delay increases due to increases V_{th} at low V_{dd}. This result indicated that the advantage of steep SS is offset by large η. Therefore, η (V_{ds} dependence of V_{th}) should be suppressed in steep subthreshold devices.

14.4.4 Negative Capacitance FET

SS is given by [18]

$$\text{SS} = \frac{\partial V_g}{\partial(\log I_d)} = \frac{\partial V_g}{\partial \Psi_s}\frac{\partial \Psi_s}{\partial\left(\log_{10} I_d\right)} = \left(1 + \frac{C_s}{C_{ins}}\right)\frac{kT}{q}\ln 10 \qquad (14.1)$$

where V_g is the gate voltage, Ψ_s is the surface potential, C_s is the semiconductor capacitance, C_{ins} is the gate insulator capacitance, and I_d is the drain current. The term $\partial V_g/\partial \Psi_s$ is the body factor and is often called the m-factor, while the term $\partial \Psi_s/\partial\left(\log_{10} I_d\right)$ is called the n-factor. It

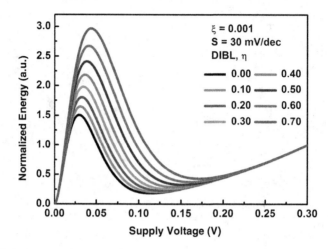

Figure 14.16 Calculated energy as a function of Vdd with fixed SS of 30 mV/dec at various η ($= |\Delta V_{th}/ \Delta V_{ds}|$) [11]

is the n-factor that is limited by kT/q due to the carrier injection in the source. The I-MOS and tunnel FETs in the previous subsections achieve SS of less than 60 mV/dec by modifying the n-factor. If the m-factor can be less than unity, the device with less than 60 mV/dec is also achievable. The negative gate insulator capacitance makes it possible.

Figure 14.17 shows a schematic of a negative capacitance FET and equivalent gate capacitance circuit [18]. The gate stack has a thin ferroelectric layer, which can introduce a positive feedback on the charge that causes polarization and amplifies the gate voltage, resulting in an m-factor of less than unity. Figure 14.18 shows measured device characteristics at room temperature [19]. SS of less than 60 mV/dec is obtained in a wide range of drain currents.

However, the average SS obtained is not small enough and the on-current is still very low for practical use. The main challenge of the negative capacitance FET is the selection of gate stack materials that can provide much smaller average SS and higher on-current. The threshold voltage adjustments, integration of nFET and pFET, and the stability, reliability, and variability will be also critical issues.

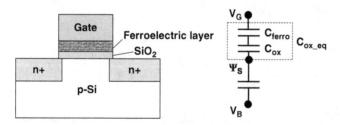

Figure 14.17 A schematic and equivalent gate capacitance circuit in the negative capacitance FET with a thin ferroelectric layer [16]

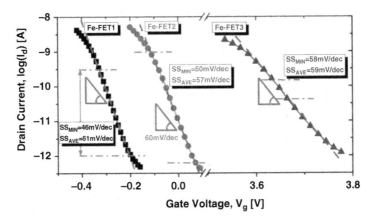

Figure 14.18 Measured I-V characteristics of negative capacitance FETs at room temperature [17]. © 2010 IEEE

14.4.5 MEMS Switch

Another approach to attain super steep subthreshold characteristics is to utilize a micro-electro-mechanical system (MEMS). The movement of a MEMS structure is affected by both elastic and electrostatic forces, and the structure suddenly moves when the balance between these forces is broken.

Figure 14.19 shows an example of a MOSFET with the MEMS structure. The device has a movable gate electrode that is fabricated by MEMS technology and suspended above the channel [20]. When the gate voltage (V_{gs}) is 0 V, the gate is located far above the gate and is not in contact with the gate oxide. Then, the inversion layer is not formed in the channel and hence, the device is in the OFF state (Figure 14.19a). When V_{gs} increases, a downward electrostatic force applied to the movable gate increases, and the movable gate is suddenly pulled down to the gate oxide at a certain voltage of V_{gs}, as shown in Figure 14.19b. This voltage is called the "pull-in" voltage ($V_{pull-in}$).

Figure 14.20 shows measured I-V characteristics of a MOSFET with a movable gate [20]. In this specific device, an additional lower gate electrode is formed in the vicinity of the channel to

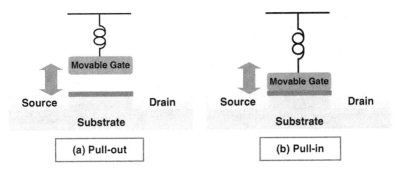

Figure 14.19 Schematics of an MOSFET with a movable gate electrode [18]. (a) OFF state at $V_{gs} = 0$ V. This state is called "pull-out." (b) ON state at sufficiently large V_{gs}. The gate electrode is pulled in and the device is at ON state

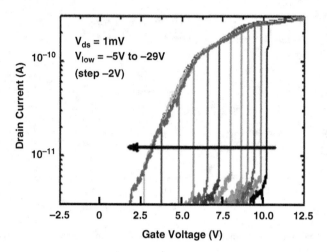

Figure 14.20 Measured I-V characteristics of an MOSFET with a movable gate electrode [18]. Very steep subthreshold characteristics are obtained

control $V_{pull-in}$ of the device. When V_{gs} increases, drain current rapidly increases at a certain V_{gs}. This voltage is $V_{pull-in}$ and the subthreshold slope is very steep. By changing the lower gate voltage (V_{low}), $V_{pull-in}$ is successfully controlled. The minimum value of SS is approximately 2 mV/dec.

The drawback of the MOSFET with a movable gate is the leakage current from drain to source that is inevitable to the MOSFET structure. In order to completely shut down the leakage current, another type of MEMS switch has been also proposed and demonstrated [21,22]. Figure 14.21 shows schematics of a MEMS relay switch with zero standby power [22]. In this structure, the source and drain are suspended and its position is controlled by the gate. Since the source and drain are completely separated mechanically in the OFF state, no leakage current flows. At $V_{gs} = 0$ V, the source electrode is located above and the device is at the OFF state. At a certain voltage of V_{gs}, the "pull-in" takes place and the drain current suddenly increases with a very steep subthreshold slope. Figure 14.22 shows a circuit composed of MEMS relay switches [21]. An inverter operation as well as NAND and NOR operations have been successfully demonstrated.

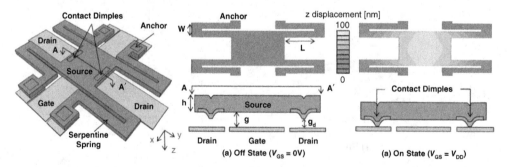

Figure 14.21 Schematics of a MEMS relay switch [20]. © 2009 IEEE (Left) Three-dimensional view of the device. (a) OFF state at $V_{gs} = 0$ V, where the source and drain are mechanically separated. (b) ON state, where the source and drain are physically contacted

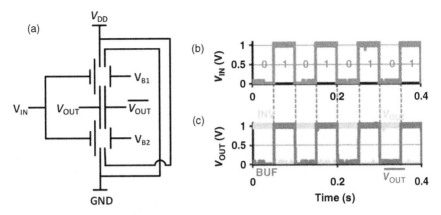

Figure 14.22 (a) A circuit schematic of a relay logic circuit. (b) Input signal. (c) Corresponding measured output signals [21]. © 2012 IEEE

14.4.6 Mott Transistor

It is well known that the abrupt resistance change by several orders of magnitude can be observed in the Mott metal–insulator transition. Mott field-effect transistor utilizes the electric field induced metal–insulator transition [23], in which the channel material shows both a metallic state and an insulating state controlled by the gate electrode. The Mott transistor could have a structure similar to conventional MOSFETs: the channel is composed of a correlated electron material, instead of semiconductor. Figure 14.23 shows schematics of the Mott metal-insulator transition and Mott transistor structure, where the channel material is VO_2 [24]. The insulating/metallic states of the channel are controlled by the gate electric field. The sharp channel resistance change by the gate voltage has been successfully demonstrated. Figure 14.24 shows measured data of a Mott transistor composed of VO_2 [24]. At low temperature, a large channel resistance change by a small change of gate voltage is observed.

Research on the Mott transistors is at a very early stage. The main challenges of the Mott transistors include a basic understanding of fundamental switching mechanisms, channel material selection, operating temperature, and stable transistor operation without hysteresis.

14.4.7 Bilayer Pseudo-spin Field-effect Transistor

In order to achieve ultimately energy-efficient information processing, a steep transition between the on state and off state in the 10 mV regime is necessary. BiSFET, bilayer pseudo-spin field-effect transistor, was recently proposed to enable the on/off transition at ultra-low voltage [25]. Figure 14.25a shows a schematic of the BiSFET structure that consists of an n-type and p-type graphene layer pair in close proximity, separated by a thin dielectric tunnel barrier. Both n-type and p-type grapheme layers have large and nearly equal carrier densities. Here, both electrons and holes are Fermions. However, under certain conditions, electrons in the n-type layer can pair with holes in the p-type layer, resulting in electron hole pairs/excitons which then condense even at room temperature in the case of the grapheme bilayer. Then, these are Bosons. This Bose–Einstein condensation alters the quantum wavefunctions in the bilayer qualitatively, resulting in a drastic reduction of the tunnel resistance between two layers for

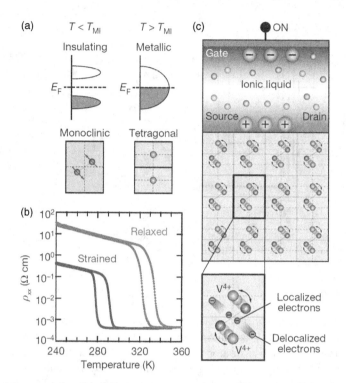

Figure 14.23 Mott metal–insulator transition and Mott transistor using VO_2 [24]. (a) The insulating state and metallic state. (b) Measured abrupt resistance change in strained 10 and relaxed 70 nm VO_2 films. (c) A schematic of Mott transistor showing localized (insulating) electrons and delocalized (metallic) electrons. Reprinted by permission from Macmillan Publishers Ltd: [Nature] [26], copyright (2012)

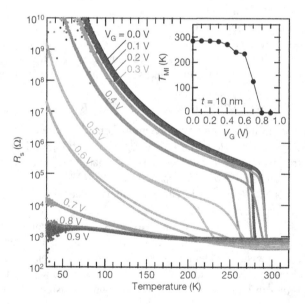

Figure 14.24 Experimental characteristics of a Mott transistor using VO_2 [24]. A large change of channel resistance by only small change of gate voltage (V_G) is observed. Reprinted by permission from Macmillan Publishers Ltd: [Nature] [26], copyright (2012)

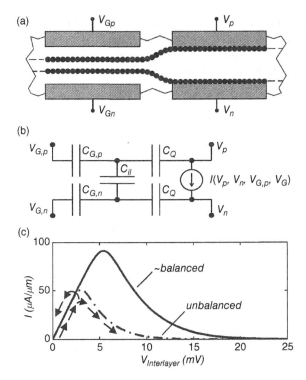

Figure 14.25 Structure, model, and I-V characteristics of BiSFET [25]. © 2009 IEEE (a) Schematic of a BiSFET. (b) Model circuit. (c) Modeled I-V characteristics

only a small interlayer bias. With a large bias, the condensation is destroyed leading to large resistance.

The device is called BiSFET because electron occupation of the top layer or bottom layer can be treated much like spin up or down, that is as a pseudospin, and the collective effects considered here are analogous to collective spin effects in a ferromagnet.

Figure 14.25b shows the circuit model and Figure 14.25c shows modeled current as a function of voltage between two grapheme layers of BiSFET [25]. A current peak appears in the I-V characteristics and then the current decreases at higher drain bias. Since the current change takes place at very low voltage ($\sim 10\,\mathrm{mV}$), very low voltage and low power operation may be possible. The simulations have shown that the energy consumption per clock cycle per BiSFET is 0.008 aJ at $100\,\mathrm{GHz}$ [25], while according to ITRS prediction MOSFET will consume 5 aJ per switching at $100\,\mathrm{GHz}$ in 2020 [1].

However, BiSFET is only a novel device concept. The main challenges include the fabrication of a perfect grapheme bilayer, a perfect gate stack with double gate structure, and an improved theory of the condensation.

14.5 Summary

In this chapter, recently proposed and/or reinvented transistors are reviewed. By closely looking at the historical trend of conventional MOSFETs in the past and comparing with the

ITRS prediction, it is pointed out that one of the most significant challenge for future integrated electron devices is to attain high on-current with keeping low off-current at very low voltage, that is, a steep subthreshold slope. Newly proposed and reinvented transistors described in this chapter have great potential to solve this grand challenge.

References

1. International Roadmap Committee (2013) International Technology Roadmap of Semiconductors (ITRS) (http://public.itrs.net/).
2. Thompson, S., Anand, N., Armstrong, M. *et al.* (2002) A 90nm logic technology featuring 50nm strained silicon channel transistors, 7 layers of Cu interconnects, low k ILD, and 1 um2 SRAM cell. IEDM Tech Dig., pp. 61–64.
3. Mistry, K., Allen, C., Auth, C. *et al.* (2007) A 45nm Logic Technology with High-k+Metal Gate Transistors, Strained Silicon, 9 Cu Interconnect Layers, 193nm Dry Patterning, and 100% Pb-free Packaging. IEDM Tech. Dig., pp. 247–250.
4. Yokoyama, M., Kim, S.H., Zhang, R. *et al.* (2011) CMOS Integration of InGaAs nMOSFETs and Ge pMOSFETs with Self-Align Ni-Based Metal S/D Using Direct Wafer Bonding. Symposium of VLSI Technology, pp. 60–61.
5. Bangsaruntip, S., Joseph, E., Klaus, D. *et al.* (2010) Gate-all-around Silicon Nanowire 25-Stage CMOS Ring Oscillators with Diameter Down to 3nm. Symposium of VLSI Technology, pp. 21–22.
6. Auth, C., Allen, C., Blattner, A. *et al.* (2012) A 22nm High Performance and Low-Power CMOS Technology Featuring Fully-Depleted Tri-Gate Transistors, Self-Aligned Contacts and High Density MIM Capacitors. Symposium of VLSI Technology, pp. 131–132.
7. Takeuchi, K., Fukai, T., Tsunomura, T. *et al.* (2007) Understanding Random Threshold Voltage Fluctuation by Comparing Multiple Fabs and Technologies. IEDM Tech. Dig., pp. 467–470.
8. Tsunomura, T., Nishida, A., and Hiramoto, T. (2009) Verification of threshold voltage variation properties in scaled transistors with ultra large-scale device matrix array test element group. *Japanese Journal of Applied Physics*, **48** (12), 124505.
9. Mizutani, T., Yamamoto, Y., Makiyama, H. *et al.* (2012) Reduced Drain Current Variability in Fully Depleted Silicon-on-Thin-BOX (SOTB) MOSFETs. IEEE Silicon Nanoelectronics Workshop, pp. 71–72.
10. Fuketa, H., Yasufuku, T., Iida, S. *et al.* (2011) Device-Circuit Interactions in Extremely Low Voltage CMOS Designs. IEDM Tech. Dig., pp. 559–562.
11. Jung, S.-M., Saraya, T., and Hiramoto, T. (2013) Effect of Drain Voltage Dependent Subthreshold Voltage Reduction on Energy Efficiency in Steep Subthreshold Slope Transistors. Japan Applied Physics Society Spring Meeting, 29a G9-3.
12. Gopalakrishnan, K., Griffin, P.B., and Plummer, J.D. (2002) I-MOS: A Novel Semiconductor Device with a Subthreshold Slope lower than kT/q. IEDM Tech. Dig., pp. 289–292.
13. Seabaugh, Alan C. and Zhang, Qin (2010) Low-voltage tunnel transistors for beyond CMOS logic. *Proceedings of the IEEE*, **98** (12), 2095–2110.
14. Ionescu, A.M. and Riel, H. (2002) Tunnel field-effect transistors as energy-efficient electronic switches. *Nature*, **479**, 329–337.
15. Huang, Q., Huang, R., Zhan, Z. *et al.* (2012) A Novel Si Tunnel FET with 36mV/dec Subthreshold Slope Based on Junction Depleted-Modulation through Striped Gate Configuration. IEDM Tech. Dig., pp. 187–190.
16. Dewey, G., Chu-Kung, B., Kotlyar, R. *et al.* (2012) III-V Field Effect Transistors for Future Ultra-Low Power Applications. VLSI Tech. Symp., pp. 45–46.
17. Tomioka, K., Yoshimura, M., and Fukui, T. (2012) Steep-slope Tunnel Field-Effect Transistors using III-V Nanowire/Si Heterojunction. VLSI Tech. Symp., pp. 47–48.
18. Salvatore, G.A., Bouvet, D., and Ionescu, A.M. (2008) Demonstration of Subthrehold Swing Smaller Than 60mV/decade in Fe-FET with P(VDF-TrFE)/SiO2 Gate Stack. IEDM Tech. Dig., p. 167–170.
19. Rusu, A., Salvatore, G.A., Jiménez, D., and Ionescu, A.M. (2010) Metal-Ferroelectric-Metal-Oxide-Semiconductor Field Effect Transistor with Sub-60mV/Decade Subthreshold Swing and Internal Voltage Amplification. IEDM Tech. Dig., pp. 395–398.
20. Park, J.S., Saraya, T., Miyaji, K. *et al.* (2008) Characteristic Modulation of Silicon MOSFETs and Single Electron Transistors with a Movable Gate Electrode. IEEE Silicon Nanoelectronics Workshop, p. S1015.
21. Liu, T.-J.K., Hutin, L., Chen, I.-R. *et al.* (2012) Recent Progress and Challenges for Relay Logic Switch Technology. VLSI Tech. Symp., pp. 43–44.

22. Kam, H., Pott, V., Nathanael, R. *et al.* (2009) Design and Reliability of a Micro-Relay Technology for Zero-Standby-Power Digital Logic Applications. IEDM Tech. Dig., pp. 809–812.

23. Newns, D.M., Misewich, J.A., Tsuei, C.C. *et al.* (1998) Mott transition field effect transistor. *Applied Physics Letters*, **73**, 780–782.

24. Nakano, M., Shibuya, K., Okuyama, D. *et al.* (2012) Collective bulk carrier delocalization driven by electrostatic surface charge accumulation. *Nature*, **487**, 459–462.

25. Banerjee, S.K., Register, L.F., Tutuc, E. *et al.* (2009) Bilayer PseudoSpin field-effect transistor (BiSFET): a proposed new logic device. *IEEE Electron Devices Letters*, **30**, 158–160.

15

Graphene Electronics

Frank Schwierz

Technical University of Ilmenau, Germany

15.1 Introduction

Over decades, semiconductor electronics, in particular IC (integrated circuit) technology, has evolved at an enormous pace and its annual growth has significantly exceeded that of other industries. Two major domains in current IC technology are *More Moore* and *More Than Moore*. Important trends in these domains are shown in Figure 15.1.

More Moore relates to digital logic that enjoyed an exponential growth of the number of transistors per chip, a continuous reduction of transistor size, and a steady increase of transistor switching speed coupled with a decrease in switching energy. The Si MOSFET (metal oxide–semiconductor field effect transistor) is the backbone of the dominating CMOS (complementary MOS) logic. In 2013, processors containing five billion MOSFETs with a gate length around 20 nm were in mass production.

More Than Moore, on the other hand, is not primarily focused on increasing circuit complexity but rather on enhancing the functionality by integrating digital logic together with nonlogic components such as high voltage modules, analog/RF (radio frequency) circuitry, sensors, actuators, and so on, on a chip. Today RF electronics is of key importance in the *More Than Moore* domain since many electronic systems use RF circuits to receive and transmit data. In particular wireless data transfer at increasing bit rates is gaining importance. Thus, *More Than Moore* needs fast transistors with excellent RF performance. Commonly two characteristic frequencies are used to assess the performance of RF transistors: (i) the cutoff frequency f_T (the frequency where the small-signal current gain has dropped to unity) and (ii) the maximum frequency of oscillation f_{max}, where the power gain has declined to unity. In RF electronics, different FET classes, most notably III-V HEMTs (high electron mobility transistor) and Si MOSFETs, are used. As shown in Figure 15.1, f_T and f_{max} of RF transistors could be enhanced continuously over decades. Currently the fastest RF FETs are InP HEMTs

---

Emerging Nanoelectronic Devices, First Edition. An Chen, James Hutchby, Victor Zhirnov and George Bourianoff.
© 2015 John Wiley & Sons, Ltd. Published 2015 by John Wiley & Sons, Ltd.

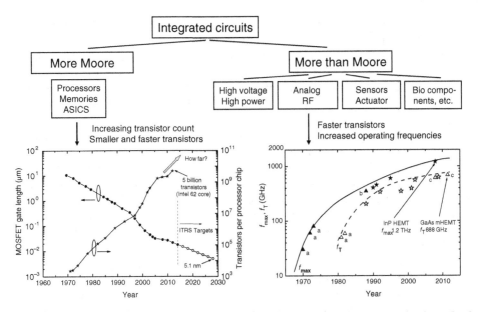

Figure 15.1 Major trends in the *More Moore* and *More Than Moore* domains of IC technology. On the left, the increase of the transistor count of processor chips and the scaling of MOSFET gate length are shown. The enhancing frequency performance of RF transistors is indicated on the right. The symbols a, b, c, and star represent the fastest FETs at certain times and the lines are a guide for the eye: a – GaAs metal–semiconductor FET, b – GaAs pHEMT (pseudomorphic HEMT), c – GaAs mHEMT, star – InP HEMT

and GaAs mHEMTs (metamorphic HEMT) that both benefit from the high mobility in their InGaAs channels.

The impressive trends shown in Figure 15.1 indicate that so far Si MOSFETs and III-V HEMTs have done an excellent job. Thus the question arises if research on new electronic materials and novel transistor concepts is needed at all. The answer is definitely yes. First, MOSFET scaling is becoming more and more challenging and will come to an end in the foreseeable future. Second, in spite of significant efforts, RF transistors are still not available for operation in the THz gap (the frequency range between 0.3 and 3.0 THz that could be used for many exceedingly attractive applications). Note that a transistor with 1 THz f_T and/or f_{max} is not fast enough for operation at 1 THz since at the operating frequency the transistor should provide a gain sufficiently higher than unity. The scaling and frequency limitations just mentioned have motivated intensive research on new material and transistor concepts.

When in 2004 two pioneering papers on the preparation of the purely two-dimensional carbon-based material graphene have been published [1,2], aspirations awakened that now a successor for the conventional semiconductors had been found that could pave the way to transistors with unprecedented performance and scaling capabilities. In particular the high mobilities observed in graphene – 10 000 cm^2/Vs demonstrated already in 2004 [1] and more than 100 000 cm^2/Vs reported later [3] – raised expectations that replacing Si by graphene could solve many problems in semiconductor electronics [4]. The community responded enthusiastically and by 2007 graphene had already found its place in the Emerging Research Devices chapter of the ITRS (International Technology Roadmap for Semiconductors) [5]. In the same year the first graphene MOSFET was demonstrated [6] and ever since many groups have become active in research on graphene transistors.

Now, only a few years later, we witness a dramatic change of mind [7–9]. The replacement of Si in logic transistors is taken off the agenda, graphene high-performance RF transistors are viewed with skepticism, and only the application of graphene transistors in some specific areas is considered promising. What is the reason for this abrupt change of the assessment of the prospects of graphene?

The early discussions on the potential of graphene in electronics almost solely focused on its outstanding mobility. Certainly a high mobility is always desirable. To be successful, however, an electronic material has to meet more requirements. A logic transistor needs two clearly distinguishable states, a highly conductive on-state and an off-state where the transistor blocks the current. For the latter, a sizeable bandgap is needed – at least in CMOS logic. RF FETs, on the other hand, should be operated in the regime of current saturation to unfold their full speed potential. To achieve satisfying saturation, again a bandgap is required. Since heat is generated in a FET that has to be removed, the channel material should show a good thermal conductivity and a small heat transfer resistance to the underlying substrate. Finally, a low contact resistance between the transistor terminals and the channel is needed. As will be shown in the next section, graphene meets only part of these requirements.

Recently several review papers were published on graphene as an electronic material and on graphene transistors, for example [10–17]. In the remainder of this chapter we will not repeat all details from [10–17] but rather emphasize relevant issues related to the requirements mentioned above, highlight the main trends in graphene transistor research, and identify promising applications for graphene transistors.

15.2 Properties of Graphene

15.2.1 Bandgap Energy

The existence of a bandgap is essential for electronic materials. Transistors for CMOS logic need a high on–off ratio I_{on}/I_{off} in the range 10^4 to 4×10^7 depending on the type of logic (high performance, low power). I_{on} is the on-state current and is not directly related to the bandgap energy E_G. The off-state current I_{off}, on the other hand, strongly depends on E_G according to $I_{off} \propto \exp -E_G/(2k_B T)$ [15] where k_B is the Boltzmann constant and T is the temperature. Thus, the on–off ratio follows the proportionality

$$\frac{I_{on}}{I_{off}} \propto \exp \frac{E_G}{2k_B T} \qquad (15.1)$$

The minimum bandgap energy required to achieve the on–off ratios for logic is estimated as 360 to 500 meV [13,15,16]. For RF FETs, on the other hand, the on–off ratio is of minor importance. Here, a bandgap is needed to achieve drain current saturation. In particular for high RF power gain and high f_{max}, current saturation is inevitable [17]. It is not exactly known how wide the bandgap for a good RF FET should be. Most likely, the requirements are less stringent than for logic transistors and a narrow bandgap could already help.

Large-area graphene is gapless and behaves like a semi-metal. Its bandstructure shows cone-shaped conduction and valence bands that touch each other at certain points of the Brillouin zone. Thus, large-area gapless graphene is not useful for CMOS logic and problematic for RF.

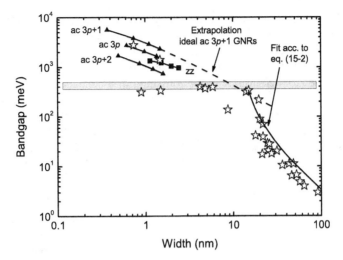

Figure 15.2 Bandgap energy versus ribbon width for GNRs (versus neck width for GNMs). Triangles with lines: simulated bandgap of ac (armchair) GNRs with ideal edges [18]. Rectangles with line: simulated bandgap of ideal zz (zigzag) GNRs [18]. Stars: measured bandgap of GNRs [19,20,22–24]. The shaded area indicates the minimum bandgap needed to achieve the on–off ratios required for CMOS logic

This does not automatically mean that graphene is useless for electronics since a bandgap can be opened in several ways, for example by

- forming narrow GNR (graphene nanoribbon) [18,19],
- biasing BLG (bilayer graphene) [10,21].

Figure 15.2 shows the simulated and measured bandgap energies of GNRs as a function of ribbon width and indicates a pronounced dependence of the bandgap energy on the width. Interestingly the qualitative trends and the actual numbers of the experimental bandgaps differ significantly from the simulated ones. According to theory, the bandgap energy critically depends on the GNR edge configuration (armchair or zigzag). If armchair GNRs are separated into three groups according to the number of carbon atoms along their width ($3p$, $3p + 1$, $3p + 2$, where p is an integer), the bandgap energies follow the order $E_G^{3p+1} > E_G^{3p} > E_G^{3p+2}$ and the bandgap energy of the GNRs of each of the three groups as well as that of zigzag GNRs increases with decreasing width due to quantum confinement.

The trend of the experimental bandgap energy is qualitatively different. Down to a width of about 15 nm, the upper limit of the measured bandgaps can be fitted by

$$E_G = \frac{a}{(w_{GNR} + b)^c} \tag{15.2}$$

where $a = 1.78$, $b = -11.7$, and $c = 1.4$ are fitting parameters, w_{GNR} is the ribbon width in nm, and E_G is the bandgap energy in eV. Below 15 nm, the experimental bandgaps scatter and do not show a clear trend. Thus, GNRs with widths below 15–20 nm fulfill the minimum bandgap requirement for logic transistors.

The reason for the difference between theory and experiment is that in reality patterned GNRs never possess well-defined edge geometries as assumed in the calculations. Instead, their edges are disordered and their width varies along their length. Moreover, it has been suggested that, in addition to pure lateral confinement [18], other effects (most notably Coulomb blockade [25]) contribute to the formation of a bandgap in GNRs.

By applying a perpendicular field to BLG, a bandgap can be opened as well. It has been shown that under high fields, bandgap energies of up to 250 meV can be achieved [10,21]. This would at least come close to the bandgap energy required for CMOS logic transistors. The bandgaps observed in experimental BLG MOSFETs, however, are limited to about 130 meV so far [26–28], which is too small for CMOS but may be sufficient for RF transistors.

15.2.2 Carrier Transport

Most frequently the mobility is used as a measure for carrier transport. It indicates how fast an average carrier moves under the influence of an electric field. Although it is argued occasionally that in nanometer FETs the mobility is no longer appropriate for assessing carrier transport, experiences show that it is very important even at the FET scaling limit. This is why significant efforts are spent on introducing high-mobility channel materials into nanometer CMOS technology. Strained Si with enhanced mobilities is already implemented into mass production [29] and research on high-mobility Ge and III-V channels for future MOSFET generations is pushed forward [30,31].

It is known that the electron mobility of semiconductors follows the general trend of decreasing mobilities for increasing bandgap energy, see, for example, Figure 8 in [17], and this is true for graphene as well. While the record mobilities of exfoliated gapless graphene clearly exceed those of the high-mobility III-V compounds InSb and InAs, opening a bandgap in GNRs and BLG modifies the bandstructure, increases the effective mass, and leads to a dramatic reduction of the mobility [32–34]. Figure 15.3 shows simulated and measured GNR mobilities as a function of ribbon width (note the good agreement between simulation and experiment). Currently the highest experimental GNR mobility is 2500 cm^2/Vs, reported for a 20-nm wide

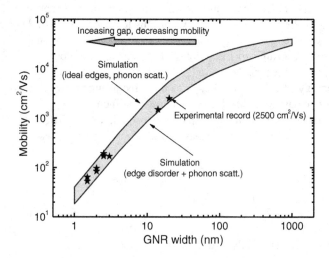

Figure 15.3 Simulated [36] and measured [35,37,38] mobility in GNRs as a function of ribbon width

ribbon [35]. This is much less compared to the mobilities of more than 10 000 cm^2/Vs in the InGaAs channels of InP HEMTs and GaAs mHEMTs. However, regarding the mobility an additional aspect should be kept in mind. For CMOS both n- and p-channel MOSFETs are needed. Since GNRs and BLG possess symmetric conduction and valence bands, nearly identical effective masses and quite similar mobilities for electrons and holes can be expected. This is certainly desirable and much different from the conventional semiconductors, in particular the III-V materials, where the hole mobility is significantly smaller than the electron mobility.

15.2.3 Heat Transport

Graphene is an excellent heat conductor with a room temperature thermal conductivity of 30–50 W/(cmK) which is much higher than that of conventional semiconductors, for example, 1.3 W/(cmK) for Si and 0.5 W/(cmK) for GaAs. A material's heat conductivity, however, does not tell the whole story of heat transport. If a transistor is located on a substrate different from the channel material, the heat generated in the transistor has to cross the channel–substrate interface that acts as a heat transfer resistance frequently called Kapitza resistance. The room-temperature Kapitza resistance of exfoliated graphene on a SiO_2/Si substrate has been measured as $(5.6–12) \times 10^{-9}$ m^2 K/W [39] and for epitaxial graphene on SiC a Kapitza resistance of 36×10^{-9} m^2 K/W has been simulated [40]. This is more than ten times as large as the Kapitza resistance of Si-SiO_2 interfaces of modern silicon on insulator structures $(0.5 \times 10^{-9}$ m^2 K/W [41]).

15.2.4 Contact Resistance

Early graphene transistors suffered from extraordinary large contact resistances between the metallic source/drain contacts and the underlying graphene channel in the range of 250–1000 $\Omega\mu$m. Meanwhile the metal–graphene contact resistance could be reduced significantly and the lowest reported values are now within the typical range of the contact resistances of state of the art Si MOSFETs and III-V HEMTs. The best graphene MOSFETs show contact resistances of 100–120 $\Omega\mu$m [42,43] compared to 20–150 $\Omega\mu$m for III-V HEMTs and 80–400 $\Omega\mu$m for Si MOSFETs. However, further improvements of the graphene contact resistance highly desirable.

15.3 Graphene MOSFETs for Mainstream Logic and RF Applications

In its early days, the GNR MOSFET has been considered as a viable candidate for replacing the Si MOSFET in future digital logic. The device has been introduced to the Emerging Research Devices chapter of the ITRS in 2007 and at about that time the first theoretical studies on the performance of GNR MOSFETs have been published. While high on–off ratios of 10^7 have been predicted for ideal GNR MOSFETs assuming ballistic transport [44], it has also been shown that edge disorder seriously degrades the performance of GNR MOSFETs [45]. In contrast to the wide body of simulation studies published since 2007, the number of papers on experimental GNR MOSFETs is rather limited. Back-gate nanoribbon transistors with ribbon widths between 2 and 5 nm and on–off ratios exceeding 10^6 [23,46] as well as top-gate GNR MOSFETs with lower on–off ratios [47] have been reported. There has been some limited activity on BLG MOSFETs but the achieved on–off ratios of 100 [27,28] at maximum are not sufficient for logic. The lack of more extensive experimental work on GNR and BLG MOSFETs is likely to be related to the processing issues and the edge problems of narrow GNRs and the limited bandgap opening in BLG, which have discouraged researchers to enhance the efforts on GNR and BLG MOSFETs.

The problems of GNR and BLG MOSFETs and the fact that RF FETs do not need to switch off have led to the situation that recently graphene logic MOSFETs receded into the background and more attention has been paid to graphene RF MOSFETs with gapless channels. Many groups have successfully demonstrated gapless graphene MOSFETs with GHz capabilities. On the other hand, reports on experimental RF graphene MOSFETs with semiconducting channels, that is, RF GNR and BLG MOSFETs, are still missing. Figure 15.4 shows the f_T and f_{max} performance of the best RF graphene MOSFETs, together with the corresponding data for three competing classes of RF FETs. The first class comprises InP HEMTs and GaAs mHEMTs having $In_xGa_{1-x}As$ channels with In contents x of 0.57–1.0, class two is constituted from GaAs pHEMTs (pseudomorphic HEMT) having InGaAs channels with a lower In content of 0.15–0.2, and the third class comprises RF Si MOSFETs. Obviously, in terms of f_T graphene MOSFETs rival successfully with InP HEMTs and GaAs mHEMTs (which are the fastest RF FETs at all) down to about 100 nm gate length and are much faster than Si MOSFETs and GaAs pHEMTs with comparable gate length [48–55].

Unfortunately, the situation is much different for the maximum frequency of oscillation f_{max}. So far, the best f_{max} data for graphene MOSFETs is below 100 GHz compared to several hundreds of GHz for the competing RF FET classes. This represents a serious limitation since for most RF applications f_{max} is more important than f_T and transistors with high f_T but low f_{max} are only of limited use. It has been shown that the disappointing f_{max} performance of graphene MOSFETs is mainly caused by the poor drain current saturation in gapless graphene channels [17], most likely combined with too large series and gate resistances. The unsatisfying drain current saturation, in turn, has its origin in the missing bandgap of the channel. This brings us to the conclusion that a semiconducting channel is needed not only for logic MOSFETs but for RF MOSFETs as well. Table 15.1 summarizes the state of the art of RF graphene MOSFETs and of the competing RF FETs.

While experimental RF data for GNR MOSFETs are still missing, the f_T performance of GNR MOSFETs has been studied by simulations. Early simulations assuming ballistic transport predicted unrealistically high cutoff frequencies in the 20–70 THz range for 10-nm gate GNR MOSFETs [56]. Later, scattering has been included and cutoff frequencies of 5 THz have been calculated for 10-nm gate transistors with 10-nm wide GNR channels [57].

Table 15.1 Frequency performance of RF FETs. $f_{T\text{-}500}$ and $f_{max\text{-}500}$ are the cutoff frequency and the maximum frequency of oscillation (in GHz) of the different FET classes according to their f_T and f_{max} trend curves for a gate length of 500 nm. Record f_T and record f_{max} are the highest reported f_T and f_{max} values for each transistor class in GHz and L is the gate length of the record transistors in nm

Transistor class	$f_{T\text{-}500}$	Record f_T	L	Ref.	$f_{max\text{-}500}$	Record	L	Ref.
Graphene MOSFET	≈100	427	67	48	≈40	70	100	49
GaAs pHEMT	65	152	100	50	153	290	100	51
InP HEMT and GaAs mHEMT	106	688	40	52	216	1200	35	53
Si MOSFET	43.5	485	29	54	70	420	29	55

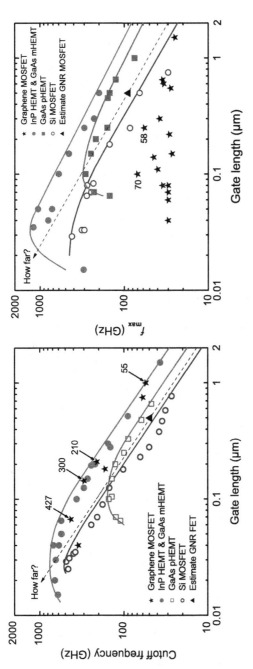

Figure 15.4 Frequency performance of experimental RF FETs as of June 2013. (a) Cutoff frequency and (b) maximum frequency of oscillation of experimental graphene RF MOSFETs (all having gapless channels) and of the three competing RF FET classes mentioned in the text versus gate length. Also shown is the projected frequency performance of GNR MOSFETs (dotted line) that will be discussed below. The numbers represent f_T and f_{max} of the fastest experimental graphene transistors in GHz and the lines are empirical trend curves. Data collected from the literature, updated from [17]

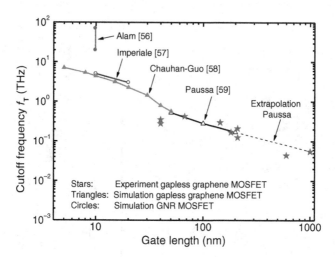

Figure 15.5 Simulated (triangles, circles) and experimental (stars) f_T performance of graphene MOSFETs. Simulation results from [58,59] are for gapless graphene MOSFETs and from [56,57] relate to GNR MOSFETs

Figure 15.5 compares the simulated cutoff frequencies of GNR MOSFETs [56,57] and gapless graphene MOSFETs [58,59] with those of the best experimental gapless graphene MOSFETs. An analysis of this data reveals to two interesting observations. First, down to 70 nm gate length the experimental f_T trend of gapless graphene MOSFETs is perfectly reproduced by simulations. Second, the simulated cutoff frequencies of GNR MOSFETs from [57] closely follow the simulated f_T trend for gapless graphene MOSFETs [58,59]. Obviously the degraded mobility in narrow GNRs is not expected to significantly affect the RF performance of short-gate GNR MOSFETs. Unfortunately, f_{max} simulations for GNR MOSFETs are still missing.

To get a reasonable idea on the RF capabilities (including f_{max}) of GNR MOSFETs, in the following an empirical-inductive approach to estimate their $f_T - f_{max}$ performance is developed. It is based on general RF performance trends observed for Si MOSFETs and III-V HEMTs and takes advantage of the scale length λ. The scale length provides useful information on the electrostatic integrity of a FET and its ability to suppress undesirable short-channel effects. The scale length concept has originally been developed for Si MOSFETs but can be applied to other FET classes as well provided the FET channel is semiconducting. A certain FET design with a small scale length can be considered to be sufficiently immune against short-channel effects down to short gate length levels. The short-channel effect relevant for RF FETs is the degradation of the drain current saturation leading to an enhanced drain conductance g_{ds} (g_{ds} is the slope of the output characteristics) that deteriorates power gain and f_{max}. Our empirical-inductive approach essentially consists of four steps.

1. Analysis of the f_T and f_{max} trend curves of the three competing RF FET classes from Figure 15.4 (long gates). In the $\log(f_T, f_{max})$-versus-$\log(L)$ plots of Figure 15.4, for $L > 400$ nm the f_T and f_{max} trend curves for the three RF FET classes are straight lines. We consider a gate length of 500 nm, extract the corresponding $f_{T\text{-}500}$ and $f_{max\text{-}500}$ from the trends curves (see Table 15.1), and plot this data versus the channel mobility as shown in Figure 15.6. As can be seen, both $f_{T\text{-}500}$ and $f_{max\text{-}500}$ are linearly dependent on the mobility.

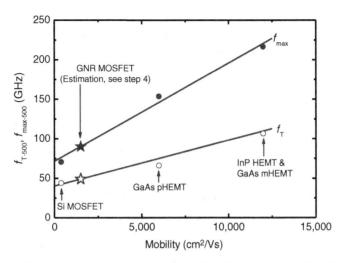

Figure 15.6 Cutoff frequency and maximum frequency of oscillation extracted from the trend curves of Figure 15.4 for 0.5-μm gate Si MOSFETs, GaAs pHEMTs, and InP HEMTs versus channel mobility

2. Analysis of the f_T and f_{max} trend curves from Figure 15.4 (short gates). At short gate lengths levels, the trend curves deviate from the long-channel behavior, reach a peak and eventually decline. Regarding f_T, for example, the deviation starts for GaAs pHEMTs at about 300 nm, for InP HEMTs and GaAs mHEMTs at 150 nm and for Si MOSFETs at 70 nm. A qualitatively similar trend can be observed for f_{max}. The main reasons for the degraded RF performance at short gate lengths levels are: (i) the increasing effect of parasitics and the external part of the device and (ii) the degraded electrostatic integrity and the enhanced short-channel effects.

3. Estimation of the scale length. As mentioned above, a measure for the robustness of a FET against short-channel effects is the scale length that can be estimated by [60]

$$\lambda = \sqrt{\frac{\varepsilon_{ch}}{\varepsilon_{bar}} t_{ch} \, t_{bar}} \tag{15.3}$$

where ε_{ch} and ε_{bar} are the dielectric constants of the channel and the barrier, t_{ch} is the channel thickness (the thickness of the Si body in fully depleted silicon on insulator MOSFETs and the thickness of the InGaAs quantum well channel in III-V HEMTs), and t_{bar} is the thickness of the barrier between gate and channel (the gate oxide in MOSFETs, the AlGaAs barrier in GaAs pHEMTs, and the InAlAs barrier in InP HEMTs and GaAs mHEMTs). Table 15.2 shows typical values for ε_{ch}, ε_{bar}, t_{ch}, and t_{bar} for the three RF FET classes from Figure 15.4 and for GNR MOSFETs, together with the scale lengths obtained from Equation 15.3. The results show a clear trend and indicate that in GNR MOSFETs short-channel effects should be much better suppressed than in the competing FET classes. Therefore, for GNR MOSFETs the linear part of the $\log(f_T, f_{max})$ versus $\log(L)$ should extend down to shorter gate length levels.

4. Estimation of the $f_T - f_{max}$ trend of GNR MOSFETs. Now we assume a fictive 0.5-μm gate GNR MOSFET with a 15 nm wide GNR channel that, according to Figure 15.2, should

Table 15.2 Typical design parameters and scale lengths of different classes of RF FETs. The much shorter scale length of GNR MOSFETs suggests that the linear shape of the log(f_T,f_{max}) versus log(L) curves can be maintained down to shorter gate lengths compared to the competing RF FETs

Transistor class	ϵ_{ch}	ϵ_{bar}	t_{ch} (nm)	t_{bar} (nm)	λ (nm)
Si MOSFET	11.9	3.9	5	0.6	3
GaAs pHEMT	13.3	12	15	35	24
InP HEMT and GaAs mHEMT	14.1	12.7	15	15	15.8
GNR MOSFET	3	3.9	0.35	2	<1

have a bandgap energy around 300 meV and, according to Figure 15.3, should show a mobility in the range of 1500–3700 cm^2/Vs. Taking the more realistic lower limit of 1500 cm^2/Vs, for this transistor a cutoff frequency of 49 GHz and a f_{max} of 90 GHz is extracted from Figure 15.6. This data is now extrapolated toward shorter gate length, see Figure 15.4. For gate length above 30 nm we expect GNR MOSFETs to show lower f_T and f_{max} than InP HEMTs and GaAs mHEMTs. For ultra-short gates, on the other hand, they might be able to outperform all competing RF FET classes provided the parasitics in GNR FETs can be minimized and the improved electrostatics suggested by the short scale length can actually be achieved in real devices.

We note that the fabrication issues and edge problems of GNRs mentioned earlier are relevant for RF GNR MOSFETs as well. However, RF ICs contain only a limited number of transistors (orders of magnitude less compared to complex logic ICs). Therefore it should be less cumbersome to develop a viable fabrication process for RF GNR MOSFETs. To conclude, it is not likely that the f_{max} problem of gapless graphene MOSFETs can be solved since it is related to the missing gap of gapless graphene. GNR MOSFETs, on the other hand, have the potential for significantly enhanced f_{max} and at ultra-short gate lengths level they possibly may outperform the conventional RF FETs. The potential of RF BLG MOSFETs is not clear at the moment. First simulations predict f_T and f_{max} in excess of 1 THz for a 40-nm gate BLG MOSFETs but experimental RF results for BLG MOSFETs are still missing [61]. More research is definitely needed to explore the RF potential of GNR and BLG MOSFETs.

15.4 Graphene MOSFETs for Nonmainstream Applications

A promising direction currently pursued is directed toward complementing the conventional semiconductors by graphene instead of trying to replace them and to use graphene in applications where other materials fail or perform poor. Intensive work is underway on graphene transistors for flexible and printable electronics.

Traditional mono-crystalline semiconductors such as Si and III-V compounds are rigid and cannot be used for flexible and printable electronics. Instead, organic semiconductors are favored in this field. Organic materials are bendable and printable, but this comes at the expense of a very low mobility. Typical mobilities of organic materials are in the 0.01–1.0 cm^2/Vs range only, which makes fast devices and low operating voltages a problem. Here, graphene definitely has an edge.

Large-area graphene is bendable and recently the deposition of graphene on flexible substrates and the use of graphene ink for printed electronics have become fields of intensive research. Impressively high mobilities of 1000–4930 cm^2/Vs for graphene on flexible substrates and printed graphene structures with mobilities of 100–365 cm^2/Vs have been reported [62,63]. Moreover, formidable results have been achieved with graphene transistors on flexible substrates [64–66]. A recent highlight is a 500-nm gate graphene MOSFET on polyethylene naphthalate substrate showing a f_T of 10.7 GHz and a f_{max} of 3.7 GHz [66]. This transistor is orders of magnitude faster than competing organic transistors on flexible substrates and shows that graphene-based GHz circuits on plastic have become a viable option. Moreover, printed graphene transistors have been reported [67] that could pave the way for graphene to the emerging field of low-cost printed circuits that can be fabricated without lithography. It should be kept in mind, however, that the flexible and printed graphene MOSFETs mentioned possess gapless large-area channels and do not switch off.

15.5 Graphene NonMOSFET Transistors

All graphene transistors discussed in the preceding sections are based on the classical MOSFET concept and suffer either from the gapless nature of large-area graphene or from the degraded mobility caused by a bandgap opening in BLG and GNRs. To exploit the full potential of graphene, however, devices taking advantage of the missing bandgap are desirable. Actually, several such transistor concepts, all based on tunneling in one way or another, have been conceived.

In 2009, the BiSFET (Bilayer Pseudospin Field Effect Transistor) has been proposed [68]. It is a vertical tunnel device concept based on the prediction that under appropriate conditions a superfluid electron-hole pair condensate forms in two graphene layers of opposite conduction type (i.e., one n-type and the other p-type conducting) that are separated by a thin dielectric [69]. Graphene is attractive for the formation of such a condensate since: (i) it is atomically thin and the missing bandgap is not a problem for proper BiSFET operation and (ii) it has a low density of states and symmetric conduction and valence bands. The active region of a BiSFET consists of the following vertical layer structure: top-gate / top-gate oxide / top graphene layer / inter-layer tunnel oxide / bottom graphene layer / bottom-gate oxide / bottom-gate. Outside the active region, the two graphene layers have contacts acting like the source and drain contacts in a FET. Under certain bias conditions (top-gate voltage, bottom-gate voltage, interlayer voltage applied across the two graphene layers), an electron-hole pair condensate forms, the resistance of the tunnel oxide dramatically reduces, and a pronounced tunnel current can flow. Beyond a certain interlayer voltage that can be small compared to the thermal voltage $k_B T/q$ at room temperature, the condensate collapses and the current drops to zero. Thus, in spite of the missing bandgap of the graphene layers, the BiSFET switches off since, in contrast to MOSFETs, not the conductivity of a graphene channel but the tunnel resistance of the interlayer tunnel oxide is controlled by the gate applied voltage.

The small interlayer voltage range within which the current reaches its maximum and then drops to zero results in extremely small switching energies. Theoretical considerations have shown that ultra-fast (100 GHz clock frequency) and ultra-low power (a few to a few tens of zeptojoules per switching event of a logic gate) logic circuits should be possible with BiSFETs [70]. While the BiSFET is an intriguing conceptual device, its experimental manifestation is still missing. One problem is certainly the proper stacking of the two graphene layers [71]. Nevertheless, in 2009 the BiSFET has been added as an emerging research device to the ITRS.

Apart from the BiSFET, several other vertical graphene-based tunnel transistor concepts are under investigation, for example [72,73]. All transistor concepts mentioned in this section have in common that at the moment their potential cannot be assessed reliably and that more work is needed to evaluate their merits and drawbacks.

15.6 Perspectives

It is always difficult to make a serious prediction on the future progress in dynamic fields such as semiconductor electronics in general and graphene electronics in particular. Thus, the following discussion should not be considered as a reliable prediction but rather as a personal view on the perspectives of graphene electronics.

15.6.1 General Remarks

Graphene is a fascinating material with outstanding properties and will undoubtedly find applications in different fields. Its prospects in electronics, in particular in digital logic and RF electronics, are vague, however. The early expectations that graphene could replace the conventional semiconductors in the foreseeable future have turned out to be wishful thinking. One should bear in mind, however, that this is not a fault of graphene – it is just a material. Instead, this is related to unrealistic performance projections based on a single property, the mobility, of graphene. One lesson we can learn from 10 years of graphene research is: Never expect too much too soon from a new material!

A second issue one should consider is the following. As mentioned above, to a large extend the discussions on future applications of graphene have been focused on high-performance transistors, in particular on the replacement of the conventional channel materials by graphene. In this regard it is helpful to recall Herbert Kroemer's *Lemma of New Technology* [74]: *The principal applications of any sufficiently new and innovative technology always have been – and will continue to be – applications created by that technology.* This means, in other words, that graphene may find/create applications in areas that have not been considered in the early days of graphene research.

Finally, graphene is by far not the only two-dimensional material. Instead, the work on graphene during the early 2000s paved the way for a wide variety of other two-dimensional materials that are currently under intensive investigation [75], and partly these new materials are very interesting for electronic applications.

15.6.2 Mid-term Perspectives

Table 15.3 summarizes our current view on the potential applications of graphene MOSFETs. The suitability of graphene in high-performance FETs for the *More Moore* (logic) and *More Than Moore* (RF) domains is considered rather critical since the conventional transistors (Si MOSFETs and III-V HEMTs) are simply too strong. While we do not see much a future for gapless graphene MOSFETs in high-performance logic and RF circuits, their application in flexible and printable electronics looks promising. The prospects of BLG MOSFETs are entirely unclear at the moment since it is much too less known on these transistors to make an even remotely reliable assessment. GNR MOSFETs provide the high on–off ratios needed for logic and excellent electrostatics due to the ultimately thin channel but are plagued by fabrication issues and degraded mobilities. In the RF field, GNR MOSFETs with ultra-short gates may be able to outperform the competing RF FETs – but this is not certain.

The BiSFET, after all, is a really intriguing device concept. Unfortunately, operating devices are still missing. Thus, an experimental confirmation that this concept really works is urgently

Table 15.3 Possible applications of graphene MOSFETs

Channel	Graphene MOSFET		
	Gapless graphene	BLG	GNRs
Bandgap (eV)	0	0.040–0.13	0.3–0.5
Channel mobility	Very high	Degraded	Degraded
On–off ratio	2–10	100	10^6
Application in logic	No	No	In principle yes
f_T (GHz)	>400	No experimental data; potentially high	No experimental data; potentially very high
f_{max} (GHz)	<100	Experimental data not available; potentially high	Experimental data not available; potentially very high
Application in RF	Limited	In principle yes	In principle yes
Further applications	Flexible and printable electronics	Unknown	Unknown

needed. Even if operating BiSFET can be realized, there will be long road ahead until scaled BiSFETs that can compete with Si CMOS and can be applied in commercial circuits will become reality.

15.6.3 Long-term Future

A view into the 2011 edition of the ITRS (the edition that was current at the time of completing this chapter) reveals that, although graphene transistors are mentioned in the Emerging Research Devices chapter, no production-stage carbon-based transistors are expected in the ITRS time frame, that is, until 2026. Instead, work on alternative channel materials with enhanced mobilities such as III-V semiconductors for n-channel MOSFETs and Ge for the p-channel counterpart is considered to be a major issue and production-stage MOSFETs with these alternative channel materials are expected before the year 2020.

Things may change, however, beyond the time frame of current ITRS when MOSFET gate lengths of 5 nm and below are expected. At such short gate length levels, direct source–drain tunneling may become a limiter for further MOSFET scaling. Recent work has shown that to suppress source-drain tunneling, channel materials having a heavier carrier effective mass (and thus by trend a lower mobility) and a wider bandgap are preferable [76–78]. This, together with the ultra-short scale length, may give two-dimensional materials such as graphene nanoribbons and transition metal chalcogenides (e.g., MoS_2, $MoSe_2$, WS_2, etc.) an edge over conventional semiconductors in MOSFETs at ≤5-nm gate length levels.

Acknowledgment

The author acknowledges financial supported from the Excellence Research Grant and the Intra-Faculty Research Grant of Technische Universität Ilmenau.

References

1. Novoselov, K.S., Geim, A.K., Morozov, S.V. *et al.* (2004) Electric field effect in atomically thin carbon films. *Science*, **306**, 666–669.

2. Berger, C., Song, Z., Li, T. *et al.* (2004) Ultrathin epitaxial graphite: 2D electron gas properties and a route toward graphene-based nanoelectronics. *Journal of Physical Chemistry B*, **108**, 19912–19916.

3. Mayorov, A.S., Gorbachev, R.V., Morozov, S.V. *et al.* (2011) Micrometer-scale ballistic transport in encapsulated graphene at room temperature. *Nano Letters*, **11**, 2396–2399.

4. Geim, A.K. and Novoselov, K.S. (2007) The rise of graphene. *Nature Materials*, **6**, 183–191.

5. International Roadmap Committee ITRS – The International Roadmap for Semiconductors, www.itrs.net/.

6. Lemme, M.C., Echtermeyer, T.J., Baus, M., and Kurz, H. (2007) A graphene field-effect device. *IEEE Electron Device Letters*, **28**, 282–284.

7. Bourzac, K. (2012) Back to analogue. *Nature*, **483**, S34–S36.

8. Segal, M. (2012) Learning from silicon. *Nature*, **483**, S43–S44.

9. Novoselov, K.S., Fal'ko, V.I., Colombo, L. *et al.* (2012) A roadmap for graphene. *Nature*, **490**, 192–200.

10. Castro Neto, A.H., Guinea, F., Peres, N.M.R. *et al.* (2009) The electronic properties of graphene. *Reviews of Modern Physics*, **81**, 109–162.

11. Weiss, N.O., Zhou, H., Liao, L. *et al.* (2012) Graphene: An emerging electronic material. *Advanced Materials*, **24**, 5782–5825.

12. Avouris, Ph. (2010) Graphene: Electronic and photonic properties and devices. *Nano Letters*, **10**, 4285–4294.

13. Schwierz, F. (2010) Graphene transistors. *Nature Nanotechnology*, **5**, 487–496.

14. Banerjee, S.K., Register, L.F., Tutuc, E. *et al.* (2010) Graphene for CMOS and beyond CMOS applications. *Proceedings of the IEEE*, **98**, 2032–2046.

15. Kim, K., Choi, J.-Y., Cho, S.-H., and Chung, H.-J. (2011) A role for graphene in silicon-based semiconductor devices. *Nature*, **479**, 338–344.

16. Reddy, D., Register, L.F., Carpenter, G.D., and Banerjee, S.K. (2011) Graphene field-effect transistors. *Journal of Physics D-Applied Physics*, **44**, 313001.

17. Schwierz, F. (2013) Graphene transistors – Status, prospects, and problems. *Proceedings of the IEEE*, **101**, 1567–1584.

18. Yang, L., Park, C.-H., Son, Y.-W. *et al.* (2007) Quasiparticle energies and band gaps in graphene nanoribbons. *Physical Review Letters*, **99**, 186801.

19. Han, M.Y., Özyilmaz, B., Zhang, Y., and Kim, P. (2007) Energy band-gap engineering of graphene nanoribbons. *Physical Review Letters*, **98**, 206805.

20. Linden, S., Zhong, D., Timmer, A. *et al.* (2012) Electronic structure of spatially aligned graphene nanoribbons on Au(788). *Physical Review Letters*, **108**, 216801.

21. Castro, E.V., Novoselov, K.S., Morozov, S.V. *et al.* (2007) Biased bilayer graphene: Semiconductor with a gap tunable by the electric field effect. *Physical Review Letters*, **99**, 216802.

22. Kim, P., Han, M.Y., Young, A.F. *et al.* (2009) Graphene nanoribbon devices and quantum heterojunction devices. *Tech. Dig. IEDM*, pp. 241–244.

23. Li, X., Wang, X., Zhang, L. *et al.* (2008) Chemically derived, ultrasmooth graphene nanoribbon semiconductors. *Science*, **319**, 1229–1232.

24. Chen, Y.-C., de Oteyza, D. G., Pedramrazi, Z. *et al.* (2013) Tuning the bandgap of graphene nanoribbons synthesized from molecular precursors. *ACS Nano*, **7**, 6123–6128.

25. Han, M., Brant, J.C., and Kim, P. (2010) Electron transport in disordered graphene nanoribbons. *Physical Review Letters*, **104**, 056801.

26. Iannaccone, G., Fiori, G., Macucci, M. *et al.* (2009) Perspectives of graphene nanoelectronics: Probing technological options with modeling. *Tech. Dig. IEDM*, pp. 245–248.

27. Xia, F., Farmer, D.B., Lin, Y.-M., and Avouris, Ph. (2010) Graphene field-effect transistors with high on/off ratio and large transport band gap at room temperature. *Nano Letters*, **10**, 715–718.

28. Szafranek, B.N., Schall, D., Otto, M. *et al.* (2011) High on/off ratios in bilayer graphene field effect transistors realized by surface dopants. *Nano Letters*, **11**, 2640–2643.

29. Thompson, S.E., Sun, G., Choi, Y.S., and Nishida, T. (2006) Uniaxial-process-induced strained-Si: Extending the CMOS roadmap. *IEEE Transactions on Electron Devices*, **53**, 1010–1020.

30. Shang, H., Frank, M.M., Gusev, E.P. *et al.* (2006) Germanium channel MOSFETs: Opportunities and challenges. *IBM Journal of Research and Development*, **50**, 377–386.

31. del Alamo, J.A. (2011) Nanometre-scale electronics with III-V compound semiconductors. *Nature*, **479**, 317–323.

32. Obradovic, B., Kotlyar, R., Heinz, F. *et al.* (2006) Analysis of graphene nanoribbons as a channel material for field-effect transistors. *Applied Physics Letters*, **88**, 142102.

33. Fang, T., Konar, A., Xing, H., and Jena, D. (2008) Mobility in semiconducting nanoribbons: Phonon, impurity, and edge roughness scattering. *Physical Review B-Condensed Matter*, **78**, 205403.

34. Li, X., Borysenko, K.M., Nardelli, M.B., and Kim, K.W. (2011) Electron transport properties of bilayer graphene. *Physical Review B-Condensed Matter*, **84**, 195453.
35. Lin, M.-W., Ling, C., Agapito, L.A. *et al.* (2011) Approaching the intrinsic band gap in suspended high-mobility graphene nanoribbons. *Physical Review B-Condensed Matter*, **84**, 125411.
36. Bresciani, M., Paussa, A., Palestri, P. *et al.* (2010) Low-field mobility and high-field drift velocity in graphene nanoribbons and graphene bilayers. Tech. Dig. IEDM, pp. 724–727.
37. Wang, X., Ouyang, Y., Li, X. *et al.* (2008) Room-temperature all-semiconducting sub-10-nm graphene nanoribbon field-effect transistors. *Physical Review Letters*, **100**, 206803.
38. Jiao, L., Wang, X., Diankov, G. *et al.* (2010) Facile synthesis of high-quality graphene nanoribbons. *Nature Nanotechnology*, **5**, 321–325.
39. Chen, Z., Jang, W., Bao, W. *et al.* (2009) Thermal contact resistance between graphene and silicon dioxide. *Applied Physics Letters*, **95**, 161910.
40. Mao, R., Kong, B.D., Kim, K.W. *et al.* (2012) Phonon engineering in nanostructures: Controlling interfacial thermal resistance in multilayer-graphene/dielectric heterojunctions. *Applied Physics Letters*, **101**, 113111.
41. Mahajan, S.S., Subbarayan, G., and Sammakia, B.G. (2011) Estimating Kapitza resistance between Si-SiO$_2$ interface using molecular dynamics simulations. *IEEE Transactions on Components, Packaging, and Manufacturing Technology*, **1**, 1132–1139.
42. Xia, F., Perebeinos, V., Lin, Y.-M. *et al.* (2011) The origin and limits of metal-graphene junction resistance. *Nature Nanotechnology*, **6**, 179–184.
43. Madan, H., Hollander, M.J., LaBella, M. *et al.* (2012) Record high conversion gain ambipolar graphene mixer at 10GHz using scaled gate oxide. Tech. Dig. IEDM, pp. 76–79.
44. Liang, G., Neophytou, N., Lundstrom, M.S., and Nikonov, D.E. (2007) Ballistic graphene nanoribbon metal-oxide-semiconductor field-effect transistors: A full real-space quantum transport simulation. *Journal of Applied Physiology*, **102**, 054307.
45. Yoon, Y. and Guo, J. (2007) Effects of edge roughness in graphene nanoribbon transistors. *Applied Physics Letters*, **91**, 073103.
46. Wang, X., Ouyang, Y., Li, X. *et al.* (2008) Room-temperature all-semiconducting sub-10-nm graphene nanoribbon field-effect transistors. *Physical Review Letters*, **100**, 206803.
47. Bai, J., Duan, X., and Huang, Y. (2009) Rational fabrication of graphene nanoribbons using a nanowire etch mask. *Nano Letters*, **9**, 2083–2087.
48. Cheng, R., Bai, J., Liao, L. *et al.* (2012) High-frequency self-aligned graphene transistors with transferred gate stacks. *PNAS*, **109**, 11588–11592.
49. Guo, Z., Dong, R., Chakraborty, P.S. *et al.* (2013) Record maximum oscillation frequency in C-face epitaxial graphene transistors. *Nano Letters*, **13**, 942–947.
50. Nguyen, L.D., Tasker, P.J., Radulescu, D.C., and Eastman, L.F. (1989) Characterization of ultra-high-speed AlGaAs/InGaAs (on GaAs) MODFETs. *IEEE Transactions on Electron Devices*, **36**, 2243–2248.
51. Tan, K.L., Dia, R.M., Streit, D.C. *et al.* (1990) 94-GHz 0.1-μm T-gate low-noise pseudomorphic InGaAs HEMTs. *IEEE Electron Device Letters*, **11**, 585–587.
52. Kim, D.-H., Brar, B., and del Alamo, J.A. (2011) $f_T = 688$ GHz and $f_{max} = 800$ GHz in $L_g = 40$ nm In$_{0.7}$Ga$_{0.3}$As MHEMTs with $g_{m\_max} > 2.7$ mS/μm. Tech. Dig. IEDM, pp. 319–322.
53. Lai, R., Mei, X.B., Deal, W.R. *et al.* (2007) Sub 50nm InP HEMT device with f_{max} greater than 1THz. Tech. Dig. IEDM, pp. 609–611.
54. Lee, S., Jagannathan, B., Narasimha, S. *et al.* (2007) Record RF performance of 45-nm SOI CMOS technology. Tech. Dig. IEDM, pp. 255–258.
55. Post, I., Akbar, M., Curello, G. *et al.* (2006) A 65 nm CMOS SOC technology featuring strained silicon transistors for RF applications. Tech. Dig. IEDM, pp. 1–3.
56. Alam, K. (2008) Gate dielectric scaling of top gate carbon nanoribbon on insulator transistors. *Journal of Applied Physiology*, **104**, 074313.
57. Imperiale, I., Bonsignore, S., Gnudi, A. *et al.* (2010) Computational study of graphene nanoribbon FETs for RF applications. Tech. Dig. IEDM, pp. 732–735.
58. Chauhan, J., Liu, L., Lu, Y., and Guo, J. (2012) A computational study of high-frequency behavior of graphene field-effect transistors. *Journal of Applied Physiology*, **111**, 094313.
59. Paussa, A., Geromel, M., Palestri, P. *et al.* (2011) Simulation of graphene nanoscale RF transistors including scattering and generation/recombination mechanisms. Tech. Dig. IEDM, pp. 271–274.
60. Yan, R.-H., Ourmazd, A., and Lee, K.F. (1992) Scaling the Si MOSFET: From bulk to SOI to bulk. *IEEE Transactions on Electron Devices*, **39**, 1704–1710.

61. Fiori, G. and Iannaccone, G. (2012) Insights on radio frequency bilayer graphene FETs. Tech. Dig. IEDM, pp. 403–406.

62. Lee, J., Tao, L., Hao, Y. *et al.* (2012) Embedded-gate graphene transistors for high-mobility detachable flexible nanoelectronics. *Applied Physics Letters*, **100**, 152104.

63. Wang, S., Ang, P.K., Wang, Z. *et al.* (2010) High mobility, printable, and solution-processed graphene electronics. *Nano Letters*, **10**, 92–98.

64. Sire, C., Ardiaca, F., Lepilliet, S. *et al.* (2012) Flexible gigahertz transistors derived from solution-based single-layer graphene. *Nano Letters*, **12**, 1184–1188.

65. Lee, J., Parrish, K.N., Chowdhury, S.F. *et al.* (2012) State-of-the-art graphene transistors on hexagonal boron nitride, high-k, and polymeric films for flexible analog nanoelectronics. Tech. Dig. IEDM, pp. 343–346.

66. Petrone, N., Meric, I., Hone, J., and Shepard, L. (2013) Graphene field-effect transistors with gigahertz-frequency power gain on flexible substrates. *Nano Letters*, **13**, 121–125.

67. Torrisi, F., Hasan, T., Wu, W. *et al.* (2012) Inkjet-printed graphene electronics. *ACS Nano*, **6**, 2992–3006.

68. Banerjee, S.K., Register, L.F., Tutuc, E. *et al.* (2009) Bilayer pseudospin field-effect transistor (BiSFET): A proposed new logic device. *IEEE Electron Device Letters*, **30**, 158–160.

69. Min, H., Bistritzer, R., Su, J.-J., and MacDonald, A.H. (2008) Room-temperature superfluidity in graphene bilayers. *Physical Review B-Condensed Matter*, **78**, 121401.

70. Reddy, D., Register, L.F., Tutuc, E., and Banerjee, S.K. (2010) Bilayer pseudospin field-effect transistor: Applications to Boolean logic. *IEEE Transactions on Electron Devices*, **57**, 755–764.

71. Basu, D., Register, L.F., MacDonald, A.H., and Banerjee, S.K. (2011) Effect of interlayer bare tunneling on electron-hole coherence in graphene bilayers. *Physical Review B-Condensed Matter*, **84**, 035449.

72. Britnell, L., Gorbatchev, R.V., Jalil, R. *et al.* (2011) Field-effect tunneling transistor based on vertical graphene heterostructures. *Science*, **335**, 947–950.

73. Mehr, W., Dabrowski, J., Scheytt, J.C. *et al.* (2012) Vertical graphene base transistor. *IEEE Electron Device Letters*, **33**, 691–693.

74. Kroemer, H. (2001) Nobel lecture: Quasielectric fields and band offsets: Teaching electrons new tricks. *Reviews of Modern Physics*, **73**, 783–793.

75. Butler, S.Z., Hollen, S.M., Cao, L. *et al.* (2013) Progress, challenges, and opportunities in two-dimensional materials beyond graphene. *ACS Nano*, **7**, 2898–2926.

76. Luisier, M., Lundstrom, M., Antoniadis, D.A., and Bokor, J. (2011) Ultimate device scaling: intrinsic performance comparison of carbon-based, InAs, and Si field-effect transistors for 5nm gate length. Tech. Dig. IEDM, pp. 251–254.

77. Sylvia, S.S., Park, H.-H., Khayer, M.A. *et al.* (2012) Material selection for minimizing direct tunnelling in nanowire transistors. *IEEE Transactions on Electron Devices*, **59**, 2064–2069.

78. Alam, K. and Lake, R.K. (2012) Monolayer MoS_2 transistors beyond the technology road map. *IEEE Transactions on Electron Devices*, **59**, 3250–3254.

16

Carbon Nanotube Electronics

Aaron D. Franklin
Department of Electrical and Computer Engineering and Department of Chemistry,
Duke University, USA

16.1 Carbon Nanotubes – The Ideal Transistor Channel

What made silicon so attractive for use as the transistor industry's foremost semiconducting material was primarily its accessibility compared to semiconductors with more ideal attributes. Compared to germanium, on which the early field-effect transistors (FETs) were demonstrated, silicon had a less desirable indirect bandgap and much lower mobility for electrons and holes. However, silicon was (and is) abundant and had a native oxide, SiO_2, which proved viable as a gate dielectric and an excellent passivation layer for surface states. After over five decades of shrinking down the size of silicon metal-oxide-semiconductor FETs (MOSFETs), the scaling limit is now effectively reached and industry is clamoring for a more scalable and more ideal replacement. This chapter presents the single-walled carbon nanotube (CNT) as the ideal semiconducting material for yielding scalable, high-performance FETs for the next generation of electronics.

16.1.1 Electronic Structure of a CNT

Physically, a CNT is composed of a single cylindrical shell of carbon atoms that are sp^2-bonded in a honeycomb lattice. It is instructive to consider the CNT as a sheet of graphene rolled into a seamless cylinder, providing both visual conception and insight into the electronic structure of a nanotube. Consider the carbon lattice for graphene in Figure 16.1, shown in both real and reciprocal space. The two unit vectors, $\mathbf{a_1}$ and $\mathbf{a_2}$ can define the location of all atoms in the lattice. A CNT is described by a chiral vector $\mathbf{c} = m\mathbf{a_1} + n\mathbf{a_2}$, which indicates the folding of one atom in the lattice onto another to form the nanotube. The "rolling" of the lattice into a CNT with a diameter d_{CNT} between 0.8 and 2.5 nm results in a quantization of the electronic states in the circumferential direction, as represented by the

Emerging Nanoelectronic Devices, First Edition. An Chen, James Hutchby, Victor Zhirnov and George Bourianoff.
© 2015 John Wiley & Sons, Ltd. Published 2015 by John Wiley & Sons, Ltd.

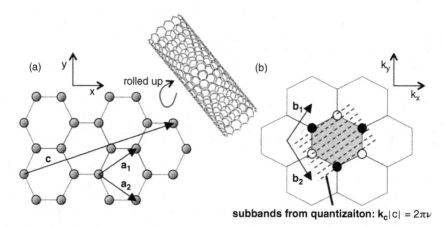

Figure 16.1 (a) Graphene's honeycomb lattice of carbon atoms in real space, with unit vectors $\mathbf{a_1}$ and $\mathbf{a_2}$, and an example chiral vector $\mathbf{c}$ illustrated as $\mathbf{c} = m\mathbf{a_1} + n\mathbf{a_2} = 3\mathbf{a_1} + \mathbf{a_2}$. (b) Graphene's lattice in reciprocal space showing unit vectors $\mathbf{b_1}$ and $\mathbf{b_2}$ with the Brillouin zone shaded. Dashed lines represent the allowed states in k_y that result from quantization in the circumferential direction from the "rolling" of a graphene sheet to form a CNT

parallel lines in Figures 16.1b and 16.2. Each of the parallel lines is a subband with its own set of 1D dispersion relations. The subbands are determined by the periodic boundary condition around the circumference of the CNT, with a circumferential wave vector $\mathbf{k_c}$ such that $\mathbf{k_c}\,\mathbf{c} = k_c|c| = k_c \pi d_{CNT} = 2\pi\nu$, which defines a series of parallel lines, each corresponding to a different integer value for ν.

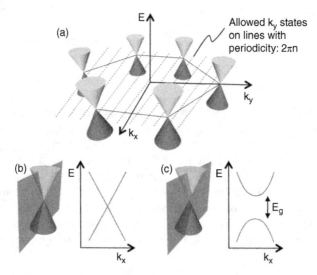

Figure 16.2 (a) Illustration of a first Brillouin zone for graphene with the conical energy dispersions shown at each K point. The quantized k-states for a nanotube's 1D subbands are depicted as parallel dashed lines. (b, c) A closer look at the dispersion relation for a subband passing (b) through a K point to form a metallic CNT and (c) away from a K point to form a semiconducting CNT

As illustrated in Figure 16.2, if a subband passes through one of the corners (known as K points in reciprocal space) of the Brillouin zone then the CNT will have no energy bandgap (E_g) and thus exhibit metallic properties. However, if none of the subbands pass through the K points, then the CNT will be semiconducting with a direct bandgap between the conduction and valence band. The bandgap is proportional to the smallest distance between a K point and the subband lines. Upon accounting for the threefold symmetry of the K points, the bandgap for a CNT becomes:

$$E_g(\text{eV}) = 2\left(\frac{\partial E}{\partial k}\right)\frac{k_c}{3} = \frac{2}{3}\hbar v_F \frac{2\pi}{|c|} = \frac{4}{3}\hbar v_F \frac{1}{d_{\text{CNT}}} \approx \frac{0.7}{d_{\text{CNT}}(\text{nm})} \tag{16.1}$$

where $\hbar$ is Planck's constant and v_F is the Fermi velocity of electrons.

Entire chapters have been written on the electronic structure of CNTs [1–3], but most of the crucial aspects can be gleaned, or extrapolated, from the information presented thus far, including:

- Semiconducting CNTs have a direct bandgap that is inversely proportional to d_{CNT}.
- "Rolling" of a 2D graphene lattice to form a CNT produces 1D subbands around the circumference of the nanotube.
- Effective mass of electrons and holes is equal (slope of conical dispersion relations is symmetric around the K points).

16.1.2 Electron Transport in CNTs

Because of quantum confinement in the circumferential direction, electron transport in CNTs is effectively 1D – carriers can either move forward or directly backward – with the carrier wave function traversing around the nanotube [3]. Carrier scattering is predominantly a result of acoustic and optical phonons. Acoustic phonon scattering is elastic and exhibits a long mean free path for CNTs on the order of 300 nm. The emission of optical phonons, however, is inelastic and has a much smaller mean free path of ∼10–15 nm, though high fields are required for these scattering events (> 0.16 eV) [4–7].

An important question for a 1D material is what the minimum resistance would be if the injection of carriers at the contacts is perfect and there is no carrier scattering? The current in such a device would simply be the product of excess carrier density (Δn), carrier velocity (v) and the elementary charge of an electron (e):

$$I = e\Delta n v = e\frac{D}{2}[eV_{ds}]v = \frac{e^2}{2}\left(\frac{4\sqrt{2}}{h}\sqrt{\frac{m}{E}}\right)V_{ds}\left(\sqrt{2}\sqrt{\frac{E}{m}}\right) = \frac{2e^2}{h}V_{ds} \tag{16.2}$$

where the application of a voltage between the drain and source contacts is V_{ds}, which opens a certain number of states in the 1D channel for transport based on the density of states, D. D and v are obtained from the parabolic dispersion relation, with dependence on Planck's constant (h) and the carrier effective mass (m). The result in Equation 16.2 is for most 1D materials with perfect transmission and no scattering; but for a CNT, the twofold band degeneracy of the

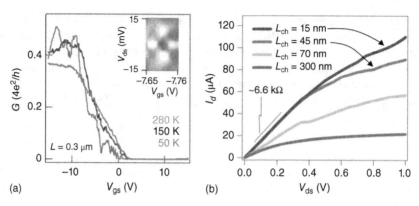

Figure 16.3 Demonstrations of reaching ballistic transport in CNTs. (a) CNT transistor with $G = 0.5$ G_Q at a 300 nm channel length [8]. (b) Metallic CNT devices down to 15 nm in length showing that below ~50 nm the device resistance at low fields is at the ballistic limit, $R_{tot} \sim R_Q$ [7]. Reprinted by permission from Macmillan Publishers Ltd: Nature [8], copyright 2003 and Nature Nanotechnology [7], copyright 2010

K points at the Fermi level yields an extra factor of two in the current and thus the following minimum resistance/conductance:

$$R_Q = \frac{h}{4e^2} = 6.5 \, k\Omega, \quad G_Q = \frac{4e^2}{h} = 154 \, \mu S \tag{16.3}$$

This quantum of resistance is the lowest possible resistance *per subband* for an electrically contacted CNT in the absence of scattering. Note that R_Q is for both of the needed contacts, source and drain, so there is effectively $3.25 \, k\Omega$ per contact. There have been several experimental demonstrations of reaching the ballistic transport limit in CNTs [7–12], two of which are shown in Figure 16.3. It is unlikely that there will ever be perfect transmission of carriers at the contacts, especially for semiconducting CNTs where Schottky barriers are typically present [13–16], and thus the best realizable resistance per nanotube in a transistor is between 6.5 and $12 \, k\Omega$.

16.1.3 The Ideal Transistor

As this book shows, there are a variety of options for a new digital switch that can replace silicon MOSFETs with higher performance, lower power, and more scalability. Aside from the dramatic platform-changing devices, such as spintronics and NEMS, most engineers can agree on what the ideal attributes would be for the channel material in a field-effect transistor. First, the ideal material should have ballistic transport for electrons and holes, be ultra-thin to maximize scalability, and have a bandgap of at least 0.5 eV. The gate should be able to wrap completely around the channel for optimal electrostatics and to eliminate the influence of stray charges or screening interactions. Finally, the semiconductor should be able to operate at low voltages while delivering high current density.

Juxtaposing CNTs against this profile of the ideal FET channel shows why they have received, and continue to receive, so much attention as an excellent candidate for silicon replacement. Their 1D nature exhibits ballistic transport, they are a mere ~1 nm in diameter/thickness, lend themselves perfectly to a gate all around geometry, and have been shown to provide high performance at low voltages. While there are certainly many material and engineering challenges that remain for CNTFETs, the impressive progress that has been made in the field combined with the CNT's ideal attributes for FETs suggest they are worth further pursuit. The proceeding sections will review the latest progress in the field and provide a brief overview of the challenges that remain.

16.2 Operation of the CNTFET

The first demonstrations of modulating a CNT channel using a gate-induced field came in 1998 [17,18]. These early CNTFETs, along with the majority of those studied since, were fabricated by using a doped silicon substrate gate to modulate the energy bands of a metal-contacted CNT, which was separated from the gate by a thermal SiO_2 gate oxide as depicted in Figure 16.4a. Since then, a variety of CNTFET structures have been demonstrated and can primarily be classified into devices with and without spacers. For a CNTFET, the spacer length (L_{spr}) is the distance between the gated channel and the source/drain metal contacts, as shown in Figure 16.4b and c.

For CNTFETs without spacers, a band diagram illustrating the basic operation is given in Figure 16.4d. Regardless of whether the gate is on bottom or top of the CNT channel, the operation of these spacer-less devices is nominally the same. Because nanotubes are intrinsic semiconductors, carriers are not induced in the channel by the gate-field, but must be injected from the contacts by the gate lowering the thermal barrier in the channel and the carriers tunneling through the Schottky barrier (SB) at the source (if there is a SB). Therefore, the contact metal work function plays a significant role in CNTFET performance, as will be discussed in Section 16.3.1. Note that the applied V_{ds} is dropped entirely at the drain-end of the device based on the ballistic transport nature of the CNT channel. Several studies have explored the impact of dropping the bias across such a small (few nanometers) length and the resultant heating of the contacts [6,19,20], which only limits the performance at very high fields where breakdown of the CNT can occur locally.

In contrast to the spacer-less devices, CNTFETs with spacers are less limited by injection at the metal-CNT source – owing to the high level of doping in the spacers from either a molecular dopant or "electrostatic doping" from a gate that thins the SB – and are controlled primarily

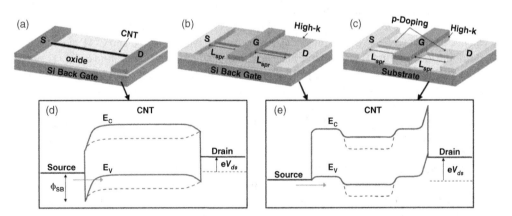

Figure 16.4 Example schematics and band diagrams of prominent CNTFET structures. Schematics of (a) substrate-gated, (b) gated-spacer, and (c) doped-spacer CNTFETs. Energy band diagrams illustrating the *on* state (solid lines) and *off* state (dashed lines) of a CNTFET (d) without spacers and (e) with gated/doped spacers

by the gate modulating the thermal barrier in the channel, like conventional MOSFETs [21]. Note that a CNTFET with spacers that are not doped in some fashion will have a large series resistance across L_{spr}, blocking carrier transport between the channel and source/drain – spacers in CNTFETs are therefore much like source/drain extensions in MOSFETs. CNTFETs with spacers will also be more resistant to ambipolar conduction because of the larger barrier to carrier injection at the drain [22,23]. Ambipolar conduction – conducting electron and hole currents in a device depending on the gate voltage – in CNTFETs has been seen as a great concern because a device would not remain in the *off* state if the gate bias is swept too far. Hence, a CNTFET that can suppress such ambipolarity is very attractive. Regardless of the CNTFET structure, movement of the energy bands by the gate in the *on* state can nominally be maintained at 1: 1 by operating the device in the quantum capacitance limit – uniquely accessible because of the 1D CNT [24–26].

16.3 Important Aspects of CNTFETs

As was introduced in the preceding section, there are a few key aspects to a CNTFET that impact the operation and performance, including the contacts, dielectrics, gate structure, and passivation treatments. Over the years, the research community has improved each of these important components, with far too much progress to cover in this short section. Therefore, emphasis herein is given to a brief summary of each aspect and the latest status in the field.

16.3.1 Contacts

Because CNTs are intrinsic semiconductors, all carriers must enter the channel by injection from the contacts, making the metal source/drain to CNT interface extremely important in CNTFET performance. For semiconductors that have dangling bonds at their surface, Fermi level pinning severely limits the effect of using metal contacts with different work functions to tune carrier injection [27]. However, CNTs have no open bonds on their honeycomb carbon lattice and hence are not impacted by such pinning of the metal Fermi level to certain mid-bandgap states [28]. Without Fermi level pinning, the choice of metals with appropriate work functions is a valuable method for maximizing performance and even tuning polarity in CNTFETs.

Many metals have been explored as source/drain contacts to CNTs [8,9,23,29–32]. Perhaps most significant was the discovery that Pd creates a nearly ideal *p*-channel contact to the valence band of CNTs [8]. Several studies on how metal work function and CNT diameter both contribute to a certain SB height have been reported, including the exploration of Pd, Ti, and Al contacts shown in Figure 16.5 [33]. While this data suggests a clear relationship between the metal work function and resulting SB height, there are other aspects to consider. For instance, Pt, which has a work function of 5.65 eV, provides a very poor contact to CNTs despite its work function being larger than Pd [8]. The reason for this inconsistency is that the physical wetting of a metal on a CNT, also referred to as coupling, is another critical factor for carrier injection [24,34,35].

A final consideration for the contacts is regarding their geometric coverage of the CNT. Most devices have used contacts that are on top of the nanotube, interfacing noncovalently with the sp^2-bonded carbon. Yet, some theoretical studies have suggested that metal contacts to the "ends" of a nanotube, where the carbon bonds are dangling, can provide many times lower contact resistance [36]. Experimental data has yet to defend or deny this claim, though Pd top contacts have already yielded CNTFETs with resistances very near R_Q [7,9], suggesting that

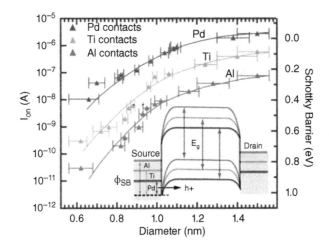

Figure 16.5 *On* current (I_{on}) and Schottky barrier height dependence on CNT diameter, which is inversely proportional to the bandgap, for contact metals with different work functions. Inset is a band diagram illustrating how the contact metal affects the SB height (ϕ_{SB}) for a CNT with the same bandgap for each metal. Adapted with permission from [33]. Copyright 2005 American Chemical Society

there is not much that can be improved by employing different contact geometries for Pd so long as the contact length is greater than the transfer length (see Section 16.4.2).

16.3.2 n-Type CNTFETs

With a work function of ~4.7 eV [37,38], the Fermi level at mid-gap, no metal-induced gap states causing Fermi level pinning, and an average bandgap of 0.6 eV, CNTs naturally lend themselves to *p*-type FETs because of the easiness of finding a robust high work function metal for injecting holes into the valence band (see Figure 16.5). Based on the common CNTFET structures in Figure 16.4, obtaining an n-type CNTFET – where electron injection is favored – can be achieved by either doping the spacer regions and/or using a low work function metal.

Doping CNTs is difficult to control. As mentioned previously, substitutional or interstitial doping is not available, so the dominant method is to apply a layer of charged molecules to the CNT surface. This has been done in gas [39,40] or solution [41,42] phase, and with both *n*- and *p*-type dopants. While this charge-transfer doping approach has enabled some high-performance CNTFETs, it is not without challenges, including: (1) low consistency/uniformity, (2) instability in air, (3) low compatibility with subsequent planarization of devices, and (4) sensitivity to high temperatures. Every molecular dopant does not necessarily have all of these challenges, but they each have at least one of them to some degree. An example of a high performance n-type CNTFET achieved by applying a layer of charged molecules to the spacer, or source/drain extension, regions of a Pd-contacted device is given in Figure 16.6a [40]. Note that a further advantage of having a CNTFET device with doped spacers is that the undesirable ambipolar conduction is largely subdued because of the sizeable barriers to carrier injection at the drain.

Establishing metal contacts with low barriers to electron injection is critical for *n*-type CNTFETs, even those with doped spacers (though in the case of doped spacers the SBs are thinned substantially by band movement from the doping). Perhaps the greatest challenge in

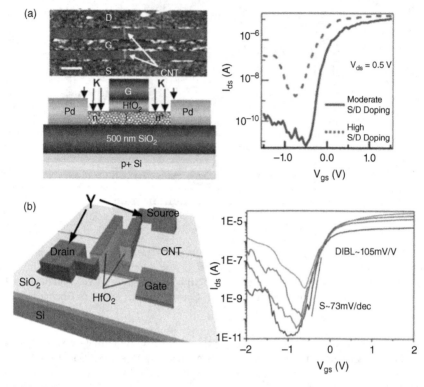

Figure 16.6 Schematic and corresponding subthreshold curves from *n*-type CNTFETs achieved by (a) doping the spacer (source/drain extension) regions with charged molecules [40] and (b) using low work function Y metal contacts for small SB to the conduction band [43]. Adapted with permission from [40] and [43]. Copyright 2005 [40] and 2009 [43] American Chemical Society

working with low work function metals is their high propensity to oxidize, which among other things compromises their air stability and consistency of material quality. However, with appropriate passivation layers used to cap the metals, there have been demonstrations of *n*-type CNTFETs having superb performance, primarily using Y [43], Sc [11], or Er [44]. The subthreshold curves from the device in Figure 16.6b with Y contacts exhibit near ideal performance for a single CNT – as good as *p*-type devices from Pd contacts. It has even been shown that, under the appropriate metal deposition conditions, the yield of *n*-type CNTFETs can be increased to the level achievable by high work function metal *p*-type devices [44]. While more difficult to achieve than *p*-type CNTFETs, high performance *n*-type devices have become accessible and no longer prove to be a bottleneck to a CNT technology.

16.3.3 Dielectrics

As pointed out previously, CNTs are made up of fully satisfied covalent carbon bonds. For this reason, choosing a dielectric for CNTFETs is not related to Fermi level pinning or passivating surface states – as is the case for most other semiconductors and the plague for III-V materials [45] – but rather what dielectric can be scalable with a compatible deposition process for the CNT. The most common fabrication method for depositing high quality high-κ dielectrics is atomic layer deposition (ALD), which is a chemical vapor deposition process that relies on gaseous precursors to react with surface states to form a dielectric one layer at a time [46]. Having no surface states, CNTs are not naturally compatible with ALD. To overcome this limitation, bottom-gate CNTFETs have been used [47,48] or adhesion layers have been applied to enable ALD nucleation on CNTs [49,50]. These adhesion layers can cause undesirable perturbations to carrier transport in the CNT without the appropriate annealing treatments applied during processing [51].

For bottom-gated devices the dielectric is completely formed prior to placing the nanotubes on the substrate, making it a great option as far as dielectric choice, scalability, and quality are concerned. But, the bottom gate geometry can also be tricky for manufacturability (see Section 16.3.5), so it is desirable to have other options. Adhesion layers for ALD nucleation have been demonstrated in the form of DNA molecules that wrap CNTs [52] or gas-phase functional layers formed in a CVD chamber prior to ALD [49]. The latter approach is more favorable in that it employs compatible oxide layers that will not compromise the needed low equivalent oxide thickness (EOT) of the final device structure. Once the adhesion layer is in place, ALD can be used to deposit the high-κ material of choice.

16.3.4 Passivation Treatments

One highly promoted application space for CNTFETs is in chemical or biological sensing [53–55]. Because of their small size and extremely high specific surface area, the presence of virtually any adsorbate is readily detected by conductance changes in a CNT. While excellent for sensing applications, the nanotube sensitivity to such stray charges is a challenge for high performance digital applications. This difficulty is manifest in the large variation of threshold voltage (V_t) among CNTFETs of the same geometry and the sizable hysteresis that is standard for most CNTFETs [56].

Transfer curves from a set of CNTFETs built on the same nanotube are shown in Figure 16.7, where the threshold voltages span a range of ~0.8 V. The simple application of a hydrophobic self-assembled monolayer in vacuum to passivate (cover) the exposed CNT channel and surrounding dielectric surface is able to reduce the range of V_t by more than 50%. This reduction in V_t variation is a result of the vacuum deposition environment driving off stray charge adsorbates (e.g., oxygen, water) and then passivating the hydroxylated surface to stave off any future adsorbates. Further, the hysteresis in the same devices was reduced on average from 500 mV to <50 mV over a gate source bias (V_{gs}) range of 4 V at $V_{ds} = -0.5$ V. Such a dramatic reduction in variability is evidence that the variation is not intrinsic to the CNTs and is primarily a result of the channel being susceptible to stray charges in the vicinity. Hence, either a technologically compatible passivation coating that can eliminate such charge, or the ability to completely wrap a CNT in the gate is needed to address the variability problem.

16.3.5 Gate Structure

Because CNTs are not substrate-bound, there is a lot of freedom in designing the gate structure. As shown in Figure 16.4, CNTFETs are typically fashioned with either bottom- or top-gate geometries. While the top-gate does provide an omega-shaped gate structure, the electrostatics are not appreciably different from the bottom gate, cylinder on plate, structure. Top gates more closely mimic the structure of Si MOSFETs, but have been challenging to realize with the difficulty of nucleating dielectrics on CNTs (see Section 16.3.3) and the inability to use reactive ion etching (RIE) in the presence of CNTs (nanotubes are readily etched away in a reactive ion environment).

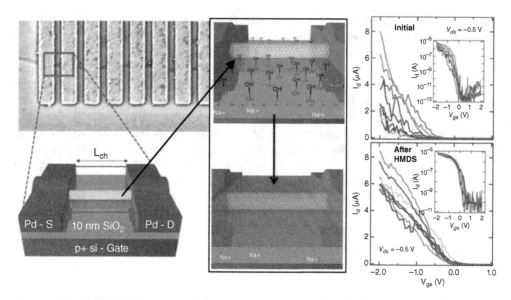

Figure 16.7 CNTFETs constructed along the same CNT showing large variation in threshold voltage. Gas-phase application of a hydrophobic monolayer passivates the hydroxylated surface and substantially lowers variation among the devices. Adapted with permission from [56]. Copyright 2012 American Chemical Society

For local bottom gate devices, there is no fabrication limit to the scalability of the dielectric thickness (t_{ox}) because it is formed prior to CNT placement, thereby minimizing the number of processing steps that the nanotubes are exposed to and preserving their quality. Another advantage of bottom gates is that they allow for any choice of passivation/planarization/capping layer to be used to cover the final device (see Section 16.3.4), rather than being restricted to the gate dielectric layer as in top-gate structures. Regardless of the gate geometry, the gate metal work function provides a 1 : 1 control over the device threshold voltage to enable the tuning of V_t in a final architecture [57].

The ultimate gate structure for CNTFETs is one that completely wraps the nanotube channel in a gate-all-around (GAA) fashion [50,58]. GAA is so natural for CNTs that it has been used for virtually all simulation studies of CNTFETs [59–61]. A demonstration of GAA-CNTFETs is given in Figure 16.8 [51]. Although the GAA provides the ideal electrostatic structure for nanotubes, it is a mistake to conclude that this marginal improvement in electrostatics is critical for achieving CNTFETs with sub-10 nm channels (see Section 16.4.1). The ultrathinness of a nanotube (<2 nm) enables excellent gate control of the channel even with only a bottom gate [62]. Actually, the primary motivation for GAA is that the gate encompasses the CNT, shielding it from stray charge, screening interactions, or other local variations. As pointed out in Section 16.3.4, nanotubes are sensitive to any charge in their vicinity, and the GAA is the only complete solution for addressing this challenge.

16.4 Scaling CNTFETs to the Sub-10 Nanometer Regime

In order for a transition from mainstream Si MOSFET technology to be worthwhile, the replacement must provide substantial benefits in performance at sub-10 nm dimensions – a

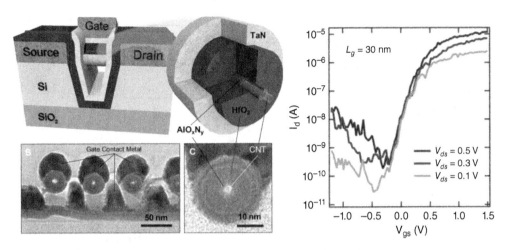

Figure 16.8 CNTFET with a gate-all-around (GAA) geometry. The transmission electron microscope images are cross-sections of the GAA-CNT showing uniform coating of the gate stack on the nanotube(s). Subthreshold curves are from a 30 nm gate length *n*-type GAA-CNTFET. Adapted with permission from [51]. Copyright 2013 American Chemical Society

regime where silicon of any dimensionality cannot compete. Therefore, while CNTFETs were shown for years to have excellent performance at hundreds of nanometers, verification of their scaled performance benefits is crucial. The two critical lengths for scaling in a CNTFET are the channel length (L_{ch}) and contact length (L_c) as illustrated in Figure 16.9.

16.4.1 Channel Length Scaling

Scaling down the channel length of a CNTFET has enabled ballistic transport and high *on* currents [7,9,63,64]. The danger with aggressive L_{ch} scaling in any FET is that the gate will lose control over the energy bands in the channel – a problem known as short channel effects (SCEs). Because of their ultrathin body, CNTs have been projected to provide excellent gate control at scaled lengths. However, with very small carrier effective mass, CNTs are also susceptible to source–drain tunneling, which is another SCE that causes an increase in subthreshold swing (*SS*) and *off* current. Some theoretical projections suggested that such tunneling would be the demise of CNTFETs at L_{ch} <15 nm [59,60].

Over the years, experimental CNTFETs with channels between 15 and 300 nm were demonstrated with little to no SCEs. It was not until 2012 that L_{ch} was scaled below 10 nm

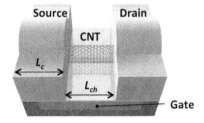

Figure 16.9 Schematic of a CNTFET with the contact and channel lengths identified

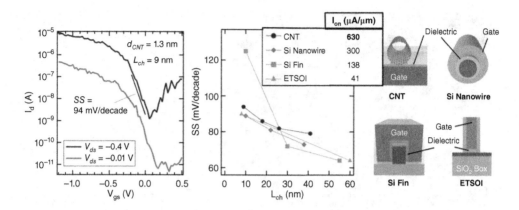

Figure 16.10 (a) Subthreshold curves from a CNTFET with a 9 nm channel length. (b) Subthreshold swing versus channel length for CNTs compared to the best experimentally demonstrated silicon-based FETs. *On* current (I_{on}) is pitch-normalized for CNTs (5 nm pitch) and fins/nanowires (20 nm pitch) and extracted at $|V_{gs} - V_t| = 0.3$ V and $|V_{ds}| = 0.5$ V (the needed voltages for a $V_{DD} = 0.5$ V technology). Adapted with permission from [62]. Copyright 2012 American Chemical Society

for a CNTFET [62], and the result defied previous theoretical expectations. The 9 nm CNTFET did not exhibit the anticipated SCEs, such as high *SS*, and instead performed better than any silicon-based device of comparable length, as shown in Figure 16.10. Importantly, the comparison of *on* currents in Figure 16.10 has been pitch-normalized, which accounts for the fact that you can only pack CNTs or fins or nanowires so close to each other and still be able to fit all of the gate stack. In the case of CNTs, the device was a local bottom gate, where there would be no gate stack to limit the pitch, though a reasonable 5 nm pitch was chosen.

16.4.2 Contact Length Scaling

It is instinctual to focus on channel length when discussing the scaling of any type of FET because SCEs are known to be a result of small L_{ch}. However, for an FET to be densely integrated in a useful digital technology, the length of the contacts (source and drain) must be scaled as aggressively as the channel. The fact that there has been less focus on L_c scaling for nanoelectronic devices is not because it has not been a challenge for Si MOSFETs – small contact lengths have led to dramatic increases in series resistance, which compromises *on* state performance. It is most probable that the lack of attention to L_c in nanomaterial-driven FETs is simply because: (1) SCEs from L_{ch} scaling is a bigger concern for Si MOSFETs and thus more popular to address and (2) there is less motivation to consider L_c when there are still SCEs present.

For CNTFETs, the first study to consider the impact of contact scaling revealed an important challenge for realizing high performance, ultrasmall devices: contact resistance (R_c) exhibits an inverse dependence on L_c, similar to MOSFETs [7,35]. While some previous studies had explored this trend for large multi-walled carbon nanotubes [65], this R_c versus L_c dependence in CNTFETs, shown in Figure 16.11, clearly defined the balance between L_{ch} and L_c scaling in these devices for achieving optimal performance.

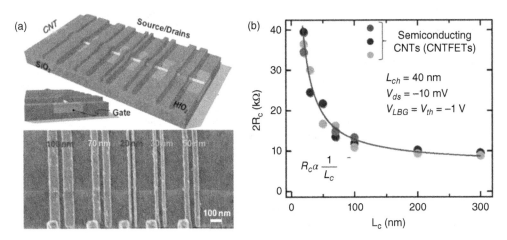

Figure 16.11 (a) Schematic and corresponding scanning electron microscope image of a set of CNTFETs with different contact lengths assembled along the same long CNT. (b) Contact resistance versus contact length from three sets of CNTFETs, each assembled on a different CNT. Adapted with permission from Macmillan Publishers Ltd: Nature Nanotechnology [7], copyright 2010

The reason for R_c showing strong dependence on L_c is different than for MOSFETs. A metal-CNT contact is a 3D-1D, metal–semiconductor interface, creating a very different scenario from the standard 3D-3D, silicide–semiconductor interface. Some had even projected short contacts to improve metal-CNT R_c by maximizing the electric field at the interface [13,66]. There have been theoretical studies that try and make sense of the observed scaling behavior, but they all deal with carrier injection between the metal-CNT and have not found a way of considering how transport in the metal-covered CNT contributes to R_c [34,35,67,68]. Overall, this result is still rather new and has only been shown for one contact metal, Pd. There are a myriad of possibilities for improving the L_c scaling behavior for CNTFETs that should come to light in the ensuing years.

16.5 Material Considerations

The focus of this chapter has predominantly been on the device considerations for a high-performance CNTFET technology. However, as with any relatively new material, CNTs have unique challenges at the materials-level that must be addressed in order for them to be viable for a densely integrated technology. The two prominent challenges are the isolation of semiconducting CNTs and the positioning of these CNTs on a substrate.

16.5.1 High-Purity Semiconducting CNTs

As discussed in Section 16.1.1, depending on the chirality of a nanotube it can either be metallic or semiconducting. The tolerance for a metallic nanotube in technologically relevant CNTFETs is very low at ~0.0001%. Statistically, the distribution of metallic versus semiconducting nanotubes is 1: 2, meaning 33% are metallic. Over the years, progress has been made in separating semiconducting CNTs from their metallic counterparts, as shown in Figure 16.12. The most promising techniques involve dispersion of CNTs into solution and then using some type of chemical technique for differentiating the semiconducting from the metallic, such as density gradient ultracentrifugation [69–71], column chromatography [72], or DNA coating [73].

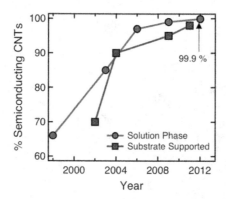

Figure 16.12   Progress in isolating semiconducting CNTs over the years, including both solution phase (CNTs dispersed in a solution) and substrate-supported (CNTs on substrate) techniques. The target for a technology is ∼99.9999%

Many reviews have been written comparing the various techniques that have been demonstrated [71,74]. The target purity is 99.9999% semiconducting [75], with the best reported as of early 2013 being 99.9% [76]. While the remaining three orders of magnitude in needed purity is daunting, it is not fundamentally impossible [77], albeit difficult to characterize.

16.5.2 Precise Placement of CNTs

One of the core differences between CNTFET fabrication processes and those of MOSFETs is that they are bottom-up versus top-down approaches, respectively. A MOSFET is created out of a bulk material – doped, annealed, patterned, coated, and so on. A CNTFET is assembled by synthesizing and placing (sometimes in one step) CNTs in desired locations and then establishing contacts, dielectrics, and gates onto them. This bottom-up approach has some advantages, including substrate independence and many contact/gate material options. The primary disadvantage is the difficulty of precisely positioning CNTs at a certain pitch (see Figure 16.13) across an entire wafer.

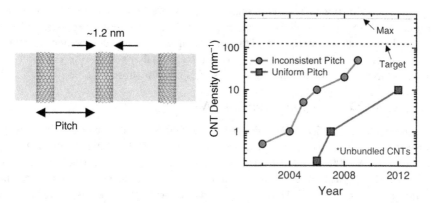

Figure 16.13   Progress in placing CNTs on a substrate at a certain linear density (number of CNTs per micron). Placing CNTs at an inconsistent pitch has reached a higher density but suffers from a lack of uniformity. The target is for at least 125 CNTs/μm (pitch ≤ 8 nm)

There has been considerable progress in the precise placement of CNTs [77], as shown in Figure 16.13. The highest reported linear density is ~55 CNTs/µm [78], though the pitch of the CNTs was inconsistent. Inconsistent pitch leads to challenges for fabricating CNTFETs with a controllable number of CNTs in each channel. Furthermore, if nanotubes are too closely spaced without being electrically isolated from each other, then charge-screening interactions will also be present and nonhomogeneous [79–82]. Placement of CNTs with uniform pitch has not progressed as rapidly, but is improving as techniques involving nanotube and/or substrate functionalization continue to advance [83–87]. The best linear density at a controlled pitch is 10 CNTs/µm [87], which is approximately an order of magnitude away from the target of at least 125 CNTs/µm [75].

16.6 Perspective

This chapter has shown that CNTFETs are aggressively scalable and have been experimentally demonstrated to outperform competing devices at low voltages. As with all "next switch" logic devices, there are challenges that remain to realizing a CNTFET technology, most of which have been reviewed in Sections 16.4 and 16.5. Should these obstacles be overcome, where would CNTFETs be expected to show up on the transistor roadmap? First it must be decided which of the roadmaps CNTFETs are best suited for. The most obvious is that of high performance digital computing, which CNTFETs have been anticipated to compete in for most of their lifetime. A technologically relevant CNTFET in this space is projected to have 5–6 CNT channels in each device, as shown in Figure 16.14a. To get an idea of how a CNTFET technology would perform compared to other options for the sub-10 nm technology nodes, a power–performance simulator was built that optimizes each

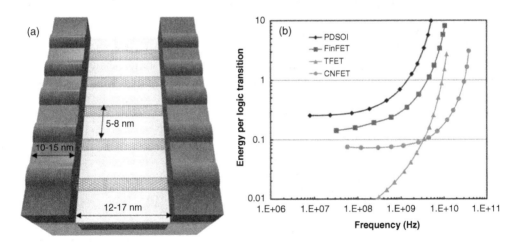

Figure 16.14 (a) Schematic of technologically relevant CNTFET device for 5 nm technology node, including 5–6 CNTs at a pitch of 5–8 nm. The noted channel length includes any spacer (source/drain extension) lengths. (b) Power versus performance simulation mimicking a P7 microprocessor architecture (1.5 M logic gates, four cores) comparing CNTFET (CNFET) to tunnel FETs (TFET), silicon FinFETs, and partially depleted silicon on insulator FETs (PDSOI) at the 7 nm technology node, with CNTFETs yielding a 3× better efficiency and 3× higher performance than FinFETs at a fixed chip power density [75]

Table 16.1 Summary parameter table of CNTFETs

Parameter		Values for the device in this chapter
Gate length L_g	Demonstrated	9 nm
	Projected	5 nm
Subthreshold Swing *SS*	Demonstrated	60 mV/dec
	Projected	60 mV/dec
Device speed, gate delay τ	Demonstrated	2 ps
	Projected	0.42 ps
Operation voltage V_{DD}	Demonstrated	0.4 V
	Projected	0.25 V
On current at $V_{DD} = 0.5$ V (normalized over contact width not CNT diameter)	Demonstrated	0.6 mA/μm
	Projected	2 mA/μm

device's dimensions for the best power–performance trade-off. As shown in Figure 16.14b, CNTFETs are projected to enable a technology with ~3× greater efficiency along with ~3× higher performance than FinFETs for a constant chip power density [75]. This level of power–performance improvement is very motivating for pursuit of a CNTFET technology in the high performance digital space.

Another way to view the type of performance that CNTFETs can offer compared to the competition is to consider relevant projected and demonstrated parameters, as shown in Table 16.1. Throughout this book, nanoelectronic devices that compete for the same application space as CNTFETs have their parameters listed in a similar fashion, providing a clear picture of the advantages available from each device and how close the promised parameters are to being demonstrated. As seen in Table 16.1, CNTFETs have very nearly demonstrated the projected performance for all of the important metrics.

While ideally suited for application in the high performance digital space, CNTFETs are also very promising for radio frequency (RF) electronics [88–92]. With the ability to carry high currents, a reportedly high linearity, and a bandgap that allows for current saturation, many see

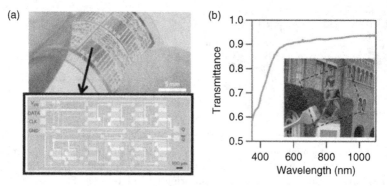

Figure 16.15 Examples of (a) flexible [94] and (b) transparent [95] CNTFET circuits. Reprinted by permission from (a) Macmillan Publishers Ltd: Nature Nanotechnology [94], copyright 2011 and (b) [95] copyright 2009 American Chemical Society

nanotubes as more suited for RF than graphene, which has received much more attention in the RF device community. The same materials challenges that face CNTFETs for digital are hedging the way for RF applications, though the needed deliverables in purity and placement density are more relaxed for RF [93].

The other roadmaps that a CNTFET technology could very well contribute to, if not completely enable, are those of flexible and/or transparent electronics. While there are a few other nanoelectronic devices that could be applicable to these fields, the CNTFET is one of the best performing and potentially the most accessible. There have already been a number of demonstrations of flexible and/or transparent circuits built using CNTFETs [94–99], as shown in Figure 16.15. These devices tend to be thin-film transistors (TFTs), which have less stringent performance, scaling, and fabrication requirements as they are traditionally assembled using large films of CNTs to make devices with large feature sizes. Ultimately, even the best performing CNTFETs would be compatible with these exotic substrates owing to their bottom-up fabrication.

16.7 Conclusion

Carbon nanotubes may very well be the most ideal semiconductor material discovered to date for high performance FETs. With demonstrations of ballistic transport, sub-10 nm performance, and complementary circuits that perform at less than 0.4 V [100], CNTFETs have proven their viability. However, as with all emerging nanoelectronic devices, there are many aspects to CNTFETs that require further exploration, understanding, and optimization in order to enable a new technology. With the ability to provide superb performance on technology roadmaps from high-performance digital to flexible electronics, it seems likely that there will eventually be a CNTFET node in the future.

References

1. Dresselhaus, M.S., Dresselhaus, G., and Avouris, P. (2001) *Carbon Nanotubes: Synthesis, Structure, Properties, and Applications*, Springer, Berlin Heidelberg.
2. Javey, A. and Kong, J. (eds) (2009) *Carbon Nanotube Electronics*, Springer.
3. Datta, S. (2005) *Quantum Transport: Atom to Transistor*, Cambridge University Press, New York.
4. Guo, J. and Lundstrom, M. (2005) Role of phonon scattering in carbon nanotube field-effect transistors. *Applied Physics Letters*, **86**(19), 193103.
5. Chandra, B., Perebeinos, V., Berciaud, S. *et al.* (2011) Low bias electron scattering in structure-identified single wall carbon nanotubes: role of substrate polar phonons. *Physical Review Letters*, **107**(14), 146601.
6. Zhou, X., Park, J.-Y., Huang, S. *et al.* (2005) Band structure, phonon scattering, and the performance limit of single-walled carbon nanotube transistors. *Physical Review Letters*, **95**(14), 146805.
7. Franklin, A.D. and Chen, Z. (2010) Length scaling of carbon nanotube transistors. *Nature Nanotechnology*, **5**(12), 858–862.
8. Javey, A., Guo, J., Wang, Q. *et al.* (2003) Ballistic carbon nanotube field-effect transistors. *Nature*, **424**, 654–657.
9. Javey, A., Guo, J., Farmer, D.B. *et al.* (2004) Self-aligned ballistic molecular transistors and electrically parallel nanotube arrays. *Nano Letters*, **4**(7), 1319–1322.
10. Mann, D., Javey, A., Kong, J. *et al.* (2003) Ballistic transport in metallic nanotubes with reliable Pd ohmic contacts. *Nano Letters*, **3**(11), 1541–1544.
11. Zhang, Z., Liang, X., Wang, S. *et al.* (2007) Doping-free fabrication of carbon nanotube based ballistic CMOS devices and circuits. *Nano Letters*, **7**(12), 3603–3607.
12. Zhang, Z., Wang, S., Ding, L. *et al.* (2008) Self-aligned ballistic n-type single-walled carbon nanotube field-effect transistors with adjustable threshold voltage. *Nano Letters*, **8**(11), 3696–3701.
13. Heinze, S., Tersoff, J., Martel, R. *et al.* (2002) Carbon nanotubes as Schottky barrier transistors. *Physical Review Letters*, **89**(10), 106801.

14. Appenzeller, J., Knoch, J., Radosavljevic, M., and Avouris, P. (2004) Multimode transport in schottky-barrier carbon-nanotube field-effect transistors. *Physical Review Letters*, **92**(22), 226802.
15. Svensson, J. and Campbell, E.E.B. (2011) Schottky barriers in carbon nanotube-metal contacts. *Journal of Applied Physiology*, **110**(11), 111101.
16. Léonard, F. and Talin, A. (2011) Electrical contacts to one-and two-dimensional nanomaterials. *Nature Nanotechnology*, **6**, 773–784.
17. Tans, S.J., Verschueren, A.R.M., and Dekker, C. (1998) Room-temperature transistor based on a single carbon nanotube. *Nature*, **393**, 49–52.
18. Martel, R., Schmidt, T., Shea, H. *et al.* (1998) Single-and multi-wall carbon nanotube field-effect transistors. *Applied Physics Letters*, **73**, 2447.
19. Pop, E. (2010) Energy dissipation and transport in nanoscale devices. *Nano Research*, **3**(3), 147–169.
20. Pop, E. (2008) The role of electrical and thermal contact resistance for Joule breakdown of single-wall carbon nanotubes. *Nanotechnology*, **19**(29), 295202.
21. Appenzeller, J., Lin, Y., Knoch, J. *et al.* (2005) Comparing carbon nanotube transistors — the ideal choice: a novel tunneling device design. *IEEE Transactions on Electron Devices*, **52**(12), 2568–2576.
22. Champlain, J. (2011) On the use of the term 'ambipolar'. *Applied Physics Letters*, **123502**, 22–24.
23. Martel, R., Derycke, V., Lavoie, C. *et al.* (2001) Ambipolar electrical transport in semiconducting single-wall carbon nanotubes. *Physical Review Letters*, **87**(25), 256805.
24. Knoch, J., Mantl, S., and Appenzeller, J. (2005) Comparison of transport properties in carbon nanotube field-effect transistors with Schottky contacts and doped source/drain contacts. *Solid-State Electronics*, **49**(1), 73–76.
25. Knoch, J. and Appenzeller, J. (2008) Tunneling phenomena in carbon nanotube field-effect transistors. *Physica Status Solidi (a)*, **205**(4), 679–694.
26. John, D.L., Castro, L.C., and Pulfrey, D.L. (2004) Quantum capacitance in nanoscale device modeling. *Journal of Applied Physiology*, **96**(9), 5180.
27. Sze, S.M. (1981) *Physics of Semiconductor Devices*, 2nd edn, John Wiley & Sons.
28. Leonard, F. and Tersoff, J. (2000) Role of fermi-level pinning in nanotube schottky diodes. *Physical Review Letters*, **84**(20), 4693–4696.
29. Appenzeller, J., Lin, Y.-M., Knoch, J., and Avouris, P. (2004) Band-to-band tunneling in carbon nanotube field-effect transistors. *Physical Review Letters*, **93**(19), 196805.
30. Tseng, Y.-C., Phoa, K., Carlton, D., and Bokor, J. (2006) Effect of diameter variation in a large set of carbon nanotube transistors. *Nano Letters*, **6**(7), 1364–1368.
31. Nosho, Y., Ohno, Y., Kishimoto, S., and Mizutani, T. (2005) n-type carbon nanotube field-effect transistors fabricated by using Ca contact electrodes. *Applied Physics Letters*, **86**(7), 073105.
32. Yaish, Y., Park, J.-Y., Rosenblatt, S. *et al.* (2004) Electrical nanoprobing of semiconducting carbon nanotubes using an atomic force microscope. *Physical Review Letters*, **92**(4), 046401.
33. Chen, Z., Lin, Y.-M., Appenzeller, J. *et al.* (2005) The role of metal-nanotube contact in the performance of carbon nanotube field-effect transistors. *Nano Letters*, **5**(7), 1497–1502.
34. Nemec, N., Tománek, D., and Cuniberti, G. (2008) Modeling extended contacts for nanotube and graphene devices. *Physical Review B-Condensed Matter*, **77**(12), 125420.
35. Nemec, N., Tománek, D., and Cuniberti, G. (2006) Contact dependence of carrier injection in carbon nanotubes: an *ab initio* study. *Physical Review Letters*, **96**(7), 076802.
36. Matsuda, Y., Deng, W.-Q., and Goddard, W.A. (2010) Contact resistance for 'end-contacted' metal– graphene and metal– nanotube interfaces from quantum mechanics. *The Journal of Physical Chemistry C*, **114**(41), 17845–17850.
37. Cui, X., Freitag, M., Martel, R. *et al.* (2003) Controlling energy-level alignments at carbon nanotube/Au contacts. *Nano Letters*, **3**(6), 783–787.
38. Javey, A., Guo, J., Paulsson, M. *et al.* (2004) High-field quasiballistic transport in short carbon nanotubes. *Physical Review Letters*, **92**(10), 106804.
39. Derycke, V., Martel, R., Appenzeller, J., and Avouris, P. (2002) Controlling doping and carrier injection in carbon nanotube transistors. *Applied Physics Letters*, **80**, 2773.
40. Javey, A., Tu, R., Farmer, D.B. *et al.* (2005) High performance n-type carbon nanotube field-effect transistors with chemically doped contacts. *Nano Letters*, **5**(2), 345–348.
41. Klinke, C., Chen, J., Afzali, A., and Avouris, P. (2005) Charge transfer induced polarity switching in carbon nanotube transistors. *Nano Letters*, **5**(3), 555–558.

42. Chen, J., Klinke, C., Afzali, A., and Avouris, P. (2005) Self-aligned carbon nanotube transistors with charge transfer doping. *Applied Physics Letters*, **86**(12), 123108.
43. Ding, L., Wang, S., Zhang, Z. *et al.* (2009) Y-contacted high-performance n-type single-walled carbon nanotube field-effect transistors: scaling and comparison with Sc-contacted devices. *Nano Letters*, **9**(12), 4209–4214.
44. Shahrjerdi, D., Franklin, A., Oida, S. *et al.* (2011) High Device Yield Carbon Nanotube NFETs for High-Performance Logic Applications. IEDM Technical Digest.
45. Engel-Herbert, R., Hwang, Y., and Stemmer, S. (2010) Quantification of trap densities at dielectric/III–V semiconductor interfaces. *Applied Physics Letters*, **97**(6), 062905.
46. Leskela, M. and Ritala, M. (2002) Atomic layer deposition (ALD): from precursors to thin film structures. *Thin Solid Films*, **409**(1), 138–146.
47. Lin, Y.M., Appenzeller, J., Chen, Z.H. *et al.* (2005) High-performance dual-gate carbon nanotube FETs with 40-nm gate length. *IEEE Electron Device Letters*, **26**(11), 823–825.
48. Franklin, A., Lin, A., Wong, H.-S., and Chen, Z. (2010) Current scaling in aligned carbon nanotube array transistors with local bottom gating. *IEEE Electron Device Letters*, **31**(7), 644–646.
49. Farmer, D.B. and Gordon, R.G. (2006) Atomic layer deposition on suspended single-walled carbon nanotubes via gas-phase noncovalent functionalization. *Nano Letters*, **6**(4), 699–703.
50. Chen, Z.H., Farmer, D., Xu, S. *et al.* (2008) Externally assembled gate-all-around carbon nanotube field-effect transistor. *IEEE Electron Device Letters*, **29**(2), 183–185.
51. Franklin, A., Koswatta, S., Farmer, D. *et al.* (2013) Carbon nanotube complementary wrap-gate transistors. *Nano Letters*, **13**, 2490–2495.
52. Lu, Y., Bangsaruntip, S., Wang, X. *et al.* (2006) DNA functionalization of carbon nanotubes for ultrathin atomic layer deposition of high kappa dielectrics for nanotube transistors with 60mV/decade switching. *Journal of the American Chemical Society*, **128**(11), 3518–3519.
53. Claussen, J.C., Franklin, A.D., Ul Haque, A. *et al.* (2009) Electrochemical biosensor of nanocube-augmented carbon nanotube networks. *ACS Nano*, **3**(1), 37–44.
54. Bradley, K., Gabriel, J.-C.P., Star, A., and Grüner, G. (2003) Short-channel effects in contact-passivated nanotube chemical sensors. *Applied Physics Letters*, **83**(18), 3821.
55. Sinha, N., Ma, J., and Yeow, J.T.W. (2006) Carbon nanotube-based sensors. *Journal of Nanoscience and Nanotechnology*, **6**(3), 573–590.
56. Franklin, A., Tulevski, G., Han, S. *et al.* (2012) Variability in carbon nanotube transistors: improving device-to-device consistency. *ACS Nano*, **6**(2), 1109–1115.
57. Chen, Z., Appenzeller, J., Lin, Y.M. *et al.* (2006) An integrated logic circuit assembled on a single carbon nanotube. *Science*, **311**(5768), 1735.
58. Franklin, A.D., Sayer, R.A., Sands, T.D. *et al.* (2009) Toward surround gates on vertical single-walled carbon nanotube devices. *Journal of Vacuum Science & Technology B*, **27**(2), 821–826.
59. Léonard, F. and Stewart, D.A. (2006) Properties of short channel ballistic carbon nanotube transistors with ohmic contacts. *Nanotechnology*, **17**(18), 4699–4705.
60. Guo, J., Datta, S., and Lundstrom, M. (2004) A numerical study of scaling issues for Schottky-barrier carbon nanotube transistors. *IEEE Transactions on Electron Devices*, **51**(2), 172–177.
61. Koswatta, S.O., Hasan, S., Lundstrom, M.S. *et al.* (2007) Nonequilibrium green's function treatment of phonon scattering in carbon-nanotube transistors. *IEEE Transactions on Electron Devices*, **54**(9), 2339–2351.
62. Franklin, A., Luisier, M., Han, S.-J. *et al.* (2012) Sub-10nm carbon nanotube transistor. *Nano Letters*, **12**(2), 758–762.
63. Choi, S.-J., Bennett, P., Takei, K. *et al.* (2013) Short-channel transistors constructed with solution-processed carbon nanotubes. *ACS Nano*, **7**, 798–803.
64. Javey, A., Qi, P., Wang, Q., and Dai, H. (2004) Ten- to 50-nm-long quasi-ballistic carbon nanotube devices obtained without complex lithography. *Proceedings of the National Academy of Sciences of the United States of America*, **101**(37), 13408–13410.
65. Lan, C., Zakharov, D.N., and Reifenberger, R.G. (2008) Determining the optimal contact length for a metal/multiwalled carbon nanotube interconnect. *Applied Physics Letters*, **92**(21), 213112.
66. Clifford, J.P., John, D.L., Castro, L.C., and Pulfrey, D.L. (2004) Electrostatics of partially gated carbon nanotube FETs. *IEEE Transactions on Nanotechnology*, **3**(2), 281–286.
67. Solomon, P.M. (2011) Contact resistance to a one-dimensional quasi-ballistic nanotube/wire. *IEEE Electron Device Letters*, **32**(3), 246–248.

68. Cummings, A.W. and Léonard, F. (2011) Electrostatic effects on contacts to carbon nanotube transistors. *Applied Physics Letters*, **98**(26), 263503.
69. Arnold, M.S., Green, A.A., Hulvat, J.F. *et al.* (2006) Sorting carbon nanotubes by electronic structure using density differentiation. *Nature Nanotechnology*, **1**, 60–65.
70. Bonaccorso, F., Hasan, T., Tan, P.H. *et al.* (2010) Density gradient ultracentrifugation of nanotubes: interplay of bundling and surfactants encapsulation. *The Journal of Physical Chemistry C*, **114**(41), 17267–17285.
71. Hersam, M.C. (2008) Progress towards monodisperse single-walled carbon nanotubes. *Nature Nanotechnology*, **3**(7), 387–394.
72. Liu, H., Feng, Y., Tanaka, T. *et al.* (2010) Diameter-selective metal/semiconductor separation of single-wall carbon nanotubes by agarose gel. *The Journal of Physical Chemistry C*, **114**(20), 9270–9276.
73. Tu, X., Manohar, S., Jagota, A., and Zheng, M. (2009) DNA sequence motifs for structure-specific recognition and separation of carbon nanotubes. *Nature*, **460**(7252), 250–253.
74. Liu, J. and Hersam, M.C. (2010) Recent developments in carbon nanotube sorting and selective growth. *MRS Bulletin*, **35**, 315–322.
75. Haensch, W., Wong, H.-S.P., and Franklin, A. (2012) Carbon Nanotubes for Digital Electronics Workshop. Workshop Summary Report.
76. Tulevski, G., Franklin, A., and Afzali, A. (2013) High purity isolation and quantification of semiconducting carbon nanotubes via column chromatography. *ACS Nano*, **7**(4), 2971–2976.
77. Franklin, A. D. (2013) The road to carbon nanotube transistors. *Nature*, **498**, 443.
78. Wang, C., Ryu, K., Arco, L.G. *et al.* (2010) Synthesis and device applications of high-density aligned carbon nanotubes using low-pressure chemical vapor deposition and stacked multiple transfer. *Nano Research*, **3**(12), 831–842.
79. Raychowdhury, A., Keshavarzi, A., Kurtin, J. *et al.* (2006) Optimal Spacing of Carbon Nanotubes in a CNFET Array for Highest Circuit Performance. 64th Device Research Conference, no. 2, pp. 129–130.
80. Deng, J. and Wong, H.-S.P. (2007) Modeling and analysis of planar-gate electrostatic capacitance of 1-D FET with multiple cylindrical conducting channels. *IEEE Transactions on Electron Devices*, **54**(9), 2377–2385.
81. Li, X.F., Chen, K.Q., Wang, L.L. *et al.* (2007) Effect of intertube interaction on the transport properties of a carbon double-nanotube device. *Journal of Applied Physiology*, **101**, 064514.
82. Léonard, F. (2006) Crosstalk between nanotube devices: contact and channel effects. *Nanotechnology*, **17**(9), 2381–2385.
83. Vijayaraghavan, A., Blatt, S., Weissenberger, D. *et al.* (2007) Ultra-large-scale directed assembly of single-walled carbon nanotube devices. *Nano Letters*, **7**(6), 1556–1560.
84. Stokes, P. and Khondaker, S.I. (2008) Local-gated single-walled carbon nanotube field effect transistors assembled by AC dielectrophoresis. *Nanotechnology*, **19**(17), 175202.
85. Jiao, L., Xian, X., Wu, Z. *et al.* (2009) Selective positioning and integration of individual single-walled carbon nanotubes. *Nano Letters*, **9**(1), 205–209.
86. Tulevski, G.S., Hannon, J., Afzali, A. *et al.* (2007) Chemically assisted directed assembly of carbon nanotubes for the fabrication of large-scale device arrays. *Journal of the American Chemical Society*, **129**(39), 11964–11968.
87. Park, H., Afzali, A., Han, S. *et al.* (2012) High-density integration of carbon nanotubes via chemical self-assembly. *Nature Nanotechnology*, **7**, 787–791.
88. Paydavosi, N., Alam, A.U., Ahmed, S. *et al.* (2011) RF performance potential of array-based carbon-nanotube transistors — Part I: intrinsic results. *IEEE Transactions on Electron Devices*, **58**(7), 1928–1940.
89. Alam, A.U., Rogers, C.M.S., Paydavosi, N. *et al.* (2013) RF linearity potential of carbon-nanotube Transistors vs. MOSFETs. *IEEE Transactions on Nanotechnology*, **12**(3), 340–351.
90. Kocabas, C., Kim, H.-S., Banks, T. *et al.* (2008) Radio frequency analog electronics based on carbon nanotube transistors. *Proceedings of the National Academy of Sciences of the United States of America*, **105**(5), 1405–1409.
91. Wang, C., Badmaev, A., Jooyaie, A. *et al.* (2011) Radio frequency and linearity performance of transistors using high-purity semiconducting carbon nanotubes. *ACS Nano*, **5**(5), 4169–4176.
92. Steiner, M., Engel, M., Lin, Y.-M. *et al.* (2012) High-frequency performance of scaled carbon nanotube array field-effect transistors. *Applied Physics Letters*, **101**(5), 053123.
93. Koswatta, S.O., Valdes-Garcia, A., Steiner, M.B. *et al.* (2011) Ultimate RF performance potential of carbon electronics. *IEEE Transactions on Microwave Theory and Techniques*, **59**(10), 2739–2750.
94. Sun, D., Timmermans, M.Y., and Tian, Y. (2011) Flexible high-performance carbon nanotube integrated circuits. *Nature Nanotechnology*, **6**(February), 156–161.
95. Ishikawa, F.N., Chang, H.-K., Ryu, K. *et al.* (2009) Transparent electronics based on transfer printed aligned carbon nanotubes on rigid and flexible substrates. *ACS Nano*, **3**(1), 73–79.

96. Chandra, B., Park, H., Maarouf, A. *et al.* (2011) Carbon nanotube thin film transistors on flexible substrates. *Applied Physics Letters*, **99**(7), 072110.

97. Cao, Q., Kim, H., Pimparkar, N. *et al.* (2008) Medium-scale carbon nanotube thin-film integrated circuits on flexible plastic substrates. *Nature*, **454**(7203), 495–500.

98. Kim, S., Ju, S., Back, J.H. *et al.* (2009) Fully transparent thin-film transistors based on aligned carbon nanotube arrays and indium tin oxide electrodes. *Advanced Materials*, **21**(5), 564–568.

99. Wang, C., Chien, J., Takei, K. *et al.* (2012) Extremely bendable, high-performance integrated circuits using semiconducting carbon nanotube networks for digital, analog, and radio-frequency applications. *Nano Letters*, **12**(3), 1527–1533.

100. Ding, L., Liang, S., Pei, T. *et al.* (2012) Carbon nanotube based ultra-low voltage integrated circuits: Scaling down to 0.4V. *Applied Physics Letters*, **100**(26), 263116.

17

Spintronics

Alexander Khitun

Material Science and Engineering, University of California, USA

17.1 Introduction

Spintronics, also known as magnetoelectronics, is an emerging technology exploiting both the intrinsic spin of the electron and its associated magnetic moment, in addition to its fundamental electronic charge, in solid-state devices [1]. Either adding the spin degree of freedom to conventional charge-based electronic devices or using the spin alone has the potential advantages of nonvolatility, increased data processing speed, decreased electric power consumption, and increased integration densities compared with conventional semiconductor devices [2]. Spintronics originates from research on the influence of spin on electrical conduction in ferromagnetic metals by Mott in 1936, Fert and Campbell in 1968, and gained further momentum with the discovery of spin-dependent electron transport phenomena in the 1980s. The notable milestones of Spintronics development include the observation of spin-polarized electron injection from a ferromagnetic metal to a normal metal by Johnson and Silsbee [3], and the discovery of giant magnetoresistance (GMR) independently by Albert Fert *et al.* [4] and Peter Grünberg *et al.* [5]. Pioneered by IBM in 1997, the GMR head enabled hard-disk drives to read smaller data bits, which led to a more than 40-fold increase in data storage density over the past seven years [6]. Today, the area of Spintronics is expanding considerably by considering spin-based phenomena, that is, spin transfer torque [7], and by introducing novel materials, that is, graphene spintronics [8] and molecular spintronics [9].

Spintronics has been recognized as one of the emerging technologies able to provide the continuation of the functional throughput enhancement [10]. On the one hand, the potential advantages of using the additional degree of freedom provided by spin are indisputable. On the other hand, the utilization of spin-based devices has to follow the mainstream of conventional electron-based devices and provide a substantial performance improvement in order to justify the technological difficulties associated with combining the two technologies. How do we utilize spin in the most efficient way? And how should we integrate novel spin-based devices with the traditional transistor-based logic circuitry? These are the main questions to be answered.

Emerging Nanoelectronic Devices, First Edition. An Chen, James Hutchby, Victor Zhirnov and George Bourianoff.
© 2015 John Wiley & Sons, Ltd. Published 2015 by John Wiley & Sons, Ltd.

Current efforts in developing spintronic logic devices involve two major approaches. The first approach is to build a spin transistor enabling lower power consumption and higher functional throughput than the scaled CMOS. This approach originates from the first proposal on the electro-optical modulator proposed by Datta and Das in 1990 [11] and has resulted in a great deal of research during the past two decades. This direction is pretty straightforward as it would continue the mainstream of conventional logic circuitry development as it worked for the past seven decades. There is a variety of spin transistors proposed so far, which can be classified into two types: the spin-FET proposed by Datta and Das (and its modified versions [12,13]) and the spin-MOSFET with several variations [14,15]. The spin-FET type transistors are aimed to take advantage of spin-polarized current modulation via spin–orbit coupling and build a transistor with unique output characteristics (i.e., output current oscillates as a function of the gate voltage). Another type of spin transistor called the spin-MOSFET resembles the conventional MOSFET with magnetic elements included. The incorporation of magnetic elements in the conventional field effect transistor structure offers advantages in both energy saving and logic functionality [15]. Besides these two major groups, there are multiple proposals to use organic materials, graphene (as the channel), and hybrid schemes combining conventional MOSFET with magnetic tunneling junction (MTJ) [16].

The second approach, which is more radical, focuses on the utilization of spin as a logic state variable. In theory, logic 0 and 1 can be assigned to the polarization of a single spin placed it in a magnetic field (so-called Single Spin Logic [17,18]). Such a logic representation promises additional functionalities beyond the capabilities of the conventional FET. A spin system can be maintained in a nonequilibrium state longer than a charge system due to the weaker spin-phonon coupling, which translates in low power consumption for low-duty cycle applications [19]. Single spin logic possesses similar problems as for single electron logic, namely high sensitivity to the structure imperfections and thermal fluctuations. More practical proposals are aimed at utilizing the collective states of a large number of spins coupled via the exchange interaction (i.e., nanomagnetic logic [20]). Theoretically, switching between the collective spin states would require the same amount of energy as for a single spin state, though the collective state would be more immune to noise. Nonvolatility (ability to preserve the result of computation without an external power source) is the key advantage offered by the nanomagnetic logic thus reducing static power consumption, which is the critical problem of the charge-based approach.

While there are promising potential advantages to using spin-based logic devices, a number of questions remain open. In this section, we describe the principle of operation, the current status of development of the spin-based logic devices and we discuss the potential advantages and key technological issues limiting the performance of these devices and Spintronics in general.

17.2 Spin Transistors

Spin transistors constitute a special type of semiconductor transistor combining conventional transistor behavior with the functionalities of a magnetoresistive device. The important features of spin transistors are: (i) variable drive current controlled by the magnetization configuration of the ferromagnetic electrodes, and (ii) nonvolatile information storage using the magnetization configuration. These features are very useful functionalities for the goal of achieving highly energy efficient, low-power circuit architectures that are inaccessible to ordinary CMOS circuits [10].

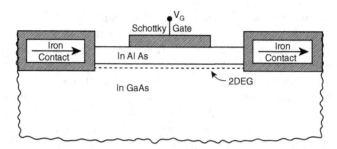

Figure 17.1 Schematics of the semiconductor electron wave analog of the electro-optic modulator – the first proposed Spin FET. Reprinted with permission from [11]. Copyright 1990, American Institute of Physics

17.2.1 Spin-FETs

It would be fair to say that the development of spin transistors has started from the proposal of the semiconductor electron wave analog of the electro-optic modulator proposed by Datta and Das [11], as shown in Figure 17.1. This is a three-terminal electronic device with source, drain, and gate contacts, where the source to drain current is controlled by applying a gate voltage. At first look, this device is very similar to the ordinary semiconductor transistor. However, the principle of operation of the spin-FET is completely different from the conventional transistors. The source and the drain are made of a ferromagnetic (or half-metallic) material working as polarizer and analyzer for the spin polarized current. The electrons are injected into a quasi one-dimensional semiconductor channel from the magnetic source. The initial spin polarization of the injected spins is supposed to be parallel to the polarization of the source. Then, the precession of the injected spins in the channel is controlled with a gate potential via the Rashba spin–orbit coupling effect. The precession angle θ is defined by the following [11]:

$$\theta = 2m^*\alpha L/\hbar^2, \tag{17.1}$$

where m^* is the effective electron mass, parameter α is the spin orbit coupling coefficient, L is the channel length, $\hbar$ is Planck's constant. The probability that the electrons will be transmitted through the channel/drain interface depends on the relative orientation of the electron's spin with the drain magnetization and the rate of precession controlled by the gate voltage. This probability is maximized when the polarization of the incoming electrons is parallel to the magnetization of the drain ($\theta = 0$), and the probability is at a minimum where the polarization of the incoming electrons and the drain are antiparallel ($\theta = \pi$). Thus, the source–drain current oscillates as a function of the gate voltage similar to the intensity modulation in the electro-optic modulators.

Efficient spin injection into the semiconductor channel, long spin coherence length in the channel, strong Rashba coupling for spin polarization control and low scattering are the key requirements for spin-FET operation. The efficiency of the spin injection is defined by the spin polarization η: $\eta = (N_\uparrow - N_\downarrow)/(N_\uparrow + N_\downarrow)$, where $N_\uparrow$ and $N_\downarrow$ is the number of the injected spin up and spin down electrons. In the ideal scenario, all of the injected electrons are of the same spin polarization $\eta = 1$. However, it is difficult to achieve high efficiency of spin injection from an ordinary ferromagnet (i.e., a ferromagnetic metal) to a semi-conductor due to the conductivity (or impedance) mismatch problem [21]. This problem originates from the large difference in electrical conductivity between a ferromagnetic metal and a semiconductor. A possible solution to this problem is via the introduction of insulating tunnel barriers between the ferromagnetic and semiconductor, or making Schottky contacts with highly doped semiconductor thin layers [22]. Spin injection in the material systems with In_xGa_{1-x}, as have been studied widely and extensively during the past decade. Typically, experimentally detected spin polarization does not exceed a few percentage points with spin diffusion length L_s of several microns (i.e., $\eta \sim 5.7\%$, $L_s \sim 5.1\,\mu m$ at $1.5\,K$ [23]).

Being injected into a semiconductor channel, the spins of conduction electrons are subject to different relaxation mechanisms (i.e., Elliott–Yafet [24], D'yakonov–Perel' [25], Bir–Aronov–Pikus [26]) which reduce spin polarization. The most critical are the Elliott–Yafet relaxation processes due to scattering via phonons, impurities, boundaries, which are inherent in all semiconductors, and the D'yakonov–Perel' mechanism in compound semiconductors without a center of inversion symmetry, which is most prominent in III-V compound semiconductors. All the scattering mechanisms tend to equalize the number of spin up and spin down electrons in a nonmagnetic semiconductor channel. In turn, the decrease of spin polarization within the ensemble of conducting electrons reduces the On/Off ratio. For spin-FETs, it is critical to suppress the spin relaxation in the channel while keeping the maximum strength of the spin–orbit interaction. The suppression of spin relaxation can be achieved via electron confinement (i.e., making one-dimensional channel structures whose characteristic size is less than the bulk spin diffusion length) [27], or by utilizing the so-called persistent spin helix (PSH) condition [28]. In the PSH condition, spin polarization is conserved even after scattering events. This conservation is predicted to be robust against all forms of spin independent scattering, including electron–electron interaction. Recently, the PSH condition has been experimentally realized and spin life time enhancement by two orders of magnitude was observed near the exact PSH point (as shown in Figure 17.2) [29]. Though these experimental data demonstrate a significant progress towards the spin control in semiconductors, the achieved spin lifetime (10–100 ps) is still not long enough for any practical application.

Besides efficient spin injection and long spin coherence length, spin-FETs require certain channel materials with a strong spin–orbit interaction such as InGaAs, InAs, and InSb. The stronger the spin–orbit coupling α, the shorter is the channel length L at which spin polarization can be π-rotated at a given gate voltage (Eq. Eq. (17.1)). There are two kinds of spin–orbit interactions known as: (i) the Dresselhaus spin–orbit interaction [30], which originates from the lack of an inversion asymmetrical lattice structure, and (ii) the Rashba spin–orbit interaction [31], which originates from the sum of the effective electric fields at the interfaces and inside the confined structure. During the past two decades, Rashba spin–orbit interaction has been widely discussed and investigated [32]. The strength of the spin–orbit coupling is usually determined from observation of beatings in the Shubnikov–de Haas (ShdH) oscillations [33–35], the measurement of magnetoresistance in the regime of weak anti-localization in low magnetic fields [36], by the observation of Zeeman splitting energies in confined structures [37], as well as by analyzing the Raman spectrum [38], spin resonance [39] and cyclotron resonance [40]. It should be noted that the voltage-controlled spin–orbit interaction may come from the competition of two or more mechanisms (e.g., the change of the carrier concentration, or the modification of the confined electron's wave function). For example, the experimental data for $In_{0.75}Ga_{0.25}As/In_{0.75}Al_{0.25}As$ hetero-structure show α to be almost

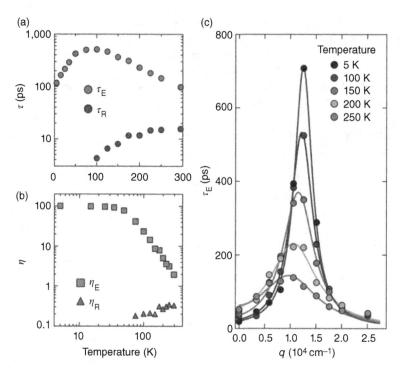

Figure 17.2 (a) Temperature dependence of the spin–orbit enhanced (τ_E) and spin–orbit reduced (τ_R) lifetimes of helix modes for the 11-nm GaAs quantum wells closest to the PSH regime. (b) Temperature dependence of the dimensionless lifetime enhancement factor η for each helix mode. (c) Temperature dependence of the PSH dispersion curves for a similar sample with a slightly reduced mobility. Reprinted by permission from [29]. Macmillan Publishers Ltd, Copyright 2009

constant near zero gate voltage and its decrease with negative Vg due to the change of the carrier concentration [33]. In contrast, the coupling constant α increases with the increase of negative gate voltage due to the modification of quantum well potential in a strained InGaAs/ InP quantum wire [34]. In all cases, the maximum observed value of spin–orbit coupling strength does not exceed the tens of 10^{-12} eVm, which implies certain limits on the gate length required for a π-phase modulation according to Equation 17.1. For example, taking $\alpha = 10 \times 10^{-12}$ eVm and the effective mass $m^* = 0.05$, the minimum gate length of spin-FETs is 260 nm [41]. Theoretically, the strength of spin–orbit coupling can be enhanced by using the PSH condition where the total effective magnetic field induced both by the Rashba spin–orbit interaction and the Dresselhaus spin–orbit interaction is doubled, reducing the length by half. It is also expected that gate length can be reduced for p-type spin FETs due to the larger spin–orbit interaction value and gate controllability [42]. It is important to note, that the channel length of spin-FETs (the length required for π-phase modulation) must be shorter than the spin scattering length, which is a grand challenge for building room temperature operating prototypes.

   The first working spin-FET prototype based on an InAs heterostructure was demonstrated in 2009 [43]. The prototype resembles a conventional lateral spin valve device for nonlocal measurements with two $Ni_{81}Fe_{19}$ electrodes on top of an InAs high-mobility channel

($\mu > 50\,000\,\text{cm}^2\,\text{V}^{-1}\,\text{s}^{-1}$) of length 1.65 µm. This device shows an oscillatory conductance as a function of the applied voltage as was originally predicted by Datta and Das [11]. However, the origin of the observed spin signals is not yet clear [44–46]. Overall, efforts to implement a Datta–Das spin transistor operating at room temperature have not yet been successful. In spite of great efforts by multiple research groups, spin-FETs are far from practical realization due to the problems associated with efficient spin injection and control, and the lack of the efficient control of the spin precession by a gate voltage. It is expected that there will be further progress in improving spin injection efficiency and spin relaxation suppression. However, the low strength of the spin–orbit coupling will remain the main fundamental problem making spin-FETs inferior to conventional CMOS.

17.2.2 Spin-MOSFETs

The problems with spin precession control by applying an electric field have stimulated further search for spin transistors the operation of which does not rely on weak spin–orbit coupling. This class of spin transistors is known as spin-MOSFETs [14]. In contrast to spin-FETs, the channel of spin-MOSFET is composed of silicon with a very weak spin–orbit interaction. There is no voltage-controlled spin polarization rotation required for spin-MOSFET operation. Current modulation through the channel is achieved in a way similar to the ordinary MOSFET by changing the carrier concentration in the channel, though the concentration of spin up and spin down electrons may differ significantly due to the spin-dependent band structure. In contrast to spin-FETs with a fixed source/drain magnetic configuration, it is assumed that the magnetization of the source and drain in spin-MOSFETs can be changed during operation and provide additional control of the output current. The latter opens a variety of ways for spin-MOSFET construction by implementing different materials for the source and drain and by engineering the spin-dependent band structure. The different types of spin-MOSFETs can be classified by the structure of the source/drain and the channel (e.g., ferromagnetic semiconductor source/drain, half-metallic source and drain, and ferromagnetic semiconductor channel and a ferromagnetic or half-metallic ferromagnetic source/drain [41]). Each type of MOSFET possesses certain advantages and constrains. For instance, the use of ferromagnetic semiconductors for source and drain forming a pn junction with the Si channel can resolve the conductivity mismatch problem. At the same time, it would require a ferromagnetic semiconductor with a large spin polarization in order to obtain a large magneto-current. Besides that, there are currently no known room temperature ferromagnetic semiconductor materials. MOSFETs with ferromagnetic/half-metallic source/drain Schottky contacts can be operated as accumulation- and inversion-type channel devices. The operation of the spin MOSFETs with Schottky junctions is similar to the Schottky barrier MOSFETs [47,48], where the current modulation is via the gate-induced modification of the Schottky barrier width [41]. A more detailed description of the spin-MOSFETs operation can be found in References [14,41,49]. The calculated output characteristics of a spin-MOSFET with a Half-Metal source and drain are presented in Reference [14]. The simulation was performed under the assumptions of ballistic transport and complete spin polarization (100%) without any spin-flip scattering. Under these rather unrealistic assumptions, the spin-MOSFET shows excellent transistor behavior.

As there is no need for spin–orbit coupling in spin-MOSFET operation, the channel can be made of silicon providing a link among the new approach with the well-established technology [49]. During the past five years, there has been a significant progress towards more

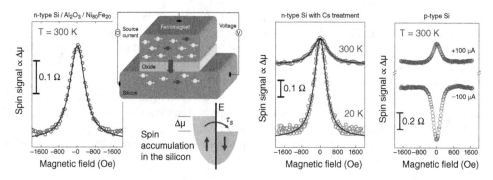

Figure 17.3 Creation of spin polarization in silicon at room temperature. Data are shown for n-type Si (left), n-type Si treated with Cs to suppress the Schottky barrier (middle) and p-type Si (right). The data show spin signal (the difference in the chemical potential ($\Delta\mu$) of the two spin populations as a function of the external magnetic field. Reprinted with permission from [49]. The Nature Publishing Group, Copyright 2012

efficient spin injection into silicon, and, finally, spin injection and detection at room temperature has been observed [50]. Figure 17.3 shows experimental data obtained for a three-terminal device comprising a $Ni_{80}Fe_{20}$ contact, a high-quality amorphous Al_2O_3 tunnel barrier, and a heavily doped Si (effective carrier density of about $2 \times 10^{19} \, cm^{-3}$) polarization [50]. The data clearly show that resistance changes with a characteristic Lorentzian shape at 300 K. These results together with several control experiments manifests room temperature spin accumulation in silicon [50]. Later on, spin injection in silicon has been confirmed by other groups [51,52]. Besides high spin injection efficiency, silicon shows a relatively long spin lifetime. In Figure 17.4, a collection of data for n-type silicon obtained via nonlocal measurements is shown. The spin lifetime at room temperature is around 0.1–1.0 ns, which is sufficient for device operation.

Though spin-MOSFETs have not yet been experimentally verified and there are many important issues to resolve, there has been important progress in increasing spin injection efficiency and observing a relative long spin lifetime. These achievements constitute an existence proof of spin-MOSFET operation at room temperature.

17.2.3 Pseudo-Spin FET

A number of problems associated with high spin injection efficiency into a semiconductor channel have stimulated a search for a hybrid device which would provide the same or similar output characteristics as a spin-MOSFET without the need for a magnetic source and drain. Such a device [referred to as a pseudo-spin-MOSFET (PS-MOSFET)] comprising an ordinary metal oxide semiconductor field-effect transistor (MOSFET) and magnetic tunnel junction has been proposed [53] and experimentally demonstrated [16]. The schematics of the PS-MOSFET are shown in Figure 17.5a. The MTJ is connected to the source of a MOSFET and feeds back its voltage drop to the gate. The degree of negative feedback depends on the resistance states of the MTJ. Thus, there are two possible values for the source–drain current at any fixed bias voltage depending on the magnetic configuration of the MTJ. The prototype PS-MOSFET was built by combining a bottom gate MOSFET and a MTJ with a full Heusler alloy (Co_2FeAl; CFA) electrode and an MgO tunnel barrier. The experimentally measured output characteristics of the

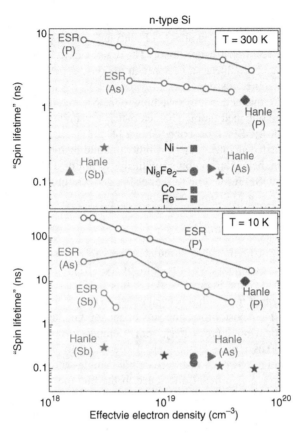

Figure 17.4 Spin lifetime in n-type Si. The spin lifetime is extracted from the Hanle measurements with electrically induced spin accumulation in silicon with different dopants (Sb, As, and P). The electron spin resonance (ESR) markers show data on bulk Si. Reprinted with permission from [49]. The Nature Publishing Group, Copyright 2012

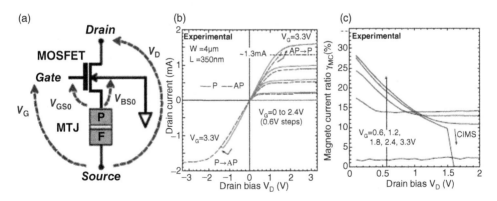

Figure 17.5 (a) Circuit configuration of the pseudo-spin-MOSFET (PS-MOSFET). (b) Experimentally measured output characteristics of the hybrid PS-MOSFET at RT. (c) Measured magneto resistance as a function of V_d for the hybrid integrated PS-MOSFET. The results show current-induced magnetization switching (CMIS) among the antiparallel (AP) and parallel (P) states. Reprinted with permission from [54]. Copyright 2012

fabricated PS-MOSFET are shown in Figure 17.5b [54]. The solid and dashed curves show the drain currents for parallel and anti-parallel MTJ configuration, respectively. Each curve follows depletion-type field effect transistor behavior. The drain current is higher for the parallel MTJ configuration over the entire linear and saturation regions. Experimental data in Figure 17.5c show the measured magnetoresistance as a function of V_d for the PS-MOSFET. The results show current-induced magnetization switching (CMIS) among the antiparallel (AP) and parallel (P) states. The obtained output characteristics are similar to the output characteristics simulated for the spin-MOSFET described above [41]. The successful demonstration of the PS-MOSFET has revealed an intriguing possibility of building hybrid logic devices possessing the standard transistor behavior combined with nonvolatility. It is highly expected that the development of PS-MOSFETs will be further elaborated within the next decade by benefiting from the rapid progress in MTJ improvement.

17.2.4 Graphene Spintronics

The quest for long spin lifetime has stimulated the study of spin transport in a variety of materials/nano-structures, which may be potentially used in spin-FETs. Here, we would like to refer to the most notable experimental works in this field.

Recently, a great deal of interest has been attracted to the potential utilization of graphene in spin-based devices. The most promising advantage of graphene is a long spin lifetime. According to theoretical predictions [55], spin lifetimes in a single and bilayer graphene sheets may exceed 100 ns at room temperature. Experimentally, a spin life time in the order of 1 ns at room temperature has been observed in both single- and bi-layer graphene [56]. The experimental data show a significant difference in the lifetime dependences for single-layer graphene (SLG) and bilayer graphene (BLG). In SLG, the spin lifetime varies linearly with the momentum scattering time as carrier concentration is varied, while the lifetime in BLG exhibit an inverse dependence [56]. Experimental data on room temperature spin lifetime in bilayer graphene are shown in Figure 17.6, indicating a strong lifetime decrease for samples with

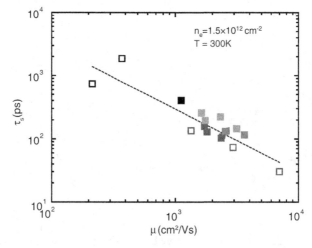

Figure 17.6 Hanle effect measurements of room temperature spin-relaxation times in a variety of bilayer graphene samples. Reprinted figure with permission from [56]. Copyright 2013 by the American Physical Society

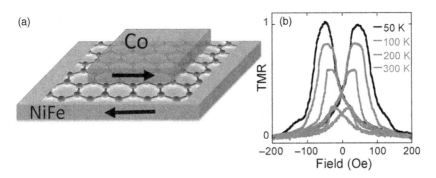

Figure 17.7 (a) Graphene tunnel junction devices. (b) TMR curves for selected temperatures to 400 K. Reprinted with permission from [57]. Copyright 2013, by the American Chemical Society

higher mobility. The major mechanisms of spin relaxation and the possibility of spin lifetime enhancement are currently under discussion and are currently not well understood.

The feasibility of using graphene as a tunneling barrier for MTJ [57] was also studied. The schematics of the device structure of SLG tunneling barrier and experimental data on the tunneling magnetoresistance are shown in Figure 17.7. Transport occurs by quantum tunneling perpendicular to the graphene plane and preserves a net spin polarization of the current from the contact so that the structures exhibit tunneling magnetoresistance to 425 K. These results demonstrate graphene capabilities as an effective tunnel barrier for both charge and spin-based devices. The unique electronic structure of graphene systems can be used to create pseudo-spin analogs of giant magnetoresistance and other established spintronics effects [8].

Despite the unique physical properties of graphene, its application in spin-based logic devices may be limited due to the following reasons. Low spin–orbit coupling (one of the reasons for long spin lifetime) makes graphene inefficient for spin-FETs. The absence of a bandgap makes the use of graphene channels in MOSFETs problematic because of the high Off current. The observed TMR in MTJ with graphene tunneling layer are orders of magnitude lower than for conventional MTJ with MgO. With all that, the long spin lifetime together with high mobility offered by graphene are most likely to be implemented in logic devices using spin polarized current for nanomagnets switching [58–60], to be discussed later in the text.

17.2.5 Organic Spintronics

The observation of long spin lifetime in organic semiconductors [61,62] has stimulated interest in organic materials as a possible medium to transport and control spin-polarized signals. Organic semiconductors are characterized by very low spin–orbit interaction, which, together with their chemical flexibility and relatively low production costs, makes them a promising materials system for spintronics applications [9]. So far, the most studied organic spintronic device is a spin valve comprising two ferromagnetic electrodes with an organic semiconductor layer in between. The typical thickness of the organic tunneling barrier is several nanometers or less allowing the observation of tunneling magnetoresistance. The first report on vertical injection organic spin valve was published in 2004 showing about 40% TMR at 11 K for (8-hydroxyquinoline)aluminum(III) (Alq3) [63]. Since then, many groups have confirmed similar MR from the devices with the same architecture, using both Alq3 and other OSC

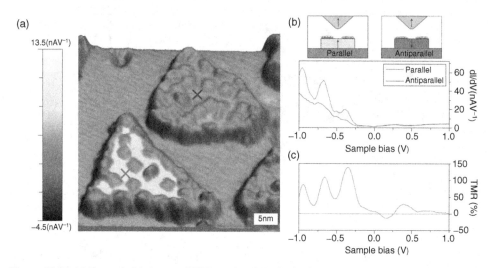

Figure 17.8 (a) Topographic image of H2Pc molecules adsorbed onto two cobalt islands on the Cu(111) surface. (b) Typical dI/dV spectra taken on parallel and antiparallel oriented cobalt islands [marked by red and blue crosses in (a)]. (c) Energy dependence of the TMR ratio calculated from the dI/dV spectra. Reprinted with permission from [66]. Copyright 2013, Nature Publishing Group

materials as a spacer layer (a review is given in Reference [64]). Remarkably, there are many discrepancies in reported results, which may be attributed to intrinsic material properties as well as to the interface quality and fabrication conditions [64,65].

Among the possible advantages offered by the Organic Spintronics is the possibility of building single-molecule spin-based devices. Recently, a giant magnetoresistance of 60% across a single, nonmagnetic hydrogen phthalocyanine molecule H_2Pc [66] was observed. The topographic image of the device and experimental data are shown in Figure 17.8. The H_2Pc molecules adsorbed onto two cobalt islands on the Cu(111) surface. The obtained I–V spectra clearly show the difference between parallel and anti-parallel oriented cobalt islands. This and other experimental reports [67] show significant progress in making organic spin-based devices.

Organic semiconductors are promising for spintronic application due to the long spin lifetime and the ability for efficient spin injection. At the same time, the spin diffusion length in organic semiconductors is limited to a range in the tens of nanometers due to the low mobility of organic crystals. Besides that, molecular spin-based devices are greatly affected by size variations and interface quality. Overall, it is difficult to expect that organic spintronics will provide any robust logic device in the foreseeable future which would be able to compete with CMOS.

17.2.6 Logic Circuits with Spin Transistors and Comparison with CMOS

The unique output characteristics provided by spin transistors opens new possibilities for logic circuit construction. For instance, the oscillation of the output current as a function of the gate voltage in spin-FETs is very useful for realization of specific logic functions such as XOR. At some point, the output characteristics of the device proposed by Datta and Das are similar to the ones of the single-electron transistor (SET) [68] and the logic circuits developed for SETs may be applicable for spin-FETs (e.g., one can compare the XOR circuit with SETs [69] and XOR

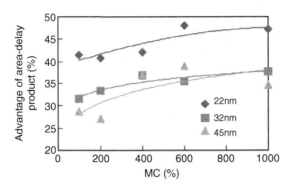

Figure 17.9 Results of numerical simulations: potential advantage of spin-MOSFET FPGA as a function of the magneto current ratio. Reprinted with permission from [77]. Copyright 2013, American Institute of Physics

circuit with spin-FETs [70]). Spin-MOSFETs possess great performance potential for reconfigurable logic by utilizing parallel/antiparallel configurations of source and drain [71]. There has been a growing interest in the MTJ-based nonvolatile logic circuits during the past five years [72–75]. However, the introduction of MTJ has certain side effects which limit the operating speed, reduce variability tolerance and static noise margin. Spin-MOSFETs and PS-MOSFETs may resolve some of the limitations inherent to the MTJ-based logic circuits [76]. Reconfigurable logic circuitry is another potential area of application for spin-MOSFETs and PS-MOSFETs [77]. Logic circuits consisting of spin-MOSFETs can be reprogramed by changing the magnetic configuration of source and drain of the individual transistors. The advantage of this approach is that the magnetization switching, which is a relatively slow process, is required only for reconfiguration, while the speed of operation itself is not limited by the magnetization switching time.

The most important question is whether or not spin transistors can outperform CMOS and provide a higher functional throughput? The optimistic projections predict significant improvement over the scaled CMOS for reconfigurable logic circuits (i.e., 6.8× improvements in delay, 32× active power, and 5.8× in area [71], or about 40% area–delay product improvement [77]). In all cases, the advantages are due to the nonvolatility provided by spin-MOSFETs. It is important to note that the potential improvements are directly related to the practically achievable magneto-current ratio. Figure 17.9 shows the estimates of the potential advantage of area–delay product as a function of the magneto current ratio, MC, defined as $MC = (I_P - I_{AP})/I_{AP}$ where I_P and I_{AP} are parallel and antiparallel currents, respectively. In turn, the magneto-current ratio is mainly defined by the spin scattering effects in the channel. Spin scattering generates a large amount of down-spin electrons, which increases the current in the antiparallel configuration, and eventually, degrades the MR ratio. According to the theoretical estimates [78], the MR ratio may exceed 1000 in the ideal channels without scattering, though the experimentally observed values are two orders of magnitude below. In Table 17.1, we summarize the key parameters of the different kinds of spin transistors, their most appealing characteristics, possible logic circuits, projected advantages over CMOS, and their current status of development.

In summary, spin transistors possess unique output characteristics, which are of potential for particular logic circuits. The oscillating output of the Spin-FET can be used for complex logic

Table 17.1 Summary of Spin-FETs

Device/material	Appealing property	Potential logic implementation	Key issue	Level of development
Spin-FET	Output current oscillates as a function of the gate voltage	Complex logic circuits (i.e., XOR)	Requires strong spin–orbit coupling	Prototype operating at 1.8 K [43]
Spin-MOSFET	Transistor + built-in memory	Reconfigurable, Nonvolatile circuits	Requires efficient spin injection	Spin injection in Si at RT [50]
PS-FET	Transistor + built-in memory	Reconfigurable, Nonvolatile circuits	—	Prototype operating at RT [81]
Graphene-based	Long spin lifetime, high mobility	As a channel for spin polarized current	Absence of bandgap	Spin injection in graphene at RT [56]
Organic-based	Long spin lifetime	One molecule – one gate, magnetostrictive memory	Short spin diffusion length	Magnetoresistance of 60% [66]

circuit construction (i.e., XOR gate) and for reducing the number of elements per circuit. Spin-MOSFETs and PS-FETs combine the characteristics of conventional transistors and a memory element, which can be utilized in nonvolatile and field programmable logic circuits [79,80]. The successful realization of room temperature operating PS-FET [81] brings the development of this device to the next circuit level. It is expected that the introduction of PS-FETs would reduce energy consumption though the use of relatively slow MTJ which would apply some restrictions on the circuit speed. The development of spin FETs may benefit from the advances in the development of novel materials (i.e., graphene- and organic-based spintronics). Table 17.1 summarizes the most appealing properties, key issues, and the level of development of Spin-FETs.

17.3 Magnetic Logic Circuits

Another (transistor-less) approach to spin-based logic devices is by exploiting magnetization as a state variable. Within this approach, logic 0 and 1 are assigned to the two ground states of the nanomagnet (i.e., magnetization along or opposite to the easy-axis) and logic functionality is associated with the switching among these states. Similar to electronic transistor-based circuits, where one transistor drives the next stage transistors by an electric field, magnetic logic circuits requires one magnet to drive next stage magnets by providing a magnetic field. There are different possible ways to interconnect the input and the output magnets (e.g., by making an array of nanomagnets sequentially switched in a domino fashion [20], by sending a spin polarized current [59], by sending a spin wave [82], or by moving a domain wall [83]). In this section, we describe the principle of operation of different magnetic logic circuits and discuss their potential advantages and drawbacks.

17.3.1 NanoMagnetic Logic

The nanomagnet logic (NML) is a variant of the edge-driven, quantum-dot cellular automata (QCA) architecture aimed to benefit from the local coupling among nano-cells. This concept

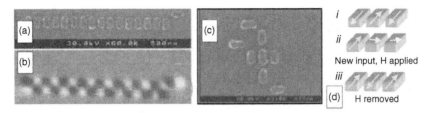

Figure 17.10 (a)Scanning electron microscope (SEM) image of a chain of 16 permalloy nanomagnets (70 × 135 nm, thickness 30 nm). (b) MFM image of the same chain shows alternating magnetization of the magnets as set by the state of the horizontal driver magnet (circled). (c) SEM image of functioning majority gate. (d) How clock works. Reprinted with permission from [93]. © 2013 IEEE

was originally proposed for the electron-based cellular structure, where an elementary cell consists of four quantum dots, with two excess electrons that can tunnel between the dots tending to occupy diagonally opposite dots due to Coulomb repulsion [84]. The electronic version of QCA appeared to be impractical, like many other single-electron logic devices due to their great sensitivity to the inevitable size variations. This issue is addressed in the magnetic version of QCA (MQCA) [85], where an elementary cell is represented by a nanomagnet of a certain shape and material to exhibit a single domain behavior. When they are connected via strong exchange interaction, the spins of the nanomagnet show collective behavior, which is much more immune to structure variation than a single electron. Later, the concept of MQCA-NML has evolved in a number of works [86–90].

Logic "0" and "1" in NML are assigned to the two ground states of the nanomagnet (i.e., magnetization along or opposite to the easy-axis). Information transfer within the NML is accomplished via dipolar coupling among nearest-neighbor nanomagnets. An example of a nanomagnetic wire is shown in Figure 17.10. It is formed from a line of permalloy islands (70 × 135 nm) separated by an air gap of 25 nm. The nanomagnets are anti-ferromagnetically ordered due to the strong dipolar interaction. The operation of NML logic gates requires a clock magnetic field to modulate the barrier between magnetization states of the coupled nano-magnets. The clocking field can be applied along the hard axis of the nanomagnets as illustrated in Figure 17.10d [91]. The initial configuration is one of two collective ground states for the line of the dipole–dipole coupled magnets. Then, an external magnetic field is applied turning the magnetic moments of the nanomagnets horizontally into a neutral logic state ("null state") against the preferred magnetic anisotropy (i.e., along the hard axes of the magnets). When the field is removed, the nanomagnets relax into the anti-ferromagnetically ordered ground state, as illustrated in Figure 17.10d. If the first magnet of the chain is influenced by an input device during relaxation, it will set the state of the whole chain via dipole–dipole interaction.

Logic functionality in the NML circuit is achieved via the specific geometric configuration of the nanomagnets. The structure of the three-input one-output majority gate is shown in Figure 17.11. The nanomagnets are arranged in two intersecting lines, such that the dipole coupling of the nanomagnets produces ferromagnetic ordering along the vertical line and antiferromagnetic ordering along the horizontal line. The nanomagnets are arranged to provide same-strength coupling to the central nanomagnet, which switches to the state corresponding to the majority of inputs. The MFM images of two possible input logic combinations are shown in Figure 17.11b–c. Such a majority gate can be reduced to a reconfigurable two-input AND/OR gate by setting the third input to a logic "0" or "1" (i.e., the MAJ gate operates as an AND gate if

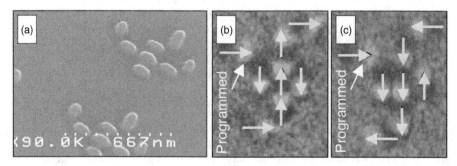

Figure 17.11 (a) SEM image of majority gates with long inputs. (b) and (c) MFM images of majority gate with "programmed" input. Reprinted with permission from [93]. © 2013 IEEE

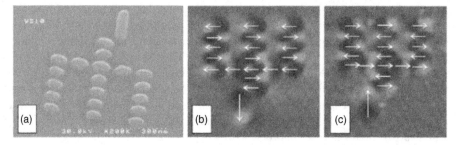

Figure 17.12 (a) SEM image of an NML ensembles one that serves as a fanout structure. (b) MFM image showing fanout operation. Reprinted with permission from [93]. © 2013 IEEE

the third input is set to 0, and the same gate operates as an OR gate if the third input is set to 1). The NOT gate in NML comprises just two anti-ferromagnetically coupled nanomagnets, as the two magnets tend to be magnetized in the opposite directions corresponding to the 01 or 10 logic states. Figure 17.12 shows a recently reported NML fan-out (1 : 3) structure with gain provided by a magnetic clocking field. All other Boolean logic gates in NML can be built via Majority gates and Inverters.

During the past five years, there has been significant progress in NML component development and experimental demonstration [89]. A summary for the experimentally demonstrated logic gates is presented in Table 17.2.

Besides that, a great deal of efforts has focused on the NML circuits modeling and application-oriented architecture development [93,96–100]. The benchmarks on the circuit area, internal delay, energy per operation, and functional throughput are shown in Table 17.2. The estimates are shown for MAJ gate, which is the base block for NM logic. The area per circuit is shown in the characteristic lithography size (DRAM's half-pitch) F. Only internal delay is taken into consideration, and the energy per operation is shown without the energy required for the clock.

Energetically efficient clocking, magneto-electric interface, and fault tolerance are the key technological issues to be addressed. As of today, all of the reported NML prototypes were clocked by the external magnetic field produced by the electric current in the conducting contours (i.e., laboratory electromagnets). Energy losses in the magnetic field generating

Table 17.2 Experimentally demonstrated NM logic circuits (all gates operate at room temperature)

Logic gate comment/ Reference	Clock wire height/width/ length (μm)	Magnetic field generating current (mA)	Structure footprint (nm)	Energy with 25 μs pulse time (nJ)	Energy with 10 μs pulse time (pJ)
Magnet[a] [92]	2.4/1.5/500	680	60×90	0.388	0.155
NM line[b] [93]	2.8/1.4/500	600	60×90×N	0.397×N	0.159×N
Two-input AND and OR gate[c] [93]	2.8/1.4/500	850	440×220	8.03	3.21
One-bit adder[b,d] [94,95]	2.8/1.4/500	760	Half-adder: 780×230 Carry gate: 320×230	11.2	4.47

[a] Magnets switched along easy axis which would not be done for clocking.

[b] (1) Higher currents/fields needed because input driver must be switched along easy axis; (2) experiments were done on a five-magnet wire, energy listed per magnet; (3) to apply current pulses, audio amplifier used in experiments which gave rise to 25 μs pulse time, this energy is reported (along with a more reasonable 10 ns pulse time); (4) assumes 50% clock wire coverage.

[c] (1) Higher currents/fields needed because input driver must be switched along easy axis; (2) assumes gate comprised of ~ 5 magnets; (3) to apply current pulses, audio amplifier used in experiments which gave rise to 25 μs pulse time – this energy is reported (along with a more reasonable 10 ns pulse time); (4) assumes 67% clock wire coverage.

[d] (1) Energy projections based on clock experiments of lines and gate summarized above; adder experiments were done with external field; (2) assumes 100% coverage.

Table 17.3 Estimates on NML MAJ gate

MAJ gate	Benchmark ($F = 15$ nm)
Circuit area ($64F^2$)	$0.0144 \, \mu m^2$
Internal delay	400 ps
Energy per operation (w/o clock)	19.31 aJ
Functional throughput	$2.51 \, \text{Pops} \, s^{-1} \, cm^{-2}$

contours may significantly exceed the internal losses within the NML circuit. According to numerical estimates [101], the Joule heat losses in a conducting loop of resistance 10 ohms, generating a 267 Oe magnetic field for magnetization reversal in a $105 \times 95 \times 6$ nm nano-magnet made of nickel will exceed 1.5 pJ. Thus, the energy losses of a single element may be 1000 times larger than the energy dissipated within the complete circuit (i.e., compare to the projected energy per operation in MAJ gate; Table 17.3). A similar problem arises for the magneto-electric interface. The input magnets in the NML circuits should be individually addressed to convert input electrical signal in the magnetization state (i.e., via spin transfer torque). The energy per switch in this case is also of the order of 0.5 pJ [101].

There are two major causes of errors in NML: static faults due to the manufacturing defects (e.g., magnet misalignment [102]) and dynamic faults due to the thermal noise. Both causes are most critical during the switching process when the nanomagnets relax from the "null" to the final state. This problem can be addressed by introducing nanomagnets with a magneto-crystalline biaxial anisotropy, which can be used to introduce a local energy minimum and to increase hard axis stability [103]. Fault tolerance can also be enhanced via adiabatic switching and/or field gradients [89]. The feasibility of building error-tolerant NML circuits is currently debated [89,101].

Overall, NML is one of the most explored approaches in the area of Spintronic Logic. NML logic gates (i.e., Inverters, Buffer lines, MAJ gates) have been experimentally demonstrated at room temperature [87]. It is expected that NML will continue to advance and more prototypes (i.e., two-bit Adder circuit) will be experimentally demonstrated within the next two years. Scalability and low power consumption are the main appealing properties of the NML approach, promising an overall power-performance advantage over the scaled CMOS. NML logic circuits might be employed for Boolean (systolic arrays) as well as for nonBoolean information processing [98,104]. The main technological issues of NML are associated with the circuit reliability and defect tolerance [105]. Thermal noise and fabrication-related imperfections can cause errors in signal transmission and logic function in NML circuits [106]. A more detailed description and analysis of NML logic will be given in Chapter 22.

17.3.2 All Spin Logic

The communication between the input and output nanomagnets can be accomplished via spin polarized currents [58]. Within this approach, nanomagnets share a common spin-coherent channel. Some of the magnets have fixed magnetization and some of them can be switched due to the spin torque produced by the spin polarized current. This approach has been further evolved in the All Spin Logic (ASL) [59]. The basic concept of switching in ASL is illustrated in Figure 17.13. There are shown two elementary magnetic cells connected via a spin-coherent channel. The cell comprises from the top to the bottom the following: metallic contact;

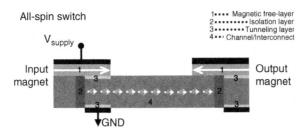

Figure 17.13 Schematics of the ASL switch. Two magnetic cells share the common spin-coherent channel. Logic 0 and 1 are assigned to the magnetic state of the free layer (1) of the cell. The cells communicate via the spin polarized current in the channel (4). Reprinted with permission from [59]. Copyright 2013, Nature Publishing Group

magnetic free layer, and a tunneling layer. It is assumed that the free layer has two thermally stable magnetic configurations and can be used as a built-in memory. The cells communicate via a spin-polarized current propagating through the channel. In order to switch the cell on the right side, the cell on the left side generates a spin polarized current which flows through the channel and provides a torque on the magnetization of the output cell. There are two possible switching scenarios, as described in [59]: (i) the magnetization of the free layer of the output cell is switched if the strength of the torque exceeds certain threshold value; (ii) a supply voltage is applied to the output cell driving the magnetization of the free layer of the output cell in a metastable state. Then, the torque provided by the current defines the relaxation direction towards one of the two stable states. The use of voltage-assisted switching has certain advantages and allows for significant reduction of the spin polarized current required for switching, as the strength of the torque has to be just above the noise level in order to define the final state. There are several possible layouts for constructing cascadable ASL gates [59]. Shown in Figure 17.14, is an example of a circuit comprising two gates (four cells per gate). Each of the gates can function as a NAND/NOR or AND/OR gate. A variety of Boolean logic

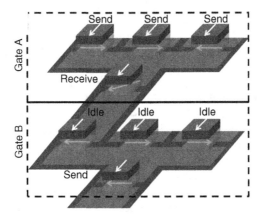

Figure 17.14 Possible layout for cascadable ASL circuit. The state of the output is determined by the spin currents injected in the channel by the inputs. Signal propagation is controlled by the voltage applied to the individual cells. Reprinted with permission from [59]. Copyright 2013, Nature Publishing Group

gates can be constructed by implementing different channel configurations and clocking schemes. The channel in ASL has to provide maximum spin coherence and can be made of semiconductors as well as metallic materials. Graphene may be the best candidate for application in ASL as it combines long coherence length with high mobility. Magnetologic gates with a hybrid graphene/ferromagnet material system have been recently proposed [60].

The ASL approach is free of constraints inherent to the schemes using dipole–dipole coupling and allows for much higher defect tolerance as the variations in the position of the elementary cells is of minor importance. It is also scalable since shorter distances between the input/output cells require less spin polarized currents for switching. The use of voltage-assisted switching may be an efficient amplification mechanism, taking energy from the conventional voltage supply. According to the theoretical estimates [107], ASL can potentially reduce the switching energy–delay product. The major issues of this scheme are associated with the electrical isolation among the electrically biased cells and the realization of the spin-coherent channel. It is desired to have spin-polarized currents flowing only between the input and the output cells though the potential difference may arise among the output cells or the cells of the different stages. No working ASL circuits have been prototyped to date.

17.3.3 Spin Wave Devices

Spin Wave Devices (SWD) are a type of magnetic logic exploiting collective spin oscillation (spin waves) for information transmission and processing [108,109]. The schematics of the SWD logic gate are shown in Figure 17.15. The circuit consists of: (i) magneto-electric cells, (ii) magnetic waveguides – spin wave buses, and (iii) a phase shifter. The magneto-electric cell (ME cell) is the element used to convert voltage pulses into the spin waves as well as to read out the voltages produced by the spin waves. The operation of the ME cell is based on the effect of magneto-electric coupling in multiferroics enabling magnetization control by applying an electric field and vice versa [82]. The principle of operation is the following. SWD receives initial information in the form of voltage pulses (e.g., $+10\,\mathrm{mV}$ corresponds to logic state 0, and $-10\,\mathrm{mV}$ corresponds to logic state 1). The polarity of the applied voltage defines the initial phase of the spin wave signal (e.g., positive voltage results in clockwise magnetization rotation and negative voltage results in counter clockwise magnetization rotation). Thus, the input information is translated into the phase of the excited wave (e.g., initial phase 0 corresponds to logic state 0, and initial phase π corresponds to logic state 1). Then, the waves propagate through the magnetic waveguides and interfere at the point of waveguide junction. For any junction with an odd number of interfering waves, there is a transmitted wave with a nonzero amplitude. The phase of the wave passing through the junction always corresponds to the majority of the phases of the interfering waves (e.g., the transmitted wave will have phase 0, if there are two or three waves with initial phase 0; otherwise, the wave will have a π-phase). As the transmitted wave passes the phase shifter, it accumulates an additional π-phase shift (e.g., phase $0 \to \pi$, and phase $\pi \to 0$). Finally, the spin wave signal reaches the output ME cell. The output cell has two stable magnetization states. At the moment of spin wave arrival, the output cell is in the metastable state (magnetization is along the hard axis perpendicular to the two stable states). The *phase* of the incoming spin wave defines the direction of the magnetization relaxation in the output cell [82,109]. The process of magnetization change in the output ME cell is associated with the change of electrical polarization in the multiferroic material and recognized by the induced voltage across the ME cell (e.g., $+10\,\mathrm{mV}$ corresponds to logic state 0, and $-10\,\mathrm{mV}$ corresponds to logic state 1).

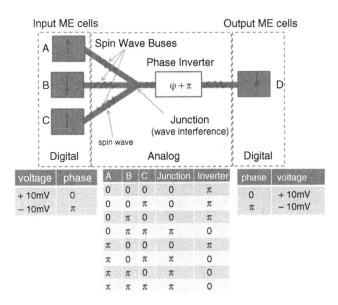

voltage	phase	A	B	C	Junction	Inverter	phase	voltage
+ 10mV	0	0	0	0	0	π	0	+ 10mV
− 10mV	π	0	0	π	0	π	π	− 10mV
		0	π	0	0	π		
		0	π	π	π	0		
		π	0	0	0	π		
		π	0	π	π	0		
		π	π	0	π	0		
		π	π	π	π	0		

Figure 17.15 Schematic view of the SWD logic circuit. There are three inputs (A,B,C) and the output D. The inputs and the output are the ME cells connected via the ferromagnetic waveguides – spin wave buses. The input cells generate spin waves of the same amplitude with initial phase 0 or π, corresponding to logic 0 and 1, respectively. The waves propagate through the waveguides and interfere at the point of junction. The phase of the wave passed the junction corresponds to the majority of the interfering waves. The phase of the transmitted wave is inverted (e.g., passing the domain wall). The table illustrates the data processing in the phase space. The phase of the transmitted wave defines the final magnetization of the output ME cell D. The circuit can operate as a NAND or NOR gate for inputs A and B depending the third input C (NOR if C = 1, NAND if C = 0). Reprinted with permission from [115]. Copyright 2013, American Institute of Physics

The operation of the SWDs is based on the spin wave interference, which provides the mechanism for phase-based data processing. During the past five years, three- and four-terminal SWDs exploiting spin wave interference have been experimentally demonstrated [110,111]. Figure 17.16a shows the top view of the four-terminal SWD. The sample consists of a silicon substrate, a 20 nm layer of $Ni_{81}Fe_{19}$, a 300 nm layer of silicon dioxide, and a set of five conducting wires on top. The distance between the wires is 2 μm. Three of the five wires are used as input ports, and the two other wires are connected in a loop to detect the inductive voltage produced by the spin wave interference. The direction of the current in the wires defines the phase difference between the excited spin waves (i.e., the waves have the same initial phase if the direction of the current is the same direction, and there is a π-phase difference if the current flows in the opposite direction). The plot in Figure 17.16b shows the output inductive voltage for different combinations of spin wave phases. The blue and red curves correspond to the case when all three waves are excited in phase, and the black and green curves correspond to the case when one of the waves has a π-phase difference with the others. The experimental data show the change of the phase of the output voltage as a function of the input combination (i.e., 000, π00, 0π0, 00π corresponds to the output phase 0, and πππ, 0ππ, π0π, ππ0 results in the change of the output phase voltage of π). This logic device operates as

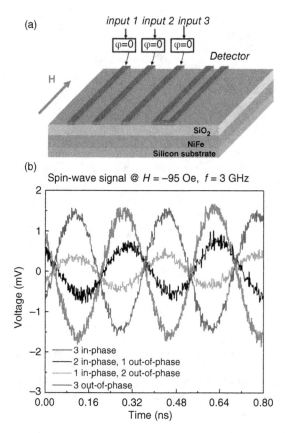

Figure 17.16 (a) Schematics of the four-terminal SWD. The device structure comprises a silicon substrate, a 20 nm layer of permalloy, a layer of silicon dioxide, and a set of five conducting wires on top (three wires to excite three spin waves, and the other two wires connected in a loop are to detect the inductive voltage). The initial phase of the excited spin wave (0 or π) is controlled by the direction of the excitation current. (b) Experimental data showing the inductive voltage as a function of time. The curves of different colors correspond to the different combinations of the phases of the interfering spin waves. Reprinted with permission from [115]

MAJ gate in the phase space. The data are taken at 95 Oe bias magnetic field (perpendicular to the spin wave propagation), 3 GHz frequency, and at room temperature.

The use of electric current wires for spin wave excitation appeared to be energetically inefficient (i.e., >pJ per spin wave), as only a small amount of energy is transferred into the spin wave. It would be much more efficient to utilize multiferroics for energy conversion between the electric and magnetic domains [112]. Recently, spin wave excitation and detection by synthetic multiferroic elements was experimentally demonstrated [113]. The schematics of the experiment and experimental data are shown in Figure 17.17. Two synthetic multiferroic elements were used to excite and detect a spin wave propagating in the permalloy waveguide (the distance among the excitation and the detection elements is 40 μm). The multiferroic element comprises a layer of piezoelectric (PZT) and a magnetostrictive material (Ni). An electric field applied across the piezoelectric produces stress, which, in turn, affects the

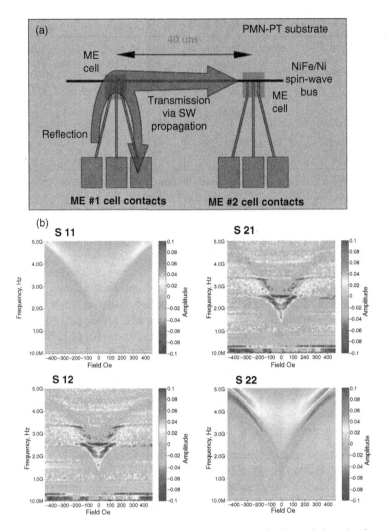

Figure 17.17 (a) Schematics of the experiment on spin wave excitation and detection by multiferroic elements (ME cells). (b) Collection of the experimental data (S11, S12, S21, and S22 parameters) obtained at different frequencies and bias magnetic field. Reprinted with permission from [113]

magnetization of the magnetostrictive material. Thus, the applying of AC voltage to the multiferroic element results in magnetization oscillation (a spin wave). Vice versa, a change of magnetization in the magnetostrictive layer results in a voltage signal due to the produced stress. The experimental data in Figure 17.17b show the collection of data obtained at different operational frequencies and bias magnetic field. The utilization of multiferroics has resulted in the energy reduction to about 10 fJ per wave [113]. The major characteristics of the experimentally demonstrated SWD prototypes are summarized in Table 17.4.

The expected advantages of SWD technology include: (i) the ability to utilize phase in addition to amplitude for building logic devices with a smaller number of elements than required for transistor-based approach [82], and (ii) parallel data processing on multiple

Table 17.4 Experimentally demonstrated SWDs (all devices operate at room temperature)

Logic gate comment/ Reference	Circuit size width/ length (μm)	Operational frequency (GHz)	Signal/ noise ratio	Energy per wave
Three-terminal SWD [110]	700/12	0.5–3.0	2 (maximum)	>10 pJ
Four-terminal SWD (MAJ gate) [111]	500/24	3.0	~10 (average)	~pJ
Two-terminal line with ME cells[a] [113,114]	500/40	0.5–5.0	N/A	10 fJ

[a] The energy per wave has been estimated based on the excitation conditions.

frequencies at the same device structure by exploiting each frequency as a distinct information channel [115]. Wave-based magnonic logic devices may complement CMOS in special task data processing (i.e., image processing), by exploiting wave interference for functional throughput enhancement. The major issues of SDW approach include the problems with energetically efficient spin wave excitation and detection, fast attenuation time (~1 ns at RT), and spin wave dispersion [102]. All of the demonstrated prototypes utilize spin waves of micrometer scale wavelength, which makes them immune with respect to the waveguide structure imperfections. It is not clear if scaling to the deep sub-micrometer range would significantly affect the signal to noise ratio as well as the speed of propagation. At some point, the success of the SWD approach depends on the ability to detect and amplify spin waves (i.e., by multiferroic elements or spin torque oscillators), which can enhance SWD noise immunity.

17.3.4 Hybrid Spintronics and Straintronics Devices

A new approach to magnetization control via applied stress was recently proposed and became known as Hybrid Spintronics and Straintronics [116,117]. The essence of this approach is in the use of two-phase composite multiferroics comprising piezoelectric and magnetostrictive materials, where an electric field applied across the piezoelectric produces stress, which, in turn, affects the magnetization of the magnetoelastic material. Although work in this area can be traced back to the 1970s [118], composite multiferroics have been in a shadow of single-phase multiferroics (i.e., $BiFeO_3$ and its derivatives [119]) for a long time. The recent resurgence of interest in composite multiferroics is due to the technological flexibility, where each phase (piezoelectric or magnetostrictive) may be independently optimized for room temperature performance. More importantly, the strength of the electro-magnetic coupling in the two-phase systems can significantly exceed the limits of the single-phase counterparts [120]. During the past two years, there have been several experimental works showing magnetization rotation in two-phase multiferroics as a function of the applied voltage [117,121]. For instance, a reversible and permanent magnetic anisotropy reorientation was reported in a magnetoelectric polycrystalline Ni thin film and (011)-oriented $[Pb(Mg_{1/3}Nb_{2/3})O_3]_{(1-x)}-[PbTiO_3]_x$ heterostructure [117]. An important feature of the magneto-electric coupling is that the changes in magnetization states are stable without the application of an electric field and can be reversibly switched by an electric field near a critical value (i.e., 0.6 MV/m for Ni/PMN-PT). Such a relatively weak electric field promises the ultra-low energy consumption required for magnetization switching [122]. The estimates

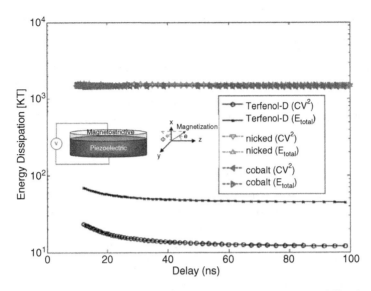

Figure 17.18 Numerical estimates on the energy dissipated in the circuit with multiferroic nanomagnets as functions of delay for three different materials used as the magnetostrictive layer in the multiferroic nanomagnet. Reprinted with permission from [116]. Copyright 2013, American Institute of Physics

of power dissipation in a two-phase multiferroic material are shown in Figure 17.18 [116]. The dissipated energy is calculated as follows:

$$E_{\text{diss}} = CV^2 + E_{\text{d}}$$

where C is the capacitance of the multiferroic element, which is defined by the dielectric properties of the piezoelectric layer, and the dimensions of the element, V is the voltage needed to generate the stress sufficient for magnetization reversal, and E_{d} is the energy dissipated in the nanomagnet due to Gilbert damping [116]. The voltage required for magnetization switching can be found by solving the Landau–Lifshitz–Gilbert (LLG) equation for particular materials or estimated, based on experimental data. The estimates predict values as low as tens of kT energy dissipated per switch at a nanosecond switching time. For example, the total energy dissipated is 45 kT for a delay of 100 ns and 70 kT for a delay of 10 ns. It is important to note that, according to the theoretical estimates [116], the energy dissipation increases sub-linearly with the switching speed. For example, in order to increase the switching speed by a factor of 10, the dissipation needs to increase by a factor of 1.6.

The ability to control magnetization by an electric field presents great potential for application of two-phase multiferroics in magnetic logic devices. Specifically, these elements may resolve the shortcoming associated with input–output isolation in the dipole-coupled NML circuits described in the previous section. In contrast to the global magnetic field provided by the electric current carrying wires, a multiferroic element allows for more local magnetization control. At the same time, stress-mediated electro-magnetic coupling eliminates the problem of the stray field. Another potential benefit of multiferroics is the possibility of building multi-state logic circuits by exploiting the biaxial magnetocrystalline anisotropy of the magnetoelastic material. An example of the four-state universal logic gate (NOR) using a linear array of three dipole-coupled magnetostrictive-piezoelectric multiferroic nanomagnets is described in Reference [123]. Similar ideas for using electric field-controlled nanomagnets can be implemented in other types of logic circuits (i.e., developed for NML).

It is expected that the development of multiferroic elements exploiting stress for magnetization switching will be intensively studied within the next two years and the estimates on ultra-low power dissipation will be experimentally validated. Another important question to be answered is whether or not stress-based multiferroic devices are capable of fast (GHz frequency range) switching.

17.3.5 Spin Torque-based Devices

The discovery of the spin torque effect [124] has resulted in the development of novel mechanism for magnetization switching, which has been successfully implemented in magnetic memory [125]. Spin torque devices (STD) comprise a stack of metallic ferromagnetic layers separated by the tunneling barrier. Passing electric current across the structure produces a spin torque affecting the magnetization of the ferromagnetic layers. Spin torque may result in the oscillation of magnetization (spin wave generation) or complete magnetization reversal of the one of the magnetic layer (i.e., the free layer) when the current density exceeds a certain threshold value. The fabrication technology is fully compatible with CMOS, which makes the spin torque device (STD) a promising candidate for implementation in logic circuits. There are two major approaches to spin torque-based devices: (i) digital devices, where logic states 0 and 1 are encoded into the magnetic state of the free layer(s) [83,126]; (ii) analog devices where logic state is associated with the frequency of the magnetic oscillation.

The schematics of the digital spin torque majority gate (STMG) are shown in Figure 17.19 [83]. There are four STD nanopillars connected via the common ferromagnetic layer. Three of the nanopillars (A, B, C in Figure 17.19) are used as the input gates, and the fourth nanopilalr (Out) is the output gate. Each of the input nanopillars and the output nanopillar has separate fixed layers which are pinned by the antiferromagnetic layers on top of them. Logic input is defined by the polarity of the voltage applied to the input gates (i.e., positive voltage corresponds to logic 1, and negative voltage corresponds to logic 0. The polarity of the applied voltage determines the direction of the spin torque affecting the magnetization of the common free ferromagnetic layer below the nanopillars. Thus, positive voltage forces the magnetization in the free layer to align parallel with the magnetization of the fixed layer, and negative voltage forces the magnetization to align in the opposite direction. As a result of the interplay between the three torque forces, the magnetization of the free layer changes according to the majority of the inputs. The numerical simulations in Figure 17.19 illustrate the process of magnetization change (domain wall motion) in the common layer. The input voltages are applied until the domain wall reaches the output cell. Once the direction of magnetization in the free layer reaches its steady-state configuration, the input voltages are turned off and the results of computation is stored in the magnetization state of the common layer. The read-out is accomplished by sensing the resistance across the output nanopillar (i.e., high resistance corresponds to logic 1 and low resistance corresponds to logic 0).

This approach looks very feasible from the technological point of view, as the material structure of the nanopillars in MAJ gate is similar to that in a typical magnetic tunnel junction [127]. The only technological challenge is associated with the electrical isolation of the STDs to prevent unnecessary electric currents, which can be possible resolved by utilizing a nonconducting ferromagnetic layer. The projected potential advantages over the CMOS are in

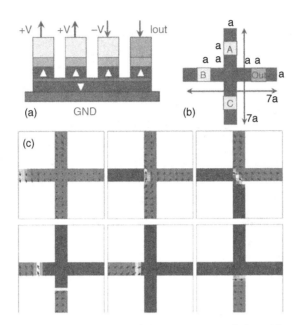

Figure 17.19 (a) Scheme of the cross section of the MAJ gate consisting of four STDs. The triangles designate the magnetization direction of the free layers. (b) Schematic of the top view of the MAJ gate. The squares designated by "A," "B," and "C" are input nanopillars. The square designated by "Out" is the output nanopillar. (c) Results of numerical simulations: several snapshots showing the magnetization of the common ferromagnetic layer. Red color denotes the out-of-plane projection of magnetization pointing up, and blue denotes that pointing down. Reprinted with permission from [83]. © 2013 IEEE

are smaller area, low power, nonvolatility, reconfigurability, and radiation hardness [83]. The estimates on a full adder circuit built of STDs from Ref. [128] are shown in Table 17.5.

There are a few factors beneficial for STDs circuit. It has a higher density than CMOS, because 28 transistors are replaced by three majority gates. The STDs circuit has lower active power and zero standby power. At the same time, the use of domain wall motion for STD coupling would limit the switching speed and the switching energy. The efficiency of the STD devices is mainly defined by the magnitude of the electric current required for magnetization switching. So far, the best experimental results show about MA/cm^2 critical current density (CoFeB free layer [129]). On the one hand, the critical switching current is proportional to the

Table 17.5 Performance estimates of a full adder circuit with STMGs

Full adder with STMGs	Benchmark ($F = 22$ nm)
Gate area	$0.13\,\mu m^2$
Switching time	2826 ps
Switching energy with external circuit	257 680 aJ
Power per gate, active	45.6 μW
Power per gate, standby	0.0 μW
Functional throughput	1.3 Mops ns^{-1} cm^{-2}

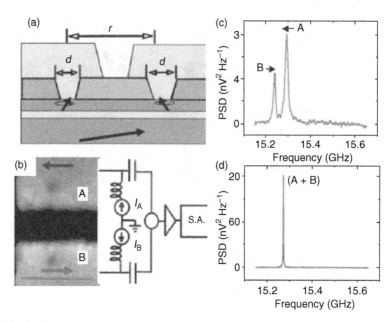

Figure 17.20 (a) Cross-sectional diagram of two STOs with a common free layer. (b) Micrograph of actual two nano-contact device with two independent leads. (c) Combined spectrum with current density below phase locking. (d) Combined spectrum with phase locking. Reprinted with permission from [130]. Copyright 2013, Nature Publishing Group

device area and can be scaled down for nanometer size structures. On the other hand, there are certain limits on the size of the free layer to be switched (i.e., due to the thermal stability requirements), which would further restrict the minimum switching current. All these fundamental issues could narrow the potential applications of the domain wall-coupled STDs.

Another approach to STD coupling is via spin waves [130]. This approach is based on the property of nanometer scale spin torque devices generate spin waves in response to a d.c. electrical current [131,132]. Electric current passing through a spin torque nano-oscillator (STNO) generates spin transfer torque and induces auto-oscillatory precession of the magnetic moment of the spin valve free layer. The frequency of the precessing magnetization is tunable by the applied dc voltage due to the strong nonlinearity of the STNO. In case of two or more STNOs sharing a common free layer, the oscillations can be frequency and phase locked via the spin wave exchange [130,133]. Figure 17.20 shows the schematics of the devices with two STNOs (marked as A and B) and the experimental data illustrating the coupling. Figure 17.20c–d plots the evolution of the combined spectrum from both STNOs as current through contact B increases. The phase-locked state is characterized by a narrowing of the signal linewidth (Figure 17.20d) as the current exceed the threshold value. This phenomenon gives rise to the proposal on the logic devices based on coupled STNOs [134]. A majority logic gate based on phase locking of STNOs consists of a common free-layer metallic ferromagnetic nanowire on a metallic nonmagnetic bottom lead with several GMR or TMR junctions patterned on top of the free layer as shown in Figure 17.21. One of these junctions can serve as the gate output while the rest of the junctions are the gate inputs. All inputs are dc current-biased at a current level above the critical current for magnetization self-oscillations. To each

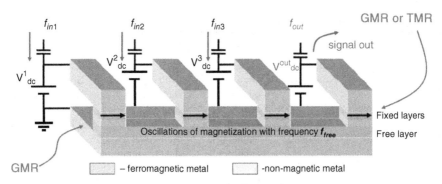

Figure 17.21 Schematics of the STO-based MAJ logic gate consisting of a metallic ferromagnetic nanowire with several injectors of spin polarized current (spin toque oscillators with a common free layer). The input frequencies (e.g., $f1, f2, f3$) assume binary values. The output frequency f_{out} in the phase locking regime is determined by the majority of the input frequencies. Reprinted with permission from [135]

input, signals of two frequencies $f1$ and $f2$ can be applied. Due to injection locking and spin wave interaction in the common free layer, the entire free layer precesses at either $f1$ or $f2$, depending on the input signal frequency applied to the majority of the inputs. Therefore, the output frequency of this logic gate is determined by the frequency applied to the majority of the input gates, and the device operates as a majority logic gate with the signal frequency as the state variable [135].

The unique properties of STDs are of great promise for future implementation. Being compatible with CMOS, STDs may serve as complementary logic units for general and special task data processing. The main challenge for the STD approach is to reduce the current required for switching and minimize active power consumption.

17.3.6 Current Status and Perspectives of Magnetic Logic Devices

In summary, magnetic devices provide an alternative approach to logic circuit construction by exploiting intrinsic nonlinearity of magnetic materials. In contrast to Spin-FETs, most magnetic devices do not require high-efficiency spin injection, and a number of room temperature operating prototypes has been demonstrated. Nonvolatility is the main advantage inherent to all schemes, which may potentially eliminate static power consumption. The future success of magnetic logic is mainly related to finding an energetically efficient mechanism for local magnetization control. Development and utilization of multiferroic materials is one of the promising approaches able to scale down energy per magnetization reversal to the atto Joule level. At the same time, the use of nanomagnets for information storage implies certain limits on the operational speed, which is restricted by the time required for magnetization reversal. It is unlikely that magnetic devices will be ever able to compete with CMOS in speed. Nevertheless, magnetic logic circuits can provide higher functional throughput or perform the same data processing tasks with lower energy consumption than CMOS by implementing complex logic functions and/or reconfigurability. For example, NML and ASL circuits are especially suited for adder and multiplier circuits as a part of arithmetic logic units (ALU). SWDs are most efficient for MAJ gate construction by exploiting wave interference and may be also promising for parallel data processing. STDs offer a variety of possible applications including nonvolatile digital circuits (i.e., MAJ gates) and analog dynamical phase-locking

Table 17.6 Summary of magnetic logic devices

Device	Coupling mechanism	Appealing property	Potential application	Key issue(s)	Level of development
NML	Dipole–dipole interaction	Nonvolatile	Digital circuits, adders, multipliers	Fault tolerance, energetically efficient clocking	Buffer, AND, OR, one-bit adder prototypes operating at RT [92,93,95]
ASL	Spin polarized current	Nonvolatile	Digital circuits, adder, multiplier circuits	Spin-coherent channel, input–output electric isolation	Numerical modeling [59]
SWD	Spin waves	Utilizes interference for data processing	Parallel data processing	Efficient mechanism for wave excitation, dispersion.	MAJ gate prototype operating at RT [121,122]
STD	Domain wall movement; spin waves	Noise immunity	CNN architectures	Electrical isolation	Two-coupled STNOs [133]

system for neuromorphic-type architectures. In Table 17.6, we summarize the most appealing properties, key issues, and the level of development of magnetic logic devices.

17.4 Summary

Adding the spin degree of freedom to conventional charge-based electronic devices appears to be a nontrivial task. So far, the tremendous efforts to exploit spin–orbit coupling for current modulation in spin-FETs have resulted in a low-temperature (1.8 K) operating prototype [43]. The attempt of building FETs without relying on spin–orbit coupling (spin-MOSFETs) has faced another grand challenge – the need for efficient spin injection. Currently, both spin-FETs and spin-MOSFETs are far from competing with CMOS technology. More encouraging is the progress in pseudo-spin FETs [54], showing the possibility of building hybrid logic devices possessing standard transistor behavior combined with nonvolatility. However, the integration of relatively slow MTJ with a scaled CMOS has certain architecture restrictions and speed-energy tradeoffs. The development of novel materials and nanostructures with long spin lifetime may further benefit the development of spin-FETs.

The idea of using spin (collection of spins) alone has resulted in a number of proposals, as described in this chapter. The principle of operation of magnetic logic devices is fundamentally different from the conventional transistor-based approach. On the one hand, the unique output characteristics of magnetic devices promise additional functionalities beyond the capabilities of the conventional FET. On the other hand, the integration of spin-based devices with conventional electron-based devices introduces a significant problem. During the past five years, there has been a continuous progress in the development of magnetic logic devices. Several room temperature operating prototypes have been demonstrated (e.g., see Table 17.4). The further development of magnetic logic is expected to show examples of specific logic gates exploiting the unique output characteristics of magnetic devices. The main challenges include the development of energetically efficient clocking, reliable magnetic interconnects, and an electro-magnetic interface for integration within the chip. The development of spin transfer- and multiferroic-based devices may significantly improve the energetic efficiency of magnetic logic. Though magnetic devices cannot complete with CMOS in speed, they hold the promise of nonvolatile and reconfigurable logic, which may provide some functional enhancement at reduced power consumption (assessment and benchmarking are presented in Chapter 21).

Overall, in spite of great efforts and research activities, spin-based logic circuitry remains in its infancy. It is very unlikely that spin-based logic devices will be able to compete with CMOS circuitry in all figures of merit. However, the unique output characteristics of magnetic devices may find applications in specific circuits and architectures aimed not to replace but to complement CMOS in special data processing tasks.

References

1. Prinz, G.A. (1998) Device physics – magnetoelectronics. *Science*, **282**, 1660–1663.
2. Wolf, S.A. *et al.* (2001) Spintronics: A spin-based electronics vision for the future. *Science*, **294**, 1488–1495.
3. Johnson, M. and Silsbee, R.H. (1985) Interfacial charge-spin coupling – injection and detection of spin magnetization in metals. *Physical Review Letters*, **55**, 1790–1793.
4. Baibich, M.N. *et al.* (1988) Giant magnetoresistance of (001)FE/(001) CR magnetic superlattices. *Physical Review Letters*, **61**, 2472–2475.
5. Binasch, G. *et al.* (1989) Enhanced magnetoresistance in layered magnetic-structures with antiferromagnetic interlayer exchange. *Physical Review B*, **39**, 4828–4830.
6. Wolf, S.A. *et al.* (2006) Spintronics – A retrospective and perspective. *Ibm Journal of Research and Development*, **50**, 101–110.
7. Slonczewski, J.C. (1996) Spintronics and spin transfer torque. *Journal of Magnetism and Magnetic Materials*, **159**, L1–L7.
8. Pesin, D. and MacDonald, A.H. (2012) Spintronics and pseudospintronics in graphene and topological insulators. *Nature Materials*, **11**, 409–416.
9. Dediu, V.A. *et al.* (2009) Spin routes in organic semiconductors. *Nature Materials*, **8**, 707–716.
10. Semiconductors, I.T.R. (2011) http://www.itrs.net vol. Chapter PIDS.
11. Datta, S. and Das, B. (1990) Electronic analog of the electro-optic modulator. *Applied Physics Letters*, **56**, 665–667.
12. Schliemann, J. *et al.* (2003) Nonballistic spin-field-effect transistor. *Physical Review Letters*, **90**, 65–67.
13. Wan, J. *et al.* (2008) Proposal for a dual-gate spin field effect transistor: A device with very small switching voltage and a large ON to OFF conductance ratio. *Physica E-Low-Dimensional Systems & Nanostructures*, **40**, 2659–2663.
14. Sugahara, S. and Tanaka, M. (2004) A spin metal-oxide-semiconductor field-effect transistor using half-metallic-ferromagnet contacts for the source and drain. *Applied Physics Letters*, **84**, 2307–2309.
15. Sugahara, S. and Tanaka, M. (2006) Spin MOSFETs as a basis for spintronics. *ACM Transaction on Storage*, **2**, 197–219.
16. Shuto, Y. *et al.* (2010) A new spin-functional metal-oxide-semiconductor field-effect transistor based on magnetic tunnel junction technology: Pseudo-Spin-MOSFET. *Applied Physics Express*, **3**, 665–667.
17. Bandyopadhyay, S. *et al.* (1994) Supercomputing with spin-polarized single electrons in a quantum coupled architecture. *Nanotechnology*, **5**, 113–133.
18. Molotkov, S.N. and Nazin, S.S. (1995) Single-electron spin logical gates. *JETP Letters*, **62**, 273–281.
19. Nikonov, D.E. *et al.* (2006) Power dissipation in spintronic devices out of thermodynamic equilibrium. *Journal of Superconductivity and Novel Magnetism*, **19**, 497–513.
20. Cowburn, R.P. and Welland, M.E. (2000) Room temperature magnetic quantum cellular automata. *Science*, **287**, 1466–1468.
21. Fert, A. and Jaffrès, H. (2001) Conditions for efficient spin injection from a ferromagnetic metal into a semiconductor. *Physical Review B*, **64**, 184420.
22. Inokuchi, T. *et al.* (2009) Electrical spin injection into n-GaAs channels and detection through MgO/CoFeB electrodes. *Applied Physics Express*, **2**, 665–667.

23. Hidaka, S. *et al.* (2012) High-efficiency long-spin-coherence electrical spin injection in CoFe/InGaAs two-dimensional electron gas lateral spin-valve devices. *Applied Physics Express*, **5**, 56–58.
24. Elliott, R.J. (1954) Theory of the effect of spin-orbit coupling on magnetic resonance in some semiconductors. *Physical Review*, **96**, 266–279.
25. D'yakonov, M.I. and Perel, V.I. (1971) Spin relaxation of conduction electrons in noncentrosymetric semiconductors. *Fiz. Tverd. Tela*, **13**, 3581–3585.
26. Bir, G.L. *et al.* (1975) Spin relaxation of electrons due to scattering by holes. *Zh. Eksp. Teor. Fiz*, **69**, 1382–1397.
27. Kettemann, S. (2007) Dimensional control of antilocalization and spin relaxation in quantum wires. *Physical Review Letters*, **98**, 665–667.
28. Bernevig, B.A. *et al.* (2006) Exact SU(2) symmetry and persistent spin helix in a spin-orbit coupled system. *Physical Review Letters*, **97**, 66–68.
29. Koralek, J.D. *et al.* (2009) Emergence of the persistent spin helix in semiconductor quantum wells. *Nature*, **458**, 610–613
30. Dresselhaus, G. (1955) Spin-orbit coupling effects in zinc blende structures. *Physical Review*, **100**, 580–586.
31. Bychkov, Y.A. and Rashba, E.I. (1984) Oscillatory effects and the magnetic-susceptibility of carriers in inversion-layers. *Journal of Physics C-Solid State Physics*, **17**, 6039–6045.
32. Huang, S.-M. (2012) Gate-controlled rashba spin-orbit interaction intensity in semiconductor heterostructures. *International Journal of Modern Physics B*, **26**, 667–669.
33. Sato, Y. *et al.* (2001) Large spontaneous spin splitting in gate-controlled two-dimensional electron gases at normal In0.75Ga0.25As/In0.75Al0.25As heterojunctions. *Journal of Applied Physics*, **89**, 8017–8021.
34. Schapers, T. *et al.* (2003) Rashba effect in strained InGaAs/InP quantum wire structures. *Science and Technology of Advanced Materials*, **4**, 19–25.
35. Engels, G. *et al.* (1997) Experimental and theoretical approach to spin splitting in modulation-doped In_{x}Ga_{1-x}As/InP quantum wells for B → 0. *Physical Review B*, **55**, R1958–R1961
36. Iordanskii, S.V. *et al.* (1994) Weak localization in quantum wells with spin-orbit interaction. *JETP Letters*, **60**, 206–211.
37. Knap, W. *et al.* (1996) Weak antilocalization and spin precession in quantum wells. *Physical Review B*, **53**, 3912–3924.
38. Jusserand, B. *et al.* (1995) Spin orientation at semiconductor heterointerfaces. *Physical Review B*, **51**, 4707–4710.
39. Wilamowski, Z. *et al.* (2002) Evidence and evaluation of the Bychkov-Rashba effect in SiGe/Si/SiGe quantum wells. *Physical Review B*, **66**, 665–667.
40. Schultz, M. *et al.* (1996) Rashba spin splitting in a gated HgTe quantum well. *Semiconductor Science and Technology*, **11**, 1168–1172.
41. Sugahara, S. and Nitta, J. (2010) Spin-transistor electronics: an overview and outlook. *Proceedings of the IEEE*, **98**, 2124–2154.
42. Gvozdic, D.M. *et al.* (2005) Comparison of performance of n- and p-type spin transistors with conventional transistors. *Journal of Superconductivity*, **18**, 349–356.
43. Cheol, K. Hyun *et al.* (2009) Control of spin precession in a spin-injected field effect transistor. *Science*, **325**, 1515–1518.
44. Bandyopadhyay, S. (2009) Comment on "Control of spin precession in a spin-injected field effect transistor" http://arxiv.org/abs/0911.0210.
45. Zainuddin, A.N.M. *et al.* (2010) Voltage Controlled Spin Precession, http://arxiv.org/abs/1001.1523 (accessed 16 July 2013).
46. Bandyopadhyay, S. (2009) Analysis of the Two Dimensional Datta-Das Spin Field Effect Transistor, http://arxiv.org/abs/1001.2705 (accessed 16 July 2013).
47. Hattori, R. *et al.* (1992) A new type of tunnel-effect transistor employing internal field emission of Schottky barrier junction. *Japanese Journal of Applied Physics, Part 2 (Letters)*, **31**, L1467–L1469.
48. Larson, J.M. and Snyder, J.P. (2006) Overview and status of metal S/D Schottky-barrier MOSFET technology. *IEEE Transactions on Electron Devices*, **53**, 1048–1058.
49. Jansen, R. (2012) Silicon spintronics. *Nature Materials*, **11**, 400–408.
50. Dash, S.P. *et al.* (2009) Electrical creation of spin polarization in silicon at room temperature. *Nature*, **462**, 491–494.
51. Li, C.H. *et al.* (2011) Electrical injection and detection of spin accumulation in silicon at 500 K with magnetic metal/silicon dioxide contacts. *Nature Communications*, **2**, 911–914.
52. Jeon, K.-R. *et al.* (2011) Electrical spin accumulation with improved bias voltage dependence in a crystalline CoFe/MgO/Si system (vol 98, 262102, 2011). *Applied Physics Letters*, **99**, 491–494.

53. Shuto, Y. *et al.* (2009) Nonvolatile static random access memory based on spin-transistor architecture. *Journal of Applied Physics*, **105**, 91–94.
54. Shuto, Y. *et al.* (2012) Design and performance of pseudo-spin-MOSFETs using nano-CMOS devices, Electron Devices Meeting (IEDM), 2012 IEEE International, 29.6.1–29.6.4.
55. Hongki, M. *et al.* (2006) Intrinsic and Rashba spin-orbit interactions in graphene sheets. *Physical Review B (Condensed Matter and Materials Physics)*, **74**, 165310, 1–5.
56. Han, W. and Kawakami, R.K. (2011) Spin relaxation in single-layer and bilayer graphene. *Physical Review Letters*, **107**, 047207.
57. Cobas, E. *et al.* (2012) Graphene as a tunnel barrier: graphene-based magnetic tunnel junctions. *Nano Letters*, **12**, 3000–3004.
58. Dery, H. *et al.* (2007) Spin-based logic in semiconductors for reconfigurable large-scale circuits. *Nature*, **447**, 573–576.
59. Behin-Aein, B. *et al.* (2010) Proposal for an all-spin logic device with built-in memory. *Nature Nanotechnology*, **5**, 266–270.
60. Dery, H. *et al.* (2012) Nanospintronics based on magnetologic gates. *IEEE Transactions on Electron Devices*, **59**, 259–262.
61. Harris, C.B. *et al.* (1973) Optically detected electron spin locking and rotary echo trains in molecular excited states. *Physical Review Letters*, **30**, 1019–1022.
62. Krinichnyi, V.I. *et al.* (1997) EPR and charge-transport studies of polyaniline. *Physical Review B*, **55**, 16233–16244.
63. Xiong, Z.H. *et al.* (2004) Giant magnetoresistance in organic spin-valves. *Nature*, **427**, 821–824.
64. Zhan, Y. and Fahlman, M. (2012) The study of organic semiconductor/ferromagnet interfaces in organic spintronics: A short review of recent progress. *Journal of Polymer Science Part B-Polymer Physics*, **50**, 1453–1462.
65. Prigodin, V.N. *et al.* (2011) New advances in organic spintronics. *Journal of Physics: Conference Series*, **292**, 012001 (8 pp.)-012001 (8 pp.).
66. Schmaus, S. *et al.* (2011) Giant magnetoresistance through a single molecule. *Nature Nanotechnology*, **6**, 185–189.
67. Sanvito, S. and Dediu, V.A. (2012) SPINTRONICS News from the organic arena. *Nature Nanotechnology*, **7**, 696–697.
68. Likharev, K.K. (1987) Single-electron transistors – electrostatic analogs of the dc squids. *IEEE Transactions on Magnetics*, **23**, 1142–1145.
69. Takahashi, Y. *et al.* (2000) Multigate single-electron transistors and their application to an exclusive-OR gate. *Applied Physics Letters*, **76**, 637–639.
70. Tan, S.G. *et al.* (2005) Single spin-FET for programmable logic gates. INTERMAG Asia 2005: Digest of the IEEE International Magnetics Conference (IEEE Cat. No. 05CH37655), pp. 1209–1210.
71. Yunfei, G. (2010) *et al.* Realistic spin-FET performance assessment for reconfigurable logic circuits. 2010 IEEE Symposium on VLSI Technology, pp. 117–118.
72. Weisheng, Z. *et al.* (2008) New non-volatile logic based on spin-MTJ. *Physica Status Solidi A*, **205**, 1373–1377.
73. Fengbo, R. and Markovic, D. (2010) True energy-performance analysis of the MTJ-based logic-in-memory architecture (1-Bit Full Adder). *IEEE Transactions on Electron Devices*, **57**, 1023–1028.
74. Hanyu, T. (2010) Special session 8B: new topic MOS/MTJ-hybrid circuit with nonvolatile logic-in-memory architecture and its impact. 2010 28th VLSI Test Symposium (VTS 2010), pp. 258–1258
75. Weisheng, Z. *et al.* (2009) High speed, high stability and low power sensing amplifier for MTJ/CMOS hybrid logic circuits. *IEEE Transactions on Magnetics*, **45**, 3784–3787.
76. Yamamoto, S.i. and Sugahara, S. (2010) Nonvolatile delay flip-flop based on spin-transistor architecture and its power-gating applications. *Japanese Journal of Applied Physics*, **49**, 491–494.
77. Tanamoto, T. *et al.* (2011) Scalability of spin field programmable gate array: A reconfigurable architecture based on spin metal-oxide-semiconductor field effect transistor (vol 109, 07C312, 2011). *Journal of Applied Physics*, **110**, 581–584.
78. Yunfei, G. *et al.* (2010) Simulation of spin field effect transistors: Effects of tunneling and spin relaxation on performance. *Journal of Applied Physics*, **108**, 083702 (10 pp.).
79. Yamamoto, S. *et al.* (2012) Nonvolatile flip-flop using pseudo-spin-transistor architecture and its power-gating applications. 2012 International Semiconductor Conference Dresden-Grenoble (ISCDG) – formerly known as the Semiconductor Conference Dresden (SCD 2012), pp. 17–20.

80. Yamamoto, S.i. *et al.* (2012) Nonvolatile power-gating field-programmable gate array using nonvolatile static random access memory and nonvolatile flip-flops based on pseudo-spin-transistor architecture with spin-transfer-torque magnetic tunnel junctions. *Japanese Journal of Applied Physics*, **51**, 491–494.
81. Shuto, Y. *et al.* Fabrication and characterization of pseudo-spin-MOSFET, *arXiv.org > cond-mat > arXiv:0910.5238*, 2009.
82. Khitun, A. and Wang, K.L. (2011) Non-volatile magnonic logic circuits engineering. *Journal of Applied Physiology*, **110**, 034306 10.
83. Nikonov, D.E. *et al.* (2011) Proposal of a spin torque majority gate logic. *IEEE Electron Device Letters*, **32**, 1128–1130.
84. Tougaw, P.D. and Lent, C.S. (1994) Logical devices implemented using quantum cellular automata. *Journal of Applied Physics*, **75**, 1818–1825.
85. Cowburn, R.P. and Welland, M.E. (2000) Room temperature magnetic quantum cellular automata. *Science*, **287**, 1466–1468.
86. Bernstein, G.H. *et al.* (2005) Magnetic QCA systems. *Microelectronics Journal*, **36**, 619–624.
87. Imre, A. *et al.* (2006) Majority logic gate for magnetic quantum-dot cellular automata. *Science*, **311**, 205–208.
88. Orlov, A. *et al.* (2008) Magnetic Quantum-dot Cellular Automata: Recent developments and prospects. *Journal of Nanoelectronics and Optoelectronics*, **3**, 55–68.
89. Niemier, M.T. *et al.* (2011) Nanomagnet logic: progress toward system-level integration. *Journal of Physics-Condensed Matter*, **23**, 491–494.
90. Varga, E. *et al.* (2010) Programmable Nanomagnet-logic Majority Gate. 2010 68th Annual Device Research Conference (DRC 2010), pp. 85–86.
91. Niemier, M.T. *et al.* (2007) Clocking structures and power analysis for nanomagnet-based logic devices. 2007 International Symposium on Low Power Electronics and Design – ISLPED, pp. 26–31.
92. Alam, M.T. *et al.* (2010) On-chip clocking for nanomagnet logic devices. *IEEE Transactions on Nanotechnology*, **9**, 348–351.
93. Alam, M.T. *et al.* (2012) On-chip clocking of nanomagnet logic lines and gates. *IEEE Transactions on Nanotechnology*, **11**, 273–286.
94. Varga, E. *et al.* (2013) Experimental realization of a nanomagnet full adder using slanted-edge input magnets. to appear at Joint MMM/Intermag Conference, Chicago, IL.
95. Varga, E. *et al.* (2013) Experimental realization of a nanomagnet full adder using slanted-edge magnets. *IEEE Transactions on Magnetics*, **10**, 60–62.
96. Carlton, D. *et al.* (2012) Investigation of defects and errors in nanomagnetic logic circuits. *IEEE Transactions on Nanotechnology*, **11**, 760–762.
97. Lambson, B. *et al.* (2011) Exploring the thermodynamic limits of computation in integrated systems: magnetic memory, nanomagnetic logic, and the landauer limit. *Physical Review Letters*, **107**, 72–76.
98. Niemier, M. *et al.* (2012) Systolic architectures and applications for nanomagnet logic. 2012 IEEE Silicon Nanoelectronics Workshop (SNW), pp. 2 pp.-2 pp.
99. Peng, L. *et al.* (2012) Power reduction in nanomagnetic logic clocking through high permeability dielectrics. 2012 70th Annual Device Research Conference (DRC), pp. 129–130.
100. Xueming, J. *et al.* (2012) Design of a systolic pattern matcher for Nanomagnet Logic. 2012 15th International Workshop on Computational Electronics (IWCE), pp. 3 pp.-3 pp.
101. Bandyopadhyay, S. (2012) Information Processing with Electron Spins. *International Scholarly Research Network*, **2012**, 1–20.
102. Bandyopadhyay, S. and Cahay, M. (2009) Electron spin for classical information processing: a brief survey of spin-based logic devices, gates and circuits. *Nanotechnology*, **20**, 10–21.
103. Carlton, D.B. *et al.* (2008) Simulation studies of nanomagnet-based logic architecture. *Nano Letters*, **8**, 4173–4178.
104. Niemier, M. *et al.* (2012) Boolean and non-boolean nearest neighbor architectures for out-of-plane nanomagnet logic. 2012 13th International Workshop on Cellular Nanoscale Networks and their Applications (CNNA 2012), pp. 6 pp.-6 pp.
105. Lambson, B. *et al.* (2012) Error immunity techniques for nanomagnetic logic. 2012 IEEE International Electron Devices Meeting (IEDM 2012), pp. 11.5 (4 pp.)-11.5 (4 pp.).
106. Spedalieri, F.M. *et al.* (2011) Performance of magnetic quantum cellular automata and limitations due to thermal noise. *IEEE Transactions on Nanotechnology*, **10**, 537–546.
107. Behin-Aein, B. *et al.* (2011) Switching energy-delay of all spin logic devices. *Applied Physics Letters*, **98**, 17–20.

108. Khitun, A. and Wang, K. (2005) Nano scale computational architectures with Spin Wave Bus. *Superlattices & Microstructures*, **38**, 184–200.
109. Khitun, A. *et al.* (2008) Spin wave magnetic NanoFabric: A new approach to spin-based logic circuitry. *IEEE Transactions on Magnetics*, **44**, 2141–2153.
110. Wu, Y. *et al.* (2009) A three-terminal spin-wave device for logic applications. *Journal of Nanoelectronics and Optoelectronics*, **4**, 394–397.
111. Shabadi, P. *et al.* (2010) Towards logic functions as the device. Proceedings of the Nanoscale Architectures (NANOARCH), 2010 IEEE/ACM International Symposium, pp. 11–16.
112. Khitun, A. *et al.* (2009) Magnetoelectric spin wave amplifier for spin wave logic circuits. *Journal of Applied Physics*, **106**, 191–200.
113. Cherepov, S. *et al.* (2011) Electric-field-induced spin wave generation using multiferroic magnetoelectric cells. Proceedings of the 56th Conference on Magnetism and Magnetic Materials (MMM 2011), DB-03, Scottsdale, Arizona.
114. Cherepov, S. *et al.* (2011) Electric-Field-Induced Spin Wave Generation Using Multiferroic Magnetoelectric Cells, vol. WIN annual report.
115. Khitun, A. (2012) Multi-frequency magnonic logic circuits for parallel data processing. *Journal of Applied Physics*, **111**, 13–20.
116. Roy, K. *et al.* (2011) Hybrid spintronics and straintronics: A magnetic technology for ultra low energy computing and signal processing. *Applied Physics Letters*, **99**, 17–21.
117. Wu, T. *et al.* (2011) Giant electric-field-induced reversible and permanent magnetization reorientation on magnetoelectric Ni/(011) [Pb(Mg1/3Nb2/3)O3](1−x)–[PbTiO3]x heterostructure. *Applied Physics Letters*, **98**, 012504 7.
118. Vanrun, A. *et al.* (1974) Insitu grown eutectic magnetoelectric composite-material .2. Physical-properties. *Journal of Materials Science*, **9**, 1710–1714.
119. Wang, J. *et al.* (2003) Epitaxial BiFeO3 multiferroic thin film heterostructures. *Science*, **299**, 1719–1722.
120. Eerenstein, W. *et al.* (2006) Multiferroic and magnetoelectric materials. *Nature*, **442**, 759–765.
121. Shabadi, P. *et al.* (2011) Spin wave functions nanofabric update. Proceedings of the IEEE/ACM International Symposium on Nanoscale Architectures (NANOARCH-11), pp. 107–113.
122. Shabadi, P. *et al.* (2010) Towards logic functions as the device. 2010 IEEE/ACM International Symposium on Nanoscale Architectures (NANOARCH 2010).
123. D'Souza, N. *et al.* (2011) Four-state nanomagnetic logic using multiferroics. *Journal of Physics D-Applied Physics*, **44**, 16–20.
124. Slonczewski, J.C. (1996) Current-driven excitation of magnetic multilayers. *Journal of Magnetism And Magnetic Materials*, **159**, L1–L7.
125. Hosomi, M. *et al.* (2005) A novel nonvolatile memory with spin torque transfer magnetization switching: spin-RAM. International Electron Devices Meeting 2005 (IEEE Cat. No.05CH37703C), pp. 4 pp.-4 pp.
126. Xiaofeng, Y. *et al.* (2012) Magnetic tunnel junction-based spintronic logic units operated by spin transfer torque. *IEEE Transactions on Nanotechnology*, **11**, 120–126.
127. Sun, J.Z. and Ralph, D.C. (2008) Magnetoresistance and spin-transfer torque in magnetic tunnel junctions. *Journal of Magnetism and Magnetic Materials*, **320**, 1227–1237.
128. Nikonov, D.E. *et al.* (2011) Nanomagnetic circuits with spin torque majority gates. 2011 IEEE 11th International Conference on Nanotechnology (IEEE-NANO), pp. 1384–1388.
129. Rahman, M.T. *et al.* (2012) Reduction of switching current density in perpendicular magnetic tunnel junctions by tuning the anisotropy of the CoFeB free layer. *Journal of Applied Physics*, **111**, 17–21.
130. Kaka, S. *et al.* (2007) Mutual phase-locking of microwave spin torque nano-oscillators. IEEE International Magnetics Conference, 2006, pp. 2–12
131. Berger, L. (1996) Emission of spin waves by a magnetic multilayer traversed by a current. *Physical Review B*, **54**, 9353–9358.
132. Katine, J.A. *et al.* (2000) Current-driven magnetization reversal and spin-wave excitations in Co/Cu/Co pillars. *Physical Review Letters*, **84**, 3149–3152.
133. Mancoff, F.B. *et al.* (2005) Phase-locking in double-point-contact spin-transfer devices. *Nature*, **437**, 393–395.
134. Csaba, G. *et al.* (2012) Spin torque oscillator models for applications in associative memories. 2012 13th International Workshop on Cellular Nanoscale Networks and their Applications (CNNA 2012), pp. 2 pp.-2 pp.
135. Krivorotov, I. (2012) Spin torque oscillator majority logic. Western Institute of Nanoelectronics, Annual Review, vol. Abstract 3.1.

18

NEMS Switch Technology

Louis Hutin and Tsu-Jae King Liu

Department of Electrical Engineering and Computer Sciences, University of California, USA

18.1 Electromechanical Switches for Digital Logic

18.1.1 Addressing the CMOS Power Crisis

For about four decades, increasing transistor density (thereby increasing chip functionality and lowering cost per function) has been carried out following a set of relatively simple transistor scaling rules [1–3]. These "constant field" device scaling guidelines were purposefully designed so that active power consumption would scale with area, hence keeping the power density constant. However, scaling the supply voltage V_{DD} with no penalty in terms of circuit speed requires the ability to maintain a sufficient gate overdrive (V_{DD}–V_T, defining the on-state resistance of a transistor). Unfortunately, there is a fundamental limit on how much the threshold voltage V_T can be reduced without causing an exponential increase in leakage current, and therefore passive power density. This limit is a direct consequence of a MOSFETs subthreshold current–voltage relationship (Figure 18.1).

With V_T pinned, the gate overdrive had to be reduced and performance sacrificed in order to keep the active power density under control. Although system throughput can be recovered through the use of parallel design (e.g., multi-core processors), a lower bound in total energy per operation is eventually reached when transistors function in the subthreshold regime [6]. Alternative field-effect transistor architectures designed to overcome the subthreshold swing limit of 60 mV/decade at room temperature have been proposed, including the impact ionization MOSFET [7], the tunnel FET [8], the negative-capacitance FET [9] (see Chapter 15). However these solid-state switches still exhibit nonzero leakage currents, hence a subsisting fundamental lower limit in energy efficiency. In contrast, a mechanical relay operating by closing and opening an air gap to switch signal on and off such as illustrated in Figure 18.2, would offer the combined advantages of zero leakage current and extremely abrupt switching behavior.

As one might expect from such an intuitive representation of an ideal switch, the use of mechanical relays for computing is not exactly a new idea, and dates back to the late 1930s [10].

Emerging Nanoelectronic Devices, First Edition. An Chen, James Hutchby, Victor Zhirnov and George Bourianoff.
© 2015 John Wiley & Sons, Ltd. Published 2015 by John Wiley & Sons, Ltd.

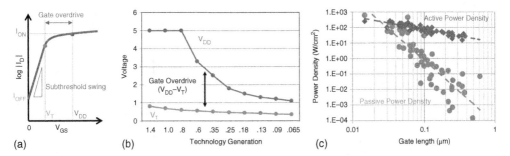

(a)   (b)   (c)

Figure 18.1   (a) Schematic typical I_D-V_{GS} of an nMOSFET. The un-scalable subthreshold swing makes I_{OFF} increase exponentially for a linear V_T shift towards 0 V. With V_T fixed, a reduction of V_{DD} provokes a decrease in I_{ON}. (b) V_{DD} and V_T scaling down to the 65 nm CMOS technology node [4]. (c) Active and passive power density in ICs for sub-micron gate lengths, and exponential increase of the latter, related to V_T reduction [5]

Konrad Zuse completed in 1941 the Z3, the world's first functional programmable, fully automated electromechanical computer, switching 2000 relays at frequencies of 5–10 Hz [11]. Almost immediately abandoned in favor of vacuum tubes and solid-state switches, it took several decades worth of technological advances in planar processing, and in particular surface micromachining, to make mechanical computing once again relevant in terms of scalability [12]. This continued progress combined to the emergence of the CMOS power crisis has recently triggered renewed interest in Micro and Nano-ElectroMechanical Systems (M/NEMS) for ultra low power integrated circuits [13,14].

18.1.2   Throughput and Area Efficiency

Various options are available to mechanically open or short an electrical connection between two electrodes, and the constraints associated to fabrication processes may vary according to the type of NEMS switch transduction. Each actuation mechanism also features distinctive advantages and limitations in terms of time and energy required for a switching event. Thus, it is useful to define rough goals in terms of energy–delay trade-off prior to assessing the adequacy of various NEMS relay families for digital logic, which will be the object of Section 18.2.

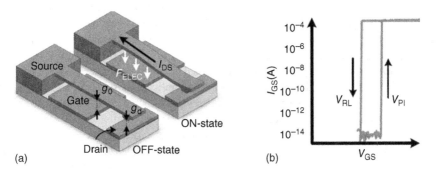

(a)   (b)

Figure 18.2   (a) Generic planar three-terminal electrostatic switch. (b) Typical measured source–drain current versus gate voltage characteristic for a three-terminal MEM switch

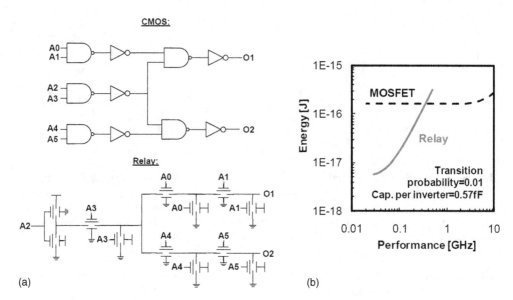

Figure 18.3 (a) CMOS to relay logic mapping [14]. (b) Simulated energy–performance trade-off curves for a 30-stage CMOS inverter chain and a 30-relay chain implemented in 65 nm equivalent technologies (based on the ITRS 2007 specifications) [19]

Today, the intrinsic delay of a single MOS transistor should be on the order of a picosecond. As a rule of thumb, this is three orders of magnitude smaller than what can be expected in terms of mechanical delay due to, for example, an electrostatically actuated relay. This may however not be a showstopper. In CMOS digital ICs, the delay is set by the electrical time constant: stacking devices comes with a quadratic delay penalty, leading circuit designers to buffer and distribute the logical effort over many stages. This is not the case for NEMS relays, for which mechanical inertia dominates over RC delay. In consequence, logic can be implemented using larger, more complex gates where all relays actuate at the same time (single mechanical delay) with in some cases little to no area penalty as device count is reduced [14,15]. Projections suggest that this relay logic could compete with CMOS at the nanoscale [16–18], with typically lower operational frequency but lower energy (Figure 18.3).

Once again, a lower throughput can be compensated by resorting to massive parallelism. But this is eventually detrimental to area efficiency and therefore cost per operation. Although the manufacturing cost of a NEM switch can arguably be predicted to be lower than, for example, a state of the art FinFET; concluding at this point that NEMS can entirely supplant CMOS for digital logic would be premature. A NEMS-only integration scheme would obviously be the least constrained in terms of process sequence, thermal budget and material selection, but at the risk of confinement to low density, moderate speed applications such as Field-Programmable Gate Arrays (FPGA). Considering the above, trying to keep the NEMS switch technology CMOS-compatible may be more prudent for high-speed digital logic. Because not all transistors in a microprocessor operate at the maximum clock frequency, a NEMS/CMOS co-integration scheme could enable saving energy using relays in the low throughput sub-blocks, while maintaining high-performance where needed.

Monolithic co-integration offers the advantage of reduced interconnect parasitics over a "system in package" approach [20]. MEMS-first integration has considerable handicaps in terms of device scalability and foundry compatibility due to the extensive preprocessing required and resulting increased wafer cost. A MEMS-last process, in which the MEMS fabrication steps are performed after completion of the CMOS back end of line, seems the most attractive. It is however critical that the underlying electronics can withstand the thermal budget of MEMS processing. Transistor characteristics must remain unaffected (avoiding dopant diffusion, equivalent oxide thickness variations), and the metal interconnects should be preserved from via resistance failure and electromigration which may arise from thermal stress and annealing effects. It was shown for example that the thermal budget limit for a 0.25 μm CMOS process was 425 °C for 10 h [21]. This significantly restricts the choice of structural materials available for NEMS fabrication (low temperature oxides, amorphous Silicon, polycrystalline SiGe, etc.).

18.2 Actuation Mechanisms

18.2.1 Electromagnetic

Electromagnetic relays convert a magnetic field into mechanical displacement based on either the Lorentz force or the magnetostrictive effect. Essentially, an inductor and a soft magnetic actuator are required, although the addition of hard magnetic films (permanent magnets) to improve energy efficiency seems unavoidable at the micro and nanoscales (Figure 18.4).

Generating a magnetic field can be achieved by flowing current across an inductor such as a Cu coil. For ease of fabrication, planar coils are preferred over their three-dimensional, multi-layered counterparts, in spite of being particularly area consuming. In addition to this, coils do not fare particularly well in terms of energy efficiency at the microscale, so that integrating auxiliary permanent magnets (PM) may be necessary to add a "background" preactuation magnetic field and hence decrease power consumption during switching. As an indicator of how poorly they scale, it can be shown that a tremendous 10^4 A/mm^2 current density is needed for a 100 μm micro-coil to generate the same magnetic moment as a 1 T magnet of same

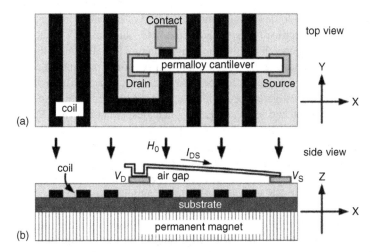

Figure 18.4 (a) Top view and (b) cross-section of an electromagnetic MEM relay, in the on-state [22]. The right side of the movable cantilever is in contact with the source electrode, forming an electrical path between drain and source

dimensions, with every size reduction by a factor of 10 resulting in an 10-fold increase of the required current density [23]. The most suitable permanent magnet materials include thin film CoPt, FePt, or rare earth-based thick film $Nd_2Fe_{14}B$, $SmCo_5$ [24]. CoPt and FePt processing requires relatively high post deposition annealing temperatures (500–700 °C) for obtaining high coercivity values, which is a major disadvantage for CMOS co-integration. On the other hand, the use of rare earth-based alloys comes with problems of easy oxidation, etching process compatibility and adhesion to Si as well as potential contamination issues.

The most straightforward way to emulate a switch is to subject a soft magnetic film to the Lorentz force. Permalloy (NiFe) cantilevers can be used [22,25] with the following advantages: well-developed technology for deposition and micromachining (owing to the longstanding use of permalloy magnetic heads in the recording industry), high saturation flux density and low hysteresis. Besides, permalloys are immune to the reverse magnetostrictive effect, which causes a change of magnetic properties under mechanical stress. Displacement can also be obtained by taking advantage of the direct magnetostrictive effect, which results in the deformation of some ferromagnetic films upon magnetization. This type of transduction can be used to provoke strain mismatch-induced deflection of a bimorph membrane or cantilever coated by magnetostrictive films [26]. Suitable candidates for such coatings in terms of sensitivity include rare earth TbFe, SmFe thin films or TbFe/FeCo multilayers [26,27].

Without specifically targeting digital logic, magnetic microrelays demonstrators have so far exhibited switching power consumption of the order of 10–100 mW, sub-kHz operating frequency and single device area measured in mm^2 [22,25]. To this day, electromagnetic switches do not seem particularly adequate for computing applications because of inherent issues related to area consumption, unconventional processing and insufficient energy efficiency.

18.2.2 Electrothermal

Thermal expansion is the driving phenomenon for electrothermal switches actuation. Two terminals are dedicated to providing power to the actuator through dissipated heat. This can simply be achieved with Joule heating caused by current flowing across a resistive actuating cantilever or beam. Specifically engineered geometry such as a V-beam [28], or thermal expansion coefficient mismatch within a bimorph structure [29,30] causes them to buckle when heated (closing a gap and making contact between two other terminals), and return to their original position once the power is turned off (breaking contact). Thermally actuated relays have been reportedly investigated for a wide range of applications which so far did not include computational logic: switched inductors [29], contact test structures [31], RF MEMS [28] and CMOS power gating for zero standby power [30].

Commonly cited advantages of such devices are high contact forces, large possible displacements and IC-compatible low actuation voltages [28]. High contact forces are typically associated to low contact resistance and therefore low insertion loss which is indeed a significant advantage for RF switches. For logic circuits however, as discussed in Section 18.1.2, the RC delay is typically orders of magnitude lower than the mechanical delay so that lowering the contact resistance is far less critical. This is especially true if its root cause is plastic deformation occurring in the contact regions, which raises additional concerns in terms of reliability (cycling, hysteresis). Large displacements can also be problematic for endurance, due to accelerated structural fatigue. Besides, because actuation is slow (hundreds of microseconds) and high currents are needed to generate sufficient heat, sub-1 V actuation voltages do not prevent thermal micro-relays from consuming around 10 mW for a single switching event [30]. Finally, from the perspective of potential for large-scale integration, the scalability of such switches seems once again fundamentally limited by the transduction mechanism.

Electrothermal NEMS relays do not appear as strong contenders for beyond CMOS computing. They are in some regards comparable to electromagnetic switches, albeit with the undeniable advantage of more conventional material processing.

18.2.3 Piezoelectric

Piezoelectric materials have the ability to charge in response to mechanical stress (this is the direct piezoelectric effect), and reciprocally to deform in response to an electric field (this is the reverse piezoelectric effect). While the former is most useful in resonators for acceleration, pressure or mass sensing [32,33], the latter is exploited in micro-actuators [34]. Because of their use for a particularly broad range of MEMS applications, and even some IC applications such as FeRAM [35], piezoelectric materials process technology has been relatively well developed. Piezoelectricity is also linear (which in itself is interesting for resonators, but of little relevance to logic switches operation) and bi-directional, meaning that displacement on both sides of the equilibrium position is possible depending on the polarity. This last property is very interesting in that it enables active turn-off of a relay, a feature which will be discussed later in Section 18.3.3, and facilitates the implementation of complementary logic.

The most commonly used piezoelectric materials for MEMS are $PbZr_xTi_{1-x}O_3$ (PZT), aluminum nitride (AlN) and zinc oxide (ZnO). PZT is known to yield the most efficient transduction, but high quality material availability is limited (need for stoichiometry control with $x = 0.53$ for maximal efficiency), and it is lead-based, which makes it incompatible with CMOS foundries [36]. AlN-based microrelays have been proposed in particular by researchers at the University of Pennsylvania, first for use as RF switches [37,38] and later for complementary logic [36,39]. These recent studies show variations of the dual beam design illustrated in Figure 18.5 featuring numerous advantages: active turn-off, complementarity, threshold adjustment via body-biasing, immunity to stress-induced out of plane bending, and

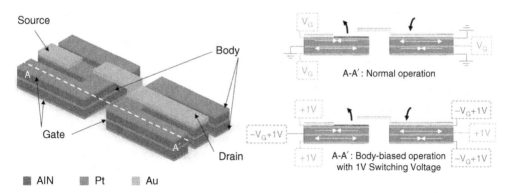

Figure 18.5 (a) 3D schematic and (b) cross-sectional views of the stress-compensated dual beam design from [36] showing body-biased operation. Note that the gap is differentially defined between two cantilevers of same dimensions and structure, hence same deflection at rest. This allows the actuation gap after release to be the same as the sacrificial thickness

an actuation mechanism (based on buckling of bimorph beams) which allows overcoming the typically small displacements generated by the reverse piezoelectric effect.

A Pt/AlN/Pt/AlN/Pt (200 nm/1 μm/200 nm/1 μm/200 nm) sandwich serves in this case as the structural stack. A gold contact is patterned on top of one of the beams, separated from the top Pt layer of the other beam by a contact air gap defined by a 200 nm amorphous Si sacrificial layer release-etched in vapor phase XeF_2. In one of the two opposing beams, the bottom AlN layer is made to longitudinally expand (respectively, contract) through reverse piezoelectric effect, while the top layer is made to contract (respectively, expand), causing the beam to buckle upward (respectively, downward). An Au/Pt contact is made as the two beams meet, transmitting the signal. The beams can be moved apart by changing biasing polarities in the Pt layers (active turn-off). With these switches, sub-1 V operation has been demonstrated thanks to body-biasing ($V_B = 14$ V), with 220 ns switching time, and endurance up to 10^7 cycles with low contact resistance ($<500 \Omega$) [39]. A complementary relay inverter was also demonstrated at 100 Hz. Remaining challenges include scaling down the vertical and lateral dimensions (the current footprint being around $200 \times 200 \mu m^2$ per switch), which would improve area efficiency and reduce parasitic capacitances limiting high frequency operation. A different Au-free contacting metal should be implemented to render processing CMOS-compatible, which may change the deal in terms of contact resistance value and stability.

Piezoelectric relays show promise for mechanical computing applications in terms of speed and energy efficiency. CMOS process compatibility and technology scalability are the next roadblocks to address. Encouragingly, recent works on resonators for gravimetric detection have led to demonstrating 50 nm thick AlN films with excellent piezoelectric properties [33].

18.2.4 Piezoresistive

The piezoresistive effect is a change of electrical resistivity in semiconductors upon application of mechanical stress. This property is widely used in strain-enhancement techniques for high-mobility CMOS. Piezoresistive elements can be incorporated in mechanical relays to generate an abrupt switching behavior when combined in series with a supplementary transducer converting an electrical signal into mechanical stress (e.g., piezoelectric, electrostatic, etc.). The appeal of such an approach with added complexity is that it circumvents the obstacle of contact adhesion forces, which may cause hysteretic characteristics or even functional failure in most air gap-based mechanical switches.

Polymers doped with carbon black or metal nanoparticles can also emulate piezoresistivity and although the fundamental mechanism is different, sharp stress-conductivity characteristics can be obtained. MIT researchers introduced a three-terminal mechanical relay consisting of a patterned polydimethylsiloxane (PDMS) pillar containing Ni nanoparticles, and becoming conductive when squeezed by an electrostatically actuated Al membrane, hence the moniker "squitch" [40]. When a sufficient pressure is applied to the nanocomposite, a conduction path is created between the conductive metallic particles and the relay turns on with a theoretical 10^{-7} R_{on}/R_{off} ratio. This architecture faces many practical problems, such as sensitivity to creep, possibility of inter-particle adhesion causing reliability issues upon cycling, and more obviously process integration and scalability.

IBMs PET concept (PiezoElectronic Transistor) relies on the inverse piezoelectric effect to transmit stress to a piezoresistive "channel" [41]. For the stress transfer to be efficient, a yoke made of a high yield material such as SiN should be preventing mechanical displacement of the free boundary of the PR film (Figure 18.6).

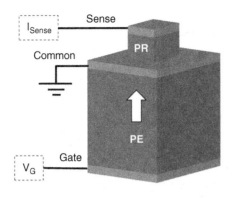 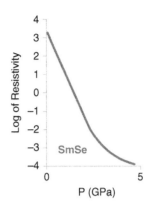

Figure 18.6 Left: PET with three terminals (gate, common, and sense), a piezoelectric element between the common and gate terminals and a piezoresistive element between the common and sense terminals. The constraining mechanical yoke is omitted for clarity. Right: Resistivity–pressure characteristic for a piezoresistive monochalcogenide: SmSe [41]

The best candidates for optimal transduction are believed to be PZN-PT $\{(1 - x)[Pb(Zn_{1/3}Nb_{2/3})O_3] - x[PbTiO_3]\}$ or PMN-PT $\{(1 - x)[Pb(Mg_{1/3}Nb_{2/3})O_3] - x[PbTiO_3]\}$ for the piezoelectric element, and SmSe, TmTe, Eu-doped SmS or Cr-doped V_2O_3 for the piezoresistive element [41]. In a best case scenario, an analysis of the device's dynamic response indicated an outstanding projected operating frequency of 20 GHz at $V_{DD} = 0.1$ V. Nevertheless, these numbers rely on many assumptions such as: (i) negligible three-dimensional effects, (ii) lossless stress transfer, and most of all that (iii) these lead-based and rare earth chalcogenide materials can be integrated down to thicknesses of a few nanometers while conserving their sensitivity. Moreover, static analysis shows a 10^{-4} R_{on}/R_{off} ratio for a gate-common bias of 0.18 V, which translates into an equivalent less than ideal subthreshold swing of 45 mV/dec.

Pending further experimental work, it is unclear at this point whether the energy efficiency benefits of piezoresistive switches can eventually be worth the implementation of an extremely disruptive technology with uncertain scalability.

18.2.5 Electrostatic

M/NEM relays utilizing electrostatic actuation may now appear excessively simple in comparison. Although not the most popular for RF switching applications due to relatively low contact forces and high contact resistance [42], these characteristics are not critical for computational purposes, as previously discussed in Section 18.1.2. These switches are essentially based on two conductive layers forming a capacitor: one being fixed and the other movable (e.g., cantilever, clamped–clamped beam, membrane, nanowire). They are therefore generally easy to fabricate with conventional planar processing techniques and conventional, inexpensive, MOS-compatible materials which do not lose their transducing properties at the nanoscale. As the capacitor charges upon applying a potential difference, the attractive electrostatic force (regardless of polarity) deflects and accelerates the movable electrode towards the fixed one. When the actuating bias disappears, the switch can be passively turned off as the spring restoring force causes the movable membrane to return to its original position, breaking the contact. Let us consider the simple model in Figure 18.7, where the top electrode

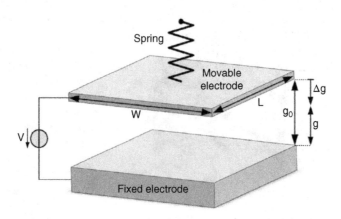

Figure 18.7 Parallel plate capacitor model for electrostatic actuation. g_0 is the as-fabricated gap corresponding to the separation at rest

of a parallel-plate capacitor of area A is suspended by a spring of elastic constant k and is separated from a fixed bottom electrode by a distance g.

Under a bias V, the electrostatic force F_{elec} and spring force F_{spring} exerted on the top electrode are [43]:

$$F_{elec} = \frac{1}{2}\frac{\varepsilon_0 A V^2}{g^2}; \quad F_{spring} = k(g_0 - g) \tag{18.1}$$

As the deflection increases, F_{elec} increases superlinearly, whereas the F_{spring} increase is only linear. Beyond a critical displacement which can be easily derived as $g = g_0/3$, the system becomes unstable, F_{elec} is always superior to F_{spring} and the switch closes abruptly: this phenomenon is called pull-in. The pull-in voltage V_{pi} is thus:

$$V_{pi} = \sqrt{\frac{8}{27}\frac{kg_0^3}{\varepsilon_0 A}} \tag{18.2}$$

In order to keep V_{pi} low and the switching energy minimal, the spring should be relatively compliant, but it also should be sufficiently stiff to overcome contact adhesion forces. Contact dimples are commonly used in order to minimize the contact area and alleviate this constraint. In this case, it is useful to distinguish between the actuation gap g_0 and the contact dimple gap g_d. If g_d is larger than $g_0/3$, the switch is operating in pull-in mode. A characteristic of pull-in mode is that although the actuated plate does not collapse onto the fixed electrode, the abrupt increase in capacitance when $V > V_{pi}$ causes an intrinsically hysteretic current–voltage characteristic. In nonpull-in mode ($g_d < g_0/3$), the adhesion forces alone may contribute to the hysteresis. It should be noted that there are ways to avoid pull-in instability beyond the $g_0/3$ displacement limit, such as leveraged bending [44] or dynamic charge control (by adding a capacitor in series) [45].

From a dynamic point of view, neglecting contact bouncing and ringing upon contact break, the mechanical delay can be approximated [16] as a function of the fundamental resonant period $[2\pi(m/k)^{1/2}]$, the dimple to actuation gap ratio (g_d/g_0), and the "gate overdrive" (V_{DD}/V_{pi}):

$$t_{delay} \cong \alpha \sqrt{\frac{m}{k}} \left(\frac{g_d}{g_0}\right)^{\gamma} \left(\frac{V_{DD}}{V_{pi}} - \chi\right)^{-\beta} \qquad (18.3)$$

α, β and γ can be numerically derived by solving Newton's second law of motion; their values primarily depend on the quality factor Q of the system and thus on the damping ratio for its first mode of oscillation. χ is an adjustment variable accounting for the dramatic increase of t_{delay} when V_{DD}/V_{pi} is very close to 1. Equation 18.3 gives us a sense that to minimize the delay for a given overdrive and gap aspect ratio, the structural material of the actuated electrode should be stiff (high Young's modulus) and light (low density). This is why speed is one of the driving arguments for investigating carbon-based electrostatic switches (Carbon Nano Tubes [46–50], graphene [51], SiC [52,53]). Increased stiffness comes however at the cost of larger V_{pi} (18.2) and therefore higher switching energy.

If the system is underdamped ($Q > 1/2$), the oscillations of the movable plate after breaking contact could interfere with the off-state electrical characteristics and therefore set an effective t_{delay} larger than the previous pull-in-based expression. A cut-off distance $g_c = 2$ nm is generally considered large enough to avoid undesirable tunneling current between electrodes during off-state. Avoiding tunneling after release translates into a condition on the logarithmic decrement of the damped oscillations, which should be larger than $\ln(g_d/g_d - g_c)$. Expressed in terms of Q:

$$Q < \sqrt{1 + \left\{2\pi/\ln\left[g_d/(g_d - g_c)\right]\right\}^2}/2 \qquad (18.4)$$

Substituting $g_c = 2$ nm, the inequality above results in a surprisingly simple but reasonably accurate rule of thumb stating that the quality factor should be designed to be smaller than the dimple gap g_d expressed in nm.

Obviously though, not all of these electrostatic relays can be modeled as parallel plate capacitors, and each design has its specific energy–delay trade-off. The next section is dedicated to the many design variations available to date in the literature for electrostatically actuated NEM relays, their assets and drawbacks.

18.3 Electrostatic Switch Designs

18.3.1 (N > 3)-Terminal Switches

Many different switch designs are reported in the literature, which can be classified by number of terminals typically ranging from two [54] to eight [55]. The majority of them are three-terminal (3-T) relays, which can probably be explained by a natural tendency to mimic the transistor configuration of a control gate allowing signal propagation between two "Source" and "Drain" terminals. However, we have already evoked (Section 18.1.2) the risks of confining NEMS relay applications to CMOS-like logic: this path should probably be avoided unless each relay can switch (mechanical and RC delay) as fast as a transistor, and its individual footprint is comparable to that of a MOSFET.

Pass-gate logic, on the other hand, comes with a reduced device count (compensating for larger individual footprint) and enables switching one complex logic gate with a single mechanical delay. Single MOS transistors are notoriously imperfect devices for pass-gate logic. This is due to their intrinsic inability to "swing full rail" at the output, that is, having the ability to propagate both a logic "0" ($V_S = 0$ V) and a logic "1" ($V_S = V_{DD}$). Because V_{GS} has to be larger than the threshold voltage V_T, an nMOSFET can only pass V_S from 0 V to V_{DD}-V_T (and conversely, a pMOSFET from V_T to V_{DD}). The fundamental issue is that the control condition is coupled to the signal to transmit. The same principle applies to 3-T electrostatic switches, but an even more basic argument is that there is no such thing as a 3-T n- or p-relay, because electrostatic actuation is ambipolar. Adding a fourth "Body" electrode permits us to:

- Differentiate "pull-down" and "pull-up" relays.
- Lower the threshold voltage by preactuating the device.
- Decouple control and signal.

An example of a 4-T electrostatic switch with insulated channel attached to the movable electrode [56] is shown in Figure 18.8.

One can note that based on the design in Figure 18.8, the number of Source and Drain outputs can easily be increased by attaching more channels to the movable plate, as long as it serves circuit design efficiency and the total equivalent parasitic gate–channels capacitance remains negligible. A 6-T variation [57] with single Gate and a double pair of Source and Drain is the most straightforward implementation, further reducing device count and reducing the area required to implement a complex logic gate by up to a factor of two [58]. Only two of these 6-T devices connected can be used to deliver simultaneously an inverted and noninverted buffered signal (input on the tied gates). The number of gate inputs can also be multiplied to implement (depending on the biasing scheme) AND, OR [59–61] or MAJORITY ($2n + 1$ inputs, $n > 1$) [55] gates with a single relay; and each of the above plus their complementary with two relays [61] (Figure 18.9).

18.3.2   In-Plane versus Out-of-Plane Actuation

Considering the substrate as the plane of reference, we have focused so far on out-of-plane deflected switches. There are nonetheless many instances of in-plane-actuated electrostatic NEM relays: [62–67]. For such devices, the gaps are defined by lithography, as opposed to a deposited sacrificial layer thickness. In terms of layout and for a given processing sequence, it means that a broad range of V_{pi} can be targeted by simply varying the gap width, which has a much lesser impact on device footprint than adjusting the actuation area.

Yet, this supposed ease of design stands under the condition that the structural material of the movable electrode (typically metal, if co-patterned with all other electrodes) can be etched vertically with good anisotropy for a high aspect ratio. This may unfortunately not always be true, and in case of a limited membrane thickness, this loss of actuation area should be recovered by designing very long beams. This would introduce nonlinear spring effects making the pull-in behavior far less predictable (compared to a parallel-plate capacitor model and Equation 18.2) and, ultimately, switch dimensioning more difficult.

Another argument to consider is that while the in-plane actuated switch approach is relatively 3-T friendly, fabricating 4-T relays (for decoupled control and signal, cf. Section 18.3.1) can prove to be more challenging. It can be achieved through the use of less trivial

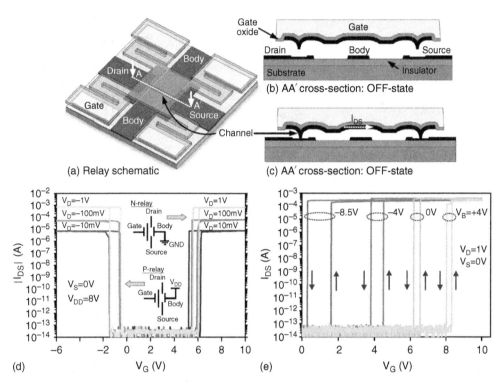

Figure 18.8 Top: (a) Isometric view of a four-terminal relay structure [56]. (b) Cross-sectional view along the channel (AA') in the off-state. (c) Cross-sectional view in the on-state. When the voltage difference between the gate and body electrodes is sufficiently large (i.e., $V_{GB} > V_{pi}$) the body electrode is actuated downward sufficiently to cause the channels – attached to the body electrode via an intermediary oxide layer – to contact their respective S/D electrodes, so that current (I_D) can flow between the S/D electrodes. Bottom: Measured I_{DS}-V_G characteristics of a single 4-T relay, with $V_S = 0$ V. (d) Operation mimicking that of either an n-channel or a p-channel MOSFET is seen by biasing the body at either 0 V or V_{DD}, respectively. (e) V_{pi} can be tuned by adjusting V_B

processing steps such as spacer definition of the insulated channel, followed by localized stringer removal to suppress shorts [67].

18.3.3 Active Turn-Off Devices

In the first section of this chapter, a premise was that relays would allow overcoming the fundamental energy limit in CMOS circuits set by subthreshold leakage. However, a legitimate concern in electrostatic NEMS relay design is that hysteretic switching behavior (coupled to the fact that the switch should turn off at 0 V) is also setting a switching energy limit. Even considering an architecture with contact dimples, operating in nonpull-in mode ($g_d < g_0/3$), counting on a passive release to break the contact means that the spring restoring force for a displacement g_d should be large enough to overcome the contact adhesion force:

$$kg_d > F_{adhesion} \qquad (18.5)$$

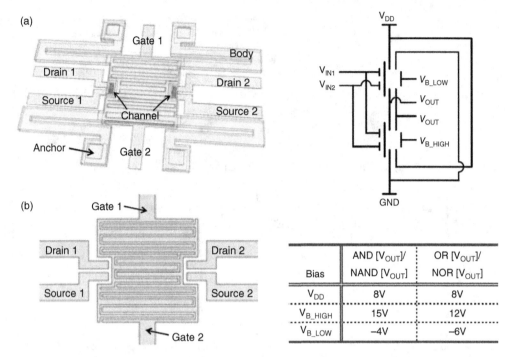

		AND [V_{OUT}]/	OR [V_{OUT}]/
Bias		NAND [V_{OUT}]	NOR [V_{OUT}]
V_{DD}		8V	8V
V_{B_HIGH}		15V	12V
V_{B_LOW}		−4V	−6V

Figure 18.9 Schematic views of a dual-gate, dual-source/drain relay [61]. (a) Isometric view. (b) Bottom electrode layout, showing interdigitated gate electrodes to ensure that they have equal influence on the body. (c) Circuit schematic of a dynamically configurable complementary relay logic circuit utilizing two dual-gate, dual-source/drain relays. (d) Corresponding AND/NAND and OR/NOR bias configurations

Unfortunately, the higher the spring constant k, the higher V_{pi} becomes (Equation 18.2). We have seen in Section 18.2.4 that piezoresistive relays could circumvent contact adhesion force issues by escaping from the conventional "physical contact make/break" scheme, but were not necessarily infinitely abrupt, zero-leakage (and depending on the choice of the piezoresistive material, not always hysteresis-free) switches.

A single-pole double-throw (SPDT) configuration is fairly common for in-plane actuated relays, because the design is straightforward. One actuating fixed electrode can be laid out on each side of the pole, and each can be used to break a contact occurring on the other side [50,63,66,67]. A notable out-of-plane SPDT variation called torsion microactuator or see-saw relay was initially designed for low voltage RF switches used for instance in passive phase shifters [68–71]. The motivation was the same as stated above: reducing the losses (power consumption) through using beams of lesser stiffness which would otherwise remain stuck down by adhesion forces.

A movable electrode is anchored along one of its coplanar symmetry axes. A fixed electrode is placed on each side of this axis under each half of the movable plate, so that it can swing on either side of the torsional axis depending on fixed electrode biasing scheme. The pull-in behavior of such devices can be predicted analytically with good accuracy [72,73]. What makes them particularly interesting for logic circuits is their built-in switching complementarity. Besides, it is possible to demonstrate switching symmetry about $V_{DD}/2$, which is

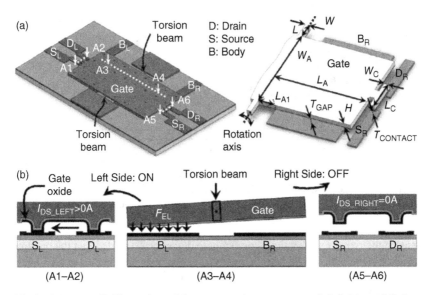

Figure 18.10 (a) Isometric illustration of the seesaw relay structure and definition of design parameters [75]. (b) Schematic cross-sectional views. ON–OFF (illustrated here) and OFF–ON are the two stable states for perfectly complementary operation. The two body electrodes (coplanar with the source/drain electrode pairs) located underneath the gate electrode, one on either side of the torsional axis, are used to apply voltages across the actuation air gap to electrostatically actuate the seesaw beam downward on only one side at a time and thereby turn on the relay on that side and turn off the relay on the other side

advantageous for symmetric noise margins [74]. An example of a 7-T design with two insulated channels [75] is shown in Figure 18.10.

Pointing out that active turn-off consumes power is a tautology; at best this feature can help to lower the energy limit set by adhesion force. Compared to a spring-driven passive release, a see-saw relay is energy-efficient in cases where the contact adhesion energy on one side is larger than the amount required for pulling the other side down. Regarding the latter, see-saws have the handicap of showing a larger gap before actuation than the nominally defined gap (as-deposited sacrificial layer thickness), because this side is deflected upward when the other is touching down. This implies that minimal excursion of the movable plate should be targeted when toggling between on and off states, with extremely scaled contact dimple gaps.

18.4 Reliability and Scalability

18.4.1 Contact Stability

Even though mechanical delay is predominant in relay switching dynamics, the electro-mechanical (EM) contact resistance remains an important figure of merit for circuit design, because it determines how many simultaneously switched stages can be cascaded with no speed penalty. It is therefore critical that its value should remain stable upon cycling. This is true as well for the contact surface topology: plastic deformation of asperities following repeated impacts can modify the adhesion force, potentially leading to permanent stiction failure. This particular failure mode may also be caused by micro welding depending on the current density when the relay is in the on-state. The reliability requirements are largely

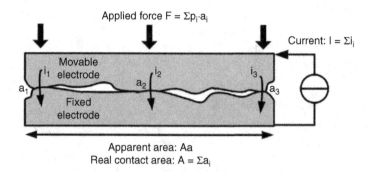

Figure 18.11 Cross-sectional schematic of an EM contact, showing three asperities (called α spots) where current can flow. Total load P and current I correspond to the sum of the contributions from every α spot

dependent on the type of application. For example, it can be estimated that the on-state contact resistance of a switch should be kept below 100 kΩ for over 10^{14} cycles so that a relay-based microcontroller for embedded sensor applications could operate reliably for 10 years at up to 100 MHz with a transition probability of 1% [17].

It is essential to properly understand the physics of electromechanical contacts at the nanometer scale in order to select the appropriate materials, design and surface treatments needed to meet the performance and stability specifications. In practice, the contacting electrodes are almost never ideally flat, and the effective contact area A_a between two conductive rough surfaces is only a fraction of the apparent dimensions (e.g., as drawn on a lithographic mask). Physical contact is only made at some asperities, each having its own associated load (p_i), local area (a_i) and local current (i_i; Figure 18.11).

The constriction resistance R_c at each microcontact (arising from the convergence and divergence of the current flow) can be described by a size-dependent expression [76]:

$$R_c = \frac{4\lambda(\rho_1 + \rho_2)}{9a} + \frac{\rho_1 + \rho_2}{2\pi r} \int_0^\infty e^{(-x\lambda/r)} \frac{\sin(\pi x)}{\pi x} dx \qquad (18.6)$$

ρ_1 and ρ_2 are the specific resistivities of the contacting surfaces 1 and 2, $\lambda = (\lambda_1 + \lambda_2)/2$ is the average electron mean free path, r is the radius of the microcontact of area a. When $r < \lambda$, the expression reduces to Sharvin's point-contact model [77]: electrons are assumed to ballistically travel across the contact without being subjected to scattering and $R_S \approx \lambda(\rho_1 + \rho_2)/2a$. If $r > \lambda$, Equation 18.6 approaches the other limit case, Holm's resistance [78], for which the constriction resistance is dominated by scattering $R_H \approx (\rho_1 + \rho_2)/4r$.

As the contact force is increased asperities can be deformed, causing the real contact area to increase, which in turn lowers the total EM contact resistance. This might seem beneficial at first glance but it should be noted that the flattening of asperities has some undesirable side-effects: increased real contact area and reduced separation of the noncontacting regions (which contribute to adhesion via van der Waals forces). As a consequence, hysteretic behavior is aggravated, possibly to the point of permanent stiction. Besides, for a fixed bias between the two contacting electrodes; a decreasing resistance leads to an increase in dissipated power ($P = V^2/R$), heating the contact as a result, which accelerates its deformation and makes it prone to micro-welding.

In order to prevent permanent mechanical wear, the contact load should be sufficiently low (typically $< 100\,\mu N$) to avoid causing plastic deformation of the electrode material. Selecting a contacting material of high Mohs hardness (>6) can also be useful. The stronger metallic bonds typically associated to hard materials [79] should not be detrimental to adhesion at this scale, because van der Waals interactions are believed to remain dominant (contacting small asperities in the elastic regime). Unlike for RF switching, the contact resistance does not have to be very low for digital logic applications, so the fact that the real contact area remains small is not a critical concern either. Iridium (Ir), rhodium (Rh), ruthenium (Ru), and tungsten (W), being hard metals, are therefore suitable candidates from a strictly electromechanical point of view.

However, extrinsic environmental contamination effects can be another important source of contact resistance instability. Late transition metals (Ir, Rh, Ru) are active catalysts and have been observed to cause the formation of friction polymers on and around the contacting surfaces [80] even in a reasonably clean, packaged environment [81]. This shows that extreme attention should be paid to removing any residual process-induced hydrocarbon contaminants (such as photoresist, or plasma etch byproduct). Tungsten has been shown to be significantly prone to oxidation, limiting the contact lifetime even under high vacuum [82].

Surface passivation techniques can be used to remedy these contamination issues. A few monolayers of conductive metal oxides such as TiO_2 can be deposited via Atomic Layer Deposition to coat W contacting electrodes after sacrificial layer release, thus improving reliability. This implies trading off some of the nominal contact resistance, but thanks to a moderate potential barrier for electrons at the W/TiO_2 interface, R_c may still fall within the $100\,k\Omega$ specifications [19]. Some other metals such as ruthenium or iridium can naturally grow a conductive oxide (RuO_2, IrO_2) when annealed in oxygen, which has the advantage of being localized as opposed to a nonselective ALD which may affect surface leakage. $2\,nm\ RuO_2$ can be thermally grown on pure sputtered Ru following a 2 h low temperature thermal oxidation at $420\,°C$ [83]. The Ru/RuO_2 system seems promising so far for enhancing contact stability, although a practical implementation of this relatively unusual metal raises various process integration considerations such as ease of deposition onto other layers, film adhesion, ease of etching, stress compatibility, and of course overall cost of fabrication.

18.4.2 Process Scalability and Perspectives

We have seen in Section 18.2 that electrostatically-actuated NEMS relays were arguably among the most scalable switches. Later (Section 18.3), basic operating principles of the simplified parallel-plate actuator model showed that the ultimate limits in terms of energy efficiency and delay can be attributed to inter-related parameters:

i. The system's spring constant – low k for low V_{pi}, high k and low mass for small delay.
ii. The actuation and contact gap – small excursion for low V_{pi}, large excursion for larger F_{spring}.
iii. The contact properties – low but stable contact resistance with adhesion force smaller than F_{spring}.

One possible route for balancing trade-offs (i) and (ii) can consist of simultaneously opting for high spring constant and small contact gap. This is the strategy motivating the use of low density, high Young's modulus, ultimately thin structural materials such as Carbon nanotubes or graphene [84,85]. Nevertheless, it should be stressed that fabricating CNT or graphene-based nanorelays with insulated control ("Gate") and transmitting electrodes ("Source and Drain") appears extremely challenging, which could be a handicap in terms of circuit design optimization. An alternative solution may be using extremely thin amorphous metal films (e.g., deposited by ALD), the associated issue being residual stress management. In the case of out-of-plane actuated switches, residual stress can cause the movable electrode to buckle, altering the nominal gap value. This can be addressed by resorting to multi-layer structural stacks for stress compensation, or scaling the lateral dimensions to prevent out-of-plane deflection from building up [57], at the respective costs of increased mass (hence t_{delay}) and reduced capacitance (thus higher V_{pi}).

Releasing nanometer-scale gaps is not trivial. One of the most common configurations is sacrificial SiO_2 selectively etched in HF. Water is a byproduct (and catalyst) of this reaction, which may cause stiction upon release due to capillarity forces. By using a critical point dryer, 15 nm gaps could be released, undercutting relatively short TiN cantilevers [86]. Anhydrous HF release at reduced pressure (75 Torr) has led to the same results on poly-SiGe membranes with lateral dimensions of several microns [88]. Although process controllability has yet to be demonstrated, some design tricks can enable to release tiny contact gaps as small as 4 nm [54]. Volume-shrinkage induced gap formation processes based on silicidation are also a promising novel alternative for high aspect-ratio structure releases [88].

Item (iii) has been addressed in the previous subsection. Yet, it is worth pointing out that the multiple-asperity-based constriction resistance model might no longer stand for ultimately scaled contact dimensions [89]. Hard contacting electrodes would still be desirable for their resistance to plastic deformation in a "single-asperity" configuration, but the adhesion force would become dominated by metallic bonding (instead of van der Waals interactions). Since hard materials are usually characterized by high metallic bonding energies, surface treatment schemes (hydrogen passivation, controlled oxidation, ALD coating, use of self-assembled monolayers) to decrease the adhesion force are likely to be required at this scale.

Considering the tremendous progress accomplished recently in the fields of material sciences and technological processes for M/NEMS fabrication dedicated to advances in device and circuit design, the odds of an emergence of nanoscale mechanical computing to alleviate the CMOS power crisis are better than ever.

References

1. Dennard, R.H., Gaensslen, F.H., Rideout, V.L. *et al.* (1974) Design of ion-implanted MOSFET's with very small physical dimensions. *Journal of Solid-State Circuits*, **SC-9**(5), 256–268.
2. Baccarani, G., Wordeman, M.R., and Dennard, R.H. (1984) Generalized scaling theory and its application to a 1/4 micrometer MOSFET design. *IEEE Transactions on Electron Devices*, **31**(4), 452–462.
3. Davari, B., Dennard, R.H., and Shahidi, G.G. (1995) CMOS scaling for high performance and low power - The next ten years. *Proceedings of the IEEE*, **83**(4), 595–606.
4. Packan, P. (2007) Device and Circuit Interactions (Short Course), Proc. Int. Electron Devices Meeting, 2007.
5. Meyerson, B. (2004) Proc. Semico Impact Conference 2004.
6. Calhoun, B.H. and Chandrakasan, A. (2004) Characterizing and modeling minimum energy operation for subthreshold circuits, Proc. Int. Symp. Low-Power Electron. Design, 2004, pp. 90–95.
7. Gopalakrishnan, K., Griffin, P.B., and Plummer, J.D. (2002) I-MOS: A novel semiconductor device with a subthreshold slope lower than kT/q, Proc. Int. Electron Devices Meeting, 2002, pp. 289–292.
8. Reddick, W.M. and Amaratunga, G.A.J. (1995) Silicon surface tunnel transistor. *Applied Physics Letters*, **67**, 494–496.
9. Salahuddin, S. and Datta, S. (2008) Can the subthreshold swing in a classical FET be lowered below 60 mV/decade?, Proc. Int. Electron Devices Meeting, 2008, pp. 693–696.

10. Shannon, C.E. (1938) A symbolic analysis of relay and switching circuits. *Transactions AIEE*, **57**, 713–723.
11. Zuse, K. (1993) *The Computer - My Life*, Springer-Verlag.
12. Petersen, K.E. (1978) Dynamic micromechanics on silicon: Techniques and devices. *IEEE Transactions on Electron Devices*, **25**(10), 1241–1250.
13. Akarvardar, K., Elata, D., Parsa, R. *et al.* (2007) Design considerations for complementary nanoelectromechanical logic gates, Proc. Int. Electron Devices Meeting, 2007, pp. 299–302.
14. Chen, F., Kam, H., Markovic, D. *et al.* (2008) Integrated circuit design with NEM relays, Proc. Int. Conf. Comput.-Aided Design, 2008, pp. 750–757.
15. Chen, F., Spencer, M., Nathanael, R. *et al.* (2010) Demonstration of integrated micro-electro-mechanical switch circuits for VLSI applications, Proc. Int. Solid State Circuits Conf., 2010, pp. 26–28.
16. Kam, H., Liu, T.-J.K., Stojanovic, V. *et al.* (2011) Design, optimization, and scaling of MEM relays for ultra-low-power digital logic. *IEEE Transactions on Electron Devices*, **58**(1), 236–250.
17. Pott, V., Kam, H., Nathanael, R. *et al.* (2010) Mechanical computing redux: relays for integrated circuit applications. *Proceedings of the IEEE*, **98**(12), 2076–2094.
18. Spencer, M., Chen, F., Wang, C.C. *et al.* (2011) Demonstration of Integrated Micro-electro-mechanical relay circuits for VLSI applications. *IEEE Transactions on Electron Devices*, **46**(1), 308–320.
19. Kam, H., Pott, V., Nathanael, R. *et al.* (2009) Design and reliability of a micro-relay technology for zero-standby-power digital logic applications, Proc. Int. Electron Devices Meeting, 2009, pp. 809–812.
20. Fedder, G.K., Howe, R.T., Liu, T.-J.K., and Quevy, E. (2008) Technologies for cofabricating MEMS and electronics. *Proceedings of the IEEE*, **96**(2), 306–322.
21. Takeuchi, H., Wun, A., Sun, X. *et al.* (2005) Thermal budget limits of quarter-micrometer foundry CMOS for post-processing MEMS devices. *IEEE Transactions on Electron Devices*, **52**(9), 2081–2086.
22. Ruan, M., Chen, J., and Wheeler, C.B. (2001) Latching microelectromagnetic relays. *Sensors and Actuators A*, **91**, 346–350.
23. Niarchos, D. (2003) Magnetic MEMS: Key issues and some applications. *Sensors and Actuators A*, **109**, 166–173.
24. Chin, T.-S. (2000) Permanent magnet films for applications in microelectromechanical systems. *Journal of Magnetism and Magnetic Materials*, **209**, 75–79.
25. Tilmans, H.A.C., Fullin, E., Ziad, H. *et al.* (1999) A fully-packaged electromagnetic microrelay, Proc. Int. Conf. Microelectromech. Syst., 1999, pp. 25–30.
26. Honda, T., Hayashi, Y., Yamaguchi, M., and Arai, K.I. (1994) Fabrication of thin-film actuators using magnetostriction. *IEEE Translation Journal on Magnetics in Japan*, **9**(6), 27–32.
27. Ludwig, A. and Quandt, E. (2000) Giant magnetostrictive thin films for applications in microelectromechanical systems. *Journal of Applied Physics*, **78**(9), 4691–4695.
28. Wang, Y., Li, Z., McCormick, D.T., and Tien, N.C. (2003) A micromachined RF microrelay with electrothermal actuation. *Sensors and Actuators A*, **103**, 231–236.
29. Zhou, S., Sun, X.-Q., and Carr, W.N. (1997) A micro variable inductor chip using MEMS relays, Proc. Int. Conf. Solid State Sens. Actuators, 1997, pp. 1137–1140.
30. Raychowdhury, A., Kim, J.I., Peroulis, D., and Roy, K. (2006) Integrated MEMS switches for leakage control of battery operated systems, Proc. Custom Integr. Circuits Conf., 2006, pp. 457–460.
31. Kruglick, E.J.J. and Pister, K.S.J. (1999) Lateral MEMS microcontact considerations. *Journal of Microelectromechanical Systems*, **8**, 264–271.
32. Karabalin, R.B., Matheny, M.H., Feng, X.L. *et al.* (2009) Piezoelectric nanoelectromechanical resonators based on aluminum nitride thin films. *Applied Physics Letters*, **95**, 103111.
33. Ivaldi, P., Abergel, J., Matheny, M.H. *et al.* (2011) 50 nm thick AlN film-based piezoelectric cantilevers for gravimetric detection. *Journal of Micromechanics and Microengineering*, **21**, 085023.
34. Baborowski, J. (2004) Microfabrication of piezoelectric MEMS. *Journal of Electroceramics*, **12**, 33–51.
35. Auciello, O., Scott, J.F., and Ramesh, R. (1998) The physics of ferroelectric memories. *Physics Today*, **51**, 22–27.
36. Sinha, N., Jones, T.S., Guo, Z., and Piazza, G. (2009) Body-biased complementary logic implemented using AlN piezoelectric MEMS switches, Proc. Int. Electron Devices Meeting, 2009, pp. 813–816.
37. Sinha, N., Mahameed, R., Zuo, C. *et al.* (2008) Dual beam actuation of piezoelectric AlN RF MEMS switches integrated with AlN contour-mode resonators, Proc. Solid-State Sens. Actuators Microsyst. Workshop, 2008, pp. 22–25.
38. Mahameed, R., Sinha, N., Pisani, M.B., and Piazza, G. (2008) Dual-beam actuation of piezoelectric AlN RF MEMS switches monolithically integrated with AlN contour-mode resonators. *Journal of Micromechanics and Microengineering*, **18**, 1–11.

39. Sinha, N., Guo, Z., Tazzoli, A. *et al.* (2012) 1 Volt digital logic circuits realized by stress-resilient AlN parallel dual-beam MEMS relays, Proc. 2012 IEEE 25th Int'l Conference on MEMS, pp. 668–671.
40. Paydavosi, S., Yaul, F.M., Wang, A.I. *et al.* (2012) MEMS switches employing active metal-polymer nano-composites, Proc. 2012 IEEE 25th Int'l Conference on MEMS, pp. 180–183.
41. Newns, D., Elmegreen, B., Liu, X.H., and Martyna, G. (2012) A low-voltage high-speed electronic switch based on piezoelectric transduction. *Journal of Applied Physics*, **111**, 084509.
42. Rebeiz, G.M. (2003) *RF MEMS: Theory, Design, and Technology*, Wiley, New York.
43. Senturia, S.D. (2001) *Microsystem Design*, Springer-Verlag., New York.
44. Seeger, J.I. and Boser, B.E. (1999) Dynamics and control of parallel-plate actuators beyond the electrostatic instability, Proc. Transducers, 1999, pp. 474–477.
45. Hung, E.S. and Senturia, S.D. (1999) Extending the travel range of analog-tuned electrostatic actuators. *Journal of Microelectromechanical Systems*, **8**(4), 497–505.
46. Lee, S.W., Lee, D.S., Morjan, R.E. *et al.* (2004) A three-terminal carbon nanorelay. *Nano Letters*, **4**, 2027–2030.
47. Rueckes, T., Kim, K., Joselevich, E. *et al.* (2000) Carbon nanotube-based nonvolatile random access memory for molecular computing. *Science*, **289**, 94–97.
48. Dujardin, E., Derycke, V., Goffman, M.F. *et al.* (2005) Self-assembled switches based on electroactuated multiwalled nanotubes. *Applied Physics Letters*, **87**, 193107.
49. Dagdour, H., Cassell, A.M., and Banerjee, K. (2008) Scaling and variability analysis of CNT-based NEMS devices and circuits with implications for process design, Proc. Int. Electron Devices Meeting, 2008, pp. 529–532.
50. Cao, J., Vitale, W.A., and Ionescu, A.M. (2012) Self-assembled nano-electro-mechanical tri-state carbon nanotube switches for reconfigurable integrated circuits, Proc. 2012 IEEE 25th Int'l Conference on MEMS, pp. 188–191.
51. Kim, S.M., Song, E.B., Lee, S. *et al.* (2011) Suspended few-layer graphene beam electromechanical switch with abrupt on-off characteristics and minimal leakage current. *Applied Physics Letters*, **99**, 023103.
52. Lee, T.-H., Bhunia, S., and Mehregany, M. (2010) Electromechanical computing at 500°C with Silicon Carbide. *Science*, **329**, 1316–1318.
53. Feng, X.L., Matheny, M.H., Zorman, C.A. *et al.* (2010) Low voltage nanoelectromechanical switches based on Silicon Carbide nanowires. *Nano Letters*, **10**, 2891–2896.
54. Lee, J.O., Song, Y.-H., Kim, M.-W. *et al.* (2013) A sub-1-volt nanoelectromechanical switching device. *Nature Nanotechnology*, **8**, 36–40.
55. Jeon, J., Hutin, L., Jevtic, R. *et al.* (2012) Multiple-Input Relay Design for More Compact Implementation of Digital Logic Circuits. *Electron Device Letters*, **33**(2), 281–283.
56. Nathanael, R., Pott, V., Kam, H. *et al.* (2009) 4-terminal relay technology for complementary logic, Proc. Int. Electron Devices Meeting, 2009, pp. 223–226.
57. Chen, I.-R., Hutin, L., Park, C. *et al.* (2012) Scaled micro-relay structure with low strain gradient for reduced operating voltage. *ECS Transactions*, **45**(6), 101–106.
58. Liu, T.-J.K., Hutin, L., Chen, I.-R. *et al.* (2012) Recent progress and challenges for relay logic switch technology, Proc 2012 VLSI Tech. Symp., pp. 43–44.
59. Hirata, A., Machida, K., Kyuragi, H., and Maeda, M. (2000) A electrostatic micromechanical switch for logic operation in multichip modules on Si. *Sensors and Actuators A*, **80**, 119–125.
60. Tsai, C.-Y., Kuo, W.-T., Lin, C.-B., and Chen, T.-L. (2008) Design and fabrication of MEMS logic gates. *Journal of Micromechanics and Microengineering*, **18**, 045001.
61. Nathanael, R., Jeon, J., Chen, I.-R. *et al.* (2012) Multi-input/multi-output relay design for more compact and versatile implementation of digital logic with zero leakage, Proc. VLSI Technology, Systems, and Applications (VLSI-TSA) 2012.
62. Chong, S., Akarvardar, K., Parsa, R. *et al.* (2009) Nanoelectromechanical (NEM) relays integrated with CMOS SRAM for improved stability and low leakage, Proc. Int. Conf. Comput.-Aided Design, 2009, pp. 478–484.
63. Czaplewski, D.A., Patrizi, G.A., Kraus, G.M. *et al.* (2009) A nanomechanical switch for integration with CMOS logic. *Journal of Micromechanics and Microengineering*, **19**, 085003.
64. Chong, S., Lee, B., Mitra, S. *et al.* (2012) Integration of nanoelectromechanical relays with Silicon nMOS. *IEEE Transactions on Electron Devices*, **59**(1), 255–258.
65. Chong, S., Lee, B., Parizi, K.B. *et al.* (2011) Integration of nanoelectromechanical (NEM) relays with Silicon CMOS with functional CMOS-NEM circuit, Proc. Int. Electron Devices Meeting, 2011, pp. 701–704.
66. Parsa, R., Shavezipur, M., Lee, W.S. *et al.* (2011) Nanoelectromechanical relays with decoupled electrode and suspension, Proc. Microelectromech. Syst. Conf., 2011, pp. 1361–1364.
67. Lee, W.S., Chong, S., Parsa, R. *et al.* (2011) Dual sidewall lateral nanoelectromechanical relays with beam isolation, Proc. Transducers, 2011, pp. 2606–2609.

68. Milanovi, V., Maharbiz, M., Singh, A. *et al.* (2000) Microrelays for batch transfer integration in RF systems, Proc. Int. Conf. Microelectromech. Syst., 2000, pp. 787–792.
69. Hah, D., Yoon, E., and Hong, S. (2000) A low-voltage actuated micromachined microwave switch using torsion springs and leverage. *IEEE Transactions on Microwave Theory and Techniques*, **48**(12), 2540–2545.
70. Kim, J., Kwon, S., Jeong, H. *et al.* (2009) A stiff and flat membrane operated DC contact type RF MEMS switch with low actuation voltage. *Sensors and Actuators A*, **153**, 114–119.
71. Plotz, F., Michaelis, S., Aigner, R. *et al.* (2001) A low-voltage torsional actuator for application in RF-microswitches. *Sensors and Actuators A*, **92**, 312–317.
72. Degani, O., Socher, E., Lipson, A. *et al.* (1998) Pull-in study of an electrostatic torsion microactuator. *Journal of Microelectromechanical Systems*, **7**, 373–379.
73. Degani, O. and Nemirovsky, Y. (2002) Design considerations of rectangular electrostatic torsion actuators based on new analytical pull-in expressions. *Journal of Microelectromechanical Systems*, **11**(1), 20–26.
74. Jeon, J., Pott, V., Kam, H. *et al.* (2010) Seesaw relay logic and memory circuits. *Journal of Microelectromechanical Systems*, **19**(4), 1012–1014.
75. Jeon, J., Pott, V., Kam, H. *et al.* (2010) Perfectly complementary relay design for digital logic applications. *Electron Device Letters*, **31**, 371–373.
76. Kogut, L. and Komvopoulos, K. (2003) Electrical contact resistance theory for conductive rough surfaces. *Journal of Applied Physics*, **94**, 3153–3162.
77. Sharvin, Y.V. (1965) A possible method for studying Fermi surfaces. *Soviet Physics - JETP*, **21**, 655.
78. Holm, R. (1999) *Electric Contacts: Theory and Application*, Springer-Verlag., Berlin, Germany.
79. Courtney, T.H. (1990) *Mechanical Behavior of Materials*, Mc-Graw Hill., New York.
80. Lee, H., Coutu, R.A., Mall, S., and Leedy, K.D. (2006) Characterization of metal and metal alloy films as contact materials in MEMS switches. *Journal of Micromechanics and Microengineering*, **16**, 557–563.
81. Czaplewski, D.A., Nordquist, C.D., Dyck, C.W. *et al.* (2012) Lifetime limitations of ohmic, contacting RF MEMS switches with Au, Pt and Ir contact materials due to the accumulation of 'friction polymer' on the contacts. *Journal of Micromechanics and Microengineering*, **22**(10), 105005.
82. Chen, Y., Nathanael, R., Jeon, J. *et al.* (2012) Characterization of contact resistance stability in MEM relays with Tungsten electrodes. *Journal of Microelectromechanical Systems*, **21**(3), 511–513.
83. de Boer, M.P., Czaplewski, D.A., Baker, M.S. *et al.* (2012) Design, fabrication, performance and reliability of Pt- and RuO$_2$-coated microrelays tested in ultra-high purity gas environments. *Journal of Micromechanics and Microengineering*, **22**(10), 105027.
84. Loh, O.Y. and Espinosa, H.D. (2012) Nanoelectromechanical contact switches. *Nature Nanotechnology*, **7**, 283–295.
85. Loh, O., Wei, X., Sullivan, J. *et al.* (2011) Carbon-Carbon contacts for robust nanoelectromechanical switches. *Advanced Materials*, **24**, 2463–2468.
86. Jang, W.W., Lee, J.O., Yoon, J.-B. *et al.* (2008) Fabrication and characterization of a nanoelectromechanical switch with 15-nm thick suspension air gap. *Applied Physics Letters*, **92**, 103110.
87. Kwon, W., Jeon, J., Hutin, L., and Liu, T.-J.K. (2012) Electromechanical diode cell for cross-point nonvolatile memory arrays. *Electron Device Letters*, **33**(2), 131–133.
88. Hung, L.-W. and Nguyen, C.T.-C. (2010) Silicide-based release of high aspect-ratio microstructures, Proc. 2010 IEEE 23rd Intl Conference on MEMS, pp. 120–123.
89. Pawashe, C., Lin, K., and Kuhn, K.J. (2013) Scaling limits of electrostatic nanorelays. *IEEE Transactions on Electron Devices*, **60**(9), 2936–2942.

19

Atomic Switch

Tsuyoshi Hasegawa and Masakazu Aono
WPI Center for Materials Nanoarchitectonics, National Institute for Materials Science, Japan

19.1 Chapter Overview

An atomic switch is a nanoionic device that controls the diffusion of metal ions and their reduction/oxidation processes in a switching operation to form/annihilate a metal atomic bridge, which in an ON state is a conductive path between two electrodes. An atomic switch works as a nonvolatile device due to an operating mechanism that requires the reduction/oxidation reaction that each occur at opposite polarities of applied bias. Since metal atoms provide a highly conductive path even if their path size is on the nanometer scale, atomic switches may enable downscaling to smaller dimensions than the 11 nm technology node.

The operation of an atomic switch was first demonstrated using a two-terminal structure, which is now widely used in applications for memories and programmable switches. Three-terminal operation was also demonstrated recently, in which the formation and annihilation of a conductive path between two electrodes is controlled by the third electrode. Three-terminal operation shows the potential for the use of atomic switches as logic devices.

The novel characteristics of atomic switches, such as their small size, low power consumption, low ON resistance and nonvolatility, will be useful in the development of future computing systems, such as nonvolatile logic systems and normally OFF computers. Novel functions have also been developed in atomic switches, including learning abilities, photo-sensing abilities, memristive operations, and synaptic functions. These novel functions are expected to contribute in the development of new types of neural computing systems.

In this chapter, the historical background, fundamentals such as operating mechanisms and characteristics, the various types and functions of atomic switches, the current state of the art, and future research directions are introduced.

19.2 Historical Background of the Atomic Switch

Precipitation of metal atoms from solid electrolytes, which was first reported more than 400 years ago [1], is a well-known phenomenon that has been studied extensively since the

---

Emerging Nanoelectronic Devices, First Edition. An Chen, James Hutchby, Victor Zhirnov and George Bourianoff.
© 2015 John Wiley & Sons, Ltd. Published 2015 by John Wiley & Sons, Ltd.

Faraday era [2]. Electronic devices that use electrochemical phenomena in their operation have been developed since the 1970s. For instance, Ag filament formation/dissolution was controlled on a surface of Ag-doped As_2Se_3 to make/annihilate a conductive path between the Ag electrode and the Au electrode formed on the surface [3]. During the same period, some devices were commercialized under the product names "memoriode" and "couliode" [4]. However, the supra-performance of semiconductor transistors in the last half-century overshadowed the atomic switches that were developed earlier. The situation is now changing as the ultimate downscaling of semiconductor transistors approaches its limits, providing various opportunities for the development of emerging nanoelectronic devices, including atomic switches. In particular, the various functions and unique characteristics of the emerging devices have attracted much attention because of their potential use in developing conceptually new computing systems.

19.3   Fundamentals of Atomic Switches

In this section we briefly introduce the operation mechanism of atomic switches. Although atomic switches have various device structures, their whole operation is based on a solid electrochemical reaction. The operating mechanisms and characteristics of two basic two-terminal atomic switches, that is, the gap-type atomic switch and the gapless-type atomic switch, are also introduced so as to understand the operation mechanism and characteristics in detail.

19.3.1   Operation Mechanism of Atomic Switches

There are several types of resistive switches, such as those based on a unipolar thermo-chemical mechanism, a bipolar valence change mechanism, and a bipolar electrochemical metallization mechanism. Atomic switches are operated by the third such mechanism, that is, bipolar electrochemical metallization. Although atomic switches have various device structures, they all have a reversible electrode, such as Ag or Cu, from which metal cations are ionized and brought to a channel region facing a counter electrode. The metal cations are reduced to neutral metal atoms at the channel region, resulting in the formation of a conductive channel. In the turning-on process, positive bias is applied to cause electrochemical reactions, that is, oxidation of metal atoms at the surface of the reversible electrode and reduction of metal cations at the channel region. When negative bias is applied to the reversible electrode, reduced metal atoms in the channel region are re-ionized and brought back to the reversible electrode, where they are reduced and subsequently introduced into the reversible electrode.

The operating bias of atomic switches, which is required for causing the electrochemical reactions, depends on the materials of both the reversible electrode and the counter electrode, as well as the material of the ionic transfer material that includes a channel region. Because the electrochemical reactions are an activation process, switching time is a function of the bias used, and becomes exponentially shorter with increasing bias. Typical operating biases are in the 1–2 V range when metal oxide is used as the ionic transfer material. Switching times in the few nano-second range have been demonstrated.

One of the characteristics that gives atomic switches an advantage over other devices is the potential for downscaling their device size to dimensions smaller than the 11 nm technology node. This is because metal atoms can make highly conductive channels, even though their size is on the nanometer scale.

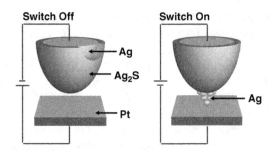

Figure 19.1 Schematic illustration of a gap-type atomic switch

In the following sections, the operation mechanism and characteristics of atomic switches are introduced in detail for each type of atomic switch.

19.3.2 Gap-type Atomic Switch

The gap-type atomic switch has a nanogap between a metal electrode facing an opposite ionic conductive material, formed on a reversible electrode. Operation of a gap-type atomic switch, that is, the controlled formation and annihilation of a metal atomic bridge in a nanogap, was demonstrated in 2001 using Ag_2S as the ionic conductive material [5], as schematically shown in Figure 19.1. This gap-type atomic switch has two key points; the first is its use of an ionic and electronic mixed conductor as the ionic conductive material, which contains metal cations to be precipitated to form a metal atomic bridge in the nanogap. The electrical conductivity of the ionic conductive material is required to achieve low resistance in the ON state, which is measured electrically by flowing a current in a circuit. The second key point is its use of a tunneling current flowing in the nanogap, which can cause the chemical reaction necessary for precipitating metal atoms. The nanogap also ensures high resistance in the OFF state.

The operating mechanism of this gap-type atomic switch is schematically shown in Figure 19.2. When a positive bias voltage is applied to the reversible electrode, metal atoms are oxidized and introduced from the electrode into the ionic conductive material as metal cations. The metal cations in the ionic conductive material migrate towards the surface, and are reduced by electrons tunneling from the counter metal electrode. The reduced metal atoms form a bridge between the ionic conductive material and the counter metal electrode, resulting in the turning on of the gap-type atomic switch.

Application of a bias voltage with an opposite polarity causes the re-ionization of the precipitated metal atoms. The ionized metal atoms, that is, cations, are re-dissolved into the ionic conductive material, resulting in annihilation of the metal atomic bridge, as a result of which the gap-type atomic switch is turned off.

As shown in Figure 19.3, the electrochemical potential changes in the switching due to the diffusion of metal cations and the applied bias, resulting in the observed precipitation and

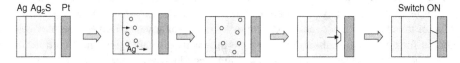

Figure 19.2 Operating mechanism of a gap-type atomic switch

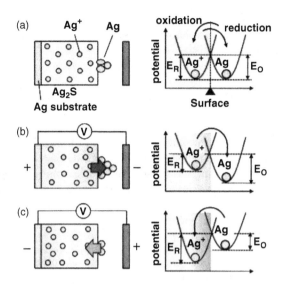

Figure 19.3 Reduction/oxidation processes induced by change in electrochemical potential due to the diffusion of metal cations in an ionic conductive material

re-ionization. This phenomenon can be caused by using various ionic conductive materials, with actual operations of gap-type atomic switches having been demonstrated using Ag_2S [5,6], Cu_2S [7], $RbAg_4I_5$ [8], and CuI.

The switching time of the gap-type atomic switch is a function of the applied bias voltage and temperature [6–8]. This becomes exponentially shorter with increases in the applied bias voltage, as shown in Figure 19.4. In most cases, two slopes are observed in the bias voltage versus the switching time, the cause of which has been explained in terms of the electrochemical nucleation of metal clusters at the surface of the ionic conductive material [8] and/or diffusion of metal cations in the ionic conductive material [7].

It would be an issue if gap-type atomic switches required an ionic conductive material, such as solid-electrolyte (Ag_2S, etc.), since these materials are not compatible with the

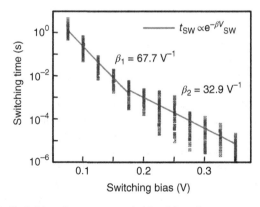

Figure 19.4 Switching time versus switching bias of a gap-type atomic switch

Cu Ta₂O₅ Pt   Switch ON

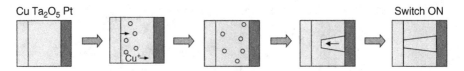

Figure 19.5 Operating mechanism of a gapless-type atomic switch

Complementary Metal Oxide Semiconductor (CMOS) process. One solution is to fabricate sulfide-based atomic switches on the top-most layer of CMOS devices [9]. Using this solution would be advantageous in such things as synaptic functions, memristive operations, and photosensing. Another unique characteristic of the gap-type atomic switch is conductance quantization, with a unit of $2e^2/h$, which occurs even at room temperature [6].

19.3.3 Gapless-type Atomic Switch

The gapless-type atomic switch operates by forming and dissolving a conductive path in an ionic conductive material, which is sandwiched by a reversible electrode and an inert electrode, as shown in Figure 19.5. The ionic conductive material of the gapless-type atomic switch should have less electrical conductivity because its OFF resistance is determined by the resistance of the ionic conductive material. In operation, the inert electrode works as a blocking electrode for diffusing metal cations, resulting in the precipitation of metal atoms at the interface between the ionic conductive material and the inert electrode. The precipitated metal atoms grow a filament towards the reversible electrode, and the gapless-type atomic switch is turned ON when the filament reaches the reversible electrode. The operating mechanism has been confirmed by observing the filaments using Transmission Electron Microscopy [10–12]. Conductance quantization at room temperature [13] has also evidenced that the conductive path consists of metal atoms. Reversing the polarity of the applied bias voltage causes dissolution of the metal filaments, resulting in the turning off of the gapless-type atomic switch.

Gapless-type atomic switches using metal oxides as the ionic conductive material are believed to use a thermal effect in their turning-off process [14]. However, they only enable bipolar operation because the subsequent supply of metal cations thickens/rebuilds the metal filament, and thermal turning-off using the same polarity of the bias voltage is not possible in the operation of this type of atomic switch. This is in contrast to the fact that unipolar operation is enabled in devices such as phase change memories, as reviewed in Chapter 5, and fuse/antifuse memories [15], where a conductive path is ruptured by a thermal effect.

Operation of the gapless-type atomic switch has been demonstrated using many types of ionic conductive materials, including Ag_2S [16], Cu_2S [17] $Ge_{0.3}Se_{0.7}$ [18], Ta_2O_5 [19], HfO_2 [20], SiO_2 [21], TiO_2 [22], CuO_x [23], GeSbTe [24], ZrO_2 [25], ZnO [26], WO_x [27], and AgI [28].

In the case of the gapless-type atomic switch, switching time is determined by applied bias voltage, the diffusion constant of cations in the ionic conductive material, the activation energies for the reduction/oxidation processes, and temperature [29]. For instance, a conductive path should be formed between both ends of an ionic conductive material, the length of which is usually longer than a few nanometers, resulting in the switching time being dependent on the cation diffusion constant [30]. This is also the cause of the initial forming process usually required for gapless-type atomic switches. Here, "forming process" means that the application of a higher bias voltage, or the application of a bias voltage for a longer period of time, is required for the initial turning-on process than in the operation thereafter [31]. After the forming process, the performance of gapless-type atomic switches is good enough for them to be commercialized. For instance, switching times on the order of nanoseconds [32–34], cyclic endurance of over 10^{10} times [35,36], and retention times of over 10 years [37] have been achieved. Gapless-type atomic switches have been used as programmable switches in programmable logic devices [38] and as switches in crossbar-type logic circuits [39–41].

19.4 Various Atomic Switches

Various atomic switches and functions have been developed based on the operating mechanism of the basic atomic switches mentioned above. For instance, three-terminal atomic switches have been developed based on the gapless-type atomic switch, for use in nonvolatile logic systems. Unique functions, such as memristive operations and synaptic functions, have been developed based on both gap-type atomic switches and gapless-type atomic switches, which has enabled the development of new types of neuromorphic computers. In this section we introduce three-terminal atomic switches, memristive operations, synaptic functions, and the photo-sensing abilities of two-terminal atomic switches.

19.4.1 Three-terminal Atomic Switches

In the operation of three-terminal atomic switches, formation and annihilation of a conductive path between the source electrode and the drain electrode is controlled by the application of a gate bias to the gate electrode. Three-terminal atomic switches have an advantage in that the signal and control lines are separated, as in a semiconductor transistor, which drastically reduces their power consumption. The nonvolatility of atomic switches, along with the advantages offered by their low power consumption, offers a solution to issues inherent in conventional semiconductor transistors, by enabling the development of nonvolatile logic systems.

There are two types of three-terminal atomic switch: one based on filament growth and another based on the nucleation of metal clusters.

19.4.1.1 Three-terminal Atomic Switch Based on Filament Growth

The three-terminal atomic switch based on filament growth uses an electrochemical technique in which the reduction/oxidation of metal cations/atoms in an electrolyte (solution) is controlled by using a gate electrode. This technique is very common in electrochemistry; it was first used in the operation of a three-terminal atomic switch in 2004 [42] using $AgNO_3 + HNO_3$ as the electrolyte. In this example, Ag filament formation and dissolution were controlled between two Au electrodes dipped in the electrolyte. The technique has also enabled three-terminal atomic switch operation in a solid-electrolyte (Cu_2S) [43], where Cu bridge formation and annihilation between the source electrode (Pt) and the drain electrode (Cu) was controlled by a gate electrode (Cu), as illustrated in Figure 19.6a. When a positive bias voltage was applied to the gate electrode, Cu^+ cations were supplied into the Cu_2S from the electrode by the oxidation of Cu atoms at the surface of the gate electrode, which is similar to the operation of a gapless-type atomic switch. Cu^+ cations migrated towards the source and drain electrodes to be reduced at the two electrodes, resulting in formation of a bridge of Cu atoms between the source and the drain. When a negative bias voltage was applied to the gate electrode, Cu atoms were extracted from the atomic bridge as a result of their oxidation to Cu^+ cations.

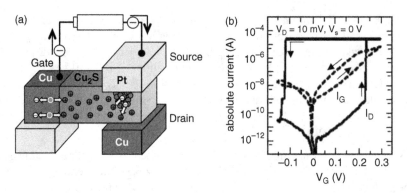

Figure 19.6 Three-terminal atomic switch controlling growth of a filament. (a) Schematic of the device. (b) Changes in drain current (I_D) and gate current (I_G) as a function of gate voltage (V_G)

The atomic bridge then became thinner and eventually disappeared. A typical operating result is shown in Figure 19.6b. In order to avoid the growth of a Cu bridge that reaches the gate electrode, the distance between the gate and the source/drain should be larger than that between the source and the drain. Although the three-terminal atomic switch uses Cu_2S, which is not CMOS process compatible, the Cu_2S-based three-terminal atomic switch can be used as a programmable switch in programmable logic devices by making them at the back end of the line of CMOS devices [44]. This circuit structure is similar to that in "CMOL" [45], which is a hybrid circuit combining CMOS stack and crossbar arrays with molecular-scale nanodevices at every crosspoint.

19.4.1.2 Three-terminal Atomic Switch Based on Nucleation Control

Another type of three-terminal atomic switch has been developed based on the nucleation control of metal clusters. Figure 19.7 shows the operating concept of this three-terminal atomic switch, in which metal cations brought from the gate electrode nucleate at the interface between the ionic conductive material and the source/drain electrodes, where the concentration of metal cations is at its highest. The key point of the development of this type of switch is the use of Ta_2O_5, which is an insulator in terms of electrical conductivity, as the

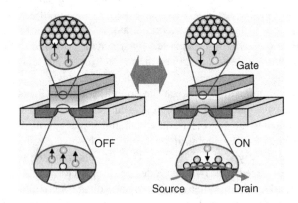

Figure 19.7 Three-terminal atomic switch controlling super-saturation induced nucleation

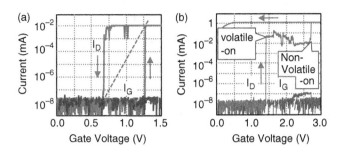

Figure 19.8 Operating results of an atom transistor: (a) volatile and (b) nonvolatile operations

ionic conductive material. Nucleation of metal clusters takes place when the concentration of metal cations, which is a function of the gate electrical field, reaches a certain threshold value, resulting in a drastic change in conductivity, that is, an insulator to conductive state transition. When the gate electrical field becomes smaller, the nuclei dissolve into metal cations moving towards the gate electrode to turn off the three-terminal atomic switch. The state variable of the three-terminal atomic switch based on nucleation control is a gate bias, which is the same as that of the conventional semiconductor transistor. Accordingly, this type of atomic switch is named the "atom transistor" [46]. This is in contrast to the three-terminal atomic switch based on filament growth, in which the state variable is a combination of a gate bias and a time.

Another unique characteristic of the atom transistor is the dual functionality of selective volatile and nonvolatile operations, depending on the sweeping range of the gate bias.

Figure 19.8 shows such selective operations when Cu was used as the gate electrode material. When a gate bias (V_G) was swept between 0 V and 1.5 V, the atom transistor showed volatile operation, where the drain current (I_D) changed by six orders of magnitude in the V_G range of 0.65–1.25 V, corresponding to a subthreshold slope of 100 mV/decade. On the other hand, two types of switching were observed when V_G was swept to 3 V. First, switching occurred at $V_G = 1.4$ V, where I_D increased to the order of 10 μA, which is similar to the result shown in Figure 19.8a. Second, switching occurred at $V_G = 2.65$ V, increasing the I_D by another two orders of magnitude. Once the second switching occurred, the ON state was kept at $V_G = 0$ V, requiring a negative V_G application to turn off the atom transistor. The ON/OFF ratio in the nonvolatile operation is eight orders of magnitude larger. In both volatile and nonvolatile operations, the gate current (I_G) remained small because Ta_2O_5 was used. The distribution of V_G required for turning-on and turning-off, both for volatile and nonvolatile operation, is shown in Figure 19.9.

Nonvolatile operation should be achieved by the nucleation of metal clusters, which accompanies the reduction/oxidation processes occurring with the application of a gate bias in the opposite polarity. One possible explanation for the volatile operation is that the metal cations themselves can make a conductive channel without nucleation when their concentration reaches a certain value. For instance, it has been reported that Ta_2O_5 lattices having a single Cu^{z+} cation in each unit cell is conductive [47]. In this case, decreasing V_G causes redistribution of the metal cations, that is, moving of some cations toward the gate electrode, resulting in the disappearance of the conductive channel.

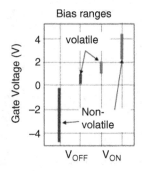

Figure 19.9 Ranges of bias voltage achieving volatile on/off and nonvolatile on/off oparations

19.4.2 Novel Functions of Two-terminal Atomic Switches

In this section we introduce the novel functions of two-terminal atomic switches, that is, memristive operations, synaptic functions, and photo-sensing functions. These newly developed functions are expected to contribute in areas such as development of neuromorphic computers, rather than in improvements to present-day von Neumann computers.

19.4.2.1 Memristive Operations

The memristor, a two-terminal device, was predicted by L. Chua from the theoretical study of symmetries in electrical devices and circuits [48]. Although it was first introduced as an element that correlates magnetic flux ϕ and charge q, it is now understood to be a resistor having a memory effect [49], and is therefore referred to as a "Memristor." In this generalized memristor, the resistance of the device changes depending on the history of input as well as the bias applied at the time. One of the operation modes of the memristor can be expressed as follows, that is, memresistance $M(q)$ is a function of time-dependent charge q.

$$M(q) = R_{OFF}\left[1 - \frac{\mu_V R_{ON}}{D^2}q(t)\right] \tag{19.1}$$

where R_{OFF}/R_{ON} are resistance in the OFF/ON states respectively, μ_V is the average ion mobility, and D is the thickness of an ionic conductive material sandwiched between two electrodes.

Memristive operation has been achieved by various types of atomic switch. The first demonstration was achieved using an Ag/solid electrolyte/AgX (gapless-type atomic switch) system, in which resistance was controlled by increasing/decreasing the Ag cations in AgX [9]. Namely, the change in Ag to X ratio in AgX caused the change in the resistance of the system. Memristive operation has also been demonstrated in a gap-type atomic switch [50], in which the gap distance and/or thickness of an atomic bridge was controlled as the state variable.

The memristor can work as a synapse in neuromorphic systems as a single device, which has been achieved by using a complex circuit consisting of CMOS devices and analog devices. Synapses in neuromorphic systems are required to change their resistance, that is, synaptic weight, according to inputs from neurons in the system. This synaptic weight change has been shown using gapless-type atomic switches with the use of neurons made of CMOS-based circuits [51], where simple neuromorphic operations, such as the so-called "Pavlov's dog" operation [52], were demonstrated.

Because the number of synapses required in neuromorphic systems is much larger than that of neurons, the achievement of synaptic functions by a single device drastically reduces the size of neuromorphic systems and enables the development of highly integrated neuromorphic systems. Thus, in recent years, a variety of emerging devices, such as anion-based ReRAMs [53], ferroelectric devices [54], and phase change memories [55], have demonstrated memristive operations.

19.4.2.2 Self-controlled Synaptic Operation by Atomic Switches

Memory is believed to occur in the human brain as a result of two types of synaptic plasticity; short-term plasticity (STP) and long-term potentiation (LTP) [56]. Here, synaptic plasticity refers to changes that occur in the organization of the brain as a result of experience. STP is achieved through the temporal enhancement of a synaptic connection, which then quickly decays to its initial state. However, repeated stimulation causes a permanent change in the connection to achieve LTP; shorter repetition intervals enable efficient LTP formation from fewer stimuli, such as is schematically shown in Figure 19.10. This synaptic behavior is achieved in biological systems by the action potential from a neuron, which generates neuro-transmitters to change the strength of a synaptic connection.

Conventional neuromorphic systems, such as those based on Spike Timing Dependent Plasticity (STDP), have been designed such that this synaptic weight change is fully controlled by signals from neurons. For instance, gradual change in short-term plasticity has been achieved by continuous or repeated application of an input bias, which is necessary because the synaptic devices developed so far require input whenever they change their resistance.

Recently, self-controlled synaptic operation was demonstrated using gap-type atomic switches, in which change in resistance occurred even after the end of signal input. Figure 19.11 shows synaptic behavior emulated by an Ag_2S-based gap-type atomic switch [57], where STP was achieved with use of a lower pulse repetition input rate, and LTP was achieved with a higher pulse repletion input rate. The key point of this achievement is the use of small input pulses that do not singly turn on/off the Ag_2S-based gap-type atomic switch. This synaptic operation was used to demonstrate concrete psychological behavior, in which new information

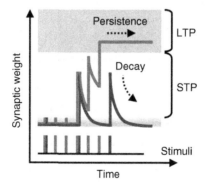

Figure 19.10 Memorization model based on the multi-store model

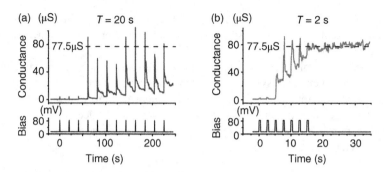

Figure 19.11 Changes in conductance of a gap-type atomic switch when input pulses were applied at intervals of 20 s (a) and 2 s (b)

from the external environment was stored as a sensory memory (SM) in the sensory register for a very short period of time, and then selected information was transferred from temporary short-term memory (STM) in the short-term store to permanent long-term memory (LTM) in the long-term store.

This self-controlled synaptic operation has also been achieved using gapless-type atomic switches [13], showing a shorter time constant than that of gap-type atomic switches. These self-controlled synaptic operations do not require complex signals from neurons to control the synaptic weight, thus enabling the development of new types of neuromorphic systems with use of simple neurons.

19.4.2.3 Photo-sensing Ability of Atomic Switches

Gap-type atomic switches operate using tunneling current, which causes the precipitation of metal atoms in a nanogap. Using a wide gap (several tens of nm), in which tunneling current does not flow, and photo-conductive molecules within the wide gap, enables photo-sensing. As shown in Figure 19.12, light irradiation causes a photocurrent to flow between an ionic conductive electrode and a counter metal electrode applied with a certain bias voltage, resulting in the growth of a metal filament from the ionic conductive electrode. The photo-assisted atomic switch is turned on when the growing metal filament reaches the counter electrode. Since filament growth does not occur without the light irradiation, the photo-assisted atomic switch works as a photo-sensor with memory function and as well as a

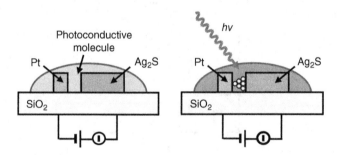

Figure 19.12 Photo-assisted atomic switch

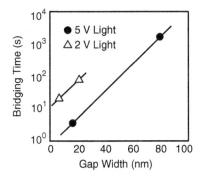

Figure 19.13 Switching time of the photo-assisted atomic switch

programmable switch, such as in an EPROM. The operation was first demonstrated using Ag_2S as the ionic conductive material and N,N'-diheptylperylenetetracarboxylic diimide (PTCDI) as the photoconductive molecule [58]. The switching time of the photo-assisted atomic switch strongly depends on the gap size used, as shown in Figure 19.13. Other characteristics, such as the switching voltage, retention time, and ON/OFF ratio are similar to those of gap-type atomic switches. Cyclic endurance is an issue that needs to be investigated, because a metal filament can grow in and may subsequently cause damage to the photoconductive molecular layer.

19.5 Perspectives

Atomic switches operate by controlling the formation and annihilation of a conductive path, made of metal atoms/cations, between two electrodes using a solid electrochemical reaction. There are several types of two-terminal atomic switch, that is, gap-type, gapless-type, and

Table 19.1 Summary parameter table of the atomic switch (two-terminal gapless-type)

Parameter	Status	Value for the device in this chapter
Cell size	Demonstrated	10 nm
	Projected	5 nm
Density	Demonstrated	$2.5E9/cm^2$
	Projected	$1E12/cm^2$
Device speed (reading/writing speed)	Demonstrated	<10 ns
	Projected	<1 ns
Operation voltages	Demonstrated	0.6 V
	Projected	<0.5 V
Switching energy (set-current/reset-current)	Demonstrated	1 μA/10 μA
	Projected	Not known
Endurance	Demonstrated	1E6
	Projected	>1E11
Retention (for memory)	Demonstrated	10 yr
	Projected	>10 yr

photo-assisted type. These have novel functions, such as memristive operations, synaptic functions, and photo-sensing with memory effect. Three-terminal atomic switches have also been developed. All such atomic switches have the potential to be downscaled beyond the 11 nm technology node. In addition to such novel functions and characteristics, development of fully CMOS compatible metal oxide-based atomic switches has attracted a great deal of attention as they gain importance as beyond-CMOS devices. For instance, because of their small size, nonvolatility and high ON/OFF ratio, two-terminal gapless-type atomic switches are expected to be the most promising candidates as programmable switches for next-generation Field Programmable Gate Arrays (FPGAs). The said characteristics are also suitable for their application to nonvolatile memory. Among the various atomic switches, the performance of two-terminal gapless-type atomic switches, shown in Table 19.1, have been much improved because of their potential for applications in the very near future. What will be more important for atomic switches is their application to beyond-von Neumann computers, such as nonvolatile logics and neural computers. Development of three-terminal atomic switches operated with a negative gate bias, corresponding to a p-type transistor, is needed to realize nonvolatile logics and to replace CMOS. The development of neuro-morphic functions will also be more important in the research of atomic switches because the realization of neural computing systems is key to achieving both a drastic reduction in power consumption and the ability to perform highly complicated computations.

References

1. Ercker, L. (1951) *Treaties on Ores and Assaying*, 1574, (translated by Sisco A. G. and Smith C. S., Univ. Chicago, pp.) 177.
2. Faraday, M. (1833) On a new law of electric conduction. *Philosophical Transactions of the Royal Society*, **123**, 507–515.
3. Hirose, Y. and Hirose, H. (1976) Polarity-dependent memory switching and behavior of Ag dendrite in Ag-photodoped amorphous As_2S_3 films. *Journal of Applied Physics*, **47**, 2767–2772.
4. Ikeda, H. and Tada, K. (1980) Sanyo's Couliodes And Memoriodes, in *Applications of Solid electrolytes* (eds T. Takahashi and A. Kozaw), Academic Press, Cleveland, pp. 40–45.
5. Terabe, K., Hasegawa, T., Nakayama, T., and Aono, M. (2001) Quantum point contact switch realized by solid electrochemical reaction. *Riken Review*, **37**, 7–8.
6. Terabe, K., Hasegawa, T., Nakayama, T., and Aono, M. (2005) Quantized conductance atomic switch. *Nature*, **433**, 47–50.
7. Nayak, A., Tsuruoka, T., Terabe, K. *et al.* (2011) Switching kinetics of a Cu_2S-based gap-type atomic switch. *Nanotechnol*, **22**, 235201-1-7.
8. Valov, I., Sapezanskaia, I., Nayak, A. *et al.* (2012) Atomically controlled electrochemical nucleation at superionic solid electrolyte surfaces. *Nature Materials*, **11**, 530–535.
9. Kaeriyama, S., Sakamoto, T., Sunamura, H. *et al.* (2005) A nonvolatile programmable solid-electrolyte nanometer switch. *IEEE Journal of Solid-State Circuits*, **40**(1), 168–176.
10. Yang, Y., Gao, P., Gaba, S. *et al.* (2012) Observation of conducting filament growth in nanoscale resistive memories. *Nature Communications*, **3**, 732.
11. Miao, F., Strachan, J.P., Yang, J.J. *et al.* (2011) Anatomy of a nanoscale conduction channel reveals the mechanism of a high-performance memristor. *Advanced Materials*, **23**, 5633–5640.
12. Liu, Q., Sun, J., Lv, H. *et al.* (2012) Real-Time Observation on Dynamic Growth/Dissolution of Conductive Filaments in Oxide-Electrolyte-Based ReRAM. *Advanced Materials*, **24**, 1844–1849.
13. Tsuruoka, T., Hasegawa, T., Terabe, K., and Aono, M. (2012) Conductance quantization and synaptic behavior in a Ta_2O_5-based atomic switch. *Nanotechnol*, **23**, 435705-1-6.
14. Tsuruoka, T., Terabe, K., Hasegawa, T., and Aono, M. (2011) Temperature effects on the switching kinetics of a Cu-Ta_2O_5-based atomic switch. *Nanotechnol*, **22**, 254013-1-1-9.
15. International Roadmap Committee (2009) International Technology Roadmap for Semiconductors, ITRS 2009 Edition, http://www.itrs.net/Links/2009ITRS/Home2009.htm (accessed 16 July 2013).

16. Kundu, M., Terabe, K., Hasegawa, T., and Aono, M. (2006) Effect of sulfurization conditions and post-deposition annealing treatment on structural and electrical properties of silver sulfide films. *Journal of Applied Physics*, **99**, 103501-1-9.
17. Sakamoto, T., Sunamura, H., Kawaura, H. *et al.* (2003) Nanometer-scale switches using copper sulfide. *Applied Physics Letters*, **82**, 3032–3034.
18. Soni, R., Meuffels, P., Petraru, A. *et al.* (2010) Probing Cu doped $Ge_{0.3}Se_{0.7}$ based resistance switching memory devices with random telegraph noise. *Journal of Applied Physics*, **107**, 024517-1-10.
19. Sakamoto, T., Lister, K., Banno, N. *et al.* (2007) Electronic transport in Ta_2O_5 resistive switch. *Applied Physics Letters*, **91**, 092110-1-3.
20. Haemori, M., Nagata, T., and Chikyow, T. (2009) Impact of Cu Electrode on Switching Behavior in a $Cu/HfO_2/Pt$ Structure and Resultant Cu Ion Diffusion. *APEX*, **2**, 061401-1-3.
21. Schindler, C., Weides, M., Kozicki, M.N., and Waser, R. (2008) Low current resistive switching in Cu–SiO_2 cells. *Applied Physics Letters*, **92**, 122910-1-3.
22. Tsunoda, K., Fukuzumi, Y., Jameson, J.R. *et al.* (2007) Bipolar resistive switching in polycrystalline TiO_2 films. *Applied Physics Letters*, **90**, 113501-1-3.
23. Chen, A., haddad, S., Wu, Y. *et al.* (2005) "Non-Volatile Resistive Switching for Advanced Memory Applications," in Tech. Dig., Int. Electron Devices Meet., pp. 765–768.
24. Aratani, K., Ohba, K., Mizuguchi, T. *et al.* (2007) "A Novel Resistance Memory with High Scalability and Nanosecond Switching," in Tech. Dig. of the IEEE Int. Electron Devices Meet. pp. 783–786.
25. Li, Q., Dou, Ch., Wang, Y. *et al.* (2009) Formation of multiple conductive filaments in the $Cu/ZrO_2:Cu/Pt$ device. *Applied Physics Letters*, **95**, 023501 1-3.
26. Yang, Y.C., Pan, F., Liu, Q., and Zeng, F. (2009) Fully Room-Temperature-Fabricated Nonvolatile Resistive Memory for Ultrafast and High-Density Memory Application. *Nano Letters*, **9**, 1636–1643.
27. Kozicki, M.N., Gopalan, C., Balakrishnan, M., and Mitkova, M. (2006) A low-power nonvolatile switching element based on copper-tungsten oxide solid electrolyte. *IEEE Transactions on Nanotechnology*, **5**, 535–544.
28. Guo, H.X., Yang, B., Chen, L. *et al.* (2007) Resistive switching devices based on nanocrystalline solid electrolyte $(AgI)_{0.5}(AgPO_3)_{0.5}$. *Applied Physics Letters*, **91**, 243513 1-3.
29. Valov, I., Waser, R., Jameson, J.R., and Kozicki, M.N. (2011) Electrochemical metallization memories—fundamentals, applications, prospects. *Nanotechnol*, **22**, 254003 1-22.
30. Banno, N., Sakamoto, T., Iguchi, N. *et al.* (2008) Diffusivity of Cu ions in solid electrolyte and its effect on the performance of nanometer-scale switch. *IEEE Transactions on Electron Devices*, **55**(11), 3283–3287.
31. Valov, I. and Kozicki, M.N. (2013) Cation-based resistance change memory. *Journal of Physics D*, **46**, 074005-1-14.
32. Tsunoda, K., Kinoshita, K., Noshiro, H. *et al.* (2007) Diffusivity of Cu ions in solid electrolyte and its effect on the performance of nanometer-scale switch, Tech. Dig., IEEE Int. Electron Device Meet. pp. 767–770.
33. Schindler, C., Weides, M., Kozicki, M.N., and Waser, R. (2008) Low current resistive switching in Cu–SiO_2 cells. *Applied Physics Letters*, **92**, 122910 1-3.
34. M. N., Kozicki., Park, M., and Mitkova, M. (2005) Nanoscale memory elements based on solid-state electrolytes. *IEEE Transactions on Nanotechnology*, **4**, 331–338.
35. Yang, J.J., Zhang, M.X., Strachan, J.P. *et al.* (2010) High switching endurance in TaOx memristive devices. *Applied Physics Letters*, **97**, 232102-1-3.
36. Lee, M., Lee, C.B., Lee, D. *et al.* (2010) A fast, high-endurance and scalable non-volatile memory device made from asymmetric Ta_2O_5−x/TaO_2−x bilayer structures. *Nature Materials*, **10**, 625–630.
37. Banno, N., Sakamoto, T., Fujieda, S., and Aono, M. (2008) On-State Reliability Of Solid-Electrolyte Switch, Proc. of the Int. Reliabil. Phys. Symp., Phoenix, pp. 707–708.
38. Tada, M., Sakamoto, T., Tsuji, Y. *et al.* (2009) Highly Scalable Nonvolatile TiOx/TaSiOy Solid-electrolyte Crossbar Switch Integrated in Local Interconnect for Low Power Reconfigurable Logic, Tech. Dig. Int. Electron Devices Meet., pp. 943–946.
39. Heath, J.R., Kuekes, P.J., Snider, G.S., and Williams, R.S. (1998) A Defect-Tolerant Computer Architecture: Opportunities for Nanotechnology. *Science*, **280**, 1716–1721.
40. Morales-Masis, M., van der Molen, S.J., Fu, W.T. *et al.* (2009) Conductance switching in Ag_2S devices fabricated by in situ sulfurization. *Nanotechnol*, **20**, 095710 1-6.
41. Borghetti, J., Snider, G.S., Kuekes, P.J. *et al.* (2010) 'Memristive' switches enable 'stateful' logic operations via material implication. *Nature*, **464**, 873–876.
42. Xie, F.-Q., Nittler, L., Obermair, C., and Schimmel, T. (2004) Gate-Controlled Atomic Quantum Switch. *Physical Review Letters*, **93**, 128303-1-4.

43. Banno, N., Sakamoto, T., Iguchi, N. *et al.* (2006) Solid-Electrolyte Nanometer Switch. *IEICE Transactions on Electronics*, **E89-C**(11), 1492–1498.

44. Sakamoto, T., Iguchi, N., and Aono, M. (2010) Nonvolatile triode switch using electrochemical reaction in copper sulfide. *Applied Physics Letters*, **96**, 252104-1-3.

45. Strukov, D.B. and Likharev, K.K. (2005) CMOL FPGA: a reconfigurable architecture for hybrid digital circuits with two-terminal nanodevices. *Nanotechnol*, **16**, 888–900.

46. Hasegawa, T., Itoh, Y., Tanaka, H. *et al.* (2011) Volatile/Nonvolatile Dual-Functional Atom Transistor. *APEX*, **4**, 015204-1-3.

47. Gu, T., Tada, T., and Watanabe, S. (2010) Conductive Path Formation in the Ta2O5 Atomic Switch: First-Principles Analyses. *ACS Nano*, **4**(11), 6477–6482.

48. Chua, L.O. (1971) Memristor - Missing Circuit Element. *IEEE Transactions on Circuit Theory*, **CT-18**, 507–519.

49. Strukov, D.B., Snider, G.S., Stewart, D.R., and Williams, R.S. (2008) The missing memristor found. *Nature*, **453**, 80–83.

50. Hasegawa, T., Nayak, A., Ohno, T. *et al.* (2011) Memristive operations demonstrated by gap-type atomic switches. *Applied Physics A*, **102**, 811–815.

51. Alibart, F., Gao, L., Hoskins, B.D., and Strukov, D.B. (2012) High precision tuning of state for memristive devices by adaptable variation-tolerant algorithm. *Nanotechnol*, **23**, 075201-1-7, 075201.

52. Ziegler, M., Soni, R., Patelczyk, T. *et al.* (2012) An Electronic Version of Pavlov's Dog. *Advanced Functional Materials*, **22**, 2744–2749.

53. Chang, T., Jo, S., and Lu, W. (2011) Short-Term Memory to Long-Term Memory Transition in a Nanoscale Memristor. *ACS Nano*, **5**, 7669–7676.

54. Chanthbouala, A., Garcia, V., Cherifi, R.O. *et al.* (2012) A ferroelectric memristor. *Nature Materials*, **11**, 860–864.

55. Wright, C.D., Hosseini, P., and Vazquez Diosdado, J.A. (2013) Beyond von-neumann computing with nanoscale phase-change memory devices. *Advanced Functional Materials*, **23**, 2248–2254.

56. Bliss, T.V.P. and Collingridge, G.L. (1993) A synaptic model of memory: long-term potentiation in the hippocampus. *Nature*, **361**, 31–39.

57. Ohno, T., Hasegawa, T., Tsuruoka, T. *et al.* (2011) Short-term plasticity and long-term potentiation mimicked in single inorganic synapses. *Nature Materials*, **10**, 591–595.

58. Hino, T., Tanaka, H., Hasegawa, T. *et al.* (2010) Photo-assisted formation of an atomic switch. *Small*, **6**, 1745–1748.

20

ITRS Assessment and Benchmarking of Emerging Logic Devices

Shamik Das

Nanosystems Group, The MITRE Corporation, USA

20.1 Introduction

This chapter provides a brief review of the assessment and benchmarking of logic devices presented in the 2011 edition of the International Technology Roadmap for Semiconductors (ITRS) [1] chapter on Emerging Research Devices (ERD) [2]. The objectives of the ITRS ERD roadmap are to review recent research in devices beyond silicon transistors and to forecast the development of novel logic switches that might replace the silicon transistor as the device driving technological development within the semiconductor industry.

Overt emphasis on emerging devices within the ITRS began with the 2003 edition, at which time it was envisioned that new technologies would be required to sustain the scaling of complementary metal-oxide semiconductor (CMOS) electronics down to the 22-nm node and beyond [3]. At present, 10 years later, 22-nm CMOS is commercially available, and further conventional scaling is envisioned that will enable transistor manufacturing down to the 7-nm or even the 5-nm node [4]. Nonetheless, it is absolutely certain that the avenues for continued conventional scaling are limited and dwindling, and therefore that alternatives to CMOS must be developed if the economic model of exponential scaling is to be sustained over the longer term. The ITRS presents a progression of such alternatives that depart increasingly from conventional device materials, structures, and mechanisms, with the goal of enabling an orderly transition of suitable emerging devices into mainstream manufacturing.

A more detailed overview of the ITRS Roadmap for Emerging Logic Devices is provided below in Section 20.2. This section enumerates the emerging devices that were examined for the 2011 edition of the ITRS and provides details of the best reported performance to date for those devices, as well as projections for their ultimate performance. Then, Section 20.3

Emerging Nanoelectronic Devices, First Edition. An Chen, James Hutchby, Victor Zhirnov and George Bourianoff.
© 2015 John Wiley & Sons, Ltd. Published 2015 by John Wiley & Sons, Ltd.

discusses recent research developments for selected emerging devices from the roadmap. These developments were chosen to highlight trends and emphases that appear likely to become evident in the 2013 edition and later editions of the roadmap. Section 20.4 explores in greater detail these important trends and potential limitations for emerging logic devices, with a view toward evaluating and prioritizing metrics for future assessments. Section 20.5 summarizes the chapter.

20.2 Overview of the ITRS Roadmap for Emerging Research Logic Devices

Within the ITRS, logic devices are categorized into one of four types: (1) conventional field-effect transistors (FETs), including bulk FETs, silicon-on-insulator (SOI) devices, and Fin FETs; (2) extensions of FET devices through novel channel materials or structures; (3) charge-based devices beyond conventional FETs; and (4) devices that use physical phenomena other than electric charge to store or carry information.

Figure 20.1 illustrates this categorization for devices considered in the 2011 edition of the ITRS. The three rightmost columns of this taxonomy constitute emerging logic devices within the ITRS viewpoint and are tracked in the ERD chapter of the roadmap [2]. For alternate channel materials/structures and charge-based devices beyond conventional FETs (middle columns of Figure 20.1), quantitative information is tracked regarding the present state of the art in performance for these devices. Devices not based on charge presently are tracked using qualitative measures, although this is anticipated to change in 2013 and beyond as more quantitative data become available (e.g., via simulation [5]).

As an example of the quantitative assessment provided by the ITRS, Table 20.1 provides performance measures for the charge-based devices tracked in the roadmap. The performance measures highlighted here are the present best demonstrations and projected maximal demonstrations of device density, speed (of individual devices as well as circuits), and energy efficiency. Supporting references for the individual entries in this table are provided in the Emerging Research Devices chapter tables [2].

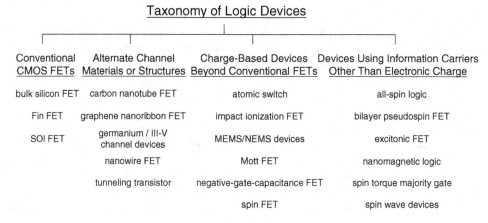

Figure 20.1 Taxonomy of logic devices considered in the 2011 ITRS Emerging Research Devices Roadmap. Devices "beyond CMOS" are categorized into one of three groups, based upon the extent of progression beyond a conventional MOSFET structure or mechanism

Importantly, the data presented for any given column of this table do not necessarily represent mutually-consistent parameters measured from a single device of the type associated with the column. For example, the best reported density, speed, and energy efficiency for carbon nanotube transistors were obtained from three distinct devices. Furthermore, it is not suggested that a single device is capable of possessing all the listed "projected" parameters in any given column.

Instead, the principal objective of the tables in the ERD chapter, as illustrated here in Table 20.1, is to provide a means of comparison between candidate emerging devices for specific applications. For example, carbon nanotube FETs or nanowire FETs appear more likely than alternative devices to enable high-performance logic, whereas nanowire FETs or tunnel FETs might be well suited for low-power circuits at lower speeds. Indeed, as conventional scaling winds down and emerging devices are adopted, some perceive that manufacturing platforms might diverge into separate lines of emphasis for high performance or low power requirements.

A key aspect of Table 20.1 is that it encapsulates the best reported results of all time. In some cases, these results may be several years old. For example, some of the measurements of highest performance for carbon nanotube transistors are from three or more years ago. In contrast, devices such as the negative gate capacitance transistor [6] were first proposed less than five years ago, with experimental demonstrations occurring even more recently. The ITRS emphasizes emerging devices for which recent results have been obtained that exhibit continued interest and focus from the research community, even if these recent results do not exceed or match the superlative demonstrations of prior years. Especially, this is the case if recent publications document the integration or manufacturing of circuits of increasing complexity based upon individual devices whose performance is captured in the ITRS ERD tables.

Some of the recent results motivating the inclusion of devices in Figure 20.1 and Table 20.1 are provided in the next section.

20.3   Recent Results for Selected Emerging Devices

In this section, recent progress is highlighted for selected devices from the ITRS roadmap for emerging logic devices. These highlights feature a representative list of technologies for which research articles over the 2011–2013 time period document substantial advances suggestive of the need for continued tracking by the ITRS.

20.3.1   Carbon Nanotube FETs

The use of carbon nanotubes (CNTs) within transistor circuits is motivated by the high performance characteristics (e.g., high mobility, high normalized drive current, minimal short-channel effects) and the small diameter of individual semiconducting CNTs [7]. The dense integration of single-CNT transistors into circuits is thought to be a promising near-term route to exceeding the performance of conventional CMOS, such that "carbon nanomaterials" were identified by the ITRS in 2008 as meriting focused emphasis for roadmapping and research investment [8].

The principal challenges to the use of carbon nanotubes (CNTs) for transistor circuits continue to be: (1) the difficulty in directed placement of individual CNTs on a substrate; (2) the inability to produce highly pure batches of CNTs with the desired electronic properties; and (3) the difficulty of addressing individual CNTs with source, drain, and gate contacts within a dense, interconnected circuit.

Table 20.1 Selected quantitative benchmark data from the ITRS Emerging Logic Devices tables [2]. References supporting individual entries are provided with the original tables [ERD12(a) and ERD12(b) in the 2011 edition of the ITRS]

Device		Conventional CMOS FET			Alternate channel materials or structures		
			CNT FET	Graphene nanoribbon FET	Ge/III-V devices		Nanowire FET
					N channel Ge FET	P channel III-V FET	
Cell size (spatial pitch)	Projected	100 nm	100 nm	100 nm	Not known	TBD	40 nm
	Demonstrated	590 nm	1.4 μm	1.4 μm	80 nm = Lg × 2	60 nm = Lg × 2	1 μm
Density (device/cm^2)	Projected	1.00E + 10	1.00E + 10	1.00E + 10	Not known	Not known	5.90E + 10
	Demonstrated	2.80E + 08	5.10E + 07	5.10E + 07	1.5E + 10 = 1/(4*Lg^2)	2.7E + 10 = 1/(4*Lg^2)	5.20E + 07
Switch speed	Projected	12 THz	6.3 THz	7 THz	Not known	Not known	6.5 THz
	Demonstrated	1.5 THz	4 GHz	300 GHz	140 GHz	601 GHz	250 GHz
Circuit speed	Projected	61 GHz	100 GHz	Not known	Not known	Not known	100 GHz
	Demonstrated	5.6 GHz	52 kHz	22 MHz	Not known	Not known	11.7 MHz
Switch energy (J)	Projected	3.00E-18	Not known	Not known	1.00E-18	Not known	4.00E-20
	Demonstrated	1.00E-16	1.00E-11	Not known	Not known	Not known	6.00E-16

Although the third challenge remains unresolved, great progress has been made recently toward resolving the first two. For example, researchers at Stanford University have developed a process named "VLSI Metallic CNT Removal" (VMR) [9], in which aligned arrays of mixed-property CNTs are fabricated on a surface and then postprocessed to form functional logic gates using multiple CNTs per transistor. This postprocessing takes the form of two steps: (1) inducing electrical breakdown of the metallic CNTs in the arrays, leaving only semi-conducting CNTs (it has been reported [9] that 99.99% of metallic CNTs are removed through this process); and (2) selective etching of CNT regions to counteract the potential for misaligned CNTs to short out connections within the logic gates. This approach recently has been used to demonstrate a one-bit microprocessor fashioned entirely out of CNT FETs [10].

Similar approaches include work at the University of Illinois to burn out metallic CNTs without electrical breakdown, as well as a "sort and place" approach by IBM [11,12] that results in a reported 99.4% of the placed CNT being semiconducting.

20.3.2 Graphene Transistors

Graphene is of great interest as a semiconductor material [13] due principally to its ultrahigh carrier mobility, as well as its likely amenability to conventional semiconductor processing techniques. However, graphene is intrinsically a zero-bandgap material, making it difficult to

	Charge-based devices beyond conventional FETs						
Tunnel FET	Atomic switch	IMOS	MEMS	Mott FET	Neg Cap Ferroelectric	Spin FET and Spin MOSFET	
20 nm	40 nm	100 nm	100 nm	10 nm	100 nm	100 nm	
Sub 60 nm	Not known	< 1 µm	Sub-1000 nm	1 µm × 150 µm	150 µm	Not known	
Channel down to 20 nm: 1E10	Not known	1.00E + 10	>1.00E10 (due to anchors)	~1E12 [1/(10 nm *10 nm)]	Similar to Si CMOS	1.00E + 10	
Not known	Not known	< 1E7	> 1E8	~6.7E5 [1/ (1 um*15 0 um)]	Not known	Not known	
Si/InAs TFET: 60 GHz/3 THz	Not known	Limited by carrier	~1 GHz	2 THz (0.5 ps)	500 GHz	10 THz or less	
Not known	~ 1 ns	CMD = 0.6 ns, SRD=1.2 ns	0.18 GHz	13.3 THz-0.1 GHz (75 fs-9 ns)	Not known	Not known	
Si/InAs TFET inverter: 20 GHz,	Not known	Limited by CRD + SRD	~1 GHz	Not known	Similar to Si CMOS	10 GHz or less	
Not known	Not known	Not known	0.18 MHz	Not known	Not known	Not known	
$C_{GG}V_{DD}^2$ (J/um)<2E-17	Not known	Similar to CMOS	<5E-17	0.1 uW	3.00E-19	~1E-17	
$C_{GG}V_{DD}^2$ (J/um) = 1E-16	Not known	Not known	Not known	Not known	Not known	Not known	

use as a digital switch [14]. Attempts to induce a bandgap in graphene have had only limited success. Furthermore, these attempts have tended to compromise the material's carrier mobility. This is true for disparate approaches to bandgap engineering, such as using bilayers or employing size-dependent effects, as is done within graphene nanoribbons.

Recent progress in the development of graphene-based field-effect devices includes, on the theoretical side, the development of greater insight into both low-field and high-field carrier transport in graphene nanoribbons and bilayers. Simulation models exist for idealized graphene FETs [14], but comparison data is lacking for graphene FETs versus other types of competing FETs. A number of graphene heterostructure devices also have been proposed and evaluated from a theoretical standpoint [15,16].

In terms of device fabrication, moderate progress has been made. Seminal results in the fabrication of nanoribbons date to 2010 [17]. There has been steady progress in prototyping FET devices using such nanoribbons, although these FETs all have widths in excess of 10 nm and lengths in excess of 0.5 µm.

20.3.3 Nanowire Transistors (NWFETs)

These devices [18], in which the conventional planar MOSFET structure is replaced by a semiconducting nanowire, potentially enable both high performance and low power due to the superior electrostatic gate control afforded by the device geometry. Along with such perform-ance increases comes the potential for ultrahigh device density (nanowires have been demonstrated with diameters as small as 0.5 nm [19]) and the prospect of integrating diverse III-V and II-VI compound semiconductor devices along with Group IV elements.

Significant recent progress has focused on the integration of nanowire devices into circuits of substantial complexity. Extended, programmable arrays of individually addressed nonvolatile nanowires have been demonstrated [20] that implement full-adder, full-subtractor, multiplexer, demultiplexer, and clocked D-latch operations. Individual logic gates also have been demon-strated with speeds up to 250 GHz [21] and a ring oscillator with 108 MHz oscillating frequency has been fabricated [22]. Further progress is needed, however, in demonstrating still larger arrays of devices, with greater densities, and clocked at much higher rates.

20.3.4 Tunnel Field Effect Transistors

The key mechanism for the tunnel field effect transistor (TFET), explored since the 1980s, is band to band tunneling in a *p-i-n* structure as modulated by gating the source and/or intrinsic layers [23]. The principal feature of such a device is that it might surpass the thermally limited subthreshold slope (STS) of conventional FETs, which is 60 mV/decade. The structure also permits a lower off-state (leakage) current than a MOSFET of comparable channel length. Both the improved STS and the lower off-state current suggest that the device might have beneficial applications for ultralow-power electronic systems.

Subthreshold slope values in the range of 20–50 mV/decade have been reported for all-silicon TFET devices [24]. However, these values are tied to the on-state current of the device, with higher "on" currents corresponding to higher STS. Furthermore, the highest of these currents is still insufficient for high-performance applications. To attempt to obtain high on-state current while maintaining advantageous STS and low off-state current, researchers presently are investigating heterostructure TFETs such as Ge-Si and InAs-Si [25,26]. In these cases, better on-state current is achieved relative to all-silicon TFETs, but the absolute current still is insufficient for desired applications.

In addition to materials research, another avenue for device engineering is in the gate overlap [27]. The specific mode of tunneling within the device (i.e., line tunneling vs. point tunneling) is dictated by the amount of gate-source overlap versus gate-channel overlap. For the best STS, line tunneling is desired, but this requires precise alignment of the gate with the source region and thus affects the scalability of the device.

Device simulation also presents a substantial challenge for TFET design. At present, no accurate compact models exist for TFET simulation. In particular, semiclassical models differ in their predictions from the slower full-quantum models by as much as an order of magnitude in drain current [28]. More accurate, fast models will be needed for iterative exploration of the device parameter space.

20.3.5 NEMS Devices

The central motivation for nanoelectromechanical system (NEMS) devices is the prospect for ultralow-power operation. This is due to two effects: (1) the possibility of effectively zero off-state (leakage) current through the device due to mechanical separation of the source and drain contacts and (2) the potential to reduce the supply voltage due to ultrasteep switching between the "off" and "on" states of the device. Three mechanisms have been proposed for using NEMS behavior for information processing: (1) using the spatial configuration of movable objects to encode information [29,30]; (2) using resonant vibrational modes for the same [31,32]; or (3) using the device as a transducer [33,34]. The principal challenge to adoption of such devices remains their scalability. Small size has been demonstrated, for example, by using suspended nanowires, but the voltage scaling has not kept pace (some devices require 10–15 V supply, for example) [35].

20.3.6 Mott FET

The Mott FET employs a Mott insulator as the device channel and exploits the insulator–metal transition as a mechanism for switching [36]. The principal advantage of exploiting such an effect is the possibility of very fast (picosecond) transitions between "on" and "off" conductivity states in the channel [37]. The rapid switching in such a device is made possible because the insulator–metal state change is effected via a purely electronic transition within a strongly correlated electron system, as opposed to an amorphous to crystalline phase transition in the channel material.

Device scaling continues to be a great challenge for Mott FETs. In particular, a very high surface carrier density under the gate is necessary to generate the Mott effect [38]. This presents an issue as the channel is miniaturized because the total number of dopants is fairly high. Recent results [39,40] using ionic liquids demonstrated that it is possible to induce very high surface electric fields using gate voltages on the order of 1 V. Furthermore, these high fields enable collective state transition throughout the device bulk, as opposed to only surface modulation, thus reducing the carrier density requirement. If the ionic liquid can replaced with a suitable solid dielectric, this may present a route to realization for Mott transistors.

20.3.7 Spin FET and Spin MOSFET

These proposed devices rely upon the integration of ferromagnetic materials to induce spin polarization of the charges in the device channel, thereby permitting information processing and transport based on electron spin [41–43]. No complete device has yet been fabricated exhibiting spin FET operation. However, recent progress has been made in the design and implementation of half-metallic ferromagnetic tunnel contacts [44,45]. These recent results demonstrate the reliable injection of spin into the device channel at room temperature. Furthermore, the probing of spin within the channel is better understood. However, spin accumulation and spin lifetime within the channel, and thereby spin gating/manipulation by the proposed devices, remain less well understood.

20.3.8 Nanomagnetic Logic and All Spin Logic

Nanomagnetic Logic (NML) and All Spin Logic (ASL) represent integrated device and interconnect approaches to computing using state variables other than electronic charge. NML [46,47] is a relatively established emerging technology with history dating back several years based upon prior investigations of quantum cellular automata (QCA) [48]. NML implements nearest neighbor cellular–automata interactions using magnetic coupling. In contrast, ASL [49], a relatively recent proposal, builds upon NML by using spin coherent channels to convey information.

Significant recent progress includes the demonstration of fanout-of-three interconnects [50] as well as one-bit full adder circuits [51]. Synchronous circuit elements also have been demonstrated [52]. An especially notable development is that of directional magnetic sensitivity: focused ion beam irradiation is used to make the nanomagnets anisotropic, such that the coupling from one nanomagnet to another can be unidirectional [53]. Both logic gates [53] and clocking elements [54] have been demonstrated. Finally, input/output conversion to the electrical domain has been proposed via the use of magnetic tunnel junctions [55].

All Spin Logic is proposed to expand upon NML by introducing interconnects that would permit logic devices to be separated spatially. A family of ASL architectures has been proposed [56], though no fabrication demonstrations have been carried out. Nonetheless, the promise of such devices and paired interconnects is that more complex computational architectures, such as neuromorphic networks, might be implemented in the magnetic/spin domain.

20.3.9 Spin Wave Devices

The principle of operation for spin wave devices is that voltage-encoded digital information is converted to analog spin waves, upon which analog information processing is performed and the output converted to voltage.

A recent example logic demonstration [57] consists of a three-input magnonic majority gate, wherein the computation is performed by analog phase cancellation of the inputs. The most recent work in this area focuses on the excitation and detection of spin waves within the device using multiferroic elements [58].

20.4 Perspective

The quantitative benchmark data collected in Table 20.1 and the recent results reported above in Section 20.3 suggest together that substantial progress has been made in demonstrating devices beyond conventional CMOS. Nonetheless, no clear "heir" to CMOS has emerged, and significant work remains to be done before any of the demonstrated or proposed devices are ready for mainstream manufacturing.

Thus, the Emerging Research Devices chapter of the ITRS continues to serve a critical role in forecasting and guiding the development of such device technologies. There are four principal ways in which the 2013 edition and future editions of the ITRS ERD Roadmap will focus efforts to identify the next logic switch beyond CMOS:

1. *Identifying new devices:* the roadmap continues to integrate and track devices as they are invented. Criteria for inclusion consist of either: (a) the publication in peer-reviewed journals of device designs or demonstrations by multiple independent groups, signifying scientific momentum behind a particular device concept; or (b) a sequence of publications by a single group over multiple years that demonstrates sustained progress in the realization and maturation of a device.
2. *Tracking additional metrics:* as the suite of devices under consideration in the roadmap matures, metrics may be tracked beyond the density, speed, and energy data delineated in Table 20.1. Examples being featured in the 2013 edition include subthreshold slope and circuit energy. The addition of circuit energy measurements to the circuit speed data tabulated in 2011 and earlier will assist in differentiating candidate devices according to their viability for either high-performance or low-power applications.
3. *Assessing quantitative performance of nonelectron devices:* devices that use information carriers or tokens other than electron charge are evaluated only qualitatively in present editions of the roadmap. This remains the case in the 2013 edition, but the advent of

systematic simulations of nonelectron devices [5] will permit future editions to compare such devices on a quantitative basis. Ultimately, the objective for ERD is to unify the assessment of all emerging devices under a uniform quantitative framework.

4. *Conducting system-level analyses:* with the maturation of experimental results demonstrating individual emerging devices, future editions of the roadmap increasingly will identify direct linkages between these devices, compatible emerging interconnects, and circuit/system architectures that might exploit these devices for beyond-CMOS architectural or computational functionality [59]. The 2013 edition initiates this linkage via the inclusion of "paired interconnects" that might operate natively with the emerging research devices already tracked in prior editions. Future editions likely will include more systematic study of the possible synergies between emerging logic devices and large-scale computational systems, to include Boolean logic as well as analog/mixed-signal circuits and nonBoolean information processing.

20.5 Summary

In this chapter, a review is provided of the assessment and benchmarking of emerging logic devices undertaken for the International Technology Roadmap for Semiconductors. This assessment quantifies the progress to date in developing devices that might eventually replace silicon CMOS as the technology underlying the electronics industry. The quantitative data suggest that suitable candidates exist for the "next switch" that will surpass CMOS transistors in ultimate performance. Furthermore, a bifurcation is observed in that some devices appear to hold promise for higher speed operation than CMOS, while others might provide a route to lower power consumption. Beyond this quantitative assessment, a qualitative evaluation of recent progress in selected devices has identified several promising opportunities for both circuit-level integration of emerging devices that are relatively mature, as well as integration of novel functions afforded by nonelectron-based information processing modalities. Thus, among the emerging device options reviewed here, or others soon to come under the purview of the ITRS, one likely will find the beyond-CMOS device that provides a basis for future decades of sustained performance improvements in information systems.

Acknowledgments

The author gratefully acknowledges his collaborators on the Emerging Research Devices Working Group of the ITRS for their sustained efforts in authoring the biannual roadmap. Thanks also are due to participants in the recent ITRS Workshop on Emerging Research Logic Devices, especially D. Frank, M. Heyns, A. Ionescu, A. Khitun, S. Mitra, W. Porod, A. Sawa, F. Schwierz, D. Strukov, and V. Sverdlov, for their contributions. The author appreciates the helpful comments of J. C. Ellenbogen, J.F. Klemic, and M.D. Taczak of The MITRE Corporation in the preparation of this chapter.

References

1. Semiconductor Industry Association (2011) International Technology Roadmap for Semiconductors (ITRS), 2011. Available at: www.itrs.net (accessed 16 July 2013).
2. Semiconductor Industry Association (2011) Emerging Research Devices Chapter of the ITRS, 2011. Available at: http://www.itrs.net/Links/2011ITRS/2011Chapters/2011ERD.pdf (accessed 16 July 2013).
3. Skotnicki, T., Hutchby, J.A., King, T.-J. *et al.* (2005) The End of CMOS Scaling. IEEE Circuits and Devices Magazine, 16–26 January/February.

4. Colwell, R. (2013) The Chip Design Game at the End of Moore's Law, in *Hot Chips 25*, Dekker, Stanford.
5. Nikonov, D.E. and Young, I.A. (2013) Overview of Beyond-CMOS Devices and a Uniform Methodology for Their Benchmarking, Proceedings of the IEEE, 7 June.
6. Salahuddin, S. and Datta, S. (2008) Use of negative capacitance to provide voltage amplification for low power nanoscale devices. *Nano Letters*, **8**(2), 405–410.
7. Bachtold, A., Hadley, P., Nakanishi, T., and Dekker, C. (2001) Logic circuits with carbon nanotube transistors. *Science*, **294** (5545), 1317–1320.
8. Hutchby, J. (2008) Emerging Research Devices, ITRS Public Conference, San Francisco, CA.
9. Patil, N., Lin, A., Zhang, J. *et al.* (2011) Scalable carbon nanotube computational and storage circuits immune to metallic and mispositioned carbon nanotubes. *IEEE Transactions on Nanotechnology*, **10** (4), 744–750.
10. Shulaker, M.M., Hills, G., Patil, N. *et al.* (2013) Carbon nanotube computer. *Nature*, **501**, 526–530.
11. Park, H., Afzali, A., Han, S.-J. *et al.* (2012) High-density integration of carbon nanotubes via chemical self-assembly. *Nature Nanotechnology*, **7**, 787–791.
12. Cao, Q., Han, S.-J., Tulevski, G.S. *et al.* (2013) Arrays of single-walled carbon nanotubes with full surface coverage for high-performance electronics. *Nature Nanotechnology*, **8**, 180–186.
13. Schwierz, F. (2010) Graphene transistors. *Nature Nanotechnology*, **5**, 487–496.
14. Luisier, M., Lundstrom, M., Antoniadis, D.A., and Bokor, J. (2011) Ultimate device scaling: intrinsic performance comparisons of carbon-based, InGaAs, and Si field-effect transistors for 5nm gate length, IEDM, Washington, D.C.
15. Fiori, G., Betti, A., Bruzzone, S. *et al.* (2011) Nanodevices in Flatland: Two-dimensional graphene-based transistors with high Ion/Ioff ratio, IEDM, Washington, DC.
16. Unluer, D., Tseng, F., Ghosh, A.W., and Stan, M.R. (2011) Monolithically patterned wide-narrow-wide all-graphene devices. *IEEE Transactions on Nanotechnology*, **10** (5), 931–939.
17. Cai, J., Ruffleux, P., Jaafar, R. *et al.* (2010) Atomically precise bottom-up fabrication of graphene nanoribbons. *Nature*, **466** (7305), 470–473.
18. Lu, W. and Lieber, C.M. (2007) Nanoelectronics from the bottom up. *Nature Materials*, **6**, 841–850.
19. Ma, D.D., Lee, C.S., Au, F.C. *et al.* (2003) Small-diameter silicon nanowire surfaces. *Science*, **299**, 1874–1877.
20. Yan, H., Choe, H.S., Nam, S. *et al.* (2011) Programmable nanowire circuits for nanoprocessors. *Nature*, **470**, 240–244.
21. Wang, R., Zhuge, J., Huang, R. *et al.* (2007) Analog/RF performance of Si Nanowire MOSFETs and the impact of process variation. *IEEE Transactions on Electron Devices*, **54** (6), 1288–1294.
22. Nam, S., Jiang, X., Xiong, Q. *et al.* (2009) Vertically integrated, three-dimensional nanowire complementary metal-oxide-semiconductor circuits. *Proceedings of the National Academy of Sciences*, **106**, 21035–21038.
23. Claeys, C., Leonelli, D., Rooyackers, R. *et al.* (2011) Trends and challenges in Si and hetero-junction tunnel field effect transistors. *ECS Transactions*, **35** (5), 15–26.
24. Gandhi, R., Chen, Z., Singh, N. *et al.* (2011) CMOS-compatible vertical-silicon-nanowire gate-all-around p-type tunneling FETs with <= 50-mV/decade subthreshold swing. *IEEE Electron Device Letters*, **32** (11), 1504–1506.
25. Koswatta, S.O., Koester, S.J., and Haensch, W. (2009) 1D broken-gap tunnel transistor with MOSFET-like on-currents and sub-60mV/dec subthreshold swing. IEDM Technical Digest, Baltimore, MD.
26. Tomioka, K., Yoshimura, M., and Fukui, T. (2012) Steep-slope tunnel field-effect transistors using III-V nanowire/Si heterojunction, Symposium on VLSI Technology, Honolulu, HI.
27. Kao, K.-H., Verhulst, A.S., Vandenberghe, W.G. *et al.* (2012) Optimization of gate-on-source-only tunnel FETs with counter-doped pockets. *IEEE Transactions on Electron Devices*, **59** (8), 2070–2077.
28. Vandenberghe, W.G., Soree, B., Magnus, W. *et al.* (2011) Two-dimensional quantum mechanical modeling of band-to-band tunneling in indirect semiconductors, IEDM Technical Digest, Washington, DC.
29. Chen, F., Spencer, M., Nathanael, R. *et al.* (2010) Demonstration of Integrated Micro-Electro-Mechanical Switch Circuits for VLSI Applications, Proceedings of ISSCC, San Francisco, CA.
30. Nathanael, R., Pott, V., Kam, H. *et al.* (2009) 4-terminal relay technology for complementary logic, IEDM Technical Digest, Baltimore, MD.
31. Weinstein, D. and Bhave, S.A. (2010) The resonant body transistor. *Nano Letters*, **10**, 1234–1237.
32. Ionescu, A.M. (2010) Resonant body transistors, Proceedings of Device Research Conference, South Bend, IN.
33. Ekinci, K.L. (2005) Electromechanical transducers at the nanoscale: actuation and sensing of motion in nano-electromechanical systems (NEMS). *Small*, **1** (8–9), 786–797.
34. Wang, X., Zhou, J., Song, J. *et al.* (2006) Piezoelectric field effect transistor and nanoforce sensor based on a single ZnO nanowire. *Nano Letters*, **6** (12), 2768–2772.

35. Jang, W.W., Lee, J.O., Yoon, J.-B. *et al.* (2008) Fabrication and characterization of a nanoelectromechanical switch with 15-nm-thick suspension air gap. *Applied Physics Letters*, **92** (10), 103110.
36. Newns, D.M., Misewich, J.A., Tsuei, C.C. *et al.* (1998) Mott transition field effect transistor. *Applied Physics Letters*, **73** (780), 15–18.
37. Matsubara, M., Okimoto, Y., Ogasawara, T. *et al.* (2007) Ultrafast photoinduced insulator-ferromagnet transition in the perovskite manganite Gd0.55Sr0.45MnO3. *Physical Review Letters*, **99** (20), 207401.
38. Ahn, C.H., Triscone, J.-M., and Mannhart, J. (2003) Electric field effect in correlated oxide systems. *Nature*, **424** (6952), 1015–1018.
39. Asanuma, S., Xiang, P.-H., Yamada, H. *et al.* (2010) Tuning of the metal-insulator transition in electrolyte-gated NdNiO3 thin films. *Applied Physics Letters*, **97** (14), 142110.
40. Nakano, M., Shibuya, K., Okuyama, D. *et al.* (2012) Collective bulk carrier delocalization driven by electrostatic surface charge accumulation. *Nature*, **487** (7408), 459–462.
41. Datta, S. and Das, B. (1990) Electronic analog of the electro-optic modulator. *Applied Physics Letters*, **56** (7), 665.
42. Osintsev, D., Sverdlov, V., Stanojevic, Z. *et al.* (2011) Properties of silicon ballistic spin fin-based field-effect transistors. *ECS Transactions*, **35** (5), 277–282.
43. Sugahara, S. and Nitta, J. (2010) Spin-transistor electronics: an overview and outlook. *Proceedings of the IEEE*, **98** (12), 2124–2154.
44. Sukegawa, H., Wen, Z., Kondou, K. *et al.* (2012) Spin-transfer switching in full-Heusler Co2FeAl-based magnetic tunnel junctions. *Applied Physics Letters*, **100** (18), 182403.
45. Takamura, Y., Hayashi, K., Shuto, Y., and Sugahara, S. (2012) Fabrication of high-quality Co2FeSi/SiOxNy/Si (100) tunnel contacts using radical-oxynitridation-formed SiOxNy barrier for Si-based spin transistors. *Journal of Electronic Materials*, **41** (5), 954–958.
46. Niemier, M.T., Bernstein, G.H., Csaba, G. *et al.* (2011) Nanomagnet logic: progress toward system-level integration. *Journal of Physics: Condensed Matter*, **23** (49), 493202.
47. Li, P., Csaba, G., Sankar, V.K. *et al.* (2012) Switching behavior of lithographically fabricated nanomagnets for logic applications. *Journal of Applied Physics*, **111** (7), 07B911.
48. Lent, C.S., Tougaw, P.D., Porod, W., and Bernstein, G.H. (1993) Quantum cellular automata. *Nanotechnology*, **4** (1), 49.
49. Behin-Aein, B., Datta, D., Salahuddin, S., and Datta, S. (2010) Proposal for an all-spin logic device with built-in memory. *Nature Nanotechnology*, **5** (4), 266–270.
50. Varga, E., Orlov, A., Niemier, M.T. *et al.* (2010) Experimental demonstration of fanout for nanomagnetic logic. *IEEE Transactions on Nanotechnology*, **9** (6), 668–670.
51. Breitkreutz, S., Kiermaier, J., Eichwald, I. *et al.* (2013) Experimental demonstration of a 1-bit full adder in perpendicular nanomagnetic logic. *IEEE Transactions on Magnetics*, **49** (7), 4464–4467.
52. Alam, M.T., Kurtz, S.J., Siddiq, M.A.J. *et al.* (2012) On-chip clocking of nanomagnet logic lines and gates. *IEEE Transactions on Nanotechnology*, **11** (2), 273–286.
53. Breitkreutz, S., Kiermaier, J., Eichwald, I. *et al.* (2012) Majority gate for nanomagnetic logic with perpendicular magnetic anisotropy. *IEEE Transactions on Magnetics*, **48** (11), 4336–4339.
54. Eichwald, I., Bartel, A., Kiermaier, J. *et al.* (2012) Nanomagnetic logic: error-free, directed signal transmission by an inverter chain. *IEEE Transactions on Magnetics*, **48** (11), 4332–4335.
55. Liu, S., Hu, X.S., Nahas, J.J. *et al.* (2011) Magnetic-electrical interface for nanomagnet logic. *IEEE Transactions on Nanotechnology*, **10** (4), 757–763.
56. Augustine, C., Panagopoulos, G., Behin-Aein, B. *et al.* (2011) Low-power functionality enhanced computation architecture using spin-based devices, Proceedings of NANOARCH, San Diego, CA.
57. Khitun, A. and Wang, K.L. (2011) Non-volatile magnonic logic circuits engineering. *Journal of Applied Physics*, **110** (3), 034306.
58. Alzate, J.G., Upadhyaya, P., Lewis, M. *et al.* (2012) Spin wave nanofabric update, Proceedings of NANOARCH, New York, NY.
59. Das, S., Picconatto, C.A., Rose, G.S. *et al.* (2007) System-level design and simulation of nanomemories and nanoprocessors, in *Nano and Molecular Electronics Handbook* (ed. S. Lyshevski), CRC, Boca Raton.

Part Four

Concepts for Emerging Architectures

21

Nanomagnet Logic: A Magnetic Implementation of Quantum-dot Cellular Automata

Michael T. Niemier, György Csaba, and Wolfgang Porod
University of Notre Dame, USA

21.1 Introduction

The system-level, information processing architectures to be discussed here are potential targets for devices that can be organized in the automata-like architecture proposed in [1] – which suggested using the position of electrons on quantum dots to represent binary state and process information.

Initially, experimentalists targeted *metal-dot* implementations of the Quantum-dot Cellular Automata (or QCA) device architecture (see device, wire, and gate schematic in Figure 21.1a, Figure 21.1b, and Figure 21.1c respectively), and individual devices [2], gates [3], latches [4], power gain [5], and fanout [6] were soon demonstrated. However, metal-dot QCA devices require cryogenic operating temperatures that severely limit the potential for deployed hardware – and hence application spaces. In an effort to raise operating temperatures, some research groups pursued/continue to pursue QCA devices that take the form of a synthesized chemical *molecule* [7]. (Scaling device size downward can increase the energy separation between states [8], and hence operating temperature.) Experimentally, a self-assembled monolayer of mixed-valence compounds appears to switch between states that could be mapped to logic 1s and 0s [9]. However, both practical devices, and a mechanism for the deterministic placement of $\sim 1 \times 1 \text{ nm}^2$ molecules with acceptable error rates, have yet to be developed. (No devices in the aforementioned switching experiment were deterministically placed.) *Semiconductor*-based devices have also been studied – and could potentially raise operating temperature [10] – but most still require subKelvin temperatures [11].

Alternatively, *magnetic* devices with nanometer feature sizes also are capable to move, process, and store information in a cellular automata-like architecture. Magnetic quantum-dot

Emerging Nanoelectronic Devices, First Edition. An Chen, James Hutchby, Victor Zhirnov and George Bourianoff.
© 2015 John Wiley & Sons, Ltd. Published 2015 by John Wiley & Sons, Ltd.

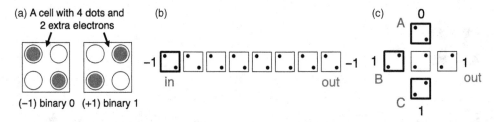

Figure 21.1 (a) QCA devices are typically comprised of four dots with two extra electrons; binary information is represented by charge configuration. (b) Electrostatic interactions propagate information from dot to dot in a QCA wire. (c) Electrostatic interactions also perform a majority voting function in a QCA majority gate

cellular automata (MQCA) devices – referred to here as Nanomagnet Logic (NML) per the latest naming convention from the ITRS [12] – will retain state without power, are radiation-hard, and are projected to dissipate less than 40 kT per switching event for a gate operation [13]. Experimentally, Cowburn performed the first *room temperature* experiments with QCA-like devices by showing that a magnetic soliton could propagate through a line of circular supermalloy discs [14]. Building on this work, researchers at the University of Notre Dame (ND) began to consider oval magnetic islands, which represent binary states due to magnetic shape anisotropy, and to explore potential approaches for clocking ensembles of NML devices – particularly focusing on how device shape could affect device to device coupling such that magnets in a circuit ensemble would settle into a logically correct state in response to a magnetic (clock) field and new inputs [15]. Additional research has since targeted – and experimentally demonstrated – the five essential "tenets" [16] that a digital device (i.e., NML) must satisfy: support for a functionally complete logic set (i.e., via majority logic gates [13] and/or 2-input AND/OR gates [17,18]), concatenability [13], nonlinear response characteristics [19], gain/fanout [20,21], and unidirectional dataflow [22]).

Given this technological state of the art, it is our belief that a magnetic implementation of the QCA device architecture (i.e., NML) represents the most viable path forward at the device level. For this reason, both our own research and this narrative considers the QCA architecture from the perspective of NML (although many of the ideas discussed here could be applied to the electrostatic QCA device architecture too).

Looking at the rest of this chapter, in Section 21.2, we will review the technological state of the art for different implementations of the NML device architecture. In Section 21.3, we address system-level architectures for NML assuming that the devices are used to implement "traditional" Boolean logic gates (adder designs and systolic arrays are discussed); Section 21.4 considers more unconventional NML-based information processing architectures (i.e., threshold logic, nonBoolean architectures, and cellular automata). Finally, in Section 21.5, we present our view on future prospects for the QCA device architecture – from a systems architecture perspective – and identify what we believe to be important and relevant future research directions.

21.2 Technology Background

21.2.1 An "In-Plane" Device Architecture

NML processes information via coupling between neighboring, nano-scale magnets. Most work with NML has focused on devices that couple in-plane (denoted iNML, see schematic in

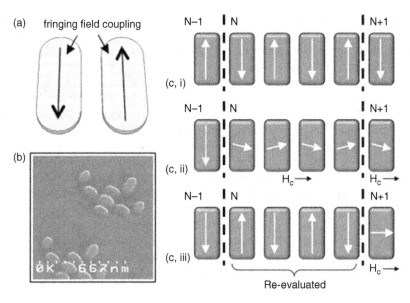

Figure 21.2 (a) iNML device schematic. (b) Scanning electron microscopy (SEM) of programmable majority gate. (c, i–iii) how an AF-line would be clocked

Figure 21.2a). (Recent work has also employed devices with out-of-plane, perpendicular magnetic anisotropy (PMA); this device architecture will be discussed in Section 21.2.2.).

Since 2010, experimental progress (driven in part by our research group at the University of Notre Dame, as well as other groups at UC-Berkeley [23], VCU [24], South Florida [25], Minnesota (UMN) [26], and so on who are now studying NML) suggests that it is possible to move beyond isolated line and gate structures (e.g., the majority gate as shown in Figure 21.2b [19]), and toward more functional, *circuit-level structures*. Notably, fanout structures [20,21], a 1-bit full adder [27], and so on have been experimentally demonstrated and successfully re-evaluated with new inputs. Both field-coupled [28] and spin transfer torque (STT) [26,29,30] electrical *inputs* have been realized. For electrical *output*, multiple NML-magnetic tunnel junction hybrids [31–33] have been proposed and simulated.

Externally supplied switching energy is needed to re-evaluate (or *clock*) a magnet ensemble with new inputs. To modulate the energy barriers of iNML device ensembles "on-chip," [34] proposed using hard axis directed magnetic fields from current driven wires. Ensembles of magnets could be grouped in clocking zones. Figure 21.2c illustrates the effects of the clocking field. From the initial state in Figure 21.2c,i, magnets in clocking zones N and N + 1 are put into a metastable logic state (i.e., along the hard/short axis) against the preferred shape anisotropy (Figure 21.2c,ii). If a driving neighbor (i.e., in zone N − 1) provides a y-directed biasing field, the driven magnet's magnetization rotates toward a preferred easy (long) axis. When the clock is removed, magnets should relax to a new, low energy ground state (Figure 21.2c,iii) in order. Clock energy could be amortized over hundreds of thousands of devices as a single clock line could control many parallel ensembles. Fringing fields from devices themselves also help with the transition to the 0/metastable state in Figure 21.2c,ii. As such, the magnitude of the required clock field (and current) need not be excessively high [34,35], and this approach to clocking could lead to performance "wins" at the application level [35–37]. Notably, proposals for line clock structures [34] were experimentally realized and used to flip the state of single magnets [38], as well as to re-evaluate lines and gates [22]. Moreover, recent experimental results [37,39,40] suggest that materials-based solutions could allow for further reductions in clock wire current.

iNML devices should also be amenable to voltage-controlled clocking, and multiferroics [41] and magnetostriction [24], have also been proposed as potential iNML clocks. Multiferroic materials (e.g., BFO) could allow for electric field control of magnetism [41]. uses multiferroic materials to demonstrate the switching of an in-plane ferromagnetic component (i.e., a nanomagnetic island). Also, [24] considers how the magnetization state of a multiferroic nanomagnet can be rotated via the coupling between a magnetostrictive layer and a piezo-electric layer. This work suggests that only 10 s of millivolts are needed to induce a rotation of 90° into a metastable state, and that stress-based clocking would lead to clock energy dissipation of just ~200 kT per device.

If voltage-controlled clocking becomes feasible, it would almost certainly be the preferred approach for iNML clocking. When looking toward circuits and systems, potential advantages include reductions in clock energy [24], and more fine-grained control of an NML ensemble (useful architecturally as discussed in [42], and Section 21.3.1.4 of this document). That said, many of the architectural ideas discussed here are by in large amenable to current or voltage-controlled clocks – particularly as both clocking approaches place devices in metastable states for re-evaluation (a potential source of error that will be considered in Section 21.5).

21.2.2 An "Out-of-Plane" Device Architecture

A pNML device schematic is shown in Figure 21.3a. Thin PMA nanomagnets switch through nucleation and domain wall propagation [43,44]. Irradiation or ion milling could be used to define a "soft spot," where a domain wall may nucleate. The nucleated wall subsequently propagates through the dot, fully reversing it. The coercivity of the magnet is determined by the coercivity of its highest-irradiated region. This reversal behavior is markedly different from in-plane permalloy nanomagnets, which remain nearly single-domain during switching and rotate coherently (through a 0/metastable state as described above).

Both computational [45] and experimental [46] results suggest that appropriately irradiated pNML structures can be controlled by a uniform, homogeneous, oscillating, global clock field. Consider the antiferromagnetically (AF) ordered line schematic in Figure 21.3b. Here, we (i) refer to the period of the sinusoidal, *out-of-plane* field as T_{pulse} with peak amplitude of H_{pulse}, and (ii) assume that dots are FIB irradiated on their left edges, so information will flow from left

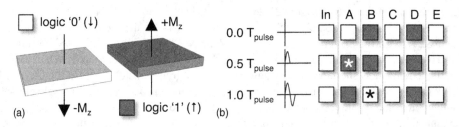

Figure 21.3 (a) Graphical representation of logic 0, 1 for pNML. (b) How information would move in an AF-ordered line (assuming the left edge of each magnet is FIB irradiated); * is used to show dataflow

to right. After the first application of H_{pulse} (in the $+z$ direction), the dot immediately adjacent to the flipped input will change state after $0.5\ T_{pulse}$. Switching occurs as H_{pulse} is of sufficient magnitude to eliminate parallel alignment in the dot pair at the input, but not of sufficiently high magnitude to alter the state of other dots in the line. (Again, irradiation location determines which dot will switch.) With successive applications of the time varying field, multiple bits of information can move through an AF-line segment simultaneously, and an inherent pipeline is created. Notably, majority gates [47], AF-lines [46], and full adders [48] comprised of multi-layered pNML devices have all been experimentally demonstrated.

When considering experimental realizations of "on-chip" clocks, large arrays of pNML devices could be controlled with on-chip inductor structures [49] that could be coupled with a capacitance in an LC oscillator – which opens the door to adiabatic energy recycling. pNML devices are also amenable to voltage-controlled clocking. Recent experiments suggest that it should be possible to change the easy axis of patterned CoFeB layers with PMA from out-of-plane to in-plane with modest voltages [50]. As such, a pNML device would transition through a metastable state with this clocking approach.

21.3 NML Circuit Design Based on Conventional, Boolean Logic Gates

21.3.1 Boolean Logic

Here, we consider the prospects for the NML device architecture when using NML devices to realize random, combinational logic blocks – for example, that might be used in a general-purpose microprocessor. We begin by discussing the pros and cons of the iNML and pNML device architectures (Sections 21.3.1.1–21.3.1.4), and conclude with an effort to quantitatively benchmark NML logic against CMOS equivalents. In Section 21.3.1, we will use an adder as a running case study.

21.3.1.1 Less Complex Signal Routing

When considering the layout of an iNML circuit, assuming a Cartesian grid, data will propagate through AF-ordered interconnect when moving in the $\pm x$ direction, and ferromagnetically (F)-ordered interconnect when moving in the $\pm y$ direction. As such, the x dimension (i.e., the width) of every device must essentially be fixed in order to allow for transitions between AF-ordered and F-ordered lines. Theoretically, this construct could be relaxed. As per Figure 21.4a, a line comprised of five devices that are each 60 nm wide, with 10 nm spacing between devices will have the same length as a line of seven devices that are 40 nm wide with 10 nm between devices. However, varying the size/aspect ratio of devices is difficult as: (i) clock field requirements can change significantly [51], (ii) very precise lithography would be required, (iii) variations to width are also limited in that we need to ensure devices remain single domain, and (iv) fringing fields from a smaller device may be insufficient to "drive" a larger device.

Assuming iNML devices of constant size, transitions between F-ordered and AF-ordered interconnect can prove to be problematic when considering concatenated gate structures. As an example, consider a majority gate-based, one-bit full adder (see schematic in Figure 21.4b). Note that in this design, the carry input signal is fed to gates M1 and M3, while the complement of the carry signal is fed to gate M2. Also, the complement of M1 is required as an input to M3.

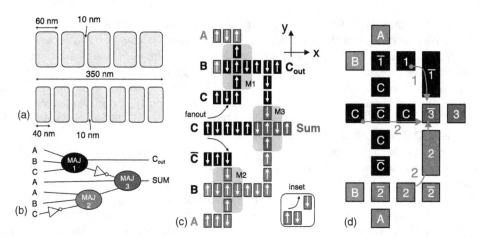

Figure 21.4 (a) With iNML, device size could be varied to create equal length line segments, but clocking requirements for the dots in each line can change significantly. (b) Majority gate-based full adder. (c) Schematic of iNML full adder. (d) Schematic of pNML full adder

Given the combination of AF- and F-ordered interconnect, obtaining the proper signal value at the desired gate input is difficult to achieve.

As a representative example, consider an iNML-based schematic of the 1-bit adder shown in Figure 21.4c. To better illustrate potential routing issues, we have made explicit copies of the A, B, and C inputs at each gate. If a single copy of the C input were to fan out to gates M1, M2, and M3, M1 and M2 would receive the wrong value. Using M2 as an example, the C input to gate M2 should have a negative y component of magnetization, but instead would have a positive y component of magnetization via F-ordered interconnect. While carry signal routing might be addressed by eliminating one magnet from the middle C input, the input to M3 would then be affected. One solution to this problem could be to create a staggered transition between AF- and F-ordered interconnect (see Figure 21.4c inset), and physical-level simulations suggest that such a structure may function properly, but this idea is unproven experimentally.

That said, this problem is much easier to address with pNML given that: (i) both x- and y-directed interconnect are AF-ordered, and (ii) that it is possible to vary the size of a given device. A potential pNML-based schematic of an adder appears in Figure 21.4d. By changing the size of the device that serves as the M1 and M2 input to M3, desired signal values can easily be routed to a given gate. While larger device sizes could impact the required length of T_{pulse}, ensembles with mixed dot sizes have been experimentally demonstrated [52], and thus no new constructs are required. Notably, magnets could be on the order of one micron long [53].

21.3.1.2 Data Races

Upon closer examination of Figure 21.4d, one might wonder whether or not "data races" could occur. Consider the input paths to M3. There is essentially just one (albeit larger) device between the output of gate M1 and the decision device in M3, there are two devices between the carry input and the decision magnet of M3, and there are two devices between the output of M2 and M3's decision magnet. (This does not even account for critical path differences given the initial evaluations of M1 and M2.) Thus, at issue is whether or not different signal arrival times will impact logical correctness.

For iNML, this is a potential concern. The success or failure of a gate is largely a function of clocking – for example, whether or not devices switch in a soliton-like fashion or devices switch in the presence of an applied clock field. With soliton-based switching, all devices in a circuit ensemble are placed into a $0°$, metastable (MS) state – for example, by a clocking field. The external field is removed, and each device in the ensemble is expected to remain in a hard-axis-biased state until set by an appropriate neighbor. (While this is an unstable state of the system, dipole fields from neighboring magnets do help to preserve it [23].) All inputs to the ensemble are set, and easy axis directed fringing fields from said inputs cause neighboring devices to switch. Fringing fields from these devices then set the state of the next neighbor, and so on.

In [23], the authors noted that thermal noise could induce premature, random, and unwanted switching. To combat this problem [23], suggested fabricating magnetic islands with a magneto-crystalline biaxial anisotropy – that is, such that $U(\theta)$ becomes $K_u \cos^2(\theta) + \frac{1}{4}K_1 \sin^2(2\theta)$, where K_1 is the biaxial anisotropy constant. The biaxial anisotropy term introduces a local minimum in the energy landscape of an individual magnetic island at $0°$ – which should further promote the hard axis stability of the ensemble. Given the effects of thermal noise at $T = 300$ K [23], reports that a line of hard-axis biased devices with a biaxial term remained in a $0°$/MS state until a device was set by an appropriate neighbor, when a device without the biaxial term did not.

In [54], Spedalieri et al. extended the work of [23] and considered the switching behavior of a suite of majority gates with biaxial anisotropy. The clocking scheme assumed in [54] was similar to the approach used in [23] (and explained above). Notably, assuming gates comprised of $30 \times 15 \times 6$ nm magnets, the gate error rate can exceed 15%. The most common errors observed were: (i) premature switching in randomized islands due to thermal noise and (ii) gates where devices essentially remained hard-axis biased (due to a biaxial constant/biaxial anisotropy that was too large).

It was also shown that sufficiently strong coupling fields in the case of pNML [68] or sufficiently small clocking zones in the case of iNML [69] reduce error rates to an acceptable level, where circuit-level solutions can be used for eliminating errors.

Alternatively, magnets can also be driven to a new, logically correct state before a clocking field is removed. [55] reports gate simulations where inputs do not arrive at a decision magnet simultaneously, but the gate still functions correctly. The clock field is applied until all signals arrive at the compute magnet, and it has time to respond. (The clock field prevents a decision magnet from relaxing prematurely.) While this approach to switching may prevent race conditions, it will also: (i) result in an increase in clock energy and (ii) make successive devices in a group weaker drivers, eventually limiting the width of a clock line.

As with signal routing, managing potential data races can be much easier with FIB-irradiated pNML. To illustrate, consider the majority gate (with some initial state) shown in Figure 21.5. As before, we will assume an out-of-plane, sinusoidal field as shown in Figure 21.3b. Now, assume that the top and middle inputs to this gate change state (see Figure 21.5b). The time evolution of the gate is shown in Figure 21.5c–j. Even though the output of the gate changes state after the new, middle input arrives (Figure 21.5f), it changes back to the logically correct state after the new, top input arrives (Figure 21.5i). As the middle input is unchanged, no metastabilities in the second input arm appear, and the compute device is allowed to react to all inputs.

Finally, consider an NML ensemble where multiple levels of logic are connected together. Using the above gate as an example, if we define T_{clock} to be NT_{pulse} (Figure 21.5j), where N is defined by the longest critical path through a bit slice, we can synchronize the application of

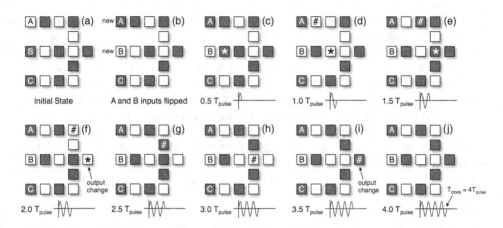

Figure 21.5 Time evolution of an pNML majority gate when subjected to a sinusoidal, out-of-plane field; data propagation from A and B inputs is illustrated with # and * respectively

new inputs in successive bit slices to ensure that a well-timed circuit (and logically correct output) is achieved.

21.3.1.3 Pipeline Depth

For iNML, a multi-phase clock is required to define dataflow directionality through a magnet ensemble (Figure 21.2c). However, controlling an iNML ensemble with multiple clock lines – for example, an extended AF-line segment like that shown in Figure 21.6a – results in an inherent pipeline, which will improve circuit throughput. More specifically, every time data propagates completely through a clock group, a new input can begin to traverse the same group. Essentially, a new value can be written into or read out of our AF-line/shift register every time the last wire cycles through all ϕ phases (where ϕ would equal three for a three-phase clock). The magnets overlaid on top of N wires (where N is the number of clock phases) represent a bit of information (see Figure 21.6a).

Assuming a line clock, the number of magnets per pipe stage will ultimately be limited by the width of a given line (see Figure 21.6b). Practically, this cannot be arbitrarily small as narrow clock wires will: (i) have a much higher resistance which will adversely affect clock energy and (ii) making very narrow, clad lines (i.e., less than 100 nm wide) with good field confinement will obviously become prohibitive. This is not to say that a bit of information could not traverse multiple gates within a given group, but only one *bit* could be stored per AF-ordered line that traverses a given group (Figure 21.6a).

We reiterate that other approaches for clocking iNML are also being studied. As noted earlier, potential advantages to electric field-based clocking include additional energy reductions, *and* more fine-grained control of an ensemble – which could lead to deeper pipelines. (It should be noted that a potential disadvantage of the aforementioned clocks is that every device may need to be contacted individually, which would increase fabrication complexity.) That said, in a pNML ensemble, pipe depth could be completely independent of clock feature size, and devices do not need to be contacted individually (i.e., assuming domain wall switching modes). Regarding fabrication and throughput, this would obviously be advantageous.

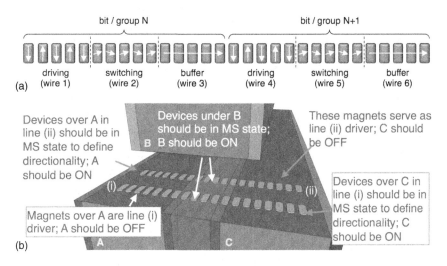

Figure 21.6   (a) Bit definition in an iNML AF-line (three-phase clock). (b) Why bi-directional dataflow is difficult to achieve with line-based, three-phase clock

21.3.1.4   Achieving Bi-directional Dataflow

Information processing hardware for countless applications requires feedback – which implicitly suggests that bi-directional dataflow is required. Assuming a line clock, with iNML, three clock wires (as shown in Figure 21.6b) *cannot* facilitate dataflow in multiple directions (e.g., in parallel, AF-ordered lines). Assuming we want data to flow from left to right in line (i) and from right to left in line (ii), the two AF-ordered lines would require different excitation patterns. For example, for line (i), wires A, B, and C should be OFF, ON, and ON respectively, while for line (ii), wires A, B, and C should be ON, ON, and OFF respectively. Thus, for a line clock, bi-directional dataflow must be accomplished at the granularity of the clock structures themselves. However, for pNML, *directionality can be achieved simply by changing the position of FIB irradiation.* Similarly, with voltage-controlled clocking, bidirectional dataflow could be achieved in a fine-grained manner for both iNML and pNML.

21.3.1.5   Benchmarking

Here, we briefly discuss energy and performance metrics for iNML and pNML adder designs. We begin by considering recent results from the NRI's architectural-level benchmarking efforts led by Nikonov and Young – where multiple devices were evaluated assuming a common set of fabrication parameters *and* circuit-level benchmarks (see [56] for more detail). Note that as only the iNML device architecture was rigorously vetted through the NRI, only results that project the energy, delay, and throughput of *iNML* current and voltage-controlled 32-bit adders are discussed below. (The adder design used in [56] is based on the design in Figure 21.4c.) That said, we do conclude this subsection by briefly comparing and contrasting the iNML and pNML adder designs in Figure 21.4c and d.

We particularly wish to highlight the impact of inherent pipelining on an iNML-based adder's performance/energy/throughput. Notably, the NRI's benchmarking analysis "[does]

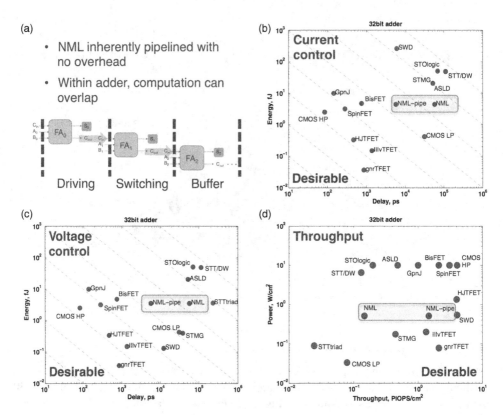

Figure 21.7 (a) Inherent pipelining in ripple carry adder. (b) NRI projections for 32-bit adder assuming current-controlled switching of spin-based devices – the impact of inherent pipelining on performance/ energy of the iNML device architecture is noted. (c) NRI projections for 32-bit adder assuming (magnetostictive) voltage controlled switching of spin-based devices – the impact of inherent pipelining on performance/energy of the iNML device architecture is noted. (d) NRI projections for throughput assuming magnetostrictive, voltage-controlled switching of spin-based devices – the impact of inherent pipelining on performance/energy of the iNML device architecture is noted. (PIOPS/cm^2 refers to the number of peta-integer operations per cm^2)

not incorporate any pipelining in [these] calculations, even though some logic technologies are not constrained by power dissipation may produce higher computational throughput by pipelining." As per the discussion above, NML circuits could (and should) be inherently pipelined with no overhead. Thus, as per Figure 21.7a, computations in an adder could overlap. If we conservatively assume that 10 adds are active simultaneously, as per Figure 21.7b and c, (generated via the code associated with [56]) the energy–delay products of iNML adders improve and become competitive with low power CMOS. Furthermore, standby power effects, intangibles regarding operating environments, and so on could represent added benefits. Moreover, as per Figure 21.7d, the throughput per unit area of an iNML adder (quantified as peta-integer operations per second, per square centimeter) moves into the most desirable area of the plot.

Finally, the pNML device architecture could offer additional advantages. When comparing the adder design schematics in Figure 21.4c (iNML) and Figure 21.4d (pNML) the number of devices in the critical path in the pNML design is 6 less than the iNML design (8 vs. 14). Furthermore, even if iNML devices with 20×30 nm footprints were employed, the area of the one-bit design would be essentially the same as a pNML design comprised of devices with 50×50 nm footprints and 50×100 nm footprints (i.e., $0.133\ \mu m^2$ for iNML vs. $0.136\ \mu m^2$ for pNML). (Differences can largely be attributed to ease in signal routing.)

21.3.2 Systolic Arrays

Here, we consider alternative, system-level architectures (e.g., nonVon Neumann architectures) that might employ NML devices to perform Boolean logic functions. Of particular interest are architectural approaches that can exploit deeply pipelined logic, and avoid large-scale interconnect.

21.3.2.1 Architecture Overview

We anticipate that NML-based hardware will be used to implement systolic architectures [57] that would in turn process information in the aforementioned application spaces. The nearest neighbor interactions and inherently pipelined logic associated with NML map extremely well to systolic architectures developed in the late 1970s and early 1980s [57]. In a systolic architecture, data flows from a computer's memory, through many (and often identical) processing elements, before returning to memory. Additional processing is done on some subset of data by each processing element. As such, this can significantly reduce the need for global signal broadcasts. This should allow us to minimize NML's drawbacks (nearest neighbor dataflow and higher latency devices when compared to CMOS) and exploit its more desirable features (inherently pipelined logic with no overhead).

As one example, consider the convolution problem where, given a sequence of weights w_1, $w_2, \ldots w_k$ and the input sequence $x_1, x_2, \ldots x_k$, the resulting sequence $y_i = w_1 x_i + w_2 x_{i+1} + w_k x_{i+k-1}$ is calculated. Also, if the multiplication and addition operations in the convolution expression above are transformed to Boolean XOR and AND operations, the convolution operation would be transformed to a pattern matching operation. In either instance, streams of data would flow from left to right while a cumulative output would flow from right to left. (And as noted in Section 21.3.1.4, this can be achieved for iNML via voltage-controlled clocking and for pNML with either a global clock *or* a voltage-controlled clock.)

Moreover, future information processing workloads will be required to perform information processing tasks that have natural systolic solutions. Systolic solutions exist for many problems including filtering, polynomial evaluation, discrete Fourier transforms, matrix arithmetic and other nonnumeric applications involving graphing algorithms and data structures. Below, we will present a more thorough case study of pattern matching (PM) hardware – which is needed at the application-level for data mining, genomics, intrusion detection, and so on.

21.3.2.2 Toward NML Implementations of Systolic Architectures

The fundamental processing element (PE) in a systolic PM circuit appears in Figure 21.8a. Individual bits of a data stream serve as one input to an XNOR gate. The other input to the XNOR is a bit (b_i) of a pattern of interest. (This bit of information could be stored directly at the gate in a nonvolatile, NML device. We are investigating the use of spin transfer torque (STT) to program the b bits.) If the bits match, the output of the XNOR gate is a logic 1; otherwise a logic 0 results. This output then becomes one input to the AND gate – the output of which captures

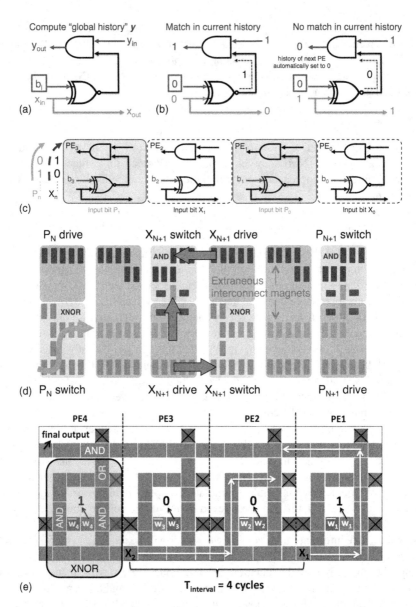

Figure 21.8 (a) Logic gates in a systolic PM PE. (b) How PE functions assuming match/no match in incoming input stream. (c) Concatenated PEs to examine a four-bit pattern; while x inputs must be staggered by one PE to ensure global history bits and new x inputs arrive at next PE simultaneously, input streams (e.g., $P_{0\text{-to-}n}$, $X_{0\text{-to-}n}$) can be interleaved to achieve full hardware usage. (d) Schematic of potential iNML-based systolic array that: (i) assumes voltage-controlled clocking (shading color denotes common clock phases) and (ii) the existence of a two-input XOR gate for simplicity (see [17] for possible realizations of this gate). (e) Concatenated pNML PEs – to reconcile pNML timing requirements, need to stagger x inputs by $4T_{\text{pulse}}$ if no interleaving

the global history of multiple, concatenated PEs. If at any point the output of an XNOR gate is a 0, the accumulated global history is set to 0, which suggests that there is no match in input stream bits $x_m \ldots x_{m+n}$ (see Figure 21.8b).

More specifically, when a computation commences, bits from the input data stream $(x_0 \ldots x_n)$ flow through the sequence of PEs (i.e., from left to right, as per Figure 21.8). When the first input bit (x_0) arrives at the rightmost PE (PE$_0$ in Figure 21.8c), initialization is complete, and streaming data analysis can begin. Thus, bit x_0 would be compared to bit w_0 via an XNOR operation, and the output of this gate would then be ANDed with the input signal y_{in} – initialized to logic "1." This "global history" is then shifted to an adjacent (left) PE (e.g., PE$_1$ in Figure 21.8c) such that we can consider the next bit of the input stream. When the shifted global history reaches the leftmost PE (PE$_3$ in Figure 21.8c), a match is detected if the initial logic "1" has been preserved.

That said, some *architectural-level* timing constraints must also be satisfied. For instance, bits of the input stream must be "spaced" two PEs apart to ensure that global history bits and streaming input data meet at the proper time. (Note the position of x_1 in Figure 21.8c). This will degrade circuit throughput. However, two different input streams could be interleaved within the systolic array – assuming that one would want to search for the same pattern in each. Interleaved dataflow (e.g., p_1, x_1, p_0, x_0 as noted in Figure 21.8c) would allow for full hardware usage and the highest possible throughput.

21.3.2.3 An iNML Implementation

A schematic representation of an iNML realization of the PEs depicted in Figure 21.8a/ Figure 21.8c appears in Figure 21.8d. Here, voltage clocking is assumed (as per Figure 21.6b, fine-grained bi-directional dataflow cannot be easily achieved with a line clock). The color of the underlying shading denotes clock phase, and arrows/labels denote how interleaved data streams would flow through the systolic array. (Note that for simplicity, this schematic assumes the existence of a two-input XNOR gate – which could be possible per [17]). By eliminating extraneous interconnect magnets (see labeled devices in Figure 21.8d) the critical path through a given PE could be reduced.

21.3.2.4 A pNML Implementation

A pNML-based design for the PM systolic array appears in Figure 21.8e. To perform the required AND function, a majority voting gate is essentially transformed to an AND gate by hard-coding one input to logic "0." Similarly, the XNOR gate is realized by performing the operation A"B + AB." Again, majority gates are transformed to AND and OR gates by hard-coding inputs to logic "0"s and logic '1's respectively. As reported in [49,58], the functionality of this design has been verified by micromagnetic simulation.[1]

By taking advantage of the ability to selectively size pNML devices,[2] it is possible to balance the critical paths through PE$_1$ and PE$_0$. This allows us to reconcile the architectural-level timing requirements of the bidirectional systolic array with those of the pNML clock – for example, to

---

[1] The NIST-based OOMMF tool was used; FIB anisotropy variations were modeled as in [45].
[2] In our simulations, for the design in Figure 21.8e, square-shaped devices were assumed to have 100×100 nm footprints, while elongated dots have 210×100 nm footprints. The distance between dots was set to be 10 nm. (To simplify this figure, overlaid arrows suggest how information should propagate through the circuit, and partially irradiated regions are not shown.)

ensure that the output of PE$_0$ is coordinated with the evaluation of the XNOR gate in PE$_1$. Thus, the output of the XNOR in PE$_1$, and the output (from the AND gate) in PE$_0$ should arrive at the AND gate in PE$_1$ simultaneously. This allows the input data stream to move through the systolic array uninterrupted to avoid stalls and maximize throughput. As highlighted in Figure 21.8e, the interval between x inputs is regular and equal to $4T_{\text{pulse}}$. This suggests that after the pipeline/systolic array is filled (i.e., the first bit of the input stream reaches PE$_0$), a new response/match check will exit the systolic array every $4T_{\text{pulse}}$ time units (if data is *not* interleaved), and every $2T_{\text{pulse}}$ time units (if data *is* interleaved). Simulations of this particular design suggest that the circuit functioned properly when subjected to an out-of-plane field where H_{pulse} was approximately $\pm 70\,\text{mT}$, and T_{pulse} was 15 ns.

21.3.2.5 Benchmarking

While a detailed discussion is beyond the scope of this document we do note that projected area–energy–delay projects could be 50–130× better than CMOS functional equivalents (even when inherent pipelining is introduced into CMOS designs) [59]. Similarly, the energy associated with pNML-based designs could be one or two orders of magnitude better than CMOS functional equivalents at iso-performance [49,60].

21.4 Alternative Circuit Design Techniques and Architectures

Here, we consider more "unconventional' approaches for processing information with the NML device architecture. Specifically, approaches for, and the impact of using NML to realize threshold logic gates (Section 21.4.1), nonBoolean information processing hardware (Section 21.4.2), and cellular automata (Section 21.4.3) are all discussed. Again, comparisons to CMOS equivalents are made whenever appropriate/possible.

21.4.1 Threshold Logic

We believe that devices with PMA (i.e., pNML) could be especially well-suited for realizing threshold logic circuits. As an example, we consider how an adder comprised of multiple layers of pNML-based devices might fare given the context of the most recent NRI benchmarking effort.

The design (Figure 21.9) is based on the two majority gate *threshold adder* from [61]. The carry output (C_{out}) can be determined simply by performing the majority voting function $M(A, B, C_{\text{in}})$ that is native to NML [13]. The sum output can then be realized by a five-input majority voter: $M(A, B, C_{\text{in}}, C'_{\text{out}}, C'_{\text{out}})$. With the in-plane device architecture, the single domain limit necessitates that device footprint be relatively small which limits fan-in. pNML devices can not only be larger (increasing fan-in), but can also couple in multiple dimensions. Note that in Figure 21.9, the AF-coupling between the A, B, and C inputs and the C_{out} target would result in the generation of C'_{out}. Additionally, by placing the Sum target *above* A, B, C_{in}, and C'_{out}, the correct signal values will be fed to Sum.[3] (The Sum target is sized such that the fringing field magnitude from C_{out} is two times that of A, B, or C_{in}.) As such, a full adder can be realized with just five devices. If we assume that $F = 15\,\text{nm}$ (as is done in the most recent NRI benchmarking), with this approach, a 32-bit adder could be ~45× smaller than an iNML-based design and ~145× smaller than a CMOS design.

---

[3] As the sum signal need not propagate, the sum magnet could be thinner to reduce interference with C_{out}.

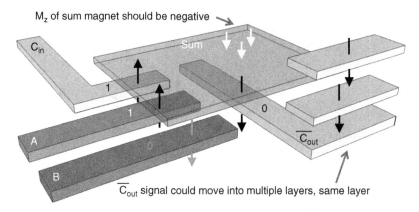

Figure 21.9 Schematic of integrated adder cell using pNML. (Assuming "0" maps to ↓, and "1" maps to ↑, the magnetization state for input combination $A = 1$, $B = 0$, $C_{in} = 1$ is shown.) Note that the C_{out} magnet (with a complemented value) could move a signal up in the vertical direction to another level, or within the same plane

Moreover, as the carry out signal and sum signals could be calculated simultaneously (see Figure 21.10b—generated via the code associated with [56]), a single, 32-bit add could be completed in the time required for 33 magnet switching events (a 900% reduction compare to iNML). as per Figure 21.10a, this could make the power-constrained throughput of

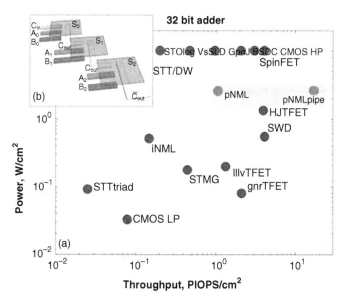

Figure 21.10 (a) Multi-layer pNML circuits offer the promise of high computational throughput (in the context of existing NRI benchmarking work). Note that power projections are based on existing NRI data for magnetostrictive switching. PIOPS/cm^2 refers to the number of peta-integer operations per cm^2. (b) Schematic showing concatenated bits. In a pipelined circuit, a new result could be completed every two clock cycles after the pipeline is filled

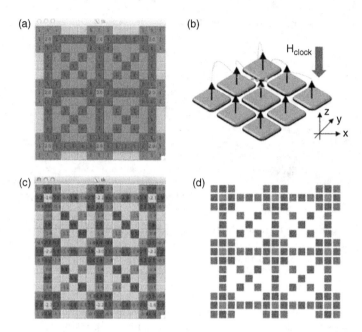

Figure 21.11 (a) Arrangement of pNML devices for edge detection. (b) Fringing fields and a clock field influence which pNML device would switch first – here, a middle device experiences the highest downward directed field $(-H_z)$. (c) Intermediate state of system shown in Figure 21.11a assuming downward directed field. (d) Micromagnetic simulations (at $0\,K$) very similar to projected states in Figure 21.11c. (Minor differences most likely due to direction of nominal x, y-directed biasing fields)

pNML-based hardware highly competitive with respect to other NRI technologies (without even considering any additional benefits from inherent pipelining associated with NML).

21.4.2 NonBoolean Computing

We also believe that NML devices could be used to create nonBoolean information processing systems. As an example, we summarize work presented in [60] that considers an arrangement of pNML devices that could be used for image edge detection.[4]

To begin, consider the arrangements of pNML devices shown in Figure 21.11a. Red boxes represent devices that are magnetized up $(+M_z)$. Later, blue boxes will represent devices that are magnetized down $(-M_z)$. Thus, to make an analogy to a black and white image, one could equate a $-M_z$ magnetization state of an input device to a white pixel, and a $+M_z$ magnetization state to a black pixel.) Input devices could presumably be set via STT [62], and are indicated by green boxes. While pNML devices are assumed, we assume that no dots are FIB irradiated. Therefore, in Figure 21.11a, we might assume that all pNML devices (in both the "interconnect" as well as the inputs) were initially magnetized such that their z-component of magnetization was positive (i.e., by an external field, voltage-based clock, and/or STT.). Additionally, this suggests that a 3×3 black square is encoded by this magnet arrangement.

---

[4] Here a design abstraction that captures how pNML devices switch when subjected to a clocking field is used for this discussion. More recently, micromagnetic simulations (at $0\,K$) also confirm the desired behavior.

Now, assume that we subject this system to an out-of-plane field that is directionally opposite to the magnetization state of "interconnect" magnets (per Figure 21.11a, a negative H_z field would be applied), and that gradually increases in magnitude. One would expect pNML devices that have more neighbors with a $+z$ component of magnetization to switch first as fringing fields from neighboring devices – also directed in the $-z$ direction – influence switching as well. (As an example, looking at Figure 21.11b, the middle device should switch first as it influenced by both the applied external field as well as neighboring fringing fields.)

In initial studies, we assumed that a given magnet is influenced by devices that are two "squares" away – that is, devices that are closer to a given device will have more influence over it than those that are farther away. (Practically speaking, the pNML design space is large, and the influence of particular device could be "tuned" by changing dot size, material, spacing between dots, irradiating dots, or combinations thereof.) We also assume that the state of the input dots does not change during the switching process – which could be achieved by varying device size, limited FIB irradiation, or combinations thereof. Thus, given different assumptions regarding neighbor to neighbor influence, we calculated how susceptible a given device is to switching when subjected to an out-of-plane field.

If we assume: (i) that the influence of a given magnet falls of as $1/r^2$, (ii) the arrangement of magnets (with the initial state) shown in Figure 21.11a, and (iii) that this magnet ensemble is subjected to a field that is directed in the opposite direction to the initial magnetization state of the devices, the final state of the system is predicted to be similar to that shown in Figure 21.11c.[5]

Note that the number of cells that surround the center input devi`ce that were initially magnetized down (opposite to the direction of the applied field) is *lower* than the number of cells that surround devices that are now magnetized up. However, the number of devices that surround the input devices on the "edge" of the 3×3 black square is lower (i.e., for the center pixel, 6 of the 8 surrounding devices have changed state, while for the edge pixels, only 4 or 5 of the surrounding devices have changed state). If one could measure the resistance of the input and its surrounding cells, one could detect the edges in a black and white image with: (i) a short field pulse per the discussion above and (ii) resistance measurements.[6] Therefore, even though NML devices have longer switching latencies than a CMOS transistor, an edge detection operation could be accomplished with just a single field pulse and simultaneous reads. The devices in the system "evolve" in parallel.

Notably, micromagnetic simulations suggest that the magnet ensemble described above will evolve almost as predicted (see Figure 21.11d). (Minor differences between Figure 21.11c and Figure 21.11d most likely result from the direction of the nominal x- and y-directed biasing fields that initiate a torque on a given pNML device.) That said, micromagnetic simulations were performed at 0 K, and multiple devices can switch simultaneously as appropriate (e.g., due to the influence of equally weighted neighbors in this environment.) However, we will ultimately need to consider how thermal noise [12] impacts ordered switching. For example, if one cell switches first, it may impact another cell that should switch at the same time, which could impact final system state.

---

[5] Once a device has changed state such that its z component of magnetization is parallel to that of the applied clock field, it should not switch back.

[6] Again, considering thermal noise, the outcome of an edge detection operation would not be impacted by switching order provided that a majority/minority of cells around black and white pixels could be obtained.

21.4.3 Cellular Automata

Cellular Automata (CA) is a classic and early example of computationally universal non-Boolean computing and ancestor of many modern cellular architectures (such as CNNs). It displays two highly desirable architectural properties of a realizable nano-scale computing system: high-level of parallelism, and nearest neighbor interconnections [63].

In the first step of a CA computation, each computing units (cells) of a CA have to initialized to a certain (possibly digital) value – the initialization pattern can be looked as an input image. Once initialized, the cell's state evolves in discrete time according to a logical rule and the result of the computation is read out after a number of computing steps elapsed. This computing scheme clearly lends itself to a natural realization in pNML architectures.

As a case study for a two-dimensional pNML-based cellular automata, we implemented an image filtering algorithm as described in [64]. Noise of a black and white image can be reduced by: (i) assigning a cell for each pixel of the image, (ii) in every CA iteration, each cell assumes the pixel value of the majority of its neighbors and itself, and (iii) the iteration is repeated for several (5–15) steps until the noise disappears but the image still retains most of its useful information. In these scheme each cell requires an external input/output (for initialization and read-out) and interconnections to the neighboring cells (to determine their majority). The results of the filtering similar to the ones obtainable by a standard Gaussian filter.

A pNML implementation we designed is shown in Figure 21.12. The design does not require any wire crossings and it is entirely tileable, that is, an arbitrary number of cells can be placed and the output (input) of each cell can directly contact the input (output) of its neighbors. Details of the design and simulation results are given in [65].

In order to arrive to this simple design, we slightly changed the used CA rule of [64] and a slightly different rule is applied to even and odd-numbered rows. For even-numbered rows the updated value of the state variable of cell i, j is:

$$\text{Next}_l(i,j) = \text{maj}[I(i,j) + 2^*I(i+1,j) + I(i,j-1) + I(i-1,j)]$$

and for rows with odd indices:

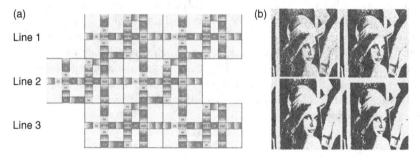

Figure 21.12 (a) Six cells of a noise-filtering CA, realized from pNML dots. The lighter side of the dots denotes their irradiated side. Note that even-index and odd-index lines obey slightly different rules and they are shifted with respect to each other, so the design is tillable. (b) Image processing results using the modified CA rule: the original image and the processed image is shown after four, eight and 12 iterations, respectively

$$\text{Next}_1(i,j) = \text{maj}[I(i,j) + 2^*I(i+1,j-1) + I(i,j-1) + I(i-1,j-1)]$$

where i, j are the column and row indices of the cell.

These cell rules ignore one neighbor (out of the four) and take into account another one with a double weight. We found that although it is a different CA rule from the one described in [64], it performs the filtering operation equally well: as it is shown in the images of Figure 21.12b. The noisy original image clears up after four, eight, and 12 iterations. The overhead of inputting and reading out the image amortized over these number of iteration steps.

Signal race conditions may cause undesired operation of this CA (or any other CA operation): depending on the data being processed some cells may pass data to neighboring cells "too early" and the computing units of the CA fall out of synchrony. To prevent this, the designs needs "transfer dots," that enable the flow of magnetic information only after each eighth clocking cycles, where all cell outputs are ready. These transfer dots are labeled as "in" in Figure 21.12a.

21.5 Retrospective, Future Challenges, and Future Research Directions

Here, we briefly review what we believe are important issues and paths to pursue in order for NML/QCA-like devices to become a viable information processing technology.

21.5.1 Retrospective: Experimental Progress to Date

When compared to other emerging information processing technologies, we believe that there has been significant progress with both iNML and pNML on the experimental front. For *iNML*, fanout structures [21], a one-bit full adder [27], and so on have been experimentally demonstrated and successfully re-evaluated with new inputs. Both field-coupled [28,66] and spin transfer torque (STT) [26,29,30] electrical inputs have been realized. For electrical output, multiple iNML-magnetic tunnel junction hybrids [33] have been proposed and simulated.

Field-based, CMOS compatible line clock structures have been used to simultaneously switch the states of multiple magnetic islands [38], as well as to re-evaluate iNML lines and gates with new inputs [22]. Additionally, recent experiments have considered materials-based solutions to further reduce field/energy requirements for line clock structures. Notably, results from [37,39] suggest that the component energy metrics could be further reduced by as much as 16×. Alternative clocking mechanisms – for example, via multiferroics [41], magnetostriction [24], or the Spin Hall Effect [67] – might also be paths to further reductions in clock energy.

Experimental results suggest that appropriately irradiated *pNML* structures (to define dataflow directionality) can be controlled by a uniform, homogeneous, oscillating, global clock field [46]. Majority gates [47], AF-lines [46], multi-plane signal crossings [68], and full adders [48,69] have all been experimentally demonstrated with this clocking approach. Field coupled inputs have also been demonstrated [70].

Looking to chip-level implementations, large arrays of pNML devices could be controlled with on-chip inductor structures [49] that could be coupled with a capacitance in an LC oscillator – which opens the door to adiabatic energy recycling. Devices with PMA are also amenable to voltage-controlled clocking [50].

That said, for both device architectures, challenges still remain – particularly with respect to efficient interconnect, reliable device switching, and fault tolerant architectures.

21.5.2 Challenge 1: Efficient Interconnect

One challenge associated with processing information using devices that communicate via fringing fields is interconnect. Signal routing and signal propagation times can be adversely affected. Devices with PMA can help to alleviate the aforementioned issues. With respect to signal routing, the large single-domain size in pNML enables the use of long (several micrometer) domain walls for information transfer. Moreover, per [63] CoPt domain walls *can cross above/ below one other*, and conducing stripes can cross a 30 nm vertical "step" without unwanted pinning. Moreover, simulations suggest that by creating a nucleation center where two stripes cross, the interaction can be switched on, and one of the walls can *stop or pass* the other wall depending on their relative orientation. (Conditional passing of domain walls may even be used to realize PLAs.) The benefits of NML/domain-wall logic hybrids are threefold: (i) the footprint of domain wall-based devices is reduced and electrical components are necessary only at the inputs and outputs, (ii) interconnect complexity and delay of dot-based interconnections in NML is also reduced, and (iii) more reliable signal transmission may ensue.

Regarding signal propagation times (as well as signal routing), consider an array multiplier – where bits of data word A must be distributed to processing elements in each column of the multiplier, and the bits of data word B must be distributed to processing elements (PE) in each row of the multiplier. If one were to move this information to individual PEs via AF-lines, both signal routing and communication delay would be adversely affected. However, domain wall motion in magnetic stripes above and below NML processing elements could be used to preprogram PEs with the A and B input bits required – vastly simplifying design layout. (The work presented in [71] could be considered to be a preliminary, experimental proof of concept.)

21.5.3 Challenge 2: Error Tolerance

Any circuit and/or architecture comprised of NML devices must ultimately be tolerant to fabrication variations. Consider an iNML device with a symmetric, rounded-rectangle shape. A given device will be in the highest energy state if magnetized along its hard axis with no applied field (see Figure 21.12). However, lower energy states are equivalent and occur at $\pm 90\,°$. [With no external stimulus to keep such a device in a hard axis-biased state, thermal noise, minor fabrication variations, etc. should determine whether or not it relaxes such that its y-component of magnetization is positive ($\uparrow$) or negative ($\downarrow$).] By changing a magnet's geometry, its energy landscape can change as well. More specifically, if the edge of a magnet becomes "slanted," the highest-energy state does *not* occur when a device is magnetized along its (geometrically) hard axis. Rather, if biased along its geometrically hard axis, a device is *already* on one side of the potential barrier and should always relax such that the sign of the y-component of magnetization (i.e., its binary state) is always the same – see Figure 21.13. If devices are sufficiently large, we can exploit the aforementioned energy barrier shift to realize nonmajority gate, two-input AND/OR logic [55]. However, as device sizes scale, the effects of minor fabrication variations on device switching behavior can become more pronounced. If energy barrier shifts like those described above occur, a stuck at fault can ensue.

In pNML, switching field distributions (SFD; from defects formed during layer growth) can impact operation. The width of SFD strongly depends on layer deposition conditions, post-growth annealing and the applied FIB dose. Thermally induced SFD also have a strong effect. Initial studies show that coupling fields are sufficiently strong to ensure device operation in the presence of SFD [72].

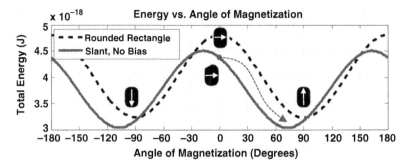

Figure 21.13   Total energy curves for slant-edge and rounded rectangle magnet – device with slanted edge already on one side of barrier in 0° state

21.5.4   Paths Forward

While NML circuits show benefits over CMOS-based logic, a "killer application" has yet to emerge. We envision that NML circuits may become especially useful in applications, which require close integration of data storage and processing functions. The combination of dense, nonvolatile, possibly three-dimensional storage capabilities of coupled nanomagnet arrays with processing capabilities (such as the systolic pattern matching or image processing) may serve as hardware for future intelligent storage devices.

A challenge with scaled NML is unwanted bit flips that can occur during the switching process due to thermal noise, lithographic variation, and so on. At the device level, improved device to device couplings [73,74] and devices with biaxial anisotropy [23] have been suggested as possible mechanisms to improve the reliability of an NML-based computation. However, even magnet ensembles that employ devices with biaxial anisotropy can be error prone [54]. Thus, fault tolerant architectures should be studied and developed. Of initial interest are stochastic computer architectures where the value of a signal is interpreted as a probability regardless of normal or faulty conditions. Stochastic computation is naturally fault tolerant since a small number of bit flips only result in small deviations from the desired value. While computation time could increase as the precision of stochastic values increases, given the inherently pipelined, streaming dataflow associated with NML, more efficient implementation of stochastic computation architectures should be possible with NML devices (e.g., as one of the main drawbacks of stochastic computation is the rapid growth of computation time with respect to precision).

References

1. Lent, C.S., Tougaw, P.D., Porod, W., and Bernstein, G.H. (1993) Quantum cellular automata. *Nanotechnology*, **4**, 49–57.
2. Orlov, A.O., Amlani, I., Bernstein, G.H. *et al.* (1997) Realization of a functional cell for quantum-dot cellular automata. *Science*, **277**, 928–930.
3. Amlani, I., Orlov, A.O., Toth, G. *et al.* (1999) Digital logic gate using quantum-dot cellular automata. *Science*, **284**, 289–291.
4. Orlov, A.O., Kummamuru, R.K., Ramasubramaniam, R. *et al.* (2001) Experimental demonstration of a latch in clocked Quantum-dot cellular automata. *Applied Physics Letters*, **78**, 1625–1627.
5. Kummamuru, R.V., Timler, J., Toth, G. *et al.* (2002) Power gain in a quantum-dot cellular automata latch. *Applied Physics Letters*, **81**, 1332–1334.

6. Yadavalli, K.K., Orlov, A.O., Timler, J.P. *et al.* (2007) Fanout gate in quantum-dot cellular automata. *Nanotechnology*, **18**, 375401.
7. Jiao, J., Long, G.J., Grandjean, F. *et al.* (2003) Building blocks for the molecular expression of quantum cellular automata: Isolation and characterization of a covalently bonded square array of two ferrocenium and two ferrocene complexes. *Journal of the American Chemical Society*, **125**, 7522–7523.
8. Lent, C.S. and Tougaw, P.D. (1997) A device architecture for computing with quantum dots. *Proceedings of the IEEE*, **84**, 541–557.
9. Qi, H., Sharma, S., Li, Z. *et al.* (2003) Molecular Quantum Cellular Automata cells: electric field driven switching of a silicon surface bound array of vertically oriented two-dot molecular quantum cellular automata. *Journal of the American Chemical Society*, **125**, 15250–15259.
10. Haider, M.B., Pitters, J.L., DiLabio, G.A. *et al.* (2009) Controlled coupling and occupation of silicon atomic quantum dots at room temperature. *Physical Review Letters*, **102**, 046805.
11. Mitic, M., Cassidy, M.C., Petersson, K.D. *et al.* (2006) Demonstration of a silicon-based quantum cellular automata cell. *Applied Physics Letters*, **89**, 013503-3.
12. ITRS (2012) International Technology Roadmap for Semiconductors. Available www.itrs.net.
13. Imre, A., Csaba, G., Ji, L. *et al.* (2006) Majority logic gate for magnetic quantum-dot cellular automata. *Science*, **311**, 205–208.
14. Cowburn, R.P. and Welland, M.E. (2000) Room temperature magnetic quantum cellular automata. *Science*, **287**, 1466–1468.
15. Bernstein, G.H., Imre, A., Metlushko, V. *et al.* (2005) Magnetic QCA systems. *Microelectronics Journal*, **36**, 619–624.
16. Waser, R. (2003) *Nanoelectronics and information technology: Advanced electronic materials and novel devices*, Wiley-VCH, Weinheim.
17. Kurtz, S., Varga, E., Niemier, M. *et al.* (2011) Two input, non-majority magnetic logic gates: experimental demonstration and future prospects. *Journal of Physics: Condensed Matter*, **23**, 053202.
18. Varga, E., Siddiq, M., Niemier, M.T. *et al.* (2010) Experimental Demonstration of Non-Majority, Nanomagnet Logic Gates, Device Research Conference, pp. 87–88.
19. Varga, E., Niemier, M.T., Bernstein, G.H. *et al.* (June 22 2009) Non-volatile and Reprogrammable MQCA-based Majority Gates, Device Research Conference, pp. 1–2.
20. Varga, E., Liu, S., Niemier, M.T. *et al.* (2010) Experimental Demonstration of Fanout for Nanomagnet Logic, Device Research Conference, Notre Dame, IN, pp. 95–96.
21. Varga, E., Orlov, A., Niemier, M.T. *et al.* (2010) Experimental demonstration of fanout for nanomagnetic logic. *IEEE Transactions on Nanotechnology*, **9**, 668–670.
22. Alam, M.T., Kurtz, S., Siddiq, M.J. *et al.* (2012) On-chip clocking of nanomagnet logic lines and gates. *IEEE Transactions on Nanotechnology*, **11**, 273–286.
23. Carlton, D.B., Emley, N.C., Tuchfeld, E., and Bokor, J. (2008) Simulation Studies of Nanomagnet-Based Logic Architecture. *Nano Letters*, **8**, 4173–4178.
24. Salehi, F.M., Roy, K., Atulasimha, J., and Bandyopadhyay, S. (2011) Magnetization dynamics, Bennett clocking and associated energy dissipation in multiferroic logic. *Nanotechnology*, **22**, 155201.
25. Pulecio, J.F. and Bhanja, S. (2010) Magnetic cellular automata coplanar cross wire systems. *Journal of APhys.*, **107**, 034308.
26. Lyle, A., Harms, J., Klein, T. *et al.* (2012) Spin transfer torque programming dipole coupled nanomagnet arrays. *Applied Physics Letters*, **100**, 012402-3.
27. Varga, E., Niemier, M.T., Csaba, G. *et al.* (2013) Experimental Realization of a Nanomagnet Full Adder Using Slanted-Edge Input Magnets, INTERMAG/MMM, Chicago, IL.
28. Siddiq, M.A.J., Niemier, M.T., Bernstein, G.H. *et al.* (2013) A Field Coupled Electrical Input for Nanomagnet Logic, accepted for publication in IEEE Transactions on Nanotechnology.
29. Lyle, A., Harms, J., Klein, T. *et al.* (2011) Integration of spintronic interface for nanomagnetic arrays. *AIP Advances*, **1**, 042177 11.
30. Lyle, A., Klemm, A., Harms, J. *et al.* (2011) Probing dipole coupled nanomagnets using magnetoresistance read. *Journal of Applied Physics*, **98**, 092502.
31. Liu, S., Hu, X.S., Nahas, J. *et al.* (2011) Design and optimization of magnetic-electrical interfaces for NML circuit output, Design Automation Conference, Work-in-progress session.
32. Liu, S., Hu, X.S., Nahas, J.J. *et al.* (2011) Magnetic-electrical interface for nanomagnet logic. *IEEE Transactions on Nanotechnology*, **10**, 757–763.

33. Liu, S., Hu, X.S., Niemier, M.T. *et al.* (2013) A design space exploration of the magnetic-electrical interfaces for nanomagnet logic. *IEEE Transactions on Nanotechnology*, **12**, 203–214.
34. Niemier, M.T., Hu, X.S., Alam, M. *et al.* (2007) Clocking Structures and Power Analysis for Nanomagnet-Based Logic Devices, International Symposium on Low Power Elec. and Design (ISLPED), Portland, OR, pp. 26–31.
35. Dingler, A., Niemier, M.T., Hu, X.S., and Lent, E. (2011) Performance and Energy Impact on Locally Controlled NML Circuits. *ACM Journal on Emerging Technologies in Computing*, **7**, 1–24.
36. Dingler, A., Niemier, M., Hu, X.S. *et al.* (2009) System-Level Energy and Performance Projections for Nano-magnet-based Logic, 2009 IEEE/ACM International Symposium on Nanoscale Architectures (NANOARCH), pp. 21–26.
37. Li, P., Csaba, G., Sankar, V.K. *et al.* (2012) Power Reduction in Nanomagnet Logic Clocking through High Permeability Dielectrics, Device Research Conference, State College, Pennsylvania, pp. 129–130.
38. Alam, M.T., Siddiq, M.J., Bernstein, G.H. *et al.* (2010) On-chip clocking for nanomagnet logic devices. *IEEE Transactions on Nanotechnology*, **9**, 348–351.
39. Li, P., Csaba, G., Sankar, V.K. *et al.* (2013) Paths to Clock Power Reduction via High Permeability Dielectrics for Nanomagnet Logic Circuits, Joint MMM/Intermag Conference, Chicago, IL.
40. Li, P., Sankar, V.K., Csaba, G. *et al.* (2012) Magnetic properties of enhanced permeability dielectrics for NML circuits. *IEEE Transactions on Magnetics*, **48**, 3292–3295.
41. Chu, Y.H., Martin, L.W., Holcomb, M.B. *et al.* (2008) Electric-field control of local ferromagnetism using a magnetoelectric multiferroic. *Nature Materials*, **7**, 478–482.
42. Crocker, M., Hu, X.S., and Niemier, M. (2010) Design and Comparison of NML Systolic Architectures, 2010 IEEE/ACM International Symposium on Nanoscale Architectures (NANOARCH), pp. 29–34.
43. Becherer, M., Kiermaier, J., Breitkreutz, S. *et al.* (2010) On-chip Extraordinary Hall-effect sensors for characterization of nanomagnetic logic devices. *Solid State Electronics*, **54**, 1027–1032.
44. Shaw, J.M., Russek, S.E., Thomson, T. *et al.* (2008) Reversal mechanisms in perpendicularly magnetized nanostructures. *Physical Review B*, **78**, 024414.
45. Ju, X., Wartenburg, S., Rezgani, J. *et al.* (2012) Nanomagnet logic from partially irradiated Co/Pt nanomagnets. *IEEE Transactions on Nanotechnology*, **11**, 97–104.
46. Eichwald, I., Bartel, A., Kiermaier, J. *et al.* (2012) Nanomagnet Logic: error-free directed signal transmission by an inverter chain. *IEEE Transactions on Magnetics*, **48**, 4332–4335.
47. Breitkreutz, S., Kiermaier, J., Eichwald, I. *et al.* (2012) Majority gate for nanomagnetic logic with perpendicular magentic anisotropy. *IEEE Transactions on Magnetics*, **48**, 4336–4339.
48. Breitkreutz, S., Kiermaier, J., Eichwald, I. *et al.* (2013) Experimental Demonstration of a 1-bit Full Adder in Perpendicular Nanomagnetic Logic, *to appear in IEEE Transactions on Magnetics*, vol. 49.
49. Ju, X., Niemier, M., Becherer, M. *et al.* (2013) Systolic pattern matching hardware with out-of-plane nanomagnet logic devices. *IEEE Transactions on Nanotechnology*, **12**, 399–407.
50. Wang, W.-G., Li, M., Hageman, S., and Chien, C.L. (2012) Electric-field-assisted switching in magnetic tunnel junctions. *Nature Materials*, **11**, 64–68 01//print.
51. Dingler, A., Kurtz, S., Niemier, M. *et al.* (2012) Making Non-Volatile Nanomagnet Logic Non-Volatile, Design Automation Conference (DAC), pp. 476–485.
52. Becherer, M., Csaba, G., Emling, R. *et al.* (2009) Field-coupled Nanomagnets for Interconnect-Free Nonvolatile Computing, International Solid-State Circuits Conference, ISSCC, pp. 474–475.
53. Hellwig, O., Berger, A., Kortright, J.B., and Fullerton, E.E. (2007) Domain structure and magnetization reversal of antiferromagnetically coupled perpendicular anisotropy films. *Journal of Magnetism and Magnetic Materials*, **319**, 13–55.
54. Spedalieri, F.M., Jacob, A.P., Nikonov, D.E., and Roychowdhury, V.P. (2011) Performance of magnetic quantum cellular automata and limitations due to thermal noise. *IEEE Transactions on Nanotechnology*, **10**, 537–546.
55. Niemier, M.T., Bernstein, G.H., Csaba, G. *et al.* (2011) Nanomagnet logic: Progress toward system-level integration. *Journal of Physics: Condensed Matter*, **23**, 493202.
56. Nikonov, D. E. and Young, I. A. (2013) Overview of Beyond-CMOS Devices and a Uniform Methodology for Their Benchmarking, *Proceedings of the IEEE*, **101**(12), 2498–2533.
57. Kung, H.T. (1982) Why systolic architectures? *Computer*, **15**, 37–46.
58. Ju, X., Niemier, M.T., Becherer, M. *et al.* (2012) Design of a Systolic Pattern Matcher for Nanomagnet Logic, IWCE 2012 (International Workshop on Computational Electronics, University of Wisconsin-Madison, May 2012).
59. Bernstein, G., Csaba, G., Hu, X.S. *et al.* (2012) Nanomagnet Logic, NRI Annual Review, Gaithersburg, Maryland.

60. Niemier, M., Ju, X., Becherer, M. *et al.* (2012) Boolean and Non-Boolean Architectures for Out-of-Plane Nanomagnet Logic, Procedings of the International Workshop on Cellular Nanoscale Networks and their Applications, August 29–31 2012, pp. 1–6.

61. Lageweg, C., Cotofana, S., and Vassiliadis, S. (2002) A full adder implementation using SET based linear threshold gates, Electronics, Circuits and Systems, 2002. 9th International Conference on, vol. 2. pp. 665–668.

62. Mangin, S., Ravelosona, D., Katine, J.A. *et al.* (2006) Current-induced magnetization reversal in nanopillars with perpendicular anisotropy. *Nature Materials*, **5**, 210–215.

63. Toffoli, T. and Margolus, N. (1987) *Cellular Automata Machines: A New Environment for Modeling*, MIT press.

64. Popovici, A. and Popovici, D. (2002) Cellular Automata in Image Processing, Fifteenth International Symposium on Mathematical Theory of Networks and Systems.

65. Haughan, K. (2013) Cellular Automata Designs for Out of Plane Nanomagnetic Logic, to be submitted for IEEE Conference on Nanotechnology.

66. Siddiq, M.A., Niemier, M., Csaba, G. *et al.* (2013). Demonstration of Field Coupled Input Scheme on Line of Nanomagnets, *accepted in IEEE Transactions on Magnetics*.

67. Bhowmik, D., Long, Y., and Salahuddin, S. (2012) Possible route to low current, high speed, dynamic switching in a perpendicular anisotropy CoFeB-MgO junction using Spin Hall Effect of Ta, Electron Devices Meeting (IEDM), 2012 IEEE International. pp. 29.7.1–29.7.4.

68. Eichwald, I., Wu, J., Kiermaier, J. *et al.* (2013) Towards a Signal Crossing in double-layer Nanomagnetic Logic, *to appear in IEEE Transactions on Magnetics*, vol. 49.

69. Breitkreutz, S., Eichwald, I., Kiermaier, J. *et al.* (2013) 1-Bit Full Adder in Perpendicular Nanomagnet Logic using a Novel 5-Input Majority Gate, accepted at the Joint European Magnetic Symposium (JEMS), Rhodos, Greece, August 25–30 2013.

70. Kiermaier, J., Breitkreutz, S., Csaba, G. *et al.* (2012) Electrical input structures for nanomagnetic logic devices. *Journal of Applied Physics*, **111, 07E341** 3.

71. Varga, E., Csaba, G., Bernstein, G.H., and Porod, W. (2012) Domain-wall assisted switching of single-domain nanomagnets. *Magnetics, IEEE Transactions on*, **48**, 3563–3566.

72. Breitkreutz, S., Kiermaier, J., Ju, X. *et al.* (2011) Nanomagnetic Logic: Demonstration of Directed Signal Flow for Field-coupled Computing Devices, ESSDERC Helsinki, Finnland.

73. Breitkreutz, S., Kiermaier, J., Yilmaz, C. *et al.* (2011) Nanomagnetic logic: compact modeling of field-coupled computing devices for system investigations. *Journal of Computational Electronics*, **10**, 352–359.

74. Csaba, G. and Porod, W. (2010) Behavior of Nanomagnet Logic in the presence of thermal noise, Computational Electronics (IWCE), 2010 14th International Workshop on, pp. 1–4.

22

Explorations in Morphic Architectures

Tetsuya Asai[1] and Ferdinand Peper[2]
[1]*Graduate School of Information Science and Technology, Hokkaido University, Japan*
[2]*Center for Information and Neural Networks, National Institute of Information and Communications Technology, USA*

22.1 Introduction

Biological systems give us examples of amorphous, unstructured devices capable of noise- and fault-tolerant information processing. They excel in massively parallel spatial problems, as opposed to digital processors, which are rather weak in that area. The *morphic* architecture was thus introduced in the Emerging Research Architecture section of ITRS 2007, to refer to biologically inspired architectures that embody a new kind of computation paradigm in which adaptation plays a key role to effectively address the particulars of problems [1]. This chapter focuses on recent progress of two morphic architectures that offer opportunities for emerging nanoelectronic devices: *neuromorphic* architectures and *cellular automaton* architectures.

22.2 Neuromorphic Architectures

The term *neuromorphic* was introduced by Carver Mead in the late 1980s to describe VLSI systems containing electronic analog circuits that mimic neuro-biological architectures in the nervous system [2]. Traditional neurocomputers employ components that are biologically rather implausible, like static threshold elements that represent neurons, whereas neuromorphic architectures are closer to biology. An example of a neuromorphic VLSI system is the silicon retina [3] (Figure 22.1) that is modeled after the neuronal structures of the vertebrate retina.

The appeal of neuromorphic architectures lies in: (i) their potential to achieve (human-like) intelligence based on the unreliable devices typically found in neuronal tissue, (ii) their strategies to deal with anomalies, emphasizing not only tolerance to noise and faults, but also the active exploitation of noise to increase the effectiveness of operations, and (iii) their potential for low-power operation. Traditional von Neumann machines are less suitable with

Emerging Nanoelectronic Devices, First Edition. An Chen, James Hutchby, Victor Zhirnov and George Bourianoff.
© 2015 John Wiley & Sons, Ltd. Published 2015 by John Wiley & Sons, Ltd.

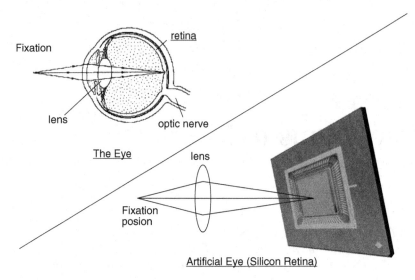

Figure 22.1   Example of neuromorphic chips (silicon retina)

regard to item (i), since for this type of tasks they require a machine complexity (the number of gates and computational power), that tends to increase exponentially with the complexity of the environment (the size of the input). Neuromorphic systems, on the other hand, exhibit a more gradual increase of their machine complexity with respect to the environmental complexity [4] (Figure 22.2). Therefore, at the level of human-like computing tasks, neuromorphic machines

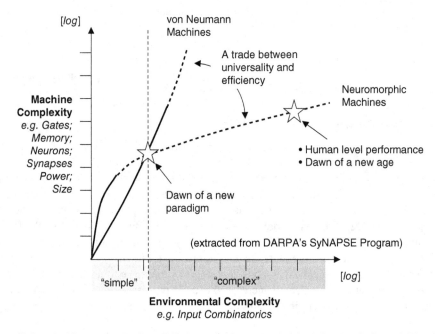

Figure 22.2   Machine versus Environmental complexity (extracted from DARPA's SyNAPSE program [4])

Table 22.1 Applications and development of neuromorphic systems

Application category		Proposed architectures
Information processing		Associative memory (CMOS [7], SET [8,9]), data mining and Inference (CMOS) [10], winner takes all selection (CMOS) [11], noise-driven computing (see Table 22.2), locomotion controller (CMOS) [12]
Intelligent sensors	Vision	Edge detection (CMOS [13], SET [14]), motion detection (CMOS [15], SET [16]), stereo vision (CMOS) [17], visual tracking (CMOS) [18], adaptive gain control (CMOS) [19], orientation selection (CMOS) [20], high-speed sensors (CMOS) [21]
	Others	Band-pass filtering (CMOS) [22], echolocation (CMOS) [23], noise canceling and figure–ground separation in auditory systems (CMOS) [24], olfactory systems (CMOS) [25]
Artificial life		Reaction–diffusion computers (CMOS, SET) [26], artificial fish [27] and octopus [28] (CMOS)
Technology		CrossNets [29], address–event representation [30], CDMA neural networks (CMOS) [31], artificial neurons (CMOS [32], SET [33]), memristive synapses [34,35], 3D implementation [36], brain–machine interface [37]

have the potential to be superior to von Neumann machines. Points (ii) and (iii) are strongly related to each other in von Neumann machines, since tolerance to noise runs counter to lowering power supply voltage and bias currents. Neuromorphic architectures, on the other hand, suffer much less from this trade-off. Unlike von Neumann machines, which can correct bit errors only to a certain extent through the use of error control techniques, neuromorphic machines tend to keep working even under high error rates.

Like the areas of the human brain, neuromorphic machines (LSIs) are application-specific. Significant performance benefits can be achieved by employing them as supplemental CMOS, as an addition to a von Neumann system, which provides universal computation ability. Neuromorphic systems thus have a diversified functionality, and consequently can be categorized as more than Moore candidates. Table 22.1 shows trends in the development of neuromorphic systems and their applications. The application category "Information processing" is only a single item in the table, which understates the potentially large benefits in terms of machine complexity of human-like information processing, since functionalities like prediction, associative memory, and inference offer new opportunities require huge machine complexity if they were implemented by von Neumann architectures. The Emerging Research Architecture section of ITRS 2009, for example, introduced an inference engine based on Bayesian neural networks [5], and in 2010 Lyric Semiconductor introduced NAND gates where the input probabilities are combined using Bayesian logic rather than the binary logic of conventional processors. Lyric Error Correction (LEC) uses this probabilistic logic to perform error detection and correction with about 3% of the circuitry and 8% of the power that would be needed for the equivalent conventional error correction scheme [6].

ITRS 2007 (Emerging Research Architecture section) did not address neuromorphic "Intelligent sensors," since they were considered ancillary to the central focus on information processing of ITRS at the time. However, intelligent sensors have been re-spotlighted, because there exist vast opportunities for high-performance architectures that combine them with emerging nanoelectronic devices. So far CMOS implementations (in the items "Vision" and "Others"; see Table 22.1) and blueprints of SET implementations (in item "Vision") have been proposed.

Another approach to build neuromorphic systems is inspired by biochemical reactions in living organisms. Reaction–diffusion computers, for example, are based on a biochemical reaction named after Belousov and Zhabotinsky [38], and they are able to efficiently solve combinatorial problems through the use of natural parallelism. Electronic implementations of this type of information processing require strong nonlinear I-V characteristics to mimic the chemical reactions involved, and there exist many opportunities for emerging nanoelectronic devices in this respect.

On the technology side, one of the key issues for neuromorphic systems is how neuronal elements are implemented. An important consideration in this respect is the level of abstraction of a biological neuron, which can range from (almost) truthful to a very simple model, such as an integrate and fire neuron. Depending on the technology used [single electron transistor (SET) neurons, RTD, memristors, etc.], this level may vary, and opportunities for emerging nanoelectronic devices exist in this respect. Another important issue is how nonvolatile analog synapses are implemented. Many attempts have been made to design analog synaptic devices based on existing flash memory technologies, but they have experienced difficulties in designing appropriate controllers for electron injection and ejection as well as increasing the limit on the number of rewriting times. Memristive devices (e.g., resistive RAMs and atomic switches) offer a promising alternative for the implementation of nonvolatile analog synapses. They are applied in the CMOL architecture, which combines memristive nano-junctions with CMOS neurons and their associated controllers. In ITRS 2007, CMOL (CrossNets) was introduced in terms of nanogrids of (ideally) single molecules fabricated on top of a traditional CMOS layer, but the concept has since been expanded to use a nanowire crossbar add on as well as memristive (two-terminal) crosspoint devices like nanowire resistive RAMs [39]. The CMOL architecture will be further expanded to include multiple stacks of CMOS layers and crossbar layers. This may result in the implementation of large-scale multi-layer neural networks, which have thus far evaded direct implementations by CMOS devices only.

A final important issue is *noise tolerance* and *noise utilization* in neural systems and their possible application in electronics. Noise and fluctuations are usually considered obstacles in the operation of both analog and digital circuits and systems, and most strategies to deal with them focus on suppression. Neural systems, on the other hand, tend to employ strategies in which the properties of noise are exploited to improve the efficiency of operations. This concept may be especially useful in the design of computing systems with noise-sensitive devices (e.g., extremely low-power devices like SET and subthreshold analog CMOS devices).

Table 22.2 shows examples of noise-driven neural processing and their possible applications in electronics. Stochastic resonance (SR) is a phenomenon where a static or dynamic threshold system responds stochastically to a subthreshold or suprathreshold input with the help of noise. In biological systems SR is utilized to detect weak signals in a noisy environment. SR on some emerging nanoelectronic devices (a SET network and GaAs nanowire FETs) has been demonstrated. SR can be observed in many bi-stable systems and will be utilized to facilitate the state transitions in emerging logic (bi-stable) memory devices. Noise-driven fast signal transmission is observed in neural networks for the vestibulo-ocular reflex, where signals are transmitted with an increased rate over a neuronal path when nonidentical neurons and

Table 22.2 Noise-driven neural processing and its possible applications

Neurophysiological phenomena	Type of noise	Applications and device examples
Stochastic resonance [40]	Dynamic/static	Sensors and logic memory (CMOS [41], SET [42], GaAs nanowire FET [43])
Fast signal transmission on slow transmission pathway [44]	Dynamic/static	Fast signal transmission, Pulse-density modulation (CMOS [45], SET [46])
Phase synchronization among isolated neurons [47]	Dynamic	Phase synchronization among isolated circuits/PLL (CMOS) [48]
Synaptic depression [49]	Static	Burst signal detection (SET) [50]
Noise shaping in inhibitory neural networks [51]	Dynamic/static	Noise shaping AD conversion (CMOS [52], SET [53])

dynamic noise are introduced. Implementation in terms of a SET circuit has demonstrated that, when several nonidentical pulse-density modulators were used as noisy neurons, performances on input–output fidelity of the population increased significantly as compared to that of a single neuron circuit. Phase synchronization among isolated neurons can be utilized for skew-free clock distribution where independent oscillators are implemented on a chip as distributed clock sources, while the oscillators are synchronized by a common temporal noise. Noise in synaptic depression can be used to facilitate the operation of a neuromorphic burst-signal detector, where the output range of the detector is significantly increased by noise. Noise shaping in inhibitory neural networks has been demonstrated in subthreshold CMOS, where static and dynamic noises can positively be taken if one could not remove a certain level of noise or device mismatches. The circuits exploit properties of device mismatches and external (temporal) noise to perform noise shaping one-bit AD conversion (pulse-density modulation).

22.3 Cellular Automata Architectures

A Cellular Automaton is an array of cells, organized in a regular grid. Each cell can be in one of a finite number of states from a predefined state set, which is usually a set of integers. The state of each cell is updated according to transition rules, which determine the cell's next state from its current state as well as from the states of the neighboring cells. The neighbors of a cell are usually the cells directly adjacent in orthogonal directions of a cell, like the north, south, east, and west neighbor in the case of a two-dimensional grid (von Neumann neighborhood), but other neighborhoods have also been used in experiments. The functionality of each cell is defined by the transition rules of the cellular automaton. The transition rules are usually the same for all cells, but heterogeneous sets of rules have also been considered, as well as programmable rules. A cell can typically be expressed in terms of a Finite Automaton, which is a model in computer science well-known for its simple but effective structure.

Cellular automata were initially proposed by von Neumann in the 1940s as a model of self-reproduction, but most of the interest they have attracted since then has been motivated by their ability to conduct computation in a distributed way. Though cellular automata have the name of their inventor in common with von Neumann architectures, they represent a radically different concept of computation.

The appeal of cellular automata as emerging research architectures lies in a number of factors. First, their regular structure has the potential for manufacturing methods that can deliver huge numbers of cells in a cost-effective way. Candidates in this respect are bottom-up manufacturing methods, such as those based on molecular self-assembly. Second, regularity also facilitates reuse of logic designs. The design of a cell is relatively simple as compared to that of a microprocessor unit, so design efforts are greatly reduced for cellular automata. Third, errors are more easily managed in the regular structures of cellular automata, since a unified approach for all cells can be followed. Fourth, wire lengths between cells are short, or wires are completely unnecessary if cells can interact with their neighboring cells through some physical mechanism. Fifth, cells can be used for multiple purposes, from logic or memory to the transfer of data. This makes cellular automata configurable in a flexible way. Sixth, cellular automata are massively parallel, offering a huge computational power for applications of which the "logical structure" fits the topology of the grid of cells.

Cellular automata may be less suitable for certain application due to the following factors. First, there is a relatively large overhead in terms of hardware. Cells tend to require a certain minimal level of complexity in order to be useful for computation [54]. In practice this means that they are configurable for logic, memory, or data transfer. The density of such functionality per unit of area tends to be lower than that of more conventional architectures. A large hardware overhead may be acceptable, though, if cells are available in huge numbers at a low cost, especially if the cellular automaton can be mapped efficiently on a certain application. Second, input and output of data to cells may be difficult. The use of the border cells of a grid for input and output is infeasible in case of huge numbers of cells, because it fails to employ all cells in parallel. Parallel input and output to cells through optical means or by wires addressing individual cells like in memory have more potential in this context. Third, it is difficult to configure cells into various patterns of states. Such configuration and reconfiguration functionality is required to give the cellular automaton its functionality for a certain computational task. Here, similar solutions as those for the input and output of data need to be employed to access cells in parallel.

There are two approaches for implementing cellular automata in hardware: fine-grained and tiny-grained. Systems that are coarse-grained are considered outside the class of cellular automata, since they are associated with multi-core architectures. Fine-grained cellular automata have cells that can be configured each as one or a few logic gates, or as a simple hub for data transfer. A cell typically contains a limited amount of memory in the order of 10–100 Bytes. Cells are usually addressed on an individual basis for input and output, or for configuration. Typically, the transition rules governing the functionality of a cell are changed during configuration. An example of a fine-grained approach is the Cell Matrix [55], which is a model capable of universal computation. Fine-grained cellular automata have a good degree of control over configuration and computation, but this comes at the price of relatively complex cells, limiting the use of cost-effective manufacturing methods that can exploit the regularity of these architectures.

The other approach to cellular automata is tiny-grained. Cells in this model have extremely simple functionality, in the order of a few states per cell and a limited number of (fixed) transition rules. The small number of states translates into memory requirements of only a few bits per cell, whereas the nonprogrammable nature of the transition rules drastically reduces the complexity of cells. The simple nature of rules poses no problems if the rules are designed to cover the functionality intended for the cellular automaton. An example of a tiny-grained approach is proposed [56] which is capable of universal computations as well as the correction of errors. Tiny-grained cellular automata have the potential of straightforward realizations of cells on nanometer scales. The challenge is to design models with as few states and transition rules per cell as possible. The theoretical minimum is two states and one transition rule. In synchronously timed models this has been approached by the Game of Life cellular automaton (two states, two rules) and in asynchronously timed models (no clock) by the Brownian Cellular Automata [57] (three states, three rules). Both models are universal. The number of states and rules should be considered only as rough yardsticks, since ultimately the most important measure is the efficiency at which cells can be realized in a technology.

Most hardware realizations of cellular automata to date are Application-Specific. In this context cellular automata are used as part of a larger system to conduct a specific set of operations with great efficiency. Applications have typically a structure that can be mapped efficiently on the hardware, and the approach followed is generally tiny-grained, since cells are optimized for one or a few simple operations. Image processing applications are the most common in such hardware realizations [58–60], since they can be mapped with great efficiency on two-dimensional cellular automata. Though the focus in the past was mostly on operations like filtering, thinning, skeletonizing, and edge detection, recent applications of cellular automata include the watermarking of images with digital image copyright [61,62]. Cellular automata have also been used for the implementation of a Dictionary Search Processor [63], memory controllers [64], and the generation of test patterns for Built-In Self-Test (BIST) of VLSI chips [65,66]. An overview of application-specific cellular automata is given in [67].

It is possible that the role of cellular automata in architectures will gradually increase with technological progress, from being merely used as dedicated subprocessor to the main part of the architecture. To do that, cellular automata would need a capacity that their application-specific cousins lack: computation universality, that is, the ability to carry out the same class of computations as our current computers do. This term is mostly used in a theoretical context, to prove equivalence to a universal Turing machine or more narrowly, to support a complete set of Boolean logic functions. The extreme inefficiency of Turing machines carries with it the misunderstanding that generallity is equivalent to inefficiency, but this is often far from the truth. A general approach to carry out operations efficiently on a cellular automaton is to configure it as a logic circuit. Cells will then be used as logic gates or for transferring data between logic gates. In fine-grained cellular automata a cell is typically sufficiently complex to be able to function as one or a few gates. In tiny-grained cellular automata, on the other hand, clusters of cells need to work together in order to obtain logic gate functionality. A cluster typically consists of up to 10 tiny-grained cells, its size depending on the functionality covered. This may seem a large overhead, but cells tend to be much less complex than in fine-grained cellular automata, making this approach feasible. Furthermore, cells used merely to transfer data – and this is the majority of the cells – see much less of their hardware unused when carrying out this simple task.

Cellular automata have seen only limited attempts at realization on a nanometer scales Molecule Cascades [68] use CO molecules on a Cu(111) grid to conduct simple logic operations. The CO molecules jump from grid point to grid point, triggering each other sequentially, like dominos. This process is quite slow and error-prone, though there appears to be potential for improvement. The mechanical nature of the operations, though, means that this cellular automaton is unlikely to reach competitive speeds. Another attempt uses layers of organic molecules on a gold grid [69]. Interactions between molecules take place via the tunneling of electrons between them. The rules that have been identified as governing those interactions appear to be influenced by the presence of electrons in the grid. This may limit the control over the operation of the cellular automaton, but it also carries the promise of efficient ways to configure the grid.

22.4   Taxonomy of Computational Ability of Architectures

Whereas von Neumann architectures generally refer to the use of memory resources separated from computational resources to store data and programs, there is an increasing need for taxonomy of those architectures that are based on different concepts. Figure 22.3 illustrates the world of "information processing" including from present Boolean-based processing of von Neumann machines to nonNeumann machines and introducing four possible architectures on it, that is, *More Neumann, More than Neumann, Less than Neumann,* and *Beyond Neumann.*

The term *More Neumann* refers to those architectures that differ from the classical von Neumann architecture only in terms of numbers (e.g., presenting multi- or many-core architectures). While the stored memory concept is still followed in *More Neumann* architectures, a certain level of parallelism is assumed, like in multi-core systems. The *More Neumann* architecture has been grown from ancient *Less than Neumann* architectures (Figure 22.4, left), and thus the performance lies in the metrics in terms of parallelism, computational ability, and programmability, as shown in Figure 22.4, right.

More than Neumann refers to architectures that do not suffer from the von Neumann bottleneck between computation and memory resources, that is, these resources are integrated to a high degree. These architectures tend to have a highly distributed character in which small elements have extremely limited memory and computation resources to the extent that each element individually is *Less than Neumann* (i.e., incapable of being used as a full-fledged von Neumann architecture, like a "logic element" unit in FPGA), yet the combination of these elements lifts them to a higher level of competence for certain applications (Figure 22.4). In *More than Neumann* architectures reorganization or reconfiguration usually plays the role that programmability has in von Neumann architectures. Programming a *More than Neumann* architecture thus involves an appropriate organization or configuration of the individual elements in order to make them perform a certain function. This reorganization may take the form of setting/adjusting the memories of the individual elements, but it may also involve a

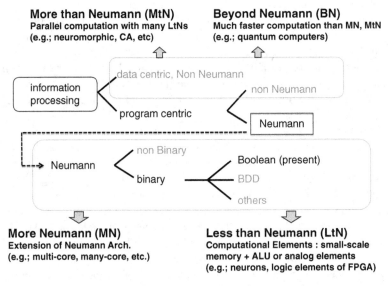

Figure 22.3 Taxonomy of information processing

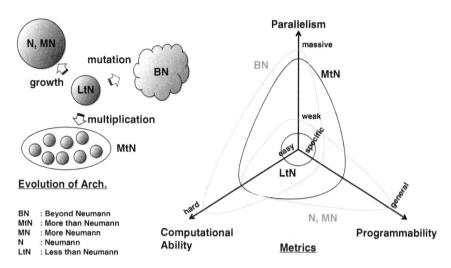

Figure 22.4 Concept of *More Neumann, More than Neumann, Less than Neumann,* and *Beyond Neumann*

relinking of interconnections between the elements. In the context of neuromorphic architectures the elements take the form of neurons and synaptic connections between them, and synapses can be adjusted based on a learning process, while in some architectures new synaptic connections can be created and old ones destroyed. In the case of cellular automata, the elements are the cells, and their functionality is changed by setting their memory states to appropriate values. *More than Neumann* architectures are typically capable of high performance on certain classes of problems, but much less so on other problems (or may be even unable to handle other problems). Neuromorphic architectures have their strengths in problems that involve learning, classification, and recognition, but they do less well on traditional computing problems. Cellular automata are strong in applications that demand a regular structure of logic or data and a huge degree of parallelism.

Beyond Neumann refers to architectures that can solve certain computational problems fundamentally faster than would be possible on the architectures outlined above (Figure 22.4), like quantum computing architectures. Problems such as these typically require computation times that are exponential as measured in terms of their input. The fundamental limits that restrict the computational power of architectures ranging from von Neumann to *More than Neumann* are exceeded in *Beyond Neumann* architectures through adopting novel operating principles. Schemes that use analog values instead of digital (neuromorphic architectures, dynamic systems, etc.), that use superposition of bit values (quantum computing schemes), or that use an analog timing scheme (asynchronous architectures) are prime candidates for this category. The flow of information in architecture may also characterize it as *Beyond Neumann*. While Turing machines embody the traditional Input–Processing–Output flow, modern computers (even von Neumann ones) are used in a more interactive mode with humans, like in gaming, or with other computers, when connected in networks. Biological brains have a somewhat related concept of input and output, but different in its implementation: their processing of information appears to be an autonomous process, that may (or may not) be modulated by the input signals in the environment [70]. This allows biological organisms to flexibly select important signals from the environment, while ignoring irrelevant ones.

Underneath this all lies an impressive neural machinery, yet to be uncovered, that can solve problems with unrivaled efficiency. Many of the above elements (analog-valued signals, asynchronous timing in combination with selective synchronization, chaotic dynamics) are thought to play an important role in neural information processing. While *Beyond Neumann* architectures are promising in principle, it needs to be emphasized that currently no practical implementations of them have been reported.

22.5 Summary

This chapter introduced a concept of the morphic architecture that refers to architectures adapted to effectively address a particular problem set, and exhibited recent progress of two morphic architectures that offer opportunities for emerging nanoelectronic devices: neuromorphic architectures and cellular automaton architectures.

Morphic architectures will be employed in a broad class of mixed-signal systems that are focused on a particular application and that draw inspiration for their structure from the application. In some cases, computation is performed in the analog manner, which may offer orders of magnitude improvement in performance and power dissipation, albeit with reduced accuracy. As an example, biologically inspired inference networks for cognition may yield to a partial analog implementation and provide substantial gains in performance relative to their digital counterparts.

Finally, taxonomy for emerging models of computation was introduced, because of an increasing need of classifying emerging architectures that are based on nonNeuman concepts. Concepts of four possible architectures were introduced: More Neumann, More than Neumann, Less than Neumann, and Beyond Neumann architectures. The morphic and cellular automaton architectures introduced in this chapter will be categorized into More than Neumann architectures that consist of huge collection of ancient Less than Neumann (i.e., incapable of being used as fully fledged von Neumann architecture) elements.

References

1. ITRS (2007) Edition, Emerging Research Devices, Chapter: Emerging Research Architectures, Section: Morphic Computational Architecture.
2. Mead, C. (1989) *Analog VLSI and Neural System*, Addison-Wesley, Reading, MA.
3. UZH (2010) http://siliconretina.ini.uzh.ch/wiki/index.php (accessed 16 July 2013).
4. DARPA (2012) http://en.wikipedia.org/wiki/SyNAPSE (accessed 16 July 2013).
5. ITRS (2009) Edition, Emerging Research Devices, Chapter: Emerging Research Architectures, Section: Inference Computing.
6. ARS (2010) http://arstechnica.com/hardware/news/2010/08/probabilistic-processors-possibly-potent.ars (accessed 16 July 2013).
7. Bui, T.T. and Shibata, T. (2008) Compact bell-shaped analog matching-cell module for digital-memory-based associative processors. *Japanese Journal of Applied Physics*, **47**(4), 2788–2796.
8. Saen, M., Morie, T., Nagata, M., and Iwata, A. (1998) A stochastic associative memory using single-electron tunneling devices. *IEICE Transactions on Electronics*, **E81-C**(1), 30–35.
9. Morie, T., Matsuura, T., Nagata, M., and Iwata, A. (2003) A multinanodot floating-gate MOSFET circuit for spiking neuron models. *IEEE Transactions on Nanotechnology*, **2**(3), 158–164.
10. Domingos, P.O., Silva, F.M., and Neto, H.C. (2005) An efficient and scalable architecture for neural networks with backpropagation learning, in Proc. Int. Conf. Field Programmable Logic and Applications, pp. 89–94.
11. Oster, M., Yingxue, W., Douglas, R., and Liu, S.-C. (2008) Quantification of a spike-based winner-take-all VLSI network. *IEEE Transactions on Circuits and Systems I*, **55**(10), 3160–3169.

12. Nakada, K., Asai, T., and Amemiya, Y. (2004) Biologically-inspired locomotion controller for a quadruped walking robot: Analog IC implementation of a CPG-based controller. *Robotics and Mechatronics*, **16**(4), 397–403.
13. Kameda, S. and Yagi, T. (2006) An analog silicon retina with multichip configuration. *IEEE Transactions on Neural Networks*, **17**(1), 197–210.
14. Kikombo, A.K., Schmid, A., Asai, T. *et al.* (2009) A bio-inspired image processor for edge detection with single-electron circuits. *Journal of Signal Processing*, **13**(2), 133–144.
15. Brinkworth, R.S.A., Shoemaker, P.A., and O'Carroll, D.C. (2009) Characterization of a neuromorphic motion detection chip based on insect visual system, in Proc. Int. Conf. Intelligent Sensors, Sensor Networks and Information Processing, pp. 289–294.
16. Kikombo, A.K., Asai, T., and Amemiya, Y. (2009) An elementary neuro-morphic circuit for visual motion detection with single-electron devices based on correlation neural networks. *Journal of Computational and Theoretical Nanoscience*, **6**(1), 89–95.
17. Tsang, E.K.C., Lam, S.Y.N., Yicong, M., and Shi, B.E. (2008) Neuromorphic implementation of active gaze and vergence control, in Proc. IEEE Int. Symp. Circuits and Systems, pp. 18–21.
18. Indiveri, G. (1999) Neuromorphic analog VLSI sensor for visual tracking: circuits and application examples. *IEEE Transactions on Circuits and Systems II*, **46**(11), 1337–1347.
19. Meng, Y. and Shi, B.E. (2008) Adaptive gain control for spike-based map communication in a neuromorphic vision system. *IEEE Transactions on Neural Networks*, **19**(6), 1010–1021.
20. Choi, T.Y.W., Merolla, P.A., Arthur, J.V. *et al.* (2005) Neuromorphic implementation of orientation hypercolumns. *IEEE Transactions on Circuits and Systems I*, **52**(6), 1049–1060.
21. IEEE (2012) http://spectrum.ieee.org/computing/hardware/a-flyeye-inspired-speed-sensor (accessed 16 July 2013).
22. Lyon, R.F. and Mead, C. (1988) An analog electronic cochlea. *IEEE Transactions on Acoustics Speech and Signal Processing*, **36**(7), 1119–1134.
23. Horiuchim, T.K. (2005) Seeing in the dark: Neuromorphic VLSI modeling of bat echolocation. *IEEE Signal Processing Magazine*, **22**(5), 134–139.
24. Park, H.M., Oh, S.H., and Lee, S.Y. (2003) A filter bank approach to independent component analysis and its application to adaptive noise cancelling. *Neurocomputing*, **55**(3–4), 755–759.
25. Koickal, T.J., Hamilton, A., Tan, S.L. *et al.* (2007) Analog VLSI circuit implementation of an adaptive neuromorphic olfaction chip. *IEEE Transactions on Circuits and Systems I*, **54**(1), 60–73.
26. Adamatzky, A., De Lacy Costello, B., and Asai, T. (2005) *Reaction-Diffusion Computers*, Elsevier, Amsterdam.
27. Fujita, D., Asai, T., and Amemiya, Y. (2011) A neuromorphic MOS circuit imitating jamming avoidance response of eigenmannia. *Nonlinear Theory and Its Applications*, **2**(2), 205–217.
28. Momeni, M. and Titus, A.H. (2006) An analog VLSI chip emulating polarization vision of octopus retina. *IEEE Transactions on Neural Networks*, **17**(1), 222–232.
29. Türel, Ö. and Likharev, K.K. (2003) CrossNets: Possible neuromorphic networks based on nanoscale components. *International Journal of Circuit Theory and Applications*, **31**(1), 37–52.
30. Costas-Santos, J., Serrano-Gotarredona, T., Serrano-Gotarredona, R., and Linares-Barranco, B. (2007) A spatial contrast retina with on-chip calibration for neuromorphic spike-based AER vision systems. *IEEE Transactions on Circuits and Systems I*, **54**(7), 1444–1458.
31. Yuminaka, Y., Sasaki, Y., Aoki, T., and Higuchi, T. (1998) Design of neural networks based on wave-parallel computing technique. *Analog Integrated Circuits and Signal Processing*, **15**(3), 315–327.
32. Scholarpedia (2012) http://www.scholarpedia.org/article/Silicon_neurons (accessed 16 July 2013).
33. Oya, T., Schmid, A., Asai, T. *et al.* (2005) On the fault tolerance of a clustered single-electron neural network for differential enhancement. *IEICE Electronics Express*, **2**(3), 76–80.
34. Jo, S.H., Chang, T., Ebong, I. *et al.* (2010) Nanoscale memristor device as synapse in neuromorphic systems. *Nano Letters*, **10**(4), 1297–1301.
35. Ohno, T., Hasegawa, T., Tsuruoka, T. *et al.* (2011) Short-term plasticity and long-term potentiation mimicked in single inorganic synapses. *Nature Materials*, **10**(8), 591–595.
36. Koyanagi, M., Nakagawa, Y., Lee, K.-W. *et al.* (2001) Neuromorphic vision chip fabricated using three-dimensional integration technology, in ISSCC Dig. Tech. Papers, pp. 270–271.
37. Yamaguchi, M., Shimada, A., Torimitsu, K., and Nakano, N. (2010) Multichannel biosensing and stimulation LSI chip using 0.18-um CMOS technology. *Japanese Journal of Applied Physics*, **49**(2), 04DL14.
38. Wikipedia (2012) http://en.wikipedia.org/wiki/Belousov–Zhabotinsky_reaction (accessed 16 July 2013).
39. Likharev, K.K. (2008) Hybrid CMOS/nanoelectronic circuits: opportunities and challenges. *Journal of Nanoelectronics and Optoelectronics*, **3**(3), 203–230.

40. Gammaitoni, L., Hanggi, P., Jung, P., and Marchesoni, F. (1998) Stochastic resonance. *Reviews of Modern Physics*, **70**(1), 223–287.

41. Utagawa, A., Asai, T., and Amemiya, Y. (2011) Stochastic resonance in simple analog circuits with a single operational amplifier having a double-well potential. *Nonlinear Theory and Its Applications*, **2**(4), 409–416.

42. Oya, T., Schmid, A., Asai, T., and Utagawa, A. (2011) Stochastic resonance in a ballanced pair of single-electron boxes. *Fluctuation and Noise Letters*, **10**(3), 267–275.

43. Kasai, S. and Asai, T. (2008) Stochastic resonance in Schottky wrap gate-controlled GaAs nanowire field effect transistors and their networks. *Applied Physics Express*, **1**(8), 083001.

44. Hospedales, T.M., van Rossum, M.C.W., Graham, B.P., and Dutia, M.B. (2008) Implications of noise and neural heterogeneity for vestibulo-ocular reflex fidelity. *Neural Computation*, **20**(3), 756–778.

45. Utagawa, A., Asai, T., and Amemiya, Y. (2011) High-fidelity pulse density modulation in neuromorphic electric circuits utilizing natural heterogeneity. *Nonlinear Theory and Its Applications*, **2**(2), 218–225.

46. Kikombo, A.K., Asai, T., and Amemiya, Y. (2011) Neuro-morphic circuit architectures employing temporal noises and device fluctuations to improve signal-to-noise ratio in a single-electron pulse-density modulator. *International Journal of Unconventional Computing*, **7**(1–2), 53–64.

47. Nakao, H., Arai, K., and Nagai, K. (2005) Synchrony of limit-cycle oscillators induced by random external impulses. *Physical Review E*, **72**(2), 026220.

48. Utagawa, A., Asai, T., Hirose, T., and Amemiya, Y. (2008) Noise-induced synchronization among sub-RF CMOS analog oscillators for skew-free clock distribution. *IEICE Transactions on Fundamentals*, **E91-A**(9), 2475–2481.

49. Senn, W., Segev, I., and Tsodyks, M. (1998) Reading neuronal synchrony with depressing synapses. *Neural Computation*, **10**(4), 815–819.

50. Oya, T., Asai, T., Kagaya, R. *et al.* (2006) Neuronal synchrony detection on signle-electron neural network. *Chaos, Solitons and Fractals*, **27**(4), 887–894.

51. Mar, D.J., Chow, C.C., Gerstner, W. *et al.* (1999) Noise shaping in populations of coupled model neurons. *Neurobiology*, **96**(18), 10450–10455.

52. Utagawa, A., Asai, T., Hirose, T., and Amemiya, Y. (2007) An inhibitory neural-network circuit exhibiting noise shaping with subthreshold MOS neuron circuits. *IEICE Transactions on Fundamentals*, **E90-A**(10), 2108–2115.

53. Kikombo, A.K., Asai, T., Oya, T. *et al.* (2009) A neuromorphic single-electron circuit for noise-shaping pulse-density modulation. *International Journal of Nanotechnology and Molecular Computation*, **1**(2), 80–92.

54. Zhirnov, V., Cavin, R., Leeming, G., and Galatsis, K. (2008) An Assessment of Integrated Digital Cellular Automata Architectures. *IEEE Computer*, **41**(1), 38–44.

55. Durbeck, L. and Macias, N. (2001) The cell matrix: An architecture for nanocomputing. *Nanotechnology*, **12**(3), 217–230.

56. Peper, F., Lee, J., Abo, F. *et al.* (2004) Fault-tolerance in nanocomputers: A cellular array approach. *IEEE Transactions on Nanotechnology*, **3**(1), 187–201.

57. Lee, J. and Peper, F. (2008) On Brownian Cellular Automata, in Proc. Automata, pp. 278–291.

58. Preston, K., Duff, M.J.B., Levialdi, S. *et al.* (1979) Basics of cellular logic with some applications in medical image processing. *Proceedings of the IEEE*, **67**(5), 826–856.

59. Sunayama, T., Ikebe, M., Asai, T., and Amemiya, Y. (2000) Cellular vMOS circuits performing edge detection with difference-of-gaussian filters. *Japanese Journal of Applied Physics*, **39**(4B), 2278–2286.

60. Asai, T., Sunayama, T., Amemiya, Y., and Ikebe, M. (2001) A vMOS vision chip based on cellular-automaton processing. *Japanese Journal of Applied Physics*, **40**(4B), 2585–2592.

61. Mankar, V.H., Das, T.S., and Sarkar, S.K. (2007) Cellular Automata Based Robust Watermarking Architecture towards the VLSI Realization, in Proc. World Acad. of Sc., Eng., and Techn, pp. 20–29.

62. Shin, J., Yoon, S., and Park, D.S. (2010) Contents-based digital image protection using 2-D cellular automata transforms. *IEICE Electronics Express*, **7**(11), 772–778.

63. Motomura, M., Yamada, H., and Enomoto, T. (1992) A 2K-word dictionary search processor (DISP) LSI with an approximate word search capability. *IEEE Journal of Solid-State Circuits*, **27**(6), 883–891.

64. Wasaki, K. (2008) Self-stabilizing model of a memory controller based on the cellular automata. *International Journal of Computer Science and Network Security*, **8**(3), 222–227.

65. Hortensius, P.D., McLeod, R.D., Pries, W. *et al.* (1989) Cellular automata-based pseudorandom number generators for built-in self-test. *IEEE Transactions on Computer-Aided Design*, **8**(8), 842–859.

66. Dasgupta, P., Chattopadhyay, S., Chaudhuri, P.P., and Sengupta, I. (2001) Cellular automata-based recursive pseudoexhaustive test pattern generator. *IEEE Transactions on Computers*, **50**(2), 177–185.

67. Ganguly, N., Sikdar, B.K., Deutsch, A. *et al.* (Dec. (2003)) A Survey on Cellular Automata, in, Technical Report, Centre for High Performance Computing, Dresden University of Technology.

68. Heinrich, A.J., Lutz, C.P., Gupta, J.A., and Eigler, D.M. (2002) Molecule cascades. *Science*, **298**(5597), 1381–1387.
69. Bandyopadhyay, A., Pati, R., Sahu, S. *et al.* (2010) Massively parallel computing on an organic molecular layer. *Nature Physics*, **6**(5), 369–375.
70. Destexhe, A. and Contreras, D. (2006) Neuronal computations with stochastic network states. *Science*, **314**(5796), 85–90.

23

Design Considerations for a Computational Architecture of Human Cognition

Narayan Srinivasa

Center for Neural and Emergent Systems, HRL Laboratories LLC, USA

23.1 Introduction

How does the brain produce cognitive behavior and is it possible to abstract cognition using computers? This question has intrigued us for a very long time. Cognition is related to processes such as thinking, reasoning, memorizing, problem-solving, analyzing, and applying. Most attempts to understand brain function from a cognitive perspective are primarily derived by describing it as the computation of behavioral responses from internal representation of stimuli and stored representations of information from past experience. The origins of this description can be traced back to two key developments in the early twentieth century. Alan Turing's pioneering work [1] in machine theory defined computation as formally equivalent to the manipulation of symbols in a temporary buffer. Similarly, the pioneering work on telephone communication by Shannon and Weaver [2] resulted in a formal definition of information where the informational content of a signal was inversely related to the probability of that signal arising from randomness.

These developments launched computer science into prominence, and as computers grew in functional complexity, the analogy between computers and the brain began to be widely recognized. The basic premise for this analogy was that computers and the brain received information from the external environment and both acted upon this information in complex ways. It soon appeared that digital computers joined human brains as the only examples of systems capable of complex reasoning. This analogy between computers and the brain (also known as the *computer metaphor*) provided a candidate mechanism to explain cognition as akin to a digital computer program that can manipulate internal representation according to a set of rules. The computer metaphor also appeared to clearly identify with memories and

Emerging Nanoelectronic Devices, First Edition. An Chen, James Hutchby, Victor Zhirnov and George Bourianoff.
© 2015 John Wiley & Sons, Ltd. Published 2015 by John Wiley & Sons, Ltd.

provided an explanation for Descartes mind-body dualism [3] where mental entities are akin to software, whereas physical mechanisms found in the brain are akin to hardware.

The extensive use of the computer metaphor has resulted in the applied notions of symbolic computations and serial processing to construct human-like adaptive behaviors. The task of brain science has become focused on answering the question of how the brain *computes* [4]. The key issues, such as serial versus parallel processing, analog versus digital coding, and symbolic versus nonsymbolic representations, are being addressed using the computer metaphor wherein perception is akin to input, action is akin to output, and cognition is akin to computation. However, traditional algorithms that are derived based on adopting the computer metaphor have yielded very limited utility in complex, real-world environments despite several decades of research to develop machines that exhibit cognitive behaviors. This impasse has resulted in a rethinking of the notion of how cognitive behavior might be realized in machines that seek to emulate the brain.

23.2   Features of Biological Computation

To make progress in understanding human cognition, it is important to realize how biological computation is very different from the digital computers of today (see Figure 23.1 for a summary). The brain is composed of very noisy analog computing elements including neurons and synapses. Neurons operate as relaxation oscillators. Synapses are implicated in memory formation in the brain and can only resolve between three to four bits of information at each synapse [5]. The dynamics of these elements are asynchronous [6] and thus clock-free [7]. Despite being clock-free, coordination can be manifest via interactions between neurons in the brain because they operate as *oscillators*. The interaction can be very weak, and sometimes hardly perceptible, but it most often causes a qualitative transition: a neuron adjusts its rhythm

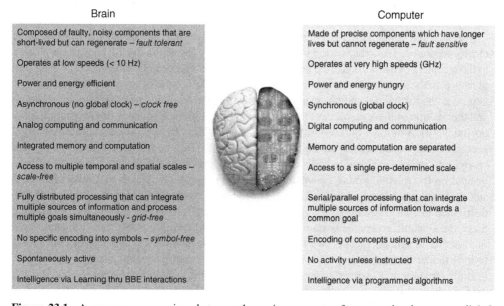

Brain	Computer
Composed of faulty, noisy components that are short-lived but can regenerate – *fault tolerant*	Made of precise components which have longer lives but cannot regenerate – *fault sensitive*
Operates at low speeds (< 10 Hz)	Operates at very high speeds (GHz)
Power and energy efficient	Power and energy hungry
Asynchronous (no global clock) – *clock free*	Synchronous (global clock)
Analog computing and communication	Digital computing and communication
Integrated memory and computation	Memory and computation are separated
Access to multiple temporal and spatial scales – *scale-free*	Access to a single pre-determined scale
Fully distributed processing that can integrate multiple sources of information and process multiple goals simultaneously - *grid-free*	Serial/parallel processing that can integrate multiple sources of information towards a common goal
No specific encoding into symbols – *symbol-free*	Encoding of concepts using symbols
Spontaneously active	No activity unless instructed
Intelligence via Learning thru BBE interactions	Intelligence via programmed algorithms

Figure 23.1   A summary comparison between the various aspects of computation between a digital computer and a brain

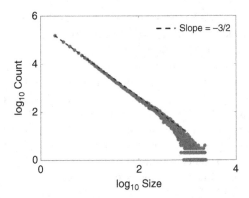

Figure 23.2 Plot showing the inverse relationship between the scale of interaction (*x* axis) and the number of neurons interacting at each scale (*y* axis) in a typical cortical laminar simulation. The slope of this relationship is at −3/2 which is characteristic of scale-free behavior in cortical networks found in the brain

in conformity with the rhythms of other neurons. This adjustment of rhythms due to an interaction is the essence of synchronization and is a universal phenomenon found in nature.

The scale of these interactions can span from a few neurons all the way to the entire network of neurons in the brain during various time instants that depend upon on the context of its interaction with its environment. This implies that the brain is a complex physical system. It is also known that the brain exhibits persistent asynchronous background activity (PABA) even in the absence of external inputs and a careful analysis has revealed rich spatio-temporal patterns of activity related to past experience that was previously neglected as background noise [8,9]. Recent insights into the relationship between PABA and brain organization suggest that the brain operates in a regime of criticality, without a single, dominant temporal or spatial scale [10–13], as reflected in power spectra with "1/f noise" (see Figure 23.2). In other words, the dynamics of brain interactions are *scale-free*.

Memory and computation are integrated at all scales, unlike digital systems where they are clearly separated. The synapses are implicated in memory formation due to biophysical changes to the receptors of the postsynaptic neuron. These changes can be both short-term and long-term in nature. In the short-term version of these changes, the synaptic efficacy at a given postsynaptic neuron is modulated depending upon the dynamics of the available neurotransmitter resources and the fraction of these resources that are utilized based on the frequency of presynaptic action potential [14]. A number of recent experimental studies [15–20] suggest that repeated pairing of pre- and postsynaptic activity in the form of action potentials, or spikes, can lead to long-term changes in synaptic efficacy as well. The sign and magnitude of the change in synaptic efficacy depend on the relative timing between the pre- and postsynaptic spikes and this is known as spike timing-dependent plasticity (STDP). The synaptic conductance represents a form of fully distributed memory with no single synapse that corresponds to any representation or encoding. Since the synaptic conductances are fully distributed and they are implicated in memory, this implies that, in general, there is no single synapse or single neural firing activity that corresponds to a particular item or concept [7]. This means that the brain is *symbol-free*.

Brain structure has evolved with the blue print of a cortex composed of laminar circuits that have six layers of neurons with a great variety in both the types of neurons and the types of synaptic receptors [7]. This basic and unique cortical structure appears to repeats itself throughout the cortex and has been established and preserved for many millions of years.

This self-similar or scale-free property of cortex enables the brain to exhibit macroscopic behavior independent of the microscopic details of the cortex. Furthermore, the brain is organized in a neither completely regular nor completely random form. To interpolate between these two extremes, an interesting concept of small-world networks was introduced [21]. In these networks, neurons are locally coupled in a dense form, but in addition, are also connected through sparse long-range connections linking physically distant brain regions. The so-called small-world networks have intermediate connectivity properties but exhibit a high degree of clustering (as in regular networks) with a small average distance between vertices (as in random networks). The connection patterns of the cerebral cortex consist of pathways linking neuronal populations across multiple levels of scale, from whole brain regions to local mini-columns [22]. There is mounting evidence that the brain is indeed organized as a small-world network [23–25]. In fact, there is also evidence for complete spatial synchronization when connections with small synaptic path lengths (i.e., number of connections between synapses) are enabled between several small worlds [26].

This organization of the brain into a small world network follows two major principles – the degree of local clustering and degree of separation (i.e., synaptic path length) which are in competition but both are needed to achieve large scale traffic with minimum wiring [7]. Local clustering is composed of strongly interacting laminar modules but sequential communication through them is highly inefficient. However, keeping synaptic path length fairly constant with brain size (Figure 23.3) is a necessity for maintaining efficient global communications.

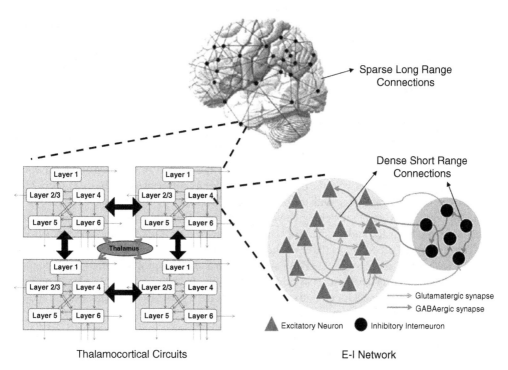

Figure 23.3 A small world network is shown (top) where the black circles represent locally densely connected thalamocortical circuits (bottom left). Each laminar layer within any thalamocortical circuit can be simply summarized as an excitatory–inhibitory (EI) network with densely connected neurons within it

This feature also allows effective integration of heterogeneous and nonlocal sources of information for multiple goals and is a hallmark of human-like cognition [7,27]. The brain thus operates in a *grid-free* fashion.

In summary, the brain exhibits clock-free, scale-free, symbol-free and grid-free dynamics that evolved to enable animals to face the challenges of a continuously changing environment. These features are very different from digital computers and thus computer metaphor-based theories of cognition are very different from what is actually found in the brain.

23.3 Evolution of Behavior as a Basis for Cognitive Architecture Design

In order to understand the process of designing an architecture inspired by mammals that can explain the emergence of complex cognitive behaviors, it is important to focus on the type of behavior. At the most basic level, there are two types of behavior. The first behavior is reflexive in nature and it is inborn, genetically determined in detail, and involves the centers in the spinal cord and base of the brain. The second behavior is learned from individual experiences via brain–body–environment interactions and is neither inborn nor genetically determined.

To survive, the animal had to develop different forms of behavior and produce different results depending on the context. The evolution of the nerve cell was important because it was uniquely positioned to influence the senses and motor parts in an extremely energy efficient fashion. The complexity of the animal's behavioral repertoire grew with the capacity of neural cells to influence each other and the body in order to realize new ways to produce action. Evolution of more complex forms of nerve cells and associated cellular mechanisms with clock-free, scale-free, symbol-free and grid-free dynamics offered a rich set of variations to the existing modes of operation of the nervous system [28]. This meant that the complex nervous system or the brain was now able to *self-organize* via nonlinear relationships between its constituent components. The more complex the brain evolved in an animal, the more it implied that the brain could generate more possible ways to self-organize and this in turn offered a rich set of behavioral repertoires that the animal exhibited in response to rapidly changing environments. This evolutionary process thus enabled the animal to survive in a far more robust manner under these changing environments.

Cognitive behavior can best be understood within the context of experience gained during continuous interactions between the brain, body, and the environment (or BBE – see Figure 23.4). This is because the brain along with the body it controls and its environment have co-evolved via learning to have extensive matching between their properties. Thus the nervous system alone *cannot be the focus* for understanding cognitive behavior. Feedback from its body movements and the dynamical properties of the environment itself with also play a vital role in the generation of cognitive behavior. The role of nervous system is *not* to direct or program behavior but to *shape it* and evoke appropriate patterns/possibilities of dynamics from the *entire* coupled system. Thus, credit or blame assignment for cognitive behavior is not assigned to any one piece of the coupled system but to the BBE system as a whole.

23.4 Considerations for a Cognitive Architecture

The dependence on BBE interactions for cognitive behavior implies that the most basic requirement for the system would be that of *timing*. In other words, the system must be capable of coordination of its actions between these three components such that proper

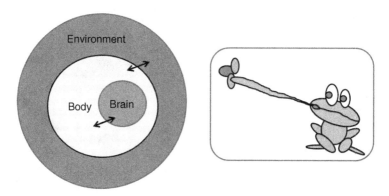

Figure 23.4 Emergence of cognitive behavior should be grounded in the idea of brain–body–environment interactions. Credit for the emergence of proper behavior given the context is not assigned to any one piece of the coupled system but to the brain–body–environment system as a whole, as in the case of the frog catching a fly with its hyper-elastic tongue with exquisite sensorimotor control and timing

cognitive behaviors are realized subject to constraints. But given that a basic feature of biological computation is its clock-free nature, the coordination has to be manifest via interactions between various elements of the architecture. This is possible if the basic elements in the architecture operate as *oscillators*. Radio communication and electrical equipment, violins in an orchestra, crickets producing chirps, and numerous man-made systems such as lasers have a common feature: they produce rhythms [29–31]. As mentioned in Section 23.2 even weak interaction between oscillators (that are self-maintaining like the neurons in the brain) can synchronize with a common phase of oscillation (also referred to as phase locking [29]). It is possible to create phase locked oscillatory clusters by weakly coupling several physical oscillators such as spin torque oscillators [32–34] and resonant body oscillators [35]. But the demands of flexibility in cognitive behaviors during interaction imply that a single clock driving everything in the architecture is not a viable solution. The alternative trick that evolution has produced in the brain is to provide the intrinsic mechanisms to *modulate* the interactions between these oscillators in a flexible fashion based on contingencies. This would be akin to modulating the coupling strength between weakly coupled physical oscillators.

The oscillators produced in the brain correspond to the neural impulse trains that serve as the carriers of the results of local interactions. For example, neurons and neural circuits exhibit endogenous oscillatory properties without any other external inputs [7,36]. Various chemicals are known to modulate the effects of neurotransmitters [37–42]. The scale-free aspect of brain computation enables a wide range of temporal and spatial variations in modulation such as gap junctions at the fast end and volume transmitters that diffuse through intercellular fluid rather than across a synapse at the slow end. These may control the graded release of transmitters rather than all or none release patterns [42–44] that result in modulatory control of population activity such that a wide range of oscillatory rhythms can be generated. In addition, small-world network connectivity aids in this process to rapidly connect distant parts of the brain to enable synchronization in disparate parts of the brain. This idea of small-world networks has been leveraged in several recent FPGA-type implementations, including reducing the delay in FPGA routing structures [45]. This in turn means that various functional forms of the same basic

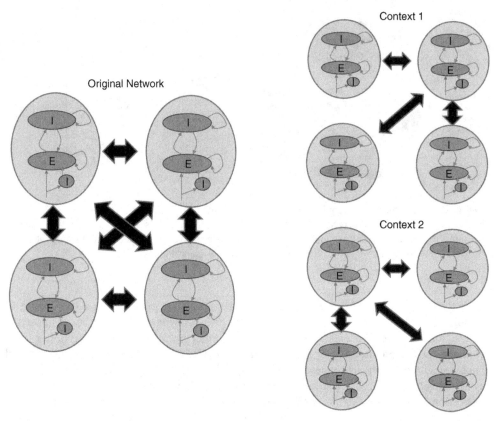

Figure 23.5 An example of a small world network composed of four EI networks is shown on the left marked "original network." Here each EI network is interacting with all other EI networks in a fully connected fashion. Depending on the context in which the animal finds itself, this architecture is flexible enough to change this fully connected behavior due to modulations of oscillations in the network that are based on context

architecture (Figure 23.5) can be realized, which is essential for the generation of cognitive behaviors under various contingencies.

The design of a cognitive architecture based on the features described above offers some key trade-offs. The first trade-off is the type of oscillators that can be used for the neuron-like basic elements. We can either choose from an assortment of CMOS like neural circuits that operate as a relaxation type oscillator [46] to nanoscale oscillators [32,35]. The nanoscale oscillators offer much higher operating frequencies with a very small footprint. This capability can enable much higher throughputs. Furthermore, these nanoscale oscillators are self-maintaining oscillators with their ground state being an energy minimizing oscillatory state with its own natural frequency. When several oscillators are weakly coupled, the system can synchronize while minimizing the energy consumed during the process, thus making the design energy efficient.

Another trade-off is to consider options for coupling neuron-like elements for both synchronization and modulatory influences on the architecture. The coupling between the neurons could be electrical (as in [46]) or through the substrate [32,35]. While electrical coupling gives more controllability, substrate-based coupling can be faster and more energy efficient. The same trade-offs come into play when we consider modulatory influences. In the brain, there seems to be a variety of mechanisms at varying scales of influence that play a role in modulatory signals. These could be broadcast either electrically or via other means, including wireless- or substrate-based coupling.

23.5   Emergent Cognition

The most followed view of cognition as promoted by cognitive psychology is the idea that there is a single executive system [47,48] that is responsible for planning and decision-making that is primarily in the frontal lobes [49–51]. This system is assumed to be independent from the systems responsible for action, such as the sensorimotor control system [52,53]. But there has been a major revision to this idea due to more compelling evidence recently from neuroscience and neurophysiological experiments that suggests a more distributed participation of various systems, including several cortical and subcortical areas of the brain during cognitive acts such as decision-making [54]. This idea of distributed participation is also closely linked to the concepts of small-world networks that enable rapid and scale-free interaction between distant brain regions.

The most recent [55] of many models [56–60] seems to offer an interesting hypothesis about how cognitive function such as decision-making emerges under the challenge of continuously changing environments and contingencies. The authors describe a multi-level model in which decisions emerge as a consensus from distributed neural activity [55] at various brain regions, with some related to sensorimotor control while others are related to more abstract aspects of behavior. These levels are reciprocally connected (as in a small-world network) and share biases that arrive from various brain regions. These biases could be influenced by the external environment related factors that assign values for actions and abstract concepts in brain regions that in turn modulate oscillations in the brain [61]. In fact, the neurons in brain areas such as the anterior cingulate cortex, orbitofrontal cortex, and lateral prefrontal cortex are known to differentially respond to the probability, magnitude, and effort associated with different options. When these oscillatory patterns do not directly produce any action, the system is said to produce cognitive behavior manifested in the form of plans or thoughts. There are several prominent models that propose that the brain is deciding among actions during this phase. When some of these patterns in effector-specific sensorimotor regions reach a threshold [62–66] they are acted upon and the system produces cognitive behavior in the form of actions. Furthermore learning during this interaction also enables stronger links between neurons in the sensory and motor brain regions that are correlated, thus enabling quicker decision making under familiar situations and contingencies.

23.6   Perspectives

In order to design a computational architecture for human cognition, it is important to ground the architecture based on brain–body–environment interactions. Using a small-world network with clock-free, scale-free, symbol-free, and grid-free computational features that allow for oscillatory interactions of its computing elements and modulations of these interactions, it may be possible in the future to design a computational architecture that could learn to plan and make decisions under constantly changing environments and contingencies in a manner reminiscent of human cognition.

References

1. Turing, A.M. (1936) On computable numbers with an application to the Entscheidungs problem. *Proceedings of the London Mathematical Society*, **2**(42), 230–265.

2. Shannon, C. and Weaver, W. (1949) *The Mathematical Theory of Information*, University of Illinois Press, Urbana, IL.

3. Dennett, D.C. (1978) Current issues in the philosophy of mind. *American Philosophical Quarterly*, **15**(4), 249–261.

4. Srinivasa, N. and Cruz-Albrecht, J. (2012) Neuromorphic adaptive plastic scalable electronics: Analog learning systems. *IEEE Pulse*, (2012), 51–56.

5. Barrett, A.B. and van Rossum, M.C.W. (2008) Optimal learning rules for discrete synapses. *PLoS Computational Biology*, **4**(11), e1000230. doi: 10.1371/journal.pcbi.1000230

6. Renart, A., De la Rocha, J., Bartho, P. *et al.* (2010) The asynchronous state in cortical circuits. *Science*, **327**, 587–590.

7. Buzsaki, G. (2009) *Rhythms of the Brain*, Oxford University Press, Oxford.

8. Freeman, W.J. (2005) A field-theoretic approach to understanding scale-free neocortical dynamics. *Biological Cybernetics*, **92**(6), 350–359.

9. Chialvo, D.R. (2010) Emergent complex neural dynamics. *Nature Physics*, **6**(10), 744–750.

10. Werner, G. (2010) Fractals in the nervous system: conceptual implications for theoretical neuroscience. *Frontiers in Physiology*, **1**. doi: 10.3389/fphys.2010.00015

11. Bak, P., Tang, C., and Wiesenfeld, K. (1987) Self-organized criticality: an explanation of 1/f noise. *Physical Review Letters*, **59**(4), 381–384.

12. Beggs, J.M. and Plenz, D. (2003) Neuronal avalanches in neocortical circuits. *Journal of Neuroscience*, **23**(35), 11167–11177.

13. Petermann, T., Thiagarajan, T.C., Lebedev, M.A. *et al.* (2009) Spontaneous cortical activity in awake monkeys composed of neuronal avalanches. *Proceedings of the National Academy of Sciences*, **106**(37), 15921–15926.

14. Markram, H. and Tsodyks, M. (1996) Redistribution of synaptic efficacy between neocortical pyramidal neurons. *Nature*, **382**, 807–810.

15. Markram, H., Lubke, J., Frotscher, M., and Sakmann, B. (1997) Regulation of synaptic efficacy by coincidence of postsynapticAPs and EPSPs. *Science*, **275**, 213–215.

16. Bi, Q.Q. and Poo, M.M. (1998) Activity-induced synaptic modification in hippocampal culture, dependence on spike timing, synaptic strength and cell type. *Journal of Neuroscience*, **18**, 10464–10472.

17. Caporale, N. and Dan, Y. (2008) Spike timing-dependent plasticity: a Hebbian learning rule. *Annual Review Neuroscience*, **31**, 25–46.

18. Debbane, D., Gahwiler, B.H., and Thompson, S. (1998) Long-term synaptic plasticity between pairs of individual CA3 pyramidal cells in rat hippocampal slice cultures. *Journal of Physiology*, **507**, 237–247.

19. Levy, W.B. and Steward, D. (1983) Temporal contiguity requirements for long-term associative potentiation/depression in the hippocampus. *Neuroscience*, **8**, 791–797.

20. Magee, J.C. and Johnston, D. (1997) A synaptically controlled, associative signal for Hebbian plasticity in hippocampal neurons. *Science*, **275**, 209–213.

21. Watts, D.J. and Strogatz, S.H. (1998) Collective dynamics of small world networks. *Nature*, **393**, 440–442.

22. Sporns, O. (2006) Small-world connectivity, motif composition, and complexity of fractal neuronal connections. *BioSystems*, (2006), 56–64.

23. Yu, S., Huang, D.B., Singer, W. *et al.* (2008) A small world of neuronal synchrony. *Cerebral Cortex*, **18**(2), 2891–2901.

24. Bassett, D.S. and Bullmore, E. (2006) Small-world brain networks. *Neuroscientist*, **12**(6), 512–523.

25. Volman, V., Baruchi, I., and Ben-Jacob, E. (2005) Manifestation of function-follow-form in cultures neuronal networks. *Physical Biology*, **2**(2), 98–110.

26. Roxin, A., Riecke, H., and Solla, S.A. (2004) Self-sustained activity in a small-world network of excitable neurons. *Physical Review Letters*, **92**, 198101.

27. Edelman, G.M. (1989) *The Remembered Present: A Biological Theory of Consciousness*, Basic Books, New York.

28. van Schaik, Carel. (2006) Why are some animals so smart? *Scientific American*, **294**(4), 64–71.

29. Andronov, A., Vitt, A.A., and Khaykin, S.E. (1966) *Theory of Oscillations*, Pergamon Press, New York.

30. Ementrout, G.B. and Kopell, N. (1984) Frequency plateaus in a chain of weakly coupled oscillators. *SIAM Journal on Mathematical Analysis*, **15**(2), 215–237.

31. Glass, L. (2001) Synchronization and rhythmic processes in physiology. *Nature*, **410**, 277–284.

32. Pufall, M.R., Rippard, W.H., Kaka, S. *et al.* (2005) Frequency modulation of spin-transfer oscillators. *Applied Physics Letters*, **86**(8), http://dx.doi.org/10.1063/1.1875762.
33. Kaka, S., Pufall, M.R., Rippard, W.H. *et al.* (2005) Mutual phase-locking of microwave spin torque nano-oscillators. *Nature*, **436**, 389–392.
34. Bertotti, G., Mayergoyz, I., and Serpico, C. (2009) *Nonlinear Magnetization Dynamics in Nanosystems*, Elsevier, Amsterdam.
35. Weinstein, D. and Bhave, S.A. (2010) The resonant body oscillator. *NanoLetters*, **10**, 1234–1237.
36. Freeman, W.J. (2005) A field theoretic approach to understanding scale free neocortical dynamics. *Biological Cybernetics*, **92**, 350–359.
37. Bloom, F.E. and Lazerson, A. (1988) *Brain, Mind and Behavior*, Freeman, London.
38. Cooper, J.R., Bloom, F.E., and Roth, R.H. (1986) *The Biochemical Basis of Neuropharmocology*, Oxford University Press, Oxford.
39. Dowling, J.E. (1992) *Neurons and Networks*, Harvard University Press, Harvard.
40. Siegelbaum, S.A. and Tsien, R.W. (1985) Modulation of gated ion channels as mode of transmitter action (ed. D. Bousfield), *Neurotransmitters in Action*, Elsevier, Amsterdam, pp. 81–93.
41. Fuxe, K. and Agnati, L.F. (1987) *Receptor-Receptor Interactions*, Plenum, New York.
42. Fuxe, K. and Aganti, L.F. (1991) *Volume Transmission in the Brain: Novel Mechanisms for Neural Transmission*, Raven, New York.
43. Bullock, T.H. (1981) Spikeless neurones: Where do we go from here? (eds A. Roberts and B.M.H. Bush), *Neurones without Impulses*, Cambridge University Press, Cambridge, pp. 269–284.
44. Krames, Elliot S., Peckham, P. Hunter, and Rezai, Ali R. (eds) (2012) *Neuromodulation*, vol. **1–2**, Academic Press, New York.
45. Nishioka, Y., Iida, M., and Sueyoshi, T. (2010) Small world network to reduce delay in FPGA routing structures. *International Journal of Innovative Computing, Information and Control*, **6**(2), 551–566.
46. Cruz-Albrecht, J., Yung, M., and Srinivasa, N. (2012) Energy-efficient, neuron, synapse and STDP integrated circuits. *IEEE Transactions on Biomedical Circuits and Systems*, **6**(3), 246–256.
47. Norman, D.A. and Shallice, T. (1980) Attention to action: willed and automatic control of behavior, Center for Human Information Processing Technical Report no. 99.
48. Baddeley, A. and Hitch, G. (1974) Working memory, in *The Psychology of Learning and Motivation* (ed. I.G.A. Bower), Academic Press, New York, pp. 47–90.
49. Baddeley, A. and Sall Della, S. (1996) Working memory and executive control. *Philosophical Transactions of the Royal Society of London. Series B, Biological Sciences*, **351**, 1397–1404.
50. Shallice, T. (1982) Specific impairments of planning. *Philosophical Transactions of the Royal Society of London. Series B, Biological Sciences*, **298**, 199–209.
51. Srinivasa, N. and Chelian, S.E. (2012) Executive control of cognitive agents using a biologically inspired model architecture of the prefrontal cortex. *Biologically Inspired Cognitive Architectures*, (2012), 13–24.
52. Fodor, J.A. (1983) *Modularity of Mind: An Essay on Faculty Psychology*, MIT Press, Cambridge, MA.
53. Pylyshyn, Z.W. (1984) *Computation and Cogntion: Toward a Foundation for Cognitive Science*, MIT Press, Cambridge, MA.
54. Cisek, P. (2007) Cortical mechanisms of action selection: the affordance competition hypothesis. *Philosophical Transactions of the Royal Society of London. Series B, Biological Sciences*, **362**, 1585–1599.
55. Cisek, P. (2012) Making decision through a distributed consensus. *Current Opinion in Neurobiology*, **22**, 927–936.
56. Padoa-Schioppa, C. (2011) Neurobiology of economic choice: a good-based model. *Annual Review of Neuroscience*, **34**, 333–359.
57. Cisek, P. (2006) Integrated neural processes for defining potential actions and deciding between them: a computational model. *Journal of Neuroscience*, **26**, 9761–9770.
58. Pastor-Bernier, A. and Cisek, P. (2011) Neural correlates of biased competition in premotor cortex. *Journal of Neuroscience*, **31**, 7083–7088.
59. Favilla, M. (1997) Reaching movements: concurrency of continuous and discrete programming. *Neuroreport*, **8**, 3973–3977.
60. Badre, G., Kayser, A.S., and D'Esposito, M. (2010) Frontal cortex and the discovery of abstract action rules. *Neuron*, **66**, 315–325.
61. Rangel, A. and Hare, T. (2010) Neural computations associated with goal-directed choice. *Current Opinion in Neurobiology*, **20**, 262–270.
62. Roitman, J.D. and Shadlen, M.N. (2002) Response of neurons in the lateral interparietal area during a combined visual discrimination reaction time task. *Journal of Neuroscience*, **22**, 9475–9489.

63. Michelet, T., Duncan, G.H., and Cisek, P. (2010) Response competition in the primary motor cortex: corticospinal excitability reflects response replacement during simple decisions. *Journal of Neurophysiology*, **104**, 119–127.
64. Pearson, B., Nelson, M.J., and Andersen, R.A. (2008) Free choice activates a decision circuit between frontal and parietal cortex. *Nature*, **453**, 406–309.
65. Lebedev, M.A., Doherty, J.E., and Nicolelis, M.A. (2008) Decoding of temporal intervals from cortical ensemble activity. *Journal of Neurophysiology*, **99**, 166–186.
66. Bennur, S. and Gold, J.I. (2011) Distinct representations of a perceptual decision and the associated occulomotor plan in the monkey lateral intraparietal area. *Journal of Neuroscience*, **31**, 913–921.

24

Alternative Architectures for NonBoolean Information Processing Systems

Yan Fang[1], Steven P. Levitan[1], Donald M. Chiarulli[1], and Denver H. Dash[2]
[1]Department of Electrical and Computer Engineering, University of Pittsburgh, USA
[2]Intel Science and Technology Center, USA

24.1 Introduction

In this chapter we explore algorithms and develop an architecture based on emerging nano-device technologies for cognitive computing tasks such as recognition, classification, and vision. Though the performance of computing systems has been continuously improving for decades, for cognitive tasks that humans are adept at, our machines are usually ineffective despite the rapid increases in computing speed and memory size. Moreover, given the fact that CMOS technology is approaching the limits of scaling, computing systems based on traditional Boolean logic may not be the best choice for these special computing tasks due to their massive parallelism and complex nonlinearities. On the other hand, recent research progress in nanotechnology provides us opportunities to access alternative devices with special nonlinear response characteristics that fit cognitive tasks better than general purpose computation. Building circuits, architectures, and algorithms around these emerging devices may lead us to a more efficient solution to cognitive computing tasks or even artificial intelligence. This chapter presents several nonBoolean architecture models and algorithms based on nano-oscillators to address image pattern recognition problems.

The idea of special hardware architectures for cognitive tasks is not new. Equivalent circuits for a neuron model go back at least to 1907, where the model was just simply composed of a resistor and a capacitor [1]. Since then, various models have been implemented based on different technologies. In the late 1980s, Carver Mead proposed the term neuromorphic system in reference to all the electronic implementations of neural systems [2]. Meanwhile, with the development of digital computing systems, theoretical and computational models for neural

Emerging Nanoelectronic Devices, First Edition. An Chen, James Hutchby, Victor Zhirnov and George Bourianoff.
© 2015 John Wiley & Sons, Ltd. Published 2015 by John Wiley & Sons, Ltd.

networks and the human brain have been improved and refined many times. From the McCulloch–Pitts neuron model to the Hodgkin and Huxley's model [3,4], computational neuroscience gives us precise mathematic descriptions of the neuron. However, models for a single neuron or synapse have not proved to be good enough to explain the cognitive ability of a whole brain. More recent research trends have focused on the hierarchical structure of human brain and vision system, like Hierarchical Temporal Memory (HTM) [5], Hierarchical Memory and X (HMAX) [6], and so on. With the new opportunities from emerging nano-devices and neuroscience, we can gain insights from both disciplines and design new architectures for cognitive computing tasks based on nonBoolean functions from a more macroscopic view of a hierarchical model of the human brain instead of single neural network.

In this chapter, we build upon this background to take advantage of the unique processing capabilities of weakly coupled nonlinear oscillators to perform the pattern recognition task. In particular, we focus on pattern matching in high dimensional vector spaces to solve the "k-nearest neighbor" search problem. We choose this problem due to its broad applicability to many machine learning and pattern recognition applications. Taking the image recognition problem as an example, we can choose whether or not to preprocess an image using feature extraction. If we choose not to perform preprocessing, the image itself can simply be considered as a long feature vector. Either way, the feature vectors of the image need to be compared with some memorized vectors and matched to the "nearest neighbor" in the feature vector space for recognition.

In the rest of this chapter, we first introduce the concept of an associative memory (AM) and Oscillatory Neural Networks and implement their function by using a traditional well-understood electronic device, the phase-locked loop (PLL). Next, we explain the structure of an AM unit based on nano-oscillators and show how to deal with patterns in high dimension vector spaces by partitioning them into segments and storing them in multiple AM units. Then, we propose two hierarchical AM models that try to address pattern matching problems in both high dimensions and for large data sets. The first hierarchical structures organize AM units into a pyramid. The image patterns are separated into different receptive fields and processed by individual AM units at the low level, and then abstracted at higher levels of the model. The second hierarchical model for the AM architecture is developed for large data sets. In this model, AM units are nodes of a tree, and patterns are classified by hierarchical k-means clustering algorithms and stored in the leaf nodes of the tree, while other nodes only memorize centroids of clusters. During the recognition process, input patterns are compared with cluster centroids and classified to subclusters iteratively until the correct pattern is retrieved.

24.1.1 Associative Memory

An associative memory is a brain-like distributed memory that learns by association. Association has been known to be a prominent feature of human memory since Aristotle, and all models of cognition use association in one form or another as the basic operation [7]. Association takes one of two forms: auto-association or hetero-association. Auto-associative memory is configured to store a set of patterns (vectors) and is required to be able to retrieve the original specific pattern when presented with a distorted or noisy version. In other words, when the input pattern is different from all the memorized patterns, the AM retrieves a best match to the input pattern given some distance metric, like the Euclidian distance. Thus, the association is a nearest neighbor searching procedure.

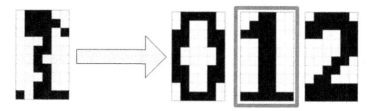

Figure 24.1   Example of image pattern retrieval

As an example shown in Figure 24.1, consider an auto-associative neural network that has memorized three binary image patterns for the numbers "0," "1," and "2." When a distorted image "1" is fed into the network, the output pattern should be the stored image "1," which is the closest pattern in this associative memory.

Hetero-associative memory is the same except its output patterns are coded differently from the input patterns. Therefore, we can consider that a mapping between input and output patterns is stored in the network for pattern retrieval. Hetero-association happens in many cognitive procedures. For instance, a group of neurons in one region of the brain activates neurons in multiple cortical regions.

In its simplest form, associative memory operations involve two phases: storage and recall. In the storage phase, we train the network with patterns by following a particular rule (e.g., a Hebbian learning rule, or a Palm rule) that can form a weight matrix which indicates the weights of connectivity between neurons. Therefore, in the training stage, the network stores the patterns we need in the interconnection weights. In the recall phase, the network performs pattern retrieval by using the interconnection weights to process a possible noisy version of one of the stored patterns. It is assumed that the input pattern is based on one of the stored patterns.

The number of patterns stored in the associative memory provides a direct measure of the storage capacity of a network. In the design of an associative memory, the tradeoff between storage capacity and correctness of retrieval is a challenge [7].

The earliest AM model explored by Steinbuch and Piske in the 1960s [8] was called Die Lernmatrix. It represents patterns with binary vectors and uses the Hebbian learning rule. Based on this model, Willshaw's group improved its information capacity with sparsely encoded binary vectors and proposed a model called an "Associative Net" [9,10]. They also tried to explain some brain network functions such as short-term memory with their model [11]. In the 1980s Palm proved the storage capacity of Willshaw's model by a different method. Palm improved his model by introducing real-value weights and an iterative structure. Brain state in a box (BSB) is another AM model proposed by Anderson's group [12,13]. In 2007 the "Ersatz Brain Project" utilized BSB in building parallel, brain-like computers in both hardware and software [14]. The Hopfield network invented by John Hopfield in the 1980s included the idea of an energy function in the computation of the recurrent networks with symmetric synaptic connections [15]. The most important contribution of the Hopfield Network is that it connects neural networks with physics and the method of analyzing network stability from the viewpoint of a dynamic system.

24.1.2   Oscillatory Neural Networks

We base our AM model design on the work of Hoppensteadt and Izhikevich that studied weakly connected networks of neural oscillators near multiple Andronov–Hopf bifurcation points [16]. They propose a canonical model for oscillatory dynamic systems which satisfy four criteria:

1. Each neural oscillator comprises two populations of neurons: excitatory neurons and inhibitory neurons.
2. The activity of each population of neurons is described by a scalar (one-dimensional) variable.
3. Each neural oscillator is near a nondegenerate supercritical Andronov–Hopf bifurcation point.
4. The synaptic connections between the neural oscillators are weak.

Given these conditions, this canonical model can be represented by the function:

$$\frac{dz_i}{dt} = (\rho_i + i\omega_i)z_i + d_i z_i |z_i|^2 + \varepsilon \sum_{j=1}^{n} C_{ij} z_j \qquad (24.1)$$

Where:

- $z = x + iy$ Complex oscillator variable
- ρ_i Damping of oscillators
- ω_i Natural frequencies of oscillators
- d_i Nonlinear factor, ensures stable amplitude
- ε Coupling parameter, typically small (weakly coupled)
- C_{ij} Coupling matrix, represent coupling strength between two oscillators similar to the weight matrix in previous models

This dynamic model was proved to be able to form attractor basins at the minima of a Lyapunov energy function by adjusting the coupling matrix though a Hopfield rule [15]. In other words, a network that consists of oscillators described by this canonical model can learn patterns as represented by phase errors and perform the functions of an associative memory.

An example of Oscillatory Neural Network that shares the same structure as a recurrent feedback Hopfield network is depicted in Figure 24.2.

The oscillator cluster (OSC1, OSC2, OSC3, . . . , OSCn) are coupled by a matrix of multipliers and adders that contain the coupling coefficients, shown as the C_{ij} box. The outputs are fed back to the network as the coupling term of Equation 24.1 for the recurrent evolution process until the network state is completely stabilized at an attractor.

Hoppensteadt and Izhikevich's Oscillatory Neural Network model demonstrates a similar functionality and dynamic behavior as the Hopfield Network [15]. An example of Oscillatory Neural Networks serving as an AM model is presented in the next part of this section.

24.1.3 Implementation of Oscillator Networks

To illustrate how to implement an Oscillatory Neural Network in hardware, we choose the example of an oscillator network composed of phase-locked loops [17]. The behavior and structure of this simple network helps to understand how Oscillatory Neural Networks can store and retrieve patterns.

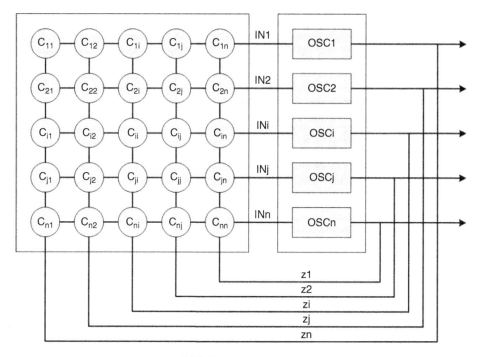

OSC Number: 1,2,...,i...,j...,n

Figure 24.2 Oscillatory neural network structure

A phase-locked loop is a device that generates an output signal whose phase is related to the phase of an input "reference" signal. As Figure 24.3 shows, it consists of a voltage-controlled oscillator and a phase detector implemented as a multiplier and low pass filter.

This circuit compares the phase of the input signal with the phase of the signal derived from its output oscillator and adjusts the frequency of its oscillator to keep the phases matched. The signal from the phase detector is used to control the oscillator in a feedback loop. Frequency is the derivative of phase, therefore, keeping the input and output phase in lock step implies keeping the input and output frequencies in lock step. Consequently, a phase-locked loop can track an input frequency, or it can generate a frequency that is a multiple of the input frequency.

A network of PLL circuits can form a structure similar to the canonical oscillator network described above. The input signal of each PLL is the weighted sum of the outputs of every PLL,

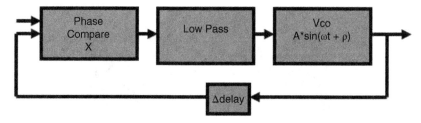

Figure 24.3 Structure of phase-locked loop

including itself. The only additional operation is that there is a $\pi/4$ delay in the feedback loop to make the input signals and reference signals inside the PLL orthogonal.

By simulation in Matlab, we can observe the properties of the PLL network more directly. We have built the model with the following parameters:

- $N = 64$ Number of oscillators
- $f = 1000$ Natural frequency of VCO (Hz)
- $d = 1/(f*100)$ Simulation step size
- $T = 0.5$ s   Simulation time
- sens $= 100$ Sensitivity of VCO
- amp $= 1$ Amplitude of VCO output
- Kd $= 1$ Gain of phase detector

$$\text{Transfer function of loop filter}: T(s) = \frac{10}{s + 10}$$

$$\text{Step iteration rule}: \Delta V_c = f_c[\text{PD}_{\text{out}} - V_c(n-1)], \ V_c(n) = V_c(n-1) + \Delta V_c$$

As an example of the operation of this model we can store two patterns, the 8×8 images of "1" and "5" and observe the convergence behaviors of the array. In Figure 24.4 we use a random binary image (as a "very noisy" input), start the simulation, and observe the evolution process of the phase behavior of the oscillators. The system converges to the closest stored pattern, which is "5" in this case.

We plot the phase error between the oscillators in terms of their cosine value. In Figure 24.5 we see the time evolution of the phase error.

In a second test, we give the system a random gray scale image, which means that the initial condition is not fixed to binary values. Similarly, Figure 24.6 and Figure 24.7 show the retrieval process.

From the experiments above, we suggest that the phase-locked loop, as both a model and an implementation for the basic neuron of oscillatory neural network, is able to behave as a canonical oscillator model. However, it is costly in terms of hardware resources compared with several novel devices presented later. It also has a slow convergence speed. In fact, based on our experiments, the PLL networks that use the Palm coding rule does not always converge.

Compared to the PLL, other oscillator circuits could be better choices for building an AM model, for instance, oscillators based on the Neuron MOS or the spin torque nano-oscillators (STNO). The Neuron MOS transistor is a type of MOS transistor that has multiple floating gates used to model a neuron [18]. In a simple CMOS ring oscillator, an inverter consisting of neuron PMOS and NMOS transistors has an input voltage capable of controlling the delay of

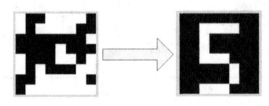

Figure 24.4 Binary image pattern retrieval in a PLL network

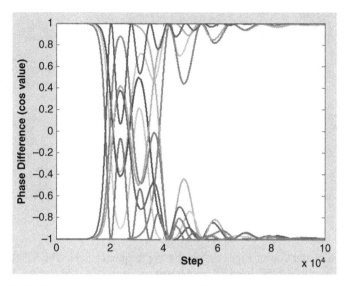

Figure 24.5 Evolution of the phase error in a binary image pattern retrieval process

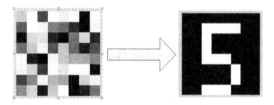

Figure 24.6 Grayscale image pattern retrieval of PLL network

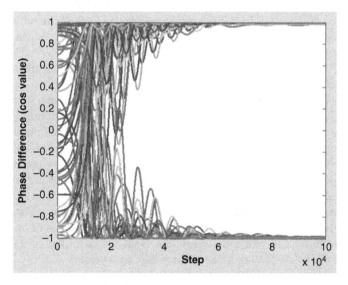

Figure 24.7 Evolution of the phase error in grayscale image pattern retrieval process

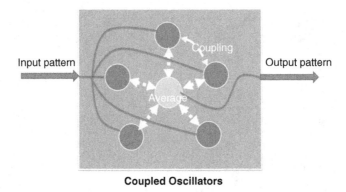

Coupled Oscillators

Figure 24.8 Coupled oscillators cluster

signal propagation and thus changes the oscillating frequency [19]. The STNO is an emerging nano-device that can also be used to implement an oscillator AM model. STNOs work on injection of high density current into multiple magnetic thin layers. Current-driven transfer of angular momentum between the magnetic layers in such a structure results in spontaneous precession of the magnetization once the current density exceeds the stability threshold. This magnetization precession generates microwave electrical signals as a result of an effect called giant magneto-resistance (GMR) that occurs in such magnetic multilayer structures [20]. STNOs demonstrate the functionality associated with VCOs, including frequency modulation and injection locking [21,22]. Given these characteristics, STNOs can couple with each other either by mutually locking on phase or by shifting each other's frequency.

Different from the structure of PLL-based oscillatory neural networks with fully connected feedback (Figure 24.2), these oscillators are designed to couple together through one node and thus form a cluster, as shown in Figure 24.8. A cluster model composed of such nano-oscillators needs no complex feedback connection matrix. Each oscillator is sensitive to all others and can simultaneously influence them without coefficients computed from multipliers and adders. Further, the number of interconnects scale as N rather than N^2. However, even for this much simpler network, the dynamic behavior of these oscillators is still determined by the energy function of the network.

A new AM unit can be designed with supportive circuits around such nano-oscillator clusters. Figure 24.9 depicts a simple structure of such an AM unit. This model's architecture is designed like a Content Addressed Memory (CAM). It is composed of multiple "Associative Words" that contain nano-clusters with local template memories and code memories, write/match control circuits, and circuits for output. As a hetero-associative memory, the template memory and code memory respectively store the memorized pattern and their code representations. Write/Match control circuits control the modes and operation of the AM unit. It has three input signals; an input pattern Xi, the code representation Ci for new stored pattern, and the model control signal. The circuits for output detect all the degrees of match (DoM) from each oscillator cluster and return the maximum DoM, d*, with the corresponding winner pattern's code, Ci, as the final outputs of the recognition process.

This AM unit has two modes, Write and Match. In the Write mode, it stores each new pattern in the local template memory of each cluster. In the Match mode, the oscillator clusters read the differences between the input pattern and stored patterns, and then outputs the (DoM). Thus,

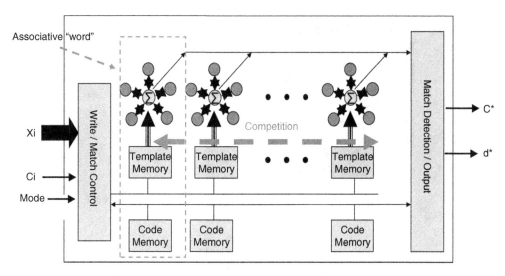

Figure 24.9   AM unit built on coupled oscillator clusters

the result for pattern recognition will be the pattern within the cluster which outputs the maximum DoM.

In the binary domain, this AM unit model can be abstracted as a simple Palm Network Model with all the patterns sparsely represented by unary codes [23]. However, in the real number domain, the oscillator clusters utilize a quite different distance metric based on the nano-oscillator's nonlinear behavior for matching.

From the perspective of nano-device technology, it may be difficult to couple a large number of nano-oscillators together, and we do not yet know about the capacity of an AM model in such a cluster structure. However, compared to the structure of traditional recurrent feedback neural networks implemented on hardware, the advantages of this AM cluster design for nano-oscillators includes fewer wires and connections, lower memory costs, and shorter convergence time.

24.2   Hierarchical Associative Memory Models

In the previous section, we have shown that weakly coupled oscillatory neural networks are capable of memorizing multiple patterns and retrieving the "closest" one to an input pattern. Taking this idea one step further, we believe that oscillatory neural networks can also perform the functions of the Palm model and form a complete associative memory system.

Nevertheless, there still exist many technical problems when novel oscillatory devices are used to form large-scale neural networks. Full connection within the associative memory requires the full and tuned connection of oscillators within the network, which is difficult for coupling between nano-scale oscillators. Even if we can build a large "flat" neural network, the maximum efficiency of a fully connected Palm model can only reach 35% [22]. Compared with a biological neural system, this type of model has a very low efficiency.

On the other hand, hierarchical structures have been identified in the neocortex of human beings and this area of the brain is believed to perform the functions of memory retrieval and

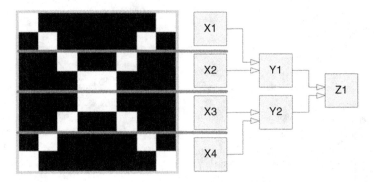

Figure 24.10   Three-level hierarchical structure and receptive field

logical inference [24]. We can hypothesize that a large number of associative memories are organized hierarchically in this structure. For both these reasons, we have investigated the design of a hierarchical associative memory system made by connecting Palm models.

24.2.1   A Simple Hierarchical AM Model

Figure 24.10 illustrates a three-level hierarchical model that can memorize 8×8 binary images (64 bits in total). As Figure 24.10 shows, each AM unit (X1 . . . X4, Y1, Y2, and Z1) represents one Palm network with 16 input neurons and eight output neurons. Every AM unit in a higher level receives the output of two units from the lower level. The training patterns and the test patterns are sliced into four pieces (shown with red lines) and are respectively fed into the four "X" modules in the bottom level.

The hierarchical model is trained from lowest level (X level) to the highest level (Z level) either using the Palm model's training rule [22] or by writing the patterns directly into the memory. After each lower level is trained, it feeds its output vectors as the training vectors to the next higher level. Therefore, eventually, every AM unit learns the patterns of its receptive field (the fraction of the pattern under it in the hierarchy) and builds the input–output mapping function to generate a corresponding output vector. The pattern retrieval task, like the training task, is processed sequentially. In this case, we put a distorted image pattern into the X level. The AM units at the X level will output unary codes to the Y level, which will then generate output codes, and the module at the Z level will recognize the whole pattern based on the Y level outputs. To more closely model the behavior of a hypothetical Palm network built from oscillatory neural networks, we adjusted the Palm model in its retrieval process. Instead of using a Heaviside function and threshold as the output, we take a "winner take all" (WTA) strategy. In this technique, the DoM is obtained by comparing two binary vectors with the bit-OR operation like a Palm model, which indicates the similarity between the input pattern and each memorized pattern. And, the pattern with the highest DoM is the winner. All other patterns will be "losers." A winner is randomly picked when multiple patterns have the same highest DoM.

In our tests, we use a unary code for the output pattern of each AM unit and train the system with four simple image patterns labeled P1 to P4 in Figure 24.11.

To test if the hierarchical model would function as expected, we input many random binary images into the model and observed the retrieval process of each AM unit. One illustrative

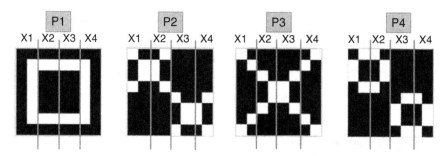

Figure 24.11   Binary image patterns for first level of hierarchical palm network

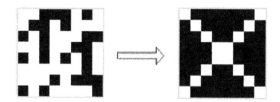

Figure 24.12   Input pattern and output pattern in function test for hierarchical model

AM	P1(dom)	P2(dom)	P3(dom)	P4(dom)	winner
X1	5	3	2	3	p1
X2	2	1	5	2	p3
X3	2	2	3	4	p4
X4	1	1	2	2	p3/p4
Y1	1	0	1	0	p1/p3
Y2	0	0	1	1	p3/p4
Z1	0	0	2	0	p3

Figure 24.13   Function test retrieval process

example of this retrieval process is shown in Figure 24.12 and Figure 24.13. In this case, the random image shown in Figure 24.12 is recognized as an "X."

Figure 24.13 explains the retrieval process of all the AM units in the hierarchy for the test in Figure 24.12. In this figure, the first column contains the name of all the AM units. The middle four columns give the DoM between the input pattern and each stored pattern from the perspective of each different AM unit. And the winner column shows the decision made by these AM units based on the DoM they generate. The circles and arrows explain how the higher levels compute the winner pattern through the information from lower levels. For example, X1 finds the local maximum DoM within its own receptive field is five, generated from pattern p1, and thus outputs p1 as the winner pattern, while X2 does the same thing but gets a different result, p3. Y1 reads the output from X1 and X2, and then puts the DoM from p1 and p3 equally

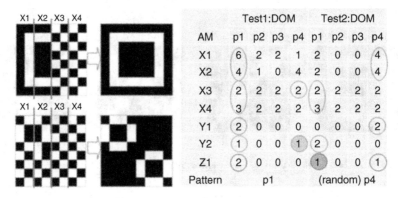

AM	Test1:DOM				Test2:DOM			
	p1	p2	p3	p4	p1	p2	p3	p4
X1	6	2	2	1	2	0	0	4
X2	4	1	0	4	2	0	0	4
X3	2	2	2	2	2	2	2	2
X4	3	2	2	2	3	2	2	2
Y1	2	0	0	0	0	0	0	2
Y2	1	0	0	1	2	0	0	0
Z1	2	0	0	0	1	0	0	1
Pattern	p1				(random) p4			

Figure 24.14 Robust test for hierarchical model

to 1. Therefore, "1010" will be Y1's output. Similarly, the top AM unit Z1 makes the final decision of the system after reading the output from Y1 and Y2. It picks the third pattern p3 as the result of this retrieval process. If we only look at the DoM output from the bottom level and sum them up for each pattern, the DoM of direct comparisons from a single AM unit can be obtained. They are 10, 7, 12, and 11, respectively for p1 to p2. Obviously p3 will also be the winning pattern with the highest DoM.

After the hierarchical model's ability to retrieve patterns was verified, we still needed to test whether the patterns retrieved met our expectations of finding the "right" pattern that is in agreement with our human perception. Hence, we developed a "robust test" where we used an input pattern with some part of the pattern exactly the same as a trained pattern and some part of the receptive field distorted. Figure 24.14 presents the recovery of two such distorted patterns and the DoM of each Palm network as the recognition process proceeds.

These simulation results show that this simple hierarchical AM model can memorize and retrieve patterns with results similar to a large single "flat" associative memory as a nearest neighbor classifier.

24.2.2 System Improvements

The simple model we discussed above illustrates the first step of our exploration of hierarchical AM architectures. However, as we explain below, there are several deficiencies in this model. We next provide an analysis of these problems and provide corresponding solutions as improvements to the AM system.

24.2.2.1 Pattern Conflict

During the pattern retrieval process, when the lower level's AM units output different patterns with the same DoM, the system sometimes performs recognition poorly due to the random choice for tie breaking. In the case shown in Figure 24.15, the correct pattern should be p1. But the AM units in the second level and top level randomly pick a pattern between p1 and p3 because they have the same DoM in the local AM unit. Since the AM units in the same layer are independent and cannot communicate with each other, the AM unit from the higher layer receives incomplete information and thus retrieves an incorrect pattern. We define this phenomenon as Pattern Conflict, which reflects the weakness of the hierarchical AM model

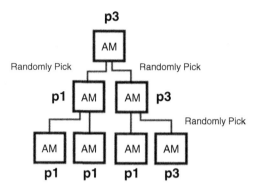

Figure 24.15   Example of the pattern conflict case

structure in passing and abstracting global information. Without more connections between networks, associative information from different receptive fields is lost from the lower levels to the higher levels.

24.2.2.2   Difficulty in Information Abstraction

In an ideal hierarchical structure, the higher level input vectors are the concatenation of lower level modules' input vectors to prevent information loss. Thus, the size of the network increases as one goes to higher levels. Unfortunately, the oscillator AM module cannot be very large due to the practical considerations of implementing large networks. Therefore, if we keep all the AM units the same size, the output patterns from the lower levels to the higher levels have to be abstracted into shorter vectors which causes a loss of the input pattern's original information.

24.2.2.3   Capacity Problem

In real applications, a hierarchical AM system may be required to memorize a large number of patterns. Since we cannot store all patterns in a single AM module, and we have fewer AM units in the higher level of a hierarchical model, the capacity problem becomes increasingly serious. Given the fact that the top level has only one AM unit, it cannot store as many patterns as the bottom level. In addition, for an oscillator network, more patterns usually mean lower speed and lower accuracy.

24.2.2.4   An Improved Hierarchical AM Model

We solve these problems described above by modifying our hierarchical design in three ways: structure, representation, and training rule. We keep the hierarchical design but change the receptive field of each block, as Figure 24.16 shows. Consider the input image as a 64×64 matrix, every module receives the input of a 2×2 square area from its lower level. The bottom Level (L1) slices the input image into 16×16 blocks and each block has 4×4 pixels. This design is better at resisting distortions and noise in the image pattern than the earlier structure.

To keep every AM module the same size, the output vectors of the four AM units are translated into a binary code and concatenated to the input vector that is fed to higher levels.

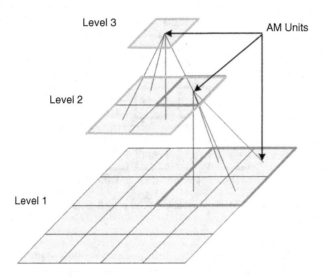

Figure 24.16 Pattern representation between layers

In this case, each AM unit can memorize 16 16-bit binary vectors. Their outputs are coded into a four-bit vector and fed to the next level, as shown in Figure 24.17.

The function of the 16-4 encoder is to convert the output of the 16-bit unary code into a four-bit binary code. The original 16-bit unary code indicates the index of the output pattern. To obtain the DoM, we use the XNOR operation instead of the multiplication in the Palm model. Figure 24.18 provides an example of how the AM unit recognizes patterns. Here the DoM is measured by the Hamming distance between the input pattern and each of the stored patterns. The unary code output has a single "1" at the index of the highest DoM.

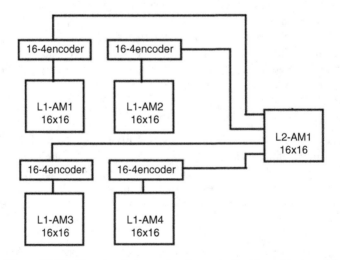

Figure 24.17 Encoders between two layers in the hierarchical model

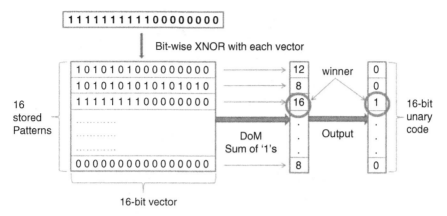

Figure 24.18 AM units function in the hierarchical model

24.2.3 Simulation and Experiments

After making these changes, we test the performance of our improved system. For these tests, we randomly pick 16 64×64 images that were converted to binary images from JPEG color pictures, as presented in Figure 24.19. Two sorts of input patterns are used for each of two tests, the first one is the original image with added noise, and the second one is a combination of different parts of two images. In these two tests, the hit rate is defined as the number of trials

Figure 24.19 Binary image set in pattern retrieval test

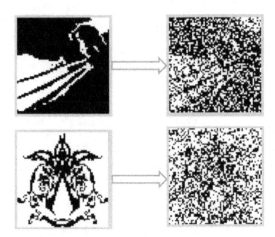

Figure 24.20 Two examples of 25% noise in the noise retrieval test

that our model retrieves the correct pattern divided by the total number of trials. The hit rate represents the ability of the model to perform pattern matching.

24.2.3.1 Noisy Pattern Retrieval Test

In the noisy pattern retrieval test, we randomly choose a percentage of image pixels and flip their value. This sort of noise is added to every image in the data set. Figure 24.20 shows the two examples of 25% noise added to an image. For this amount of noise, it is even hard for a human to recognize these patterns.

We vary the noise percentage from 0 to 50% and run each simulation 500 times for each pattern under different noisy inputs. We observe the hit rate of each simulation. Figure 24.21 gives the hit rate for all 16 patterns. We can see that even for a relatively high degree of noise, the hierarchical AM model performs quite well.

24.2.3.2 Pattern Conflict Test

To test if the Hierarchical AM model solves the pattern conflict problem, we intentionally combine two images by replacing some columns of one image by another's. In the simulation, we vary the replacement ratio from 10 to 50% and use 16 different patterns to interfere with the fourth pattern. The curves of hit rates and replacement ratios are shown in Figure 24.22. From this result we find that this model reduces pattern conflicts as we expect.

24.2.3.3 Pattern Storage

Figure 24.23a–d presents us with the utilization of storage space from the bottom level of hierarchy to the second level from the top. We did not show the top level because we know the top level has only one AM unit and stores all 16 patterns. The heights and colors of the 3-D bars indicate how many patterns each AM unit memorized for this data set. The color bar gives the color code of these four graphs from violet (0) to red (16). The maximum bar height is 16, which is the total capacity of the AM unit, represented with red. The x and y axis show the position of a module in the AM array of each layer. As we expected, AM units in the lower levels store fewer vectors than AM units in a higher level because some images share the same

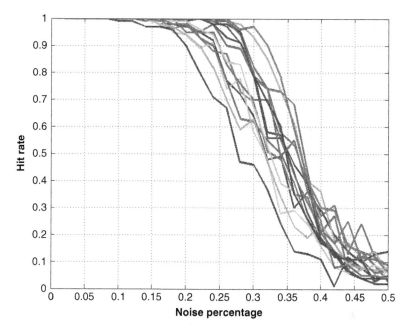

Figure 24.21   Hit rate versus noise degree in noise retrieval test

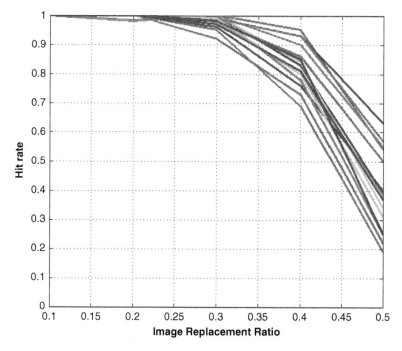

Figure 24.22   Replacement ratio versus hit rate in patten conflict test

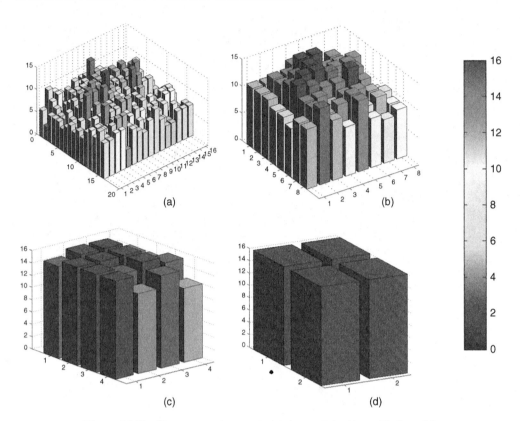

Figure 24.23   The memory usage of each layer of the hierarchical model

pattern in a small receptive field. For this test, both level 1 and level 2 run out of storage, we cannot memorize more patterns without changing the size of the AM unit. This illustrates a fundamental problem of the associative memory. If we choose to perform a direct comparison between a set of memorized patterns and an input pattern, the system must scale to accommodate all the patterns. However, it is not practical to build such a large monolithic AM unit. Therefore, in the next section, we investigate a new hierarchical architecture and algorithm which allows us to support large data sets.

24.3   N-Tree Model

To address the fundamental capacity problem that stems from the need for direct comparisons between the input and the data set, we introduce the N-Tree model [25]. In this model, we perform indirect comparisons between the input and representative patterns for subsets of the data. We also investigate the use of branch and bound search to improve the performance of this model.

24.3.1 N-Tree Model Description

The Hierarchical AM model we investigated above does not solve the capacity problem because it only partitions the pattern vectors and compresses them. The input pattern still has to be compared with all the memorized patterns for the nearest neighbor search. Therefore, it is difficult to scale and cannot effectively handle large data sets in high dimension spaces, since both the number and length of vectors grow. Given the fact that a tree structure always contains the least information in its root node and the most information in the totality of its leaf nodes, we reconsider the organization of the data set and propose another model called the N-Tree model.

24.3.1.1 Model Structure

Assume there is a pattern set with m patterns, $\{p_1, p_2, \ldots, p_m\}$, required to be memorized by an AM system. When the number of patterns, m, is very large, it becomes difficult for a single AM unit based on oscillators to hold all the patterns. As discussed previously, this is due to both fabrication and interconnects symmetry constraints. In order for the oscillators to work correctly as a pattern matching circuit, the oscillators need to be matched in performance and the interconnect needs to be symmetrical. Both of these goals can only be reliably achieved if we keep the clusters of oscillators relatively small. Therefore, as a solution to this problem we use a hierarchical AM model that organizes multiple AM units into a tree structure.

If each AM unit can only store n patterns, while $m > n$, then the m patterns can be clustered hierarchically into an N-Tree structure and stored in an AM model with the corresponding architecture. In this tree structure, every node is an AM unit and has at most n children nodes. Only leaf nodes (nodes that have no children) store specific patterns from the original input set: $\{p_1, p_2, \ldots, p_m\}$. The higher nodes are required to memorize the information associated with different levels of pattern clusters. Thus, every search operation, to retrieve one pattern, results in a path in the tree from the root to a leaf. During the retrieval process, the key pattern is input recursively into the AM nodes at a different level of the tree and finally the pattern with the highest DoM is output.

For example, if we take the simple case where $m = 8$, $n = 2$, this case forms the binary tree AM model shown in Figure 24.24. At the root node, patterns are clustered into two groups, $\{p_1, p_2, p_3, p_4\}$ and $\{p_5, p_6, p_7, p_8\}$, respectively, associated with two representative patterns: C_{11} and C_{12}. There are several ways that the C patterns can be generated during training, as explained below. From the second level, these two groups of patterns are split again into four

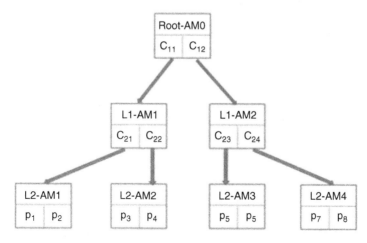

Figure 24.24 Three layers N-Tree model with fanout = 2

groups and are stored in four AM units in the third level. In effect, each AM node has an associated pattern, C, stored in its parent node which represents the patterns in its local group.

24.3.1.2 Training Process

Once the hierarchical structure of the AM model is fixed, the next question is how to organize the patterns into a tree structure and which features can be used for the recognition at each node? These questions determine the method of training and pattern retrieval. Notice the fact that the AM unit always outputs the pattern with the highest DoM, namely the one nearest to the input pattern in the data space. For this work, we use a hierarchical k-means clustering algorithm.

k-Means clustering [26] is a method of cluster analysis which aims to partition n patterns into k clusters in which each pattern belongs to the cluster with the nearest mean. Since these are multi-dimensional means, we also refer to them as "centroids."

The steps of the algorithm can be summarized as:

1. Randomly select k data points (vectors) as initial means (centroids).
2. Assign every other data point in the training set to its closest mean (Euclidian distance) and thus form k clusters.
3. Recalculate the current mean of each cluster based on all the data points that have been assigned.
4. Repeat (2) and (3) until there are no changes in the means.

Hierarchical k-means clustering is a "top-down" divisive clustering strategy. All patterns start in one cluster and are divided recursively into clusters by clustering subsets within the current cluster as one moves down the hierarchy. The clustering process ends when all the current subsets have less than k elements, and they are nondividable. This algorithm generates a k-tree structure that properly matches our AM structure. The internal nodes are centroids of each cluster in different levels and the external nodes (leaf nodes) are exact patterns. During the clustering process, some clusters may become nondividable earlier than others and, therefore, have higher positions in the tree than the other leaf nodes.

Consider the previous example, the first clustering generates two clusters, $\{p_1, p_2, p_3, p_4\}$ and $\{p_5, p_6, p_7, p_8\}$, with centroids C_{11} and C_{12}. Then these two clusters are split into four subclusters, $\{p_1, p_2\}$, $\{p_3, p_4\}$, $\{p_5, p_6\}$, and $\{p_7, p_8\}$. Their centroids are C_{21}, C_{22}, C_{23}, and C_{24}. Since the number of patterns in one cluster is less than the threshold, we set ($n = 2$) and the clustering process ends. After the centroids and patterns are written into each AM node correctly, the training process is finished.

24.3.1.3 Recognition Process

The pattern retrieval process starts from the root node. When we input the pattern that needs to be recognized, the AM unit at the root node will output the nearest centroid as the winner. This centroid will lead to a corresponding node in the next level and the system will repeat the recognition process until the current node is a leaf node. The final output is a stored pattern rather than a centroid. Considering the same example as above, assume we have a test pattern p_t, which is closest to p_5. The retrieval process is:

1. Input p_t to the root node AM0, the output pattern is C_{12}, which points to the AM2 of level 1, with the cluster $\{p_5, p_6, p_7, p_8\}$.

2. Input p_t to L1-AM2, the output pattern is C_{23}, which points to the AM3 of level 2, with the cluster $\{p_5, p_6\}$.
3. Input p_t to L2-AM3, the final output pattern is p_5.

In general, the N-Tree model reduces the steps of comparison by locating the input pattern in different clusters. Also, this architecture makes an AM system capable of scaling up or down as the data set changes. We only need to vary the tree structure of the data set.

24.3.2 Testing and Performance

In this section, we use two image data sets to evaluate the performance of the N-Tree model. The first data set is from the ATT Cambridge Image Database [27]. This data set contains 400 grayscale images of human faces. These images are photos of 40 people with 10 views for each person. The original images have 92×112 pixels with 256 gray levels.

We convert the image size into 32×32 and normalize the intensity of all the images by the following method:

1. Calculate the means of the grayscale values of all pixels for each image, $(x_1, x_2, \ldots, x_{400})$.

2. Calculate the mean of all images' pixels, $(x_1, x_2, \ldots, x_{400})$, $M = \frac{\sum_{i=1}^{400} x_i}{400}$.

3. For the i-th $(1 <= i <= 400)$ image, every pixel is scaled by multiplying the factor $\alpha = \frac{M}{x_i}$ so that all the images in the data set are normalized to the same intensity.

For the tests in this section, we do not use any image processing techniques, like feature extraction, to preprocess these images. Rather, the pixel values of an image, or centroid, are directly stored as one vector in our AM model. The purpose of the normalization of size and intensity is to make all pattern vectors fit our model.

In these tests we define the size of our AM unit as 16×1024. One AM unit can store 16 images and each image has 1024 (32×32) pixels. Thus the nodes of the tree structure can have at most 16 child nodes, instead of two in the example above. The definition of the algorithms in terms of both the training and recognition process remain unchanged.

In this task, the N-Tree model is required to recognize the person in the input image after training. We use a hierarchical version of the ARENA algorithm, a simple, memory-based algorithm for face recognition algorithm [28], which performs nearest neighbor search with a reduced-resolution face image. For evaluation, first, we split the data set into two parts, a training data set and a test data set. The partitioning is based on the subjects. For each subject, we pick nine images for training and one for test. Thus there are 360 images in the training data set and 40 images in the test data set.

After the tree structure has been built from the 360 images, we input images of the test data set sequentially. If the output image and the input image belong to the same subject (from different views) we say the recognition is successful, otherwise we call it a failure. The "hit rate" of this simulation is computed based on the 40 test results, one per subject.

In order to evaluate the system's performance, we use a cross-validation technique [29]. For each run, we randomly pick images for training and for test of each subject such that the model is trained and tested on different data set every time. One example of this partitioning is shown in Figure 24.25. We performed 1000 runs each with a different partition.

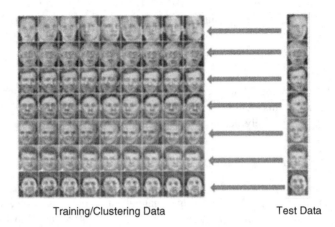

Training/Clustering Data   Test Data

Figure 24.25   Part of ATT data set partitioned for training and testing

This testing environment allowed us to examine two questions. First, what is the influence on performance brought by different norms used in the nearest neighbor search? And second, how good is our hierarchical model's performance compared with a single AM module? Each of these questions is fundamentally important to understand the capability of the N-Tree. Because we have no precise model of a distance metric for the behavior of coupled nano-oscillators, we use several different norms to test in our model. Moreover, since the retrieval performance depends on both the data set and the algorithm we use, our goal is to achieve the performance as an ideal single AM.

To gain some insight into the qualities of different norms, we tested different distance metrics for both the k-means clustering and the searching processes. The model was tested using an L2 norm, L1 norm, and L0 norm respectively.

The L2 norm is the Euclidean distance. On an n-dimensional vector space, the distance between two vectors x and y can be defined as:

$$d(x,y) = \sqrt{\sum_{i=1}^{n} (x_i - y_i)^2} \tag{24.2}$$

The L1 norm is also known as city block distance, or Manhattan distance, which is defined as:

$$d(x,y) = \sum_{i=1}^{n} |x_i - y_i| \tag{24.3}$$

Similarly, the L0 norm is defined as:

$$d(x,y) = \sum_{i=1}^{n} \delta_i, \; where \; \delta_i = \begin{cases} 1, & if \; x_i - y_i > \theta \\ 0, & otherwise \end{cases} \tag{24.4}$$

θ is a customized threshold that is determined by the data set in this case.

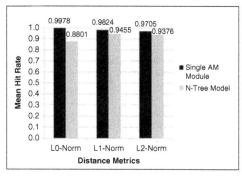 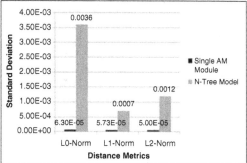

Figure 24.26   Retrieval performance on ATT data set

For comparison between our hierarchical methods and a direct implementation of ARENA, we assume we have an ideal single large AM unit that can hold all 360 images and can classify each input image by directly comparing the DoM to every memorized image. This model represents the best performance possible using pixel by pixel distance to classify the face image data.

The results of these tests are summarized in Figure 24.26. It gives the average hit rates and standard deviations of the N-Tree model and the ideal single AM module under different norms. While the standard deviations of all these tests are very small compared to the average hit rates, the N-Tree model's standard deviations are larger than the single AM module because different training data sets generate different trees of data clusters, which directly influences the performance of retrieval. Considering the three different norms, the L0 norm is the best distance metric for nearest neighbor search, while the L1 norm is the second, and the L2 norm is the worst. However, the L0 norm performs the worst for N-Tree model, because it is not a good norm for the k-mean clustering algorithm and thus leads to a bad data clustering.

For the ATT data set, when compared to an ideal "flat" single AM model, our N-Tree model is very close to the best performance with a higher efficiency. Even though the AM performs its comparisons in parallel for each case, the large single AM model has to compute the DoM for all of the 360 stored images, while the N-Tree hierarchical model with a three-level depth performs at most $3 \times 16 = 48$ operations of comparison with a time penalty of only 3×.

After verifying our model's functionality as an AM system, we employ the same noise test as we used in the previous section to examine the performance of the N-Tree model on a different real application. Salt and pepper noise is a form of noise typically seen on images. It appears as randomly occurring white and black pixels. We keep the training data set and test data set unchanged but make noisy input images by adding salt and pepper noise to test images with a density ranging from 0 to 0.5. Figure 24.27 shows different levels of noise density.

We run the simulation 1000 times with each image under each noise density and compute the hit rate. Figure 24.28 shows the 40 hit rate curves representing each of the 40 images' retrieval tasks. Though a noise density larger than 0.1 can make the image hard to recognize for human eyes, the N-Tree model can still achieve a high hit rate for most of these images.

Although the tests above show good performance for the hierarchical AM model on the ATT data set, we note that our model still fails to achieve the exact performance of the large single AM model (ARENA's direct implementation). In order to test our performance on a larger and more challenging problem, we used a second data set from The Facial Recognition Technology (FERET) Database [30]. The test subset we used has 2015 grayscale images of human faces

Density from 0 to 0.5, Increment = 0.02

Figure 24.27   Images with salt and pepper noise of different densities

from 724 people. For this data set, each subject has a different number of images, which range from 1 to 20. These images also include various facial expressions.

For the second set of tests, we use the same technique and same system specification as the ATT data set tests. That is, for each person, pick one image for testing and the rest for training. This gives 1291 images in the training data set and 724 images in the test data set. However, among these 724 subjects, the training set may have very few or even no images for some people. For example, if there is only one image for a person, this image will be used as test data and never be recognized correctly because the cluster analysis will have no instances of the person to train with. Hence, the recognition task on this data set is more difficult than that of the ATT data set due to it having more classes, unbalanced training data, and images with diverse features.

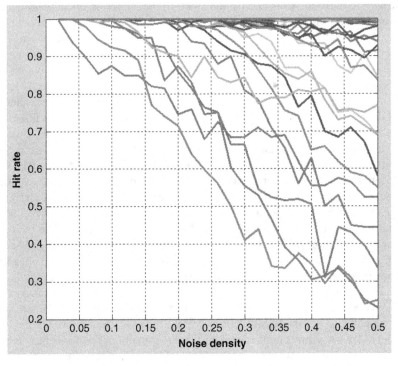

Figure 24.28   Noise test results of N-Tree model

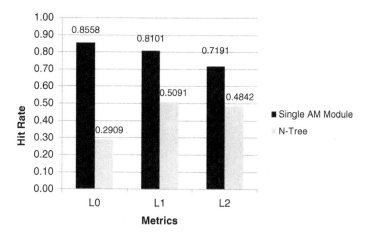

Figure 24.29   Retrieval performances on FERET data set

The results are presented in Figure 24.29, which indicates that hit rates of both the single flat AM model and the N-Tree decreased when compared to the ATT data set. Further, for this test data, the N-Tree model performs much worse than the single AM model with hit rates of only 29–51%.

24.3.3   Fan-out versus Performance

Since our goal is to obtain the same retrieval performance as a single flat AM, we explore the correlation between the clustering tree's fan-out and performance. The fan-out of the N-Tree model is the size of each AM node. For example, if the fan-out is four, each data cluster splits into four subclusters during hierarchical k-means clustering, and each AM node can only store four subcluster centroids or image vectors. For a tree of a data set generated from hierarchical k-means clustering, fewer children for each node usually means smaller clusters, which may increase the possibility of making a mistake during the retrieval procedure. This is because once an AM unit outputs the "closest" centroid of a subcluster, the input pattern is compared with data in this subcluster only and may miss the real closest patterns. However, a larger cluster means more comparison operations and lower speed, to give a higher possibility of finding the right answer. We repeat the same simulation on the FERET data set but vary the fan-out number from 2 to 30. The results are all plotted in Figure 24.30. As we expect, generally a higher fan-out number improves the hit rate to some extent. However, it is still far worse than our goal.

24.3.4   Branch and Bound Search

The simulation results above tell us that the clustered data structure generated from the hierarchical k-means algorithm cannot provide us the same performance as the flat single AM model. The reason is that the hierarchical k-means clustering is initialized randomly and might not always cluster all the data to its nearest neighbor centroid.

To improve the performance of the hierarchical architecture, we adopt an algorithm called branch and bound search [31]. The basic idea is to search additional nodes that possibly contain a better answer, after we finish a routine N-Tree search. In other words, when a leaf node is

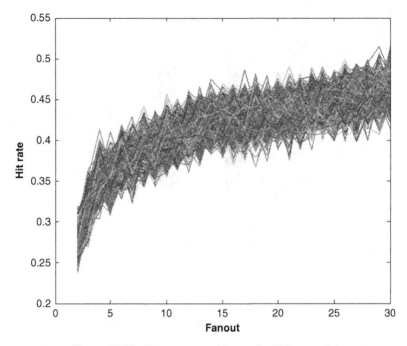

Figure 24.30 Fan-out versus hit rate for N-Tree model

visited at the end of the walk down of the tree using associative processing, we backtrack on the tree and check other nodes. The branch and bound search algorithm provides a method for generating a list of other nodes that need to be searched in order to optimize the result.

In branch and bound search, a simple rule can be used to check if a node contains a possible candidate for the nearest neighbor to the input vector. For a cluster with centroid C_i, we define its radius r_i as the farthest distance between the centroid and any element in the cluster. Here we define X as the input vector and B as the distance between X and the best nearest neighbor found so far. Thus, if a cluster satisfies the condition:

$$B + r_i < d(X, C_i) \tag{24.5}$$

where d is the distance function, then no element in cluster C_i can be nearer to X than the current bound B. Initially, B is set to be $+\infty$. The distance function d gives the distance between two vectors as defined by a norm, for example, L0, L1, or L2. This rule is illustrated in Figure 24.31.

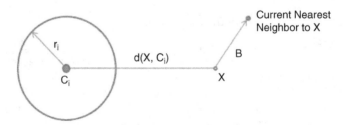

Figure 24.31 Branch and bound search rule

The full branch and bound search algorithm can be summarized as:

1. Set $B = +\infty$, current level $L = 1$, and current node to be root node.
2. Place all child nodes of root node into the active list.
3. For each node in the active list at the current level, if it satisfies the check rule, remove it from the active list.
4. If the active list at the current level is empty, backtrack to the previous level, set $L = L - 1$, terminate the algorithm if $L = 0$, go to (2) if L is not 0. If there is a node in the list, go to (5).
5. In the active list at the current level, choose the node that has the centroid nearest to input vector and call it the current node. If the current level is the final level, go to (6). Otherwise set $L = L + 1$, go to (1).
6. In the final level, choose the stored vector which is nearest to the input vector X, and update B. Back track to previous level, set $L = L - 1$, go to (2).

Theoretically, branch and bound search can be guaranteed to find the nearest neighbor of an input vector from a hierarchical k-means clustered data set, which means it should have the same recognition performance as the flat single AM model.

Therefore, we enhance the N-Tree AM search with the branch and bound search algorithm and repeat the face recognition task. With the same training model, we see that branch and bound search always achieves the same performance as the ARENA algorithm, regardless of data set and fan-out for the N-Tree model. Figure 24.32 compares the hit rates of the flat (ARENA) model, the pure N-Tree model, and the branch and bound algorithms when the fan-out is four. The results indicate that the N-Tree model with branch and bound search can achieve the same performance of the ideal single AM module.

Table 24.1 gives us the number of nodes visited for the branch and bound search and indicates the efficiency of this technique compared to the simple N-Tree search with the L1 norm. However, the price of higher performance is lower efficiency. Without branch and bound search, the number of nodes visited is always just the depth of N-Tree structure, on average 9.3. To test all the possible candidates for nearest neighbor, the algorithm has to access additional nodes in the tree, on average 159.9, with a standard deviation of 86.2. This extra work is what is

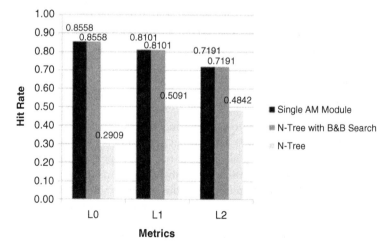

Figure 24.32 Retrieval performances on FERET data set

Table 24.1 Nodes visited in branch and bound search

B&B search	Total nodes	Visited nodes	Visited leaf nodes
Maximum	1199	454	293
Minimum	795	9	1
Mean	956.7	159.9	59.6
Standard deviation	34.3	86.28	47.4
Average N-Tree size	Total nodes	Depth	Leaf nodes
	956.7	9.3	846.7

required to achieve the comparable performance of the flat network shown in Figure 24.32. Thus, there is an obvious tradeoff between hit rate and processing speed. The original search routine for the N-Tree model is "depth-only" one-way from the root node to a leaf node. In other words, if one AM unit in the middle level makes a wrong decision due to an imprecise clustered data set, the final output vector will be incorrect. In contrast, the branch and bound search provides opportunities to backtrack and search more possible clusters in other AM units at the cost of total search speed. As Table 24.1 shows, the average number of nodes that branch and bound search visits is 17 times as many as the original search of N-Tree model.

To speed up the search, we adjust the branch and bound search by fixing the condition of the checking rule as follows:

$$B + \alpha \cdot r_i < d(X, C_i) \tag{24.6}$$

where α is a factor used to reduce the effective radius of the clusters, and ranges from 0 to 1. For some clusters, most elements are far away from X, only very few of them are in the bound. These clusters rarely contain the nearest neighbor. However, by the previous rule, we have to search them to guarantee optimality. Reducing the cluster's radius can help us avoid these less likely clusters and access significantly fewer nodes in the process.

In a final test, we sweep the value of α from 1 to 0 and observe the trends of hit rates and number of nodes visited during the search. The result is shown in Figure 24.33. We notice that the slopes of the three curves are different. Nodes searched decreases as radius is reduced. But the hit rate stays the same until α gets to 0.4. Thus, by reducing the cluster radius to some extent, we can improve the efficiency of searching without significantly sacrificing the hit rate.

24.4 Summary and Conclusion

24.4.1 Summary

In this chapter, we introduce the basic concept and brief history of associative memories both from the perspective of a pattern matching memory and in terms of their relationship to Oscillatory Neural Networks and develop a hierarchical associative memory for pattern matching.

We start by presenting the simulation of a hardware-based model of an Oscillatory Neural Network using phase-locked loops and simulate this model to show the functionality of these systems for pattern matching. We show that they perform this function by minimizing the energy of the system as they phase and frequency lock. We develop the design of an Associative Unit as a basic architectural building block and discuss several alternative emerging technology-based devices as candidates for efficient implementations of these blocks.

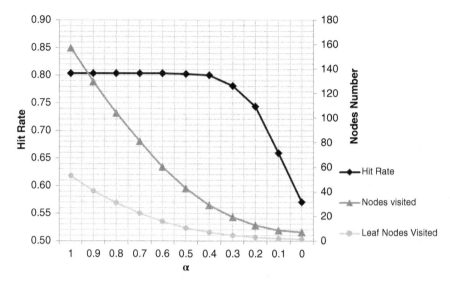

Figure 24.33 Performance versus speed of modified branch and bound search

We then show a set of evolving designs for a hierarchical associative memory system. Our first hierarchical AM model is a simple system containing three levels of AM units. We then identify three problems with this simple design when compared to an ideal fully connected "flat" AM: Pattern Conflicts, Capacity Limits, and Representation Limits. To solve these problems we develop an improved version of the hierarchical model, which can solve the pattern conflict and representation limits but cannot solve the capacity limit problem.

We solve this final problem with a fundamentally different architecture: the N-Tree hierarchical model. For this model we change both the architecture and the recognition algorithm. We first show how the stored patterns are clustered and represented. Based on this data structure, we build a corresponding hierarchical tree structure from the same AM units. We test this architecture on two human face recognition problems. To further increase the recognition performance of the system, we add a branch and bound search algorithm which increases performance with little impact on search times.

24.4.2 Conclusion

Based on the design explorations in this chapter, we foresee several important research directions that can fundamentally change our notion of computing in the next decade.

First, we need to pursue both technologies and new computational functions at the same time, and not be tied to Boolean operations as the common building block of all computation. We see that for oscillatory neural networks, we can compute the DoM in an associative memory based on the attractor basins of the system's energy function. These can be formed by adjusting the coupling strength weights between oscillators or by using offsets on the frequency of the oscillators themselves. These networks can be implemented in one of several new technologies, including spin torque oscillators and resonant body transistor oscillators. Thus, associative pattern matching and retrieval operations can be done without using Boolean primitives.

Second, this is an opportunity to re-examine our notions of "memory" and "processing" as two distinct functions. Rather the incorporation of local processing together with storage might be a more promising avenue for the application of future nano-technologies. However, we need to address at the architectural level both the advantages and deficiencies of new technologies. We show that the scaling limitations of the oscillatory networks can be overcome by the use of hierarchy and partitioning.

Third, we must consider the application domain since we are not building general purpose processors. However, understanding the application can allow us to tailor the algorithms to more closely match the capabilities of the underling architecture. We show that the N-Tree model can realize an efficient associative memory for high dimension real number vectors and can approach the performance of an ideal model of nearest neighbor search. However, this is only possible because we first perform a clustering of the data. Thus, we partition the application into two phases – clustering and pattern retrieval – to optimize our use of the limited size of the AM modules.

Finally, we must be creative in our solutions. We see that the limitations on the size of the AM modules leads to a partitioning of the data that is less than ideal. Therefore, we add the use of a branch and bound search algorithm to improve the hit rate at the expense of multiple probes of the memory. This enlarges the search area and thus increases the possibility of finding the nearest pattern. Branch and bound search algorithms slow down the retrieval process but this tradeoff can be optimized by varying the search rules. Therefore, by moderately increasing the complexity of our algorithm, we can approach an optimum solution for our particular application.

Acknowledgments

The authors would like to thank the Intel University Research Office; George Bourianoff, Youssry Botros, and Dmitri Nikonov from Intel; Mircea Stan from the University of Virginia; Dan Hammerstrom from Portland State University; Dana Weinstein from MIT; Mathew Puffal and William Rippard from NIST; Tamas Roska from PPCU, Hungary; and Valeriu Beiu from UAEU, United Arab Emirates. This work was funded, in part, by the National Science Foundation under the grants INSPIRE "Track 1: Sensing and Computing with Oscillating Chemical Reactions," DMR-1344178 and NSF "Expeditions in Computing Collaborative Research: Visual Cortex on Silicon," CCF-1317373.

References

1. Lapique, L. (1907) Sur l'Excitation Electrique Des Nerfs. *Journal of Physiology*, **1907**, 620–635.
2. Mead, C. (1989) *Analog VLSI and Neural Systems*, Addison-Wesley, Reading, MA.
3. McCulloch, W.S. and Pitts, W. (1943) A logical caculus of ideas immanet in nervous activity. *Bulletin of Mathmatical Biophysics*, **5**, 59–62.
4. Hodgkin, A.L. and Huxley, A.F. (1952) Currents carried by sodium and potassium ions through the membrane of giant squid axon of loligo. *Journal of Physiology*, **116**, 449–472.
5. Hawkins, J. and Dileep, G. (2006) *Hierarchical Temporal Memory: Concepts, Theory and Terminology*, Numenta Inc., www.numenta.com.
6. Riesenhuber, M. and Poggio, T. (1999) Hierarchical models of object recognition in cortex. *Nature Neuroscience*, **2**, 1019–1025.
7. Hakin, S. (1999) *Neural Networks: A Comprehensive Foundation*, Prentice Hall, New Jersey.
8. Steinbuch, K. and Piske, U.A.W. (1963) Learning matrices and their applications. *IEEE Transactions on Computers*, **EC-12**, 846–862.
9. Willshaw, D.J., Buneman, O.P. *et al.* (1969) Non-holographic associative memory. *Nature*, **222**, 960–962.

10. Buckingham, J. and Willshaw, D. (1992) Performance characteristics of the associative net. *Network: Computation in Neural Systems*, **3**(4), 407–414.
11. Henson, R.N.A. and Willshaw, D. (1995) Short-term associative memory. *Proceedings of the INNS World Congress on Neural Networks*, **1**, 438–441.
12. Anderson, J.A., Silverstein, J.W. *et al.* (1977) Distinctive features, categorical, perception, and probability learning: some applications of a neural model. *Psychological Review*, **84**, 413–451.
13. Anderson, J.A. (1993) *Associative Neural Memories: Theory and Implementation* (ed. M.H. Hassoun), Oxford University Press.
14. Anderson, J.A., Allopenna, P. *et al.* (2007) Programming a Parallel Computer: The Ersatz Brain Project, in *In Challenges for Computational Intelligence*, Springer, Berlin Heidelberg, pp. 61–98.
15. Hopfield, J.J. (1982) Neural networks and physical systems with emergent collective computational abilities. *Proceedings of the National Academy of Sciences*, **79**(8), 2554–2558.
16. Hoppensteadt, F.C. and Izhikevich, E.M. (1996) Synaptic organizations and dynamical properties of weakly connected neural oscillators: I. Analysis of canonical model. *Biological Cybernetics*, **75**, 129–135.
17. Hoppensteadt, F.C. and Izhikevich, E.M. (2000) Pattern recognition via synchronization in phase-locked loop neural networks. *IEEE Transactions on Neural Networks*, **11**, 734–738.
18. Shibata, T. (1993) Neuron MOS binary-logic integrated circuits: I. Design fundamentals and soft-hardware-logic circuit implementation. *IEEE Transactions on Electron Devices*, **40**(3), 570–576.
19. Shibata, T., Zhang, R., Levitan, S.P. *et al.* (2012) CMOS supporting circuitries for nano-oscillator-based associative memories, 13th International Workshop on Cellular Nanoscale Networks and Their Applications (CNNA).
20. Tsoi, M. *et al.* (1998) Excitation of a magnetic multilayer by an electric current. *Physical Review Letters*, **80**, 4281–4284.
21. Pufall, M.R., Rippard, W.H., Kaka, S. *et al.* (2005) Frequency modulation of spin-transfer oscillators. *Applied Physics Letters*, **86**(8), 082506–082506-3.
22. Rippard, W.H., Pufall, M.R., Kaka, S. *et al.* (2005) Injection locking and phase control of spin transfer nano-oscillators. *Physical Review Letters*, **95**(6), 067203.
23. Palm, G. (1980) On associative memory. *Biological Cybernetics*, **36**, 19–31.
24. George, D. (2008) *How the Brain Might Work: A Hierarchical and Temporal Model for Learning and Recognition*, Electrical Engineering, Stanford University.
25. Levitan, S.P., Fang, Y., Dash, D.H. *et al.* (2012) Non-Boolean Associative Architectures Based on Nano-oscillators, 13th International Workshop on Cellular Nanoscale Networks and Their Applications (CNNA).
26. Duda, R.O., Hart, P.E., and Stork, D.G. (2000) *Pattern Classification*, John Wiley & Sons.
27. Cambridge University (2013) http://www.cl.cam.ac.uk/research/dtg/attarchive/facedatabase.html (accessed 16 July 2013).
28. Sim, T., Sukthankar, R., Mullin, M.D., and Baluja, S. (2000) High-Performance Memory-based Face Recognition for Visitor Identification, Proceedings of IEEE Conference on Automatic Face and Gesture Recognition, Grenoble, pp. 214–220.
29. Devijver, P.A. and Kittler, J. (1982) *Pattern Recognition: A Statistical Approach*, Prentice-Hall, London, GB.
30. NIST (2012) http://www.itl.nist.gov/iad/humanid/feret/feret_master.html (accessed 16 July 2013).
31. Fukunaga, K. (1975) A branch and bound algorithm for computing k-nearest neighbors. *IEEE Transactions on Computers*, **C-24**(7), 750–753.

25

Storage Class Memory

Geoffrey W. Burr[1] and Paul Franzon[2]
[1]*IBM, USA*
[2]*North Carolina State University, USA*

25.1 Introduction

One challenge in computing systems is the need for new memory technologies that can improve overall performance. More and more frequently, the ability of a CPU to rapidly execute programs is being limited by the rate at which data can arrive at the processor. Unfortunately, scaling does not automatically solve this problem. One evolutionary solution has been to increase the size of cache memory and thus the floor space that SRAM occupies on a CPU chip. However, this trend eventually leads to a decrease of the net information throughput.

While DRAM offers higher density than SRAM, auxiliary circuitry is required to maintain the stored data. True nonvolatility has conventionally required external storage media (e.g., magnetic Hard Disk Drives (HDDs), optical CDs, etc.), with access times that are slower than the volatile memory by many orders of magnitude. Solid-State Disks (SSDs) based on NAND Flash have recently offered nonvolatility at significantly lower latencies than HDDs, but are block-based, much slower to erase than to program, and offer fairly poor cycle endurance.

The development of an *electrically accessible nonvolatile* memory with *high speed, high density*, and *high endurance*, referred to as "Storage Class Memory" or SCM, would initiate a revolution in computer architecture. By using CMOS-compatible fabrication technology scaled beyond the present limits of SRAM and FLASH, such a memory technology could enable breakthroughs in both stand-alone and embedded memory applications. This development could potentially provide a significant increase in information throughput beyond the benefits traditionally associated with scaling CMOS devices into the nanoscale.

This chapter is structured as follows. Sections 25.2 and 25.3 motivate and define SCMs. Sections 25.4 through 25.6 discuss target specifications and potential device solutions. Sections 25.7 through 25.10 discuss architectural implications of SCMs and open issues related to architectural exploitation.

Emerging Nanoelectronic Devices, First Edition. An Chen, James Hutchby, Victor Zhirnov and George Bourianoff.
© 2015 John Wiley & Sons, Ltd. Published 2015 by John Wiley & Sons, Ltd.

25.2   Traditional Storage: HDD and Flash Solid-state Drives

Conventionally, magnetic hard-disk drives are used for nonvolatile data storage. The cost of HDD storage in US$/GB is extremely low and continues to decrease. Although the bandwidth with which contiguous data can be streamed is high, the poor random access time of HDDs limits the maximum number of I/O requests per second (IOPs). Also, HDDs have relatively high energy consumption, large form factor, and are subject to mechanical reliability failures in ways that solid state technologies are not. However, the sheer number and growth in HDD shipments per year (380 000 Petabytes in 2012, growing at 32% per year) means that magnetic disk storage is highly unlikely to be "replaced" by solid-state drives at any time in the foreseeable future [1].

Nonvolatile semiconductor memory in the form of NAND Flash has become a widely used alternative storage technology, offering faster access times, smaller size and lower energy consumption when compared to HDD. However, there are several serious limitations of NAND Flash for storage applications, such as poor endurance (10^4–10^5 erase cycles), modest retention (typically 10 years on a new device, but only 1 year at the end of rated endurance lifetime), long erase time (~ms), and high operating voltage (~15 V). Another difficult challenge of NAND Flash SSD is due to its page/block-based architecture. By not allowing for direct overwrite of data, sophisticated garbage collection, wear-leveling and bulk erase procedures are required. This in turn requires additional computation (thus limiting perform-ance and increasing cost and power associated with a local processor, RAM, and logic), and over-provisioning which further increases cost per effective user-bit of data [2].

Although Flash memory technology continues to project for further density scaling, inherent performance characteristics such as read, write, and erase latencies have been nearly constant for over a decade [3]. While the introduction of multi-level cell (MLC) Flash devices extended Flash memory capacities by a small integral factor (2–4), the combination of scaling and MLC have resulted in the degradation of both retention time and endurance, two parameters critical for storage applications. The migration of NAND Flash into the vertical dimension above the silicon is expected to continue this trend of improving bit density (and thus cost per bit) while maintaining or even slightly degrading the latency, retention, and endurance characteristics of present-day NAND Flash.

This outlook for existing technologies has opened interesting opportunities for prototypical and emerging research memory technologies to enter the nonvolatile solid-state memory space.

25.3   What is Storage Class Memory?

Storage class memory (SCM) describes a device category that combines the benefits of solid-state memory, such as high performance and robustness, with the archival capabilities and low cost of conventional hard-disk magnetic storage [4,5]. Such a device requires a nonvolatile memory (NVM) technology that could be manufactured at a very low cost per bit.

A number of suitable NVM candidate technologies have long received research attention, originally under the motivation of readying a "replacement" for NAND Flash, should that prove necessary. Yet the scaling roadmap for NAND Flash has progressed steadily so far, without needing any replacement by such technologies. So long as the established commodity continues to scale successfully, there would seem to be little need to gamble instead on implementing an unproven replacement technology.

However, while these NVM candidate technologies are still relatively unproven compared to Flash, there is a strong opportunity for one or more of them to find success in applications that

Table 25.1 Target device and system specifications for SCM

Parameter	Benchmark			Target	
	HDD	NAND flash	DRAM	Memory-type SCM	Storage-type SCM
Read/write latency	3–10 ms	~100 μs (block erase ~1 ms)	<100 ns	<200 ns	1–5 μs
Endurance (cycles)	Unlimited	10^3–10^5	Unlimited	$>10^9$	$>10^6$
Retention	>10 yr	~10 yr	64 ms	>5 d	~10 yr
ON power (W/GB)	0.003–0.05	~0.01–0.04	0.4	<0.4	<0.10
Standby power	~52–69% of ON power	<10% ON power	~25% ON power	<5% ON power	<5% ON power
Areal density	~10^{11} bit/cm^2	~10^{11} bit/cm^2	~10^9 bit/cm^2	$>10^{10}$ bit/cm^2	$>10^{10}$ bit/cm^2
Cost (US$/GB)	~0.1–1.0	2	10	<10	<3–4

do not involve simply "replacing" NAND Flash. Storage Class Memory can be thought of as the realization that many of these emerging alternative nonvolatile memory technologies can potentially offer significantly *more* than Flash, in terms of higher endurance, significantly faster performance, and direct-byte access capabilities. In principle, Storage Class Memory could engender two entirely new and distinct levels within the memory and storage hierarchy. These levels would be differentiated from each other by access time, with both levels located within the more than two orders of magnitude between the latencies of off-chip DRAM (~80 ns) and NAND Flash (20 μs).

Table 25.1 lists a representative set of *target* specifications for SCM devices and systems, which are compared against benchmark parameters offered by existing technologies (HDD, NAND Flash, and DRAM). Two columns are shown, one for slower S-class Storage Class Memory, and one for fast M-class SCM.

25.3.1 Storage-type SCM

The first new level, identified as S-type storage-class memory (S-SCM), would serve as a high-performance solid-state drive, accessed by the system I/O controller much like an HDD. S-SCM would need to provide at least the same data retention as Flash, allowing S-SCM modules to be stored offline, while offering new direct overwrite and random access capabilities (which can lead to improved performance and simpler systems) that NAND Flash devices cannot provide. However, it would be absolutely critical that the eventual device cost for S-SCM be no more than 1.5–2.0× higher than NAND Flash. While such costs need not be realized immediately at first introduction, it would need to be very clear early on that costs would steadily approach such a level relative to Flash, in order to guarantee large unit volumes and justify the sizeable up-front capital investment in an unproven new technology. Note however that such system cost reduction can come from sources other than the raw cost of the device technology: a slightly higher-cost NVM technology that enables a simple, low-cost SSD by eliminating or simplifying costly and/or performance-degrading overhead components would achieve the same overall goal. If the cost per bit could be driven low enough through ultrahigh memory density, ultimately such an S-SCM device could potentially replace magnetic hard-disk drives in enterprise storage server systems as well as in mobile computers.

25.3.2 Memory-type SCM

The second new level within the memory and storage hierarchy, termed M-type storage-class memory (M-SCM), should offer a read/write latency of less than ~200 ns. These specifications would allow it to remain synchronous with a memory system, allowing direct connection from a memory controller and bypassing the inefficiencies of access through the I/O controller. The role of M-SCM would be to augment a small amount of DRAM to provide the same overall system performance as a DRAM-only system, while providing moderate retention, lower power/GB and lower cost/GB than DRAM. Again, as with S-SCM, the cost target is critical. It would be desirable to have cross-use of the same technology in either embedded applications or as a standalone S-SCM, in order to spread out the development risk of an M-SCM technology. The retention requirements for M-SCM are less stringent, since the role of nonvolatility might be primarily to provide full recovery from crashes or short-term power outages.

Particularly critical for M-SCM will be device endurance, since the time available for wear-leveling, error correction, and other similar techniques is limited. The volatile portion of the memory hierarchy will have effectively infinite endurance compared to any of the nonvolatile memory candidates that could become an M-SCM. Even if device endurance can be pushed well over 10^9 cycles, it is quite likely that the role of M-SCM will need to be carefully engineered within a cascaded cache or other Hybrid Memory approach [6]. That said, M-SCM offers a host of new opportunities to system designers, opening up the possibility of programming with truly persistent data, committing critical transactions to M-SCM rather than to HDD, and performing commit-in-place database operations.

25.4 Target Specifications for SCM

Since the density and cost requirements of SCM transcend the straightforward scaling application of Moore's Law, additional techniques will be needed to achieve the ultrahigh memory densities and extremely low cost demanded by SCM, such as: (1) 3-D integration of multiple layers of memory, currently implemented commercially for write-once solid-state memory [7] and/or (2) multiple bits per cell (MLC) techniques.

Table 25.1 lists a representative set of *target* specifications for SCM devices and systems compared with benchmark parameters of existing technologies (HDD and NAND Flash). As described above, SCM applications can be expected to naturally separate based on latency. Although S-class SCM is the slower of these two targeted specifications, read and write latencies should be in the 1–5 μsec regime in order to provide sufficient performance advantage over NAND Flash. Similarly, endurance of S-class SCM should offer at least one million program-erase cycles, offering a distinct advantage over NAND Flash. In order to support offline storage, 10 year retention at 85 °C should be available.

To make overall system power usage competitive with NAND Flash and HDD, and since faster I/O interfaces can be expected to consume considerable power, the device-level power requirements must be extremely minimal. This is particularly important since low latency is necessary but not sufficient for enabling high bandwidth–high parallelism is also required. This in turn mandates a sufficiently low power per bit access, both in terms of peripheral circuitry and device-level write and read power requirements. Finally, standby power should be made extremely low, offering opportunities for significant system power savings without loss of performance through rapid switching between active and standby states.

To achieve the desired cost target of within 1.5–2.0× of the cost of NAND Flash, the effective areal density will similarly need to be very close to NAND Flash. This low-cost structure would then need to be maintained by subsequent SCM generations, through some combination of further scaling in lateral dimension, by increasing the number of multiple layers, or by increasing the number of bits per cell.

Also shown in Table 25.1 are the target specifications for M-type SCM devices. Given the faster latency target (which enables coherent access through a memory controller), program-erase cycle endurance must be higher, so that the overall nonvolatile memory system can offer a sufficiently large lifetime before needing replacement or upgrade. Although some studies have shown that a device endurance of 10^7 is sufficient to enable device lifetimes on the order of 3–10 years [8], we anticipate that the need for sufficient engineering margin would suggest a minimum cycle endurance of 10^9 cycles. While such endurance levels support the use of M-class SCM in memory support roles, significantly higher endurance values would allow M-class SCM to be used in more varied memory applications, where the total number of memory accesses may become very large.

As was discussed above, the expected use of M-class SCM within an online system reduces the requirement of nonvolatility greatly. Here the minimum retention that would be required would be sufficient to allow recovery of transaction or other critical data from the system, within a week or so of loss of power either due to a localized or wide-scale loss of power to the overall system. Obviously, a larger retention time would support even more varied applications, including even long-term offline storage. Active power should be competitive with or better than DRAM, with standby power again being significantly lower. While M-class SCM will not need to be refreshed, it should be noted that this may not necessarily lead to large power and efficiency improvements – in existing DRAM systems, the power used and overhead associated with refresh has typically been fairly modest. However, this trend may change as DRAM itself scales to more aggressive technology nodes. As with S-class SCM, the areal density of M-class SCM will need to be high in order to support a low cost as compared with DRAM.

Note that S-class SCM could potentially fail to match NAND Flash in cost yet still "succeed" in the market, because it can offer inherently higher performance, higher endurance, and the additional benefits of direct byte-accessibility. In contrast, M-class SCM will almost certainly have to offer slower performance and lower cycle endurance than DRAM. While nonvolatility is an attractive feature that DRAM cannot provide, it is unlikely that M-class SCM will meet with widespread success unless it can also prove attractive in some combination of lower cost, higher bit density, and/or lower power usage, with respect to DRAM.

25.5   Device Candidates for SCM

As discussed above, numerous nonvolatile memory candidates have been researched as potential replacements for NAND Flash, or more recently, as potential enablers of Storage Class Memory. Some of these memory candidates have been successfully commercialized, yet are still unsuitable for enabling SCM. This is not surprising since the required combination of attributes – high nonvolatility (ranging from 1 week to 10 years), very low latencies (ranging from hundreds of nanoseconds up to tens of microseconds), physical durability during practical use, and most important, ultra-low cost per bit – is nontrivial to attain.

Necessary attributes of a memory device for the storage-class memory applications, mainly driven by the requirement to minimize the cost per bit, are *Scalability*, the potential for high areal density through either *MLC* (Multilevel Cells) and/or *3D integration*, low *Fabrication cost*, long-term *Retention*, low *Latency*, low *Power*, high cycle *Endurance*, and low *Variability*.

Table 25.2 shows the potential of the *prototypical* memory entries and the current emerging research memory entries for storage-class memory applications based on the above parameters.

Table 25.2 Potential of prototypical and emerging research memory candidates for SCM applications.

| Parameter | Prototypical | | | | Emerging | | Redox RRAM | |
	FeRAM	STT-MRAM	PCRAM	Emerging ferroelectric memory	Conducting bridge	Metal oxide: bipolar filament	Metal oxide: unipolar filament	Metal oxide: bipolar interface effects
Scalability								
MLC								
3D integration								
Fabrication cost								
Retention								
Latency								
Power								
Endurance								
Variability								

(Ratings in each cell are shown graphically as face icons indicating relative potential.)

Each of the above-mentioned categories is qualitatively described as either a strength of that technology (represented with a green face), a nonstrength (yellow face), or a decided weakness (red face), in terms of suitability for SCM applications.

The three prototypical memory candidates are FeRAM, STT-MRAM, and PCRAM. While FeRAM was the first to be commercialized, its difficulties in scalability, MLC, and 3D integration make it a poor candidate for SCM, despite its excellent latency, power, and endurance characteristics. STT-MRAM can be expected to be particularly good in latency and endurance, but achieving scalability while maintaining thermal stability, low write power, and sharply defined resistance distributions are significant challenges. In particular, implementing MLC – by stacking multiple STT-MRAM cells with different and carefully tuned characteristics – can be expected to be quite difficult to implement. While PCRAM has been shown to be scalable, capable of MLC, and only requires a unipolar selection device for 3D integration, reducing power and maintaining good endurance, retention, latency, and low fabrication costs are still a work in progress. However, the recent release of PCRAM as NOR replacement products could potentially provide an opportunity to improve these aspects for SCM applications.

Of the emerging research memory entries, the emerging ferroelectric memories such as FeFETs are quite similar to FeRAM but can be expected to be more scalable, while offering lower endurance due to charge-trapping effects. The large amount of work over the past 3–5 years in the area of Redox Memories have made it clear that these memories need to be further subcategorized into Conducting bridge memories, and Metal Oxide memories based on Bipolar Filaments, Unipolar Filaments, or Bipolar Interface effects. The metallic filaments of copper or silver in conducting bridge devices provide a large resistance contrast suitable for MLC, but also tend to lead to poor retention characteristics. While a unipolar Metal Oxide memory (such as NiO) can be more suitable for 3D integration, these memories also tend to exhibit high switching power and poor endurance. Since three of these new subcategories involve filaments, the potential for scalability should be strong. However, while Bipolar Metal Oxide memories (such as HfO_x or TaO_x) have no other strong weaknesses, it is not yet clear whether the broad cycle to cycle variations in both resistances and switching voltages will support aggressive scaling to future technology nodes. So far, the low switching currents required at such nodes have tended to lead to increased *variability*. Finally, Bipolar Interface Effects in Metal Oxides such as PCMO (PrCaMnO) avoid problems related to filaments. The need to move ions across the entire device aperture tends to set up an unpleasant tradeoff between writing speed and long-term data retention.

The next section will discuss architectural issues in Storage Class Memory further. This discussion will touch on the system design and potential application spaces for SCM.

25.6 Architectural Issues in SCM

In traditional computing, SRAM is used as a series of caches, which DRAM tries to refill as fast as possible. The entire system image is stored in a nonvolatile medium, traditionally a hard drive, and is then swapped to and from memory as needed. However, this situation has been changing rapidly. Application needs are both scaling in size and evolving in scope, and thus are rapidly exhausting the capabilities of the traditional memory hierarchy.

By combining the reliability, fast access, and endurance of a solid-state memory together with the low-cost archival capabilities and vast capacity of a magnetic hard disk drive, Storage Class Memory (SCM) offers several interesting opportunities for creating new levels in the memory hierarchies that could help with these problems. SCM offers compact yet robust nonvolatile memory systems with greatly improved cost/performance ratios relative to other technologies. S-class SCM represents ultra-fast, long-term storage, similar to an SSD but with higher endurance, lower latencies, and byte-addressable access. M-class represents dense and low-power nonvolatile memory at speeds close to DRAM.

In order to implement SCM, both the emerging memory technologies as well as new interfaces and architectures will be needed, in order to fully use the potential, and to compensate for the weaknesses of various new memory technologies. In this section, we explore the Emerging Research Architecture implications and challenges associated with Storage Class Memory.

25.6.1 Challenges in Memory Systems

Current memory systems range in size from Gigabytes (low-volume ASIC systems, FPGAs, and mobile systems) through Terabytes (multicore systems that manage execution of many threads for personal or departmental computing), to Petabytes (for database, Big Data, cloud computing, and other data analytics applications), and up to Exabytes (next-generation, exascale scientific computing). In all cases, speed (both in terms of latency of data reads and writes as well as bandwidth), power consumption, and cost are absolutely critical. However, the importance of other system aspects can vary across these different application spaces.

Some applications such as data analytics and ASIC systems can benefit from having associative memories or content addressability, while other applications might gain little. Mobile systems can become even more compact if many different memory tiers can be combined on the same chip or package, including nonvolatile M-class or even S-class Storage Class Memory.

Many computer systems are not running at peak load continuously. Such systems (including mobile or data analytics) become much more efficient if power can be turned off rapidly while maintaining persistent stored data, since power usage can then become proportional to the computational load. This provides additional incentive for the nonvolatile storage aspect of SCM.

Access patterns in data-intensive computing can vary substantially. While some companies continue to use relational databases, others have switched to flat databases that must be separately indexed to create connections among entries. In general, database accesses tend to be fairly atomic (as small as a few Bytes) and can be widely distributed across the entire database. This is true for both reads and writes, and since the relative ratio of reads and writes varies widely by application, the optimality of any one design can depend strongly on the particular workload.

Total cost of ownership is influenced by cost to purchase, cost to maintain, and system lifetime. Current cost to purchase trends are that Hard Disk Drives (HDD) costs roughly an order of magnitude less per bit than Flash memory, which in turn costs almost an order of magnitude less per bit than DRAM. However, cost to purchase is not the only consideration. It is anticipated that S-class SCM will consume considerably less power than HDD (both directly and in terms of required cooling), and will take up considerably less floorspace. By 2020, if the main storage system of a data center is still built solely from HDD, the target performance of 8.4 G-SIO/s could consume as much as 93 MW and require 98 568 square feet ($9157\,m^2$) of floor space [5]. In contrast, the improved performance of emerging memories could supply this performance with only 4 kW and 12 square feet $(1.11 \, \text{m}^2)$ [5]. Given the cost of energy, this differential can easily shift the total cost advantage to emerging memory, away from HDD, even if a cost per bit differential still exists.

Roughly one-third of the power in a large computer system is consumed in the memory subsystem [9]. Some portion of this is refresh power, required by the volatile nature of DRAM. As a result, modern data servers consume considerable power even when operating at low utilization rates. For example, Google [10] reports that servers are typically operating at over 50% of their peak power consumption even at very low utilization rates. The requirement for rapid transition to full operation precludes using a hibernate mode. Thus a persistent memory that did not require constant refresh would be valuable.

These requirements have led to considerable early investigation into new memory architectures, exploiting emerging memory devices, often in conjunction with DRAM and HDDs in novel architectures. These new Storage Class Memories (SCM) are differentiated as whether they are intended to be close to the CPU (M-class), or to largely supplement the hard-drives and SSDs (S-class).

Because of the inherent speed in SCMs, the software can easily limit the system performance. Thus IO software from the file system, through the operating system and up to applications will have to be redesigned in order to best leverage SCMs. The number of software interactions must be reduced and disk-centric features will need to be removed. Conventional software can account for anywhere from 70 to 94% of the total IO latency [11]. It is likely to be valuable to give application software direct access to the SCM interface, although this can then require additional considerations to protect the SCM device from malicious software.

25.6.2 Emerging Memory Architectures for M-Class SCM

Storage Class Memory architectures that are intended to replace, merge with, or support DRAM, and be close to the CPU, are referred to as M-type or Memory-type SCM (M-SCM). The required properties of this memory have many similarities to DRAM, including its interfaces, architecture, endurance, and read and write speed. Since write endurance of an emerging research memory is likely to be inferior to DRAM, considerable scope exists for architectural innovation. It will necessary to choose how to integrate multiple memory technologies to optimize performance and power while maximizing lifetime. In addition, advanced load leveling that preserves the word level interface and suitable error correction will be needed.

The interface is likely to be a word addressable bus, treating the entire memory system as one flat address space. Since the cost of adapting to a new memory interfaces is sizeable, an interface standard that could support multiple generations of M-SCM devices would be highly preferred. Many systems (such as in automobiles) might be deployed for a long time, so any new standard should be backward-compatible. Such a standard should be compatible with DRAM interfaces (though with simpler control commands), and should reuse existing controllers and PHY (physical layers), as well as power supplies, as much as possible. It should be power efficient, for example, supporting small page sizes, and should support future directions, such as 3D master/slave configurations. The M-SCM device should indicate when writes have been completed successfully. Finally, an M-SCM standard might have to support multiple data rates, such as a DDR-like speed for the DRAM and a slower rate for the NVRAM [12].

While wear-leveling in a block-based architecture requires significant overhead to track the number of writes to each block, simple techniques such as "Start-Gap" Wear-Leveling are available for direct byte access memories such as PCM (Phase Change Memory) [13]. In this technique, a pair of registers is used to identify the location of the start point and an empty gap within a region of memory. After some threshold number of write accesses, the gap register is moved through the region, with the start register incrementing each time the gap register passes through the entire region [13]. Additional considerations can be added to defend against detrimental attacks intended to intentionally wear out the memory [13].

With such techniques, even an M-class SCM that is markedly slower than DRAM can offer improved performance by increasing available capacity and by reducing the occurrence of costly cache misses [8]. The presence of a small DRAM cache helps keep the slower speed of the M-class SCM from affecting overall system performance in many common workloads. Even with an endurance of 10^7, device lifetime has been shown to be on the order of three years [8]. Techniques for reducing the write traffic back to the SCM device can help improve this by as much as a factor of three under realistic workloads [8].

Direct replacement of DRAM with a slightly slower M-class SCM has also been considered, for the particular example of STT-MRAM [14]. Since individual byte-level writes to STT-MRAM consume more power than in DRAM, a direct replacement is not competitive in terms of energy or performance. However, by re-architecting the interaction between the output buffer and the STT-MRAM, unnecessary writes back to the NVM can be eliminated, producing a sizeable energy improvement at almost no loss in performance [14]. However, the use of write buffers means that the device must be able to complete all writes back to nonvolatile memory in the event of power loss. Integrating PCM into the mobile environment, together with a redesigned memory management controller, is predicted to deliver a six times improvement in speed and also extends the memory lifetime six times [15].

Caches are intended to ensure that frequently needed data is located close to the processor in nearby, low-latency memory. In storage architectures, "hot" or frequently accessed data is identified and then moved to faster tiers of storage. However, as the number of tiers or caches increases, a significant amount of time and energy is being spent moving data. An alternative approach is to completely rethink the hardware/software interface. By organizing the computational system around the data, data is not brought to the processor but instead processing is performed in proximity to the stored data. One such emerging data-centric chip architecture, termed "Nanostores" [16], is predicted to offer 10–60× improvements in energy efficiency [17].

25.6.3 Emerging Storage Architectures for S-Class SCM

S (Storage) type SCMs are intended to replace or supplement the hard-disk drive as main storage. Their main advantage will be speed, avoiding the seek time penalty of main drives. However, to succeed, their total cost of ownership needs to approach that of HDDs. Research issues include whether the SCM serves as a disk cache or is directly managed, how load leveling is implemented while retaining a sufficiently fast and flexible interface, how error correction is implemented, and what is the optimal mix of fast yet expensive and slow yet inexpensive storage technologies.

The effective performance of Flash SSD, itself slower than S-SCM, has been strongly affected by interface performance. The standard SATA (Serial Advanced Technology Attachment) interface, which is a commonly used interface for SSD, was originally designed for HDD and is not optimized for Flash SSD [18]. There are several approaches for novel interfaces or architectures that can take advantage of the native Flash SSD performance [18–20], including PCIe, Thunderbolt, and Infiniband [11].

A likely introduction of these new memory devices to the market would be as *hybrid* solid-state discs, where the new memory technology complements the traditional Flash memory to boost the SSD performance. Experimental implementations of FeRAM/Flash [21] and PCRAM/Flash [22] have been explored. It was shown that the PCRAM/Flash hybrid improves SSD operations by decreasing the energy consumption and increasing the lifetime of Flash memory [22].

Additional open questions for S-SCM include storage management, interface, and architectural integration, including whether such a system should be treated like a fast disk drive, or as a managed extension of main memory. To date, disk-like systems built using nonvolatile memories have had disk-like interfaces, with fixed-sized blocks and a translation layer used to obtain block addresses. However, since the filesystem also performs a table lookup, some portion of SCM performance is sacrificed. In addition, nonNAND Flash SCMs have randomly accessible bits and do not need to be organized as fixed-size blocks [23].

While preserving this two-table structure means that no changes to the operating system are required to use, or switch between, new S-SCM technologies, the full advantages of such fast storage devices cannot be realized. There are two alternative approaches to eliminate one of these lookup tables. In Direct Access mode, the translation table is removed, so that the operating system must then understand how to address the SCM devices. However, any change in how table entries are calculated (such as improvements in garbage collection or wear-leveling) would require changes in the operating system [23].

In contrast, in an Object-Based Access model, the file system is organized as a series of (key, value) objects. While this requires a one-time change to operating systems, all specific details of the SCM could be implemented at a low level. This model leads to greater efficiency in terms of both speed and effective "file" density, and also offers potential for enhanced reliability [23].

Even first-generation PCM chips, although implemented without a DRAM cache, compare favorably with state of the art SSDs implemented with NAND Flash, particularly for small (<2 KB) writes and for reads of all sizes [24]. The CPU overhead per input–output operation is also greatly reduced [24]. Another observation for even first-generation PCM chips is that, while the average read latency is similar to NAND Flash, the worst-case latency outliers for NAND Flash can be many orders of magnitude slower than the worst-case PCM access. This is particularly important considering that such S-class SCM systems will typically be used to increase system performance by improving the delivery of urgently needed "hot" data.

Another new software consideration for both S- and M-class SCM is the increased importance of avoiding memory corruption, either through memory leaks, pointer errors, or other issues related to memory allocation and deallocation [25]. Since part of the memory system is now nonvolatile, such issues are now pervasive and may be difficult to detect and remove without affecting stored user data.

25.7 Conclusions

Storage-class memory (SCM) describes a device category that combines the benefits of solid-state memory, such as high performance and robustness, with the archival capabilities and low cost of conventional hard-disk magnetic storage. Memory class SCM is intended to be almost as fast as DRAM while providing at least five days of nonvolatility (for data recovery) at a similar or lower cost per bit. Storage class SCM is intended to outperform NAND Flash at a comparable or slightly higher cost per bit. Promising device candidates for SCM include STT-MRAM, PCM and some types of RRAM. Architecturally, Memory-class SCM is likely to be organized like DRAM, while Storage-class SCM is likely to be accessed through disk type interfaces. Challenges exist in error management and wear-leveling as these steps will have to be performed at much faster latencies and data rates than is done today for NAND Flash.

References

1. Fontana, R.E. Jr., Decad, G.M., and Hetzler, S.R. (2013) The Impact of Areal Density and Millions of Square Inches (MSI) of Produced Memory on Petabyte Shipments of TAPE, NAND Flash, and HDD, MSS&T 2013 Conference Proceedings, May 2013.
2. Deng, Y. and Zhou, J. (2011) Architectures and optimization methods of Flash memory based storage systems. *Journal of Systems Architecture*, **57**, 214–227.
3. Grupp, L.M., Caulfield, A.M., Coburn, J. *et al.* (2009) Characterizing Flash Memory: Anomalies, Observations, and Applications, MICRO'09, Dec. 12–16, 2009, New York, NY, USA, p. 24–33.
4. Burr, G.W., Kurdi, B.N., Scott, J.C. *et al.* (2008) Overview of Candidate Device Technologies for Storage-Class Memory. *IBM Journal of Research and Development*, **52**(4/5), 449–464.
5. Freitas, R.F. and Wilcke, W.W. (2008) Storage-class memory: The next storage system technology. *IBM Journal of Research and Development*, **52**(4/5), 439–447.
6. Franceschini, M., Qureshi, M., Karidis, J. *et al.* (2010) Architectural Solutions for Storage-Class Memory in Main Memory, CMRR Non-volatile Memories Workshop, April 2010, http://cmrr.ucsd.edu/education/workshops/documents/Franceschini_Michael.pdf.
7. Johnson, M., Al-Shamma, A., Bosch, D. *et al.* (2003) 512-Mb PROM with a three-dimensional array of diode/antifuse memory cells. *IEEE Journal of Solid-State Circuits*, **38**(11), 1920–1928.
8. Qureshi, M.K., Srinivasan, V., and Rivers, J.A. (2009) Scalable high performance main memory system using phase-change memory technology, ISCA '09 - Proceedings of the 36th annual International Symposium on Computer Architecture, ACM, pages 24–33.
9. "Final Report, Exascale Study Group: Technology Challenges in Advanced Exascale Systems" (DARPA) (2007).
10. Barroso, L.A. and Holzle, U. (2007) The case for energy-proportional computing. *IEEE Computer*, **40**(12), 33–37.
11. Swanson, S. (2012) System architecture implications for M/S-class SCMs, ITRS SCM workshop, July 2012, http://www.itrs.net/ITWG/ERD_files.html.
12. Kim, K.H. (2012) Memory Interfaces for M-Class SCMs, ITRS SCM workshop, July 2012, http://www.itrs.net/ITWG/ERD_files.html.
13. Qureshi, M.K., Karidis, J., Franceschini, M. *et al.* (2009) Enhancing lifetime and security of pcm-based main memory with start-gap wear leveling, MICRO 42: Proceedings of the 42nd Annual IEEE/ACM International Symposium on Microarchitecture, ACM, pages 14–23.
14. Kultursay, E., Kandemir, M., Sivasubramaniam, A., and Mutlu, O. (2013) Evaluating STT-RAM as an energy-efficient main memory alternative, Proceedings of the 2013 IEEE International Symposium on Performance Analysis of Systems and Software (ISPASS).
15. Lee, H. (2010) High-performance NAND and PRAM hybrid storage design for consumer electronics. *IEEE Transactions on Consumer Electronics*, **56**(1), 112–118.
16. Ranganathan, P. (2011) From microprocessors to nanostores: Rethinking data-centric systems. *COMPUTER*, **44**, 39–48.
17. Chang, J. (2012) Data-centric computing and Nanostores, ITRS SCM workshop, July 2012, http://www.itrs.net/ITWG/ERD_files.html.
18. Kim, D., Bang, K., Ha, S.-H. *et al.* (2010) Architecture exploration of high-performance PCs with a solid-state disk. *IEEE Transactions on Computers*, **59**, 879–890.
19. Fusion (2013) www.fusionio.com (accessed 16 July 2013).
20. NVM (2013) http://download.intel.com/standards/nvmhci/NVM_Express_Explained.pdf (accessed 16 July 2013).
21. Yoon, J.H., Nam, E.H., Seong, Y.J. *et al.* (2008) Chameleon: A high performance Flash/FRAM hybrid solid state disk architecture. *IEEE Computer Architecture Letters*, **7**, 17–20.
22. Lee, H.G. (2010) High-performance NAND and PRAM hybrid storage design for consumer electronics. *IEEE Transactions on Consumer Electronics*, **56**, 112–118.

23. Miller, E.L. (2012) Object-based interfaces for efficient and portable access to S-class SCMs, ITRS SCM workshop, July 2012, http://www.itrs.net/ITWG/ERD_files.html.
24. Akel, A., Caulfield, A.M., Mollov, T.I. *et al.* (2011) Onyx: a protoype phase change memory storage array. *Hot Storage*, **2011**, 10–19.
25. Coburn, J., Caulfield, A.M., Akel, A. *et al.* (2012) Nv-Heaps: Making persistent objects fast and safe with next-generation, nonvolatile memories. *ACM Sigplan Notices*, **47**(4), 105–117.

Part Five

Summary, Conclusions, and Outlook for Nanoelectronic Devices

26

Outlook for Nanoelectronic Devices

An Chen[1], James Hutchby[2], Victor V. Zhirnov[2], and George Bourianoff[3]
[1]GLOBALFOUNDRIES Inc., USA
[2]Semiconductor Research Corporation, USA
[3]Components Research Group, Intel Corporation, USA

26.1 Introduction

The purpose of this chapter is to summarize the potential of emerging research devices to perform their intended memory or information processing function benchmarked against current memory or CMOS technologies. These targeted functions include: (1) to extend and/or eventually replace CMOS with a highly scalable, high performance, low power information processing device technology and (2) to provide a memory or storage technology capable of scaling either volatile or nonvolatile memory technology beyond the 15 nm generation.

Two independent methods have been used to assess emerging devices. In a so-called "quantitative logic benchmarking," emerging logic devices are evaluated by their operations in conventional Boolean Logic circuits, for example, a unity gain inverter, a two-input NAND gate, and a 32-bit adder. Metrics evaluated include speed, areal footprint, power dissipation, and so on. Each parameter is compared with that of the projected high-performance and low-power 15 nm CMOS. In the second method, referred to as "survey-based benchmarking," ITRS Emerging Research Device (ERD) group conducts a survey among international experts to evaluate each emerging device technology against eight criteria normalized to high-performance CMOS for logic or to the memory technology targeted for replacement. The rating for each criterion reflects consensus opinions on the potential and maturity of these technology entries.

An important issue regarding emerging charge-based nanoelectronic switch elements is related to the fundamental scaling limits of these devices, and how they compare with CMOS technology at its projected end of scaling. An analysis [1] concludes that the fundamental limit of scaling an electronic charge-based switch is only 3× smaller than the physical gate length of silicon MOSFETs in 2024. Furthermore, the density of these switches is limited by maximum

Emerging Nanoelectronic Devices, First Edition. An Chen, James Hutchby, Victor Zhirnov and George Bourianoff.
© 2015 John Wiley & Sons, Ltd. Published 2015 by John Wiley & Sons, Ltd.

practical allowable power dissipation of approximately $100\,\mathrm{W/cm^2}$, and not by their size. The conclusion of this work is that MOSFET technology scaled to its practical limit in terms of size and power density will asymptotically reach the theoretical limits of scaling for charge-based devices.

Most of the proposed beyond-CMOS replacement devices are very different from their CMOS counterparts, and often pass computational state variables (or tokens) other than charge. Alternative state variables include collective or single spins, excitons, plasmons, photons, magnetic domains, qubits, and various ordering parameters observed in solid state phase dynamics (e.g., ferromagnetic state). With the multiplicity of computational state variables and physical mechanisms employed in the new structures, it is necessary to find new ways to benchmark the technologies effectively. This requires a combination of existing benchmarks used for CMOS and new benchmarks which take into account the idiosyncrasies of the new device behavior. Even more challenging is to extend this process to consider new circuits and architectures beyond the Boolean architecture used by CMOS today.

26.2 Quantitative Logic Benchmarking for Beyond CMOS Technologies

The Nanoelectronics Research Initiative (NRI; nri.src.org) has been benchmarking several diverse beyond-CMOS technologies over the past six years, trying to balance the need for quantitative metrics to assess a new device concept's potential with the need to allow device research to progress in new directions which might not lend themselves to existing metrics. Several of the more promising NRI devices have been described in detail in this book. The intermediate results on the benchmarking efforts were outlined in a 2010 IEEE Proceedings article [2]. The benchmark results were updated again based on refined device data in 2011 [3]. Another independent benchmark effort reported in 2012 pursued uniform methodology for benchmark for an expanded set of devices using similar reference circuits and parameters [4].

While all these efforts are still very much a work in progress – and no concrete decisions have been made on which devices should be chosen or eliminated as candidates to significantly extend or augment CMOS – this chapter summarizes some of the data and insights gained from these studies. Further benchmarking may alter some of the conclusions here and the outlook on some of these devices, but the overall message on the challenge of finding a beyond-CMOS device which can compete well across the full spectrum of benchmarks of interest has not changed. Such a beyond-CMOS device has not yet been identified.

26.2.1 Architectural Requirements for a Competitive Logic Device

Circuit designers and architects depend on the logic switch to exhibit specific desired characteristics to realize a wide range of applications, including [5]:

- Inversion and flexibility (can form an infinite number of logic functions).
- Isolation (output does not affect input).
- Logic gain (output may drive more than one following gate with a high I_{on}/I_{off} ratio).
- Logical completeness (the device can be used to realize any arbitrary logic function).
- Self-restoring/stable (signal quality restored in each gate).
- Low cost manufacturability (acceptable process tolerance).
- Reliability (aging, wear-out, radiation immunity).

- Performance (transaction throughput improvement).
- Span of control [6] (measures how other devices may be contacted within a characteristic delay of a switch, and is dependent on not only switch delay but also switch area, as well as communication speed).

Devices with intrinsic properties supporting the above features will be adopted more readily by the industry. Moreover, devices which enable architectures that address emerging concerns such as computational efficiency, complexity management, self-organized reliability and serviceability, and intrinsic cyber-security [7] are particularly valuable.

26.2.2 Quantitative Results

Preliminary analyses sponsored by NRI [2,3] and an independent study [4] surveyed the potential logic opportunities afforded by major emerging research switches using a variety of information tokens and communication transport mechanisms. Specifically, the projected effectiveness of these devices used in a number of logic gate configurations was evaluated and normalized to CMOS at the 15 nm generation, as captured by the ITRS. Many devices were assessed based on simulations only, since they had not yet been built or optimized. So the assessment should be considered only a "snapshot in time" of their potential, as the research on all of them is at a very early stage and hence the data is evolving.

At a high level, the data from these studies corroborates qualitative insights from earlier works, suggesting that many new logic switch structures are superior to CMOS in energy but inferior to CMOS in delay, as shown in the plot of median data for the device in Figure 26.1. This is not surprising, since the primary goal for NRI is to find a lower power device [8] to address the primary concern of increasing power density for future CMOS scaling. The power–speed tradeoffs commonly observed in CMOS is also extended into the emerging devices. In the area–energy–delay characteristics of a NAND2 gate in Figure 26.2, several devices have significantly lower power (even lower than that of low-voltage CMOS), while maintaining a reasonable delay.

Moving beyond the logic gate, it is important to understand the potential impact of the transport delay for the different information tokens these devices employ. As shown in Figure 26.3, communication with many of the noncharge tokens can be significantly slower than moving charge, although this may be balanced in some cases with significantly lower energy for transport. Moreover, the combination of the new balance between switch speed, switch area, and

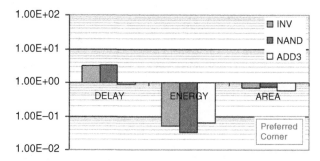

Figure 26.1 Median delay, energy, and area of proposed devices in NRI benchmark, normalized to ITRS 15 nm CMOS (courtesy of Kerry Bernstein) [3]

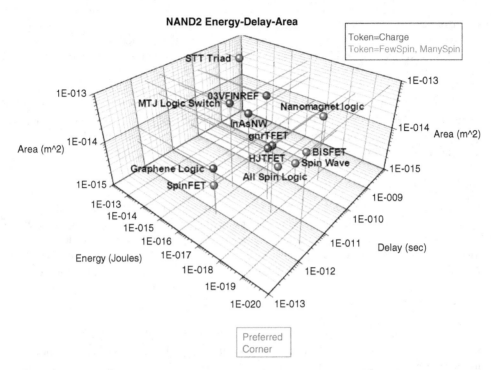

Figure 26.2 Area, energy, and delay of NAND2 gate of various postCMOS technologies from 2011 NRI benchmark (courtesy of Kerry Bernstein) [3]

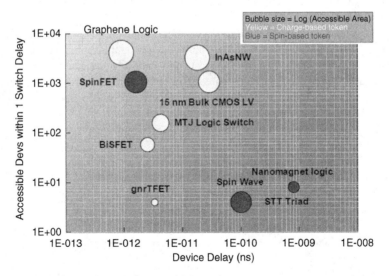

Figure 26.3 Transport impact on switch delay, size, and area of control. Circle size is logarithmically proportional to physically accessible area in one delay. Projections for 15 nm CMOS included as reference (courtesy of Kerry Bernstein) [3]

interconnect speed can lead to advantages in the span of control for a given technology. Finally, for some of the technologies, such as Nano-Magnetic Logic (NML), there is no strong distinction between the switch and the interconnect, indicating opportunities of novel architecture to exploit some of these attributes.

An independent study [4] showed disadvantages of spin-based devices in terms of speed without obvious energy advantage. However, it should be emphasized that the study is only based on active switching energy without considering other unique advantages of these devices for creative designs to reduce overall power consumption, for example, nonvolatility to eliminate standby power. This study also evaluated power-constrained throughput as an important parameter for low-power design.

At the architecture level, the ability to speculate how these devices will perform is still in its infancy. While the ultimate goal is to compare at a very high level – for example, how many MIPS can be produced for 100 mW in 1 mm^2 – the current work must extrapolate from only very primitive gate structures. One initial attempt to start this process has been to look at the relative "logical effort" [9] for these technologies, a figure of merit which ties fundamental technology to a resulting logic transaction. Several of the devices appear to offer advantage over CMOS in logical effort, particularly for more complex functions, which increases the urgency of joint device–architecture co-design for emerging technologies.

26.2.3 Observations and Perspectives

A number of common themes have emerged from these benchmark studies [10,11].

1. The energy–delay tradeoff conundrum will continue to be a challenge for all devices. Getting to low voltage remains a priority for achieving low power, but new approaches to improve throughput with "slow" devices need to be developed.
2. Most of the architectures considered to date in the context of new devices utilize binary logic to implement von Neumann architecture. In this area, CMOS implementations are very competitive across the spectrum of energy, delay and area – not surprising since these architectures have evolved over several decades to exploit the properties and limitations of CMOS. Novel electron-based devices – which can include devices that take advantage of collective and nonequilibrium effects – appear to be the best candidates as a drop-in replacement for CMOS for binary logic applications.
3. As the behavior of other emerging research devices becomes better understood, work on novel architectures that leverage these features will be increasingly important. A device that may not be competitive at doing a simple NAND function may have advantages in doing a complex adder or multiplier instead. Understanding the right building blocks for each device to maximize throughput of the system will be critical. This may be best accomplished by thinking about the high-level metric a system or core is designed to achieve (e.g., computation, pattern recognition, FFT, etc.) and finding the best match between the device and circuit for maximizing this metric.
4. Increasing functional integration and on-chip switch count will continue to grow.
5. Patterning, precise layer deposition, material purity, dopant placement, and alignment precision critical to CMOS will continue to be important in new device architectures.
6. Assessment of novel switches must also include the transport mechanism for the information tokens. Fundamental relationships connecting information generation with information communication spatially and temporally will dictate CMOS' successor.

Based on the current data and observations, it is clear that CMOS will remain the primary basis for IC chips for the coming decade. While it is unlikely that any of the current emerging devices could entirely replace CMOS, several do seem to offer advantages, such as ultra-low power or nonvolatility, which could be utilized to augment CMOS or to enable better performance in specific application spaces. One potential area for entry is special-purpose cores or accelerators that could off-load specific computations from the primary general-purpose processor and provide overall improvement in system performance. This is particularly attractive given the move to multi-core chips. These would include specific, custom-designed cores dedicated to accelerate high-value functions, such as accelerators already widely used today in CMOS (e.g., encryption/decryption, compression/decompression, floating point units, digital signal processors, etc.), as well as potentially new, higher-level functions (e.g., voice recognition). While integrating dissimilar technologies and materials is a big challenge, advances in packaging and 3D integration may make this more feasible over time.

An accelerator using a non-CMOS technology would likely need to offer an order of magnitude performance improvement relative to its CMOS implementation to be considered worthwhile. That is a high bar, but there may be instances where the unique characteristics of emerging devices, combined with a complementary architecture, could be used in implementing a particular function. At the same time, the changing landscape of electronics (moving from uniform, general purpose computing devices to a spectrum of devices with varying purposes, power constraints, and environments spanning servers in data centers to smart phones to embedded sensors) and the changing landscape of workloads and processing needs (Big Data, unstructured information, real-time computing, 3D-rich graphics) are increasing the need for new computing solutions. One of the primary goals then for future beyond-CMOS work should be to focus on specific emerging functions and optimize between the device and architecture to achieve solutions that can break through the current power/performance limits.

26.3 Survey-based Critical Assessment of Emerging Devices

26.3.1 Overall Technology Requirements and Relevance Criteria

The second method for benchmarking emerging memory and information processing devices is based on a survey in the ERD group. Some emerging devices are charge-based structures proposed to extend CMOS to the end of the current roadmap, also known as "CMOS extension." Other emerging devices use charge as the information token but in a non-MOSFET device structure or even use an information token other than charge, known as "beyond-CMOS." A set of relevance or evaluation criteria are used to parameterize the extent to which proposed "CMOS extension" and "beyond-CMOS" technologies are applicable to memory or information processing applications. The criteria include: (1) scalability, (2) speed, (3) energy efficiency, (4) gain (logic) or ON/OFF ratio (memory), (5) operational reliability, (6) operational temperature, (7) CMOS technological compatibility, and (8) CMOS architectural compatibility.

26.3.2 Methodology

Each CMOS extension and beyond-CMOS device is evaluated against these criteria. Performance potential for each criterion is assigned a value from 1 to 3, with "3" substantially exceeding ultimately scaled CMOS, and "1" substantially inferior to CMOS or a comparable existing memory technology. These numbers are more precisely defined in Table 26.1.

Table 26.1 Individual potential for emerging logic and memory devices related to each technology relevance criterion

3	Substantially exceeds ultimately scaled CMOS or baseline memory (relevance criteria 1–5).
	(6) *or* is compatible with CMOS or baseline memory operating temperature.
	(7) *or* is monolithically integrable with CMOS or baseline memory wafer technology.
	(8) *or* is compatible with CMOS or baseline memory wafer technology.
2	Comparable to ultimately scaled CMOS or baseline memory (relevance criteria 1–5).
	(6) *or* requires a very aggressive forced air cooling technology.
	(7) *or* is functionally integrable (easily) with CMOS or baseline memory wafer technology.
	(8) *or* can be integrated with CMOS or baseline memory architecture with some difficulty.
1	Substantially (2×) inferior to ultimately scaled CMOS or baseline memory (relevance criteria 1–5).
	(6) *or* requires very aggressive liquid cooling technology.
	(7) *or* is not integrable with CMOS or baseline memory wafer technology.
	(8) *or* cannot be integrated with CMOS or baseline memory architecture.

26.3.3 Results of the Survey-based Critical Assessment

The critical review results for the memory and logic devices produced both numerical evaluations based on the criteria above and more qualitative evaluations of the current critical issues and ultimate potential. The review results are plotted in spider charts in Figures 26.4 to 26.7. Some of the qualitative factors are interpreted briefly below.

26.3.3.1 Emerging Research Memory Technologies

In Ferroelectric FET memory, a ferroelectric dielectric forms the gate insulator of a FET. The main concern for FeFET memory lies in operation reliability. Operational reliability of the FeFET memory is limited by the time dependent remnant polarization of the ferroelectric gate dielectric, reflected in retention loss. Control of the ferroelectric–semiconductor interface is critical for FeFET properties. The scalability of FeFET memory beyond the 22 nm generation is uncertain (please refer to Chapter 6 in this book for details).

In ferroelectric tunnel junction (FTJ) memory, the device is controlled by the ferroelectric polarization of a tunnel barrier. With a simple two-terminal structure, FTJ is expected to have better scalability than FeFET. Ferroelectric thin film seems to maintain ferroelectric properties down to a couple of nm. Overall, FTJ has very similar performance rating as FeFET, because of the common underlying material property and device physics (please refer to Chapter 9 in this book for details).

RRAM includes multiple device types and mechanisms with varying level of maturity. The survey is based on rating of the general field rather than specific types. Some recent breakthroughs in RRAM significantly enhanced perceived potential of this technology, for example, 32 Gb array demonstration [12]. Overall RRAM assessment is similar or better than existing CMOS-based nonvolatile memories (Flash). An advantage of RRAM is scalability owing to the filamentary conduction and switching mechanisms. The simple device structure and fabrication-friendly materials also contribute to high rating in CMOS compatibility. One of the major concerns of RRAM is the operation reliability due to stochastic nature and defect-related mechanisms. Large variation of RRAM switching parameters has been commonly observed and is considered an intrinsic feature of RRAM mechanisms (please refer to Chapter 8 in this book for details).

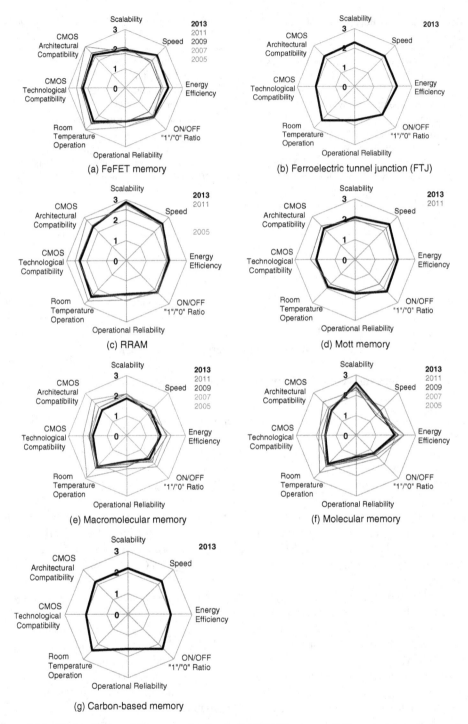

Figure 26.4 Technology performance evaluation for (a) FeFET memory, (b) ferroelectric tunnel junction (FTJ), (c) RRAM, (d) Mott memory, (e) macromolecular memory, (f) molecular memory, and (g) carbon-based memory

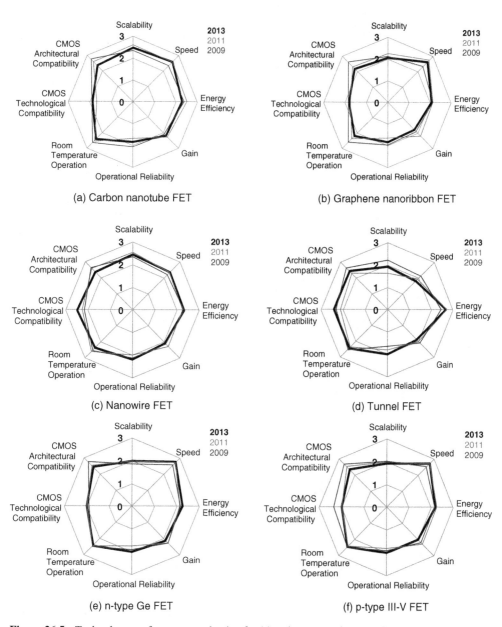

Figure 26.5 Technology performance evaluation for (a) carbon nanotube FET, (b) graphene nanoribbon FET, (c) nanowire FET, (d) tunnel FET, (e) n-type Ge FET, and (f) p-type III-V FET

Mott memory is one of the electronic-effect resistive switching memories, which is believed to have a fast switching speed. Important challenges include operation temperature and reliability. Depending on whether the conditions that triggers the transition are sustained after switching, the resistance change may or may not be nonvolatile (please refer to Chapter 9 in this book for details).

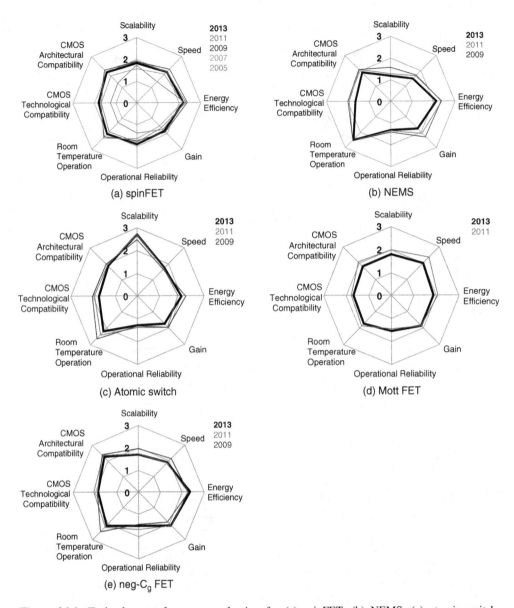

Figure 26.6 Technology performance evaluation for (a) spinFET, (b) NEMS, (c) atomic switch, (d) Mott FET, and (e) neg-C_g FET

Carbon-based memory includes different types of carbon materials (carbon nanotube, graphene, amorphous carbon, etc.), which may also vary widely in maturity. Carbon-based memory is generally perceived to have similar or better performance than Flash memory except for reliability. Lack of clear understanding of the operation mechanisms in these devices makes it difficult to predict or improve their reliability.

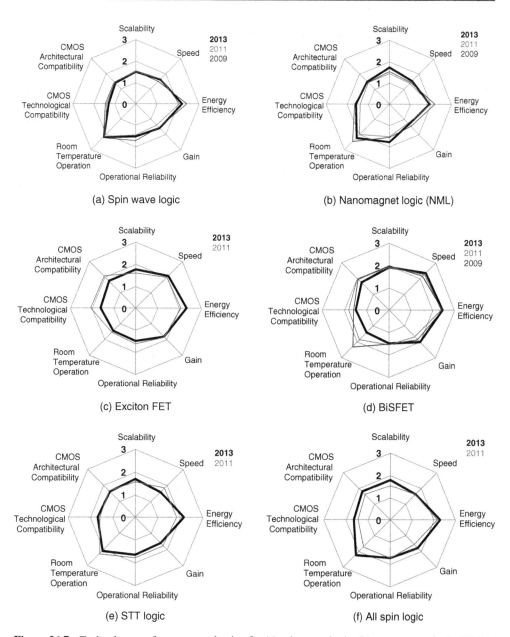

Figure 26.7 Technology performance evaluation for (a) spin wave logic, (b) nanomagnet logic (NML), (c) exciton FET, (d) BiSFET, (e) STT logic, and (f) all spin logic

Neither of the last two memory technologies evaluated, Macromolecular Memory and Molecular Memory, is considered to have long range potential for high-performance computing due to low expectations for speed, operational reliability, and I_{on}/I_{off} ratio (please refer to Chapters 10 and 11 in this book for details).

26.3.3.2   Emerging Research Logic Technologies

Nanowire FETs offer an appealing approach to scale CMOS with attributes similar to Carbon nanotube (CNT) FETs, including excellent gate control of the channel and reduced short channel effects. Nanowire FETs also have challenges similar to those for CNTs related to well controlled growth, placement and fabrication as well as parasitic resistances and capacitances.

CNT FETs offer high carrier mobility, high quasi-ballistic charge carrier velocity, and a tubular structure ideally suited for minimized short channel effects (i.e., abrupt turnoff of channel current), if a gate all around process is developed. Other challenges include the ability to obtain single wall semiconducting nanotubes, control the bandgap energy, control growth of nanotube position/direction, and control the carrier type and density (please refer to Chapter 16 in this book for details).

The tunnel FET offers an appealing concept for substantially lowering the energy dissipated in a switching device by substituting a tunneling process for a thermionic process for injecting charge carriers into the channel of a MOSFET. The major challenge is to simultaneously obtain a sharp subthreshold slope (much less than 60 mV/decade) with a high on-current, I_{on}. Furthermore the tunnel FET may have a problem with operational reliability due to high sensitivity to slight variations of the tunnel structure and the resulting tunnel barrier (please refer to Chapter 14 in this book for details).

While graphene nanoribbon (GNR) as a channel replacement material offers an attractive alternative, it faces several important challenges. Even assuming a solution to these materials and process challenges (including development of a viable epitaxial growth technology), GNR may not offer sufficient device gain to be competitive (please refer to Chapter 15 in this book for details).

The negative C_g MOSFET offers another approach to lower the energy dissipated in switching a MOSFET by utilizing negative capacitance in a MOSFET gate stack. The major challenge concerns identification of appropriate materials (ferroelectrics and/or oxides) that can provide the best voltage swing with minimal hysteresis. Another challenge comes from the integration of high-quality single-crystalline ferroelectric oxides on Si. Operational reliability may also be a concern for this device concept (please refer to Chapter 14 in this book for details).

The spin transistor category represents two different device structures. One is the spinFET and the other is the spin MOSFET. In both instances the device complements the usual field effect behavior of a MOSFET with additional functionality of magnetoresistive devices. Consequently, the spin transistors may enable more complex transfer functions with fewer devices than CMOS implementation. However, in spite of considerable focused research on these devices, none have been realized experimentally. In addition, there are concerns with the scalability, gain, operational reliability, and CMOS technological compatibility of these devices (please refer to Chapter 17 in this book for details).

The atomic switch is classified as an electrochemical switch using the diffusion of metal cations and their reduction/oxidation processes to form/dissolve a metallic conductive path. Both two-terminal and three-terminal devices have been developed. Switching speed, cyclic endurance, and uniformity need to be improved for general usage as a logic device. Establishment of clear device physics is one of the most important issues. In addition, development of the architecture to utilize the nonvolatility of these devices is desired.

The bilayer pseudo-spin field effect transistor (BiSFET) is a recently proposed concept for an ultra-low-power and fast transistor based on the possibility of a room temperature exciton (paired electron and hole) superfluid condensate in two oppositely charged (n-type and p-type) layers of graphene separated by a thin dielectric. So far, the device has remained at conceptual stage. Fabrication of BiSFETs with the necessary degree of control of graphene, dielectric and surface quality, and alignment imposes great challenges.

All spin logic (ASL) utilizes magnets to encode binary information and operates via spin current. Spin injected by a magnetic contact drives the switching of another magnet in the next stage. The concept has attracted attention due to progress in spin-based materials and devices (please refer to Chapter 17 in this book for details).

The Mott field effect transistor (Mott FET) utilizes a phase change in a correlated electron system induced by a gate voltage as the fundamental switching paradigm. It could have a similar structure as conventional semiconductor FETs, with the semiconductor channel replaced by correlated electron materials. Besides electric field excitation, the Mott phase transition can also be triggered by photo- and thermal excitations for potential optical and thermal switches. Challenges with Mott FETs include fundamental understanding of gate oxide-functional oxide interfaces and local energy bandstructure changes in the presence of electric fields (please refer to Chapter 14 in this book for details).

Micro/Nano-Electro-Mechanical (M/NEM) switches (or relays) are based on the displacement of a solid beam under the influence of electrostatic force in order to create a conducting path between two electrodes. Ideally, M/NEM relays feature two key properties for logic computation which are unavailable in MOSFETs: zero leakage and zero subthreshold swing. The most important issue with M/NEM relays is nanoscale contact reliability. High impact velocity at the end of pull-in and the resultant "tip bouncing" (which also increases the effective switching delay) can aggravate the problem. The presence of surface forces (van der Waals or Casimir) can cause sticking and switching failure. Scaling is another concern for M/NEM relays. At gaps of a few nm, the subthreshold swing is expected to degrade due to the onset of tunneling current. With sufficiently small dimensions and gaps, the effect of Brownian beam motion may become significant (please refer to Chapter 18 in this book for details).

The other spintronics devices, spin wave devices and nanomagnetic logic, are viewed to be very early in their development or as being limited by important challenges related to their projected speed, gain, operational reliability, and CMOS technological compatibility.

26.3.3.3 Observations and Perspectives Based on the ERD Survey

Several new, nonvolatile memory technologies are emerging as attractive candidates to supplement or perhaps to eventually replace Flash memories. These memory device technologies are also stimulating research efforts to explore new applications.

1. Charge-based FET and FET-like devices (e.g., nanowire FETs, III-V and/or Ge alternate channel material MOSFETs, carbon-based alternate channel MOSFETs, homojunction and heterojunction tunneling FETs) may enable extension of CMOS scaling in standard binary digital von Neumann applications.
2. Spin-based devices offer lower dynamic switching energy and smaller device area compared to silicon CMOS. However these advantages come at a cost of slower speed. An important advantage of some spin devices is their ability to perform complex circuit functions with fewer devices than CMOS implementations. Successful execution of this strategy could allow increased device densities on the die.

3. Nonvolatility of some emerging logic devices may be utilized in low-power designs. This advantage is not yet quantified in the benchmark results. Innovative architecture designs are needed to fully utilize this unique property.

26.4   Retrospective Assessment of ERD Tracked Technologies

The value of ERD's technology evaluation activities is estimated by the extent to which ERD transfers new Potential Solutions to other ITRS topical areas [for example, PIDS (Process, Integration, Devices, and Structures) and FEP (Front End Processes)]. Impact is also judged by transfer of evaluated technologies entries to one or more semiconductor companies.

Contributions of value to the semiconductor industry can be both direct and indirect. Direct influence is achieved through the transfer of maturing technologies from ERD to PIDS and to FEP (Tables 26.2–26.5). The acceptance by PIDS and/or FEP of a technology entry recommended by ERD sends an important message to the tool supplier community that they may need to begin R&D activities in preparation to meet their customers' eventual needs.

Indirect impact is also achieved through other avenues. For example, the chapter is used by several universities in teaching courses in nanoelectronics. University researchers are also

Table 26.2 Emerging memory devices evaluated by ERD

Memory device	Year accepted	Year transitioned/ removed	Transitioned to PIDS?		Transitioned to Manufacture?
			Yes	No	
Phase change	2001	2005	X		Yes
Magnetic RAM	2001	2011 (STT-MRAM)	X		Yes
Floating body DRAM	2001	2005	X		
Floating nano-gate NVRAM	2001	2007		X	
Engineered tunnel barrier NVRAM	2005	2009	X		
Single/few electron transistor DRAM	2001	2005		X	
Insulator resistance change	2003	2007		X	
ReRAM	2011				
– Electrochemical (chemical bridge)	2007	2011		X	
– Fuse/anti-fuse (nanothermal)	2007	2011		X	
– Valence change (nanoionic)	2007	2011		X	
Electronic effects RAM	2007	2011		X	
Nanomechanical NVRAM	2007				
Molecular RAM	2001				
Polymer RAM	2005				
Ferroelectric FET RAM	2005	2011		X	
Ferroelectric RAM	2011				
Mott memory	2011				

Table 26.3 Extending nonclassical CMOS to the end of the roadmap – technology entries evaluated by ERD

Logic or information processing device	Year accepted	Year transitioned/ removed	Transitioned to PIDS?		Transitioned to manufacture?
			Yes	No	
Ultrathin body SOI	2001	2005	X		Yes
Strained silicon (band engineered CMOS)	2001	2005	X		Yes
Vertical CMOS	2001	2005		X	
Multiple gate CMOS	2001	2005	X		Yes
Schottky source/drain	2003	2005		X	
CNT FET	2001				
GNR FET	2009				
Nanowire FET	2003				
III-V channel MOSFET	2007	2011–2013	X		
Ge channel MOSFET	2007	2011–2013	X		
Tunnel MOSFET	2009				

aware of the key scientific and technological issues identified by the industry and ERD as gating acceptance of their new technology for further exploration and potential commercialization. In addition some international government agencies use the findings and recommendations of ERD as input to their research planning, funding, and execution decisions.

Tables 26.2–26.5 list all technology entries evaluated by ERD from 2001 to 2013, and illustrate their outcomes as to whether or not they were recommended for transfer to PIDS and FEP. For example, of the 12 memory technologies, shown in Table 26.2 as having been transitioned, four were transferred to PIDS, three re-classified under ReRAM, one re-classified as Mott RAM, and four were dropped. Two of the four technologies transferred to PIDS and FEP are in manufacture. An additional five memory technologies, including ReRAM and its component technologies, remain under active evaluation.

Table 26.4 Charge-based nonconventional beyond CMOS – nonFET devices and other charge-based information carrier devices – technology entries evaluated by ERD

Logic or information processing device	Year accepted	Year transitioned/ removed	Transitioned to PIDS?		Transitioned to manufacture?
			Yes	No	
Spin FET/spin MOSFET	2001				
IMOS	2009				
NEMS	2009	2013		X	
Mott FET	2011				
Atomic switch	2009				
Negative Cg FET	2009				
RTD FET	2001	2009		X	
E:QCA	2001	2009		X	
SET	2001	2011		X	

Table 26.5 Alternative information processing devices – technology entries evaluated by ERD

Logic or information processing device	Year accepted	Year transitioned/ removed	Transitioned to PIDS?		Transitioned to manufacture?
			Yes	No	
RSFQ	2011	2007		X	
Nanomagnetic logic (coherent spin devices)	2009				
Pseudospintronic (BiSFET)	2009				
Spin wave (collective spin devices)	2009				
Excitonic FET	2011				
All spin logic	2011				
Spin torque majority gate	2011				
Moving domain wall	2007	2011		X	

Tables 26.3–26.5 represent technology entries for logic and information processing that fall into three different classifications: (1) extending nonclassical CMOS, (2) charge-based non-FET devices, and (3) noncharge-based alternative information processing devices. Of the 28 technology entries in these three tables, a total of 13 have been transitioned out. Five of them, including ultra-thin body SOI, strained Si, multiple-gate CMOS (trigate, MUGFET, FinFET), III-V n-channel MOSFET, and Ge p-channel MOSFET, were transferred to PIDS and FEP; three of these technologies are currently in manufacture.

References

1. Zhirnov, V.V., Cavin, R.K., Hutchby, J.A., and Bourianoff, G.I. (2003) Limits to binary logic scaling – a gedanken model. *Proceedings of the IEEE*, **91**(11), 1934–1939.
2. Bernstein, K., Cavin, R.K., Porod, W. *et al.* (2010) Device and architecture outlook for beyond CMOS switches. *Proceedings of the IEEE*, **98**(12), 2169–2184.
3. Bernstein, K. (2011) NRI architecture benchmarking study phase 1.5 metrics readout – overview. *NRI Annual Review*.
4. Nikonov, D.E. and Young, I.A. (2013) Overview of beyond-CMOS devices and a uniform methodology for their Benchmarking. *Proceedings of the IEEE*, **101**(12), 2498.
5. Keyes, R.W. (1979) The evolution of digital electronics towards VLSI. *IEEE Transactions on Electron Devices*, **26** (4), 271–279.
6. Matzke, Doug. (1997) Will physical scalability sabotage performance gains? *IEEE Computer*, **30**(9), 37–39.
7. Welser, J. and Bernstein, K. (Jun 2011) Challenges for post-CMOS devices & architectures, IEEE Device Research Conference Technical Digest (Santa Barbara, CA), pp. 183–186.
8. Theis, T.N. and Solomon, P.M. (2010) Quest of the 'Next switch': prospects for greatly reduced power dissipation in a successor to the silicon field-effect transistor. *Proceedings of the IEEE*, **98**(12), 2005–2014.
9. Sutherland, I. *et al.* (Feb. 1999) *Logical effort: design fast CMOS circuits*, 1st edn, Morgan Kaufmann, San Mateo, CA, ISBN: 10:1558605576.
10. Bourianoff, G. *et al.* (May 2008) "Boolean logic and alternative information-processing devices," Computer, pp. 38–46.
11. Bourianoff, G., Brillouët, M., Cavin, R.K., Jr III *et al.* (2010) An extremely valuable collection of different approaches to post-CMOS technology. *Proceedings of the IEEE*, **98**(12), 441–448.
12. Liu, T.Y. *et al.* (2013) "A 130.7 mm^2 2-layer 32 Gb ReRAM memory device in 24nm technology," ISSCC, pp. 210–212.

Index

Emerging Nanoelectronic Devices, First Edition. An Chen, James Hutchby, Victor Zhirnov and George Bourianoff.
© 2015 John Wiley & Sons, Ltd. Published 2015 by John Wiley & Sons, Ltd.

Printed in the United States
By Bookmasters